T3-BUM-091

Friedrich G. Barth *A Spider's World: Senses and Behavior*

Springer
*Berlin
Heidelberg
New York
Barcelona
Hong Kong
London
Milan
Paris
Tokyo*

Friedrich G. Barth

A Spider's World
Senses and Behavior

Translated by Ann Biederman-Thorson
with financial support from
the Austrian Academy of Sciences

With 309 Figures,
including 16 Color Plates

Springer

Professor Dr. Friedrich G. Barth
Universität Wien
Institut für Zoologie
Biozentrum
Althanstraße 14
1090 Wien, Österreich

Translator
Dr. M. Ann Biederman-Thorson
The Old Marlborough Arms
Church Walk
Combe, Witney
Oxfordshire OX29 8NQ, UK

Title of the Original German Edition
Friedrich G. Barth:
Sinne und Verhalten: aus dem Leben einer Spinne
ISBN 3-540-67716-X

ISBN 3-540-42046-0 Springer-Verlag Berlin Heidelberg New York

Library of Congress Cataloging-in-Publication Data
Barth, Friedrich G., 1940–
[Sinne und Verhalten. English]
A spider's world: senses and behavior/Friedrich G. Barth; translated by M. A. Biederman-Thorson.
p. cm.
Includes bibliographical references (p.)
ISBN 3540420460 (hardcover)
1. Spiders – Nervous system. 2. Spiders – Behavior. I. title.
QL458.4 .B36413 2001
573.8′1544–dc21

Springer-Verlag Berlin Heidelberg New York
a member of BertelsmannSpringer Science+Business Media GmbH

http://www.springer.de

Printed in Germany

Typesetting: K+V Fotosatz GmbH, Beerfelden
Cover design: Design & Production, Heidelberg
Cover photograph: F. G. Barth; *Cupiennius coccineus*, Costa Rica

SPIN 10794287 31/3130-5 4 3 2 1 0 – Printed on acid-free paper

For the next generation
of sensory biologists
and arachnophilic zoologists

Prologue

You must neither be discontented with yourself, which is weak spirited, nor self-satisfied, which is folly.

Balthasar Gracián 1637

Writing is a hard job. Long experience with the production of scientific texts leaves me with no doubt: selection pressure toward developing this skill has obviously been very low; evolution was much better at teaching our brains other things. When an author nevertheless undertakes the labor of writing a scientific book, a special *motivation* must be behind it.

Spiders are wonderful creatures, and do not deserve the bad reputation they often have. Their varied and complex behavior relies on highly developed sensory systems and is eminently well matched to the conditions in the habitat – as is evident from their evolutionary success. Since receiving my doctorate in 1967 I have never really stopped being preoccupied with the sensory systems of spiders. The main object of my attention has been *Cupiennius salei*, a large spider of Central America, and it is also the leading character in this book. I have been acquainted with *Cupiennius* for over three decades now. That is a long time. It seems an awfully long time if one places the work of all these years before the bar of public opinion, but also a delightfully long time when I think of the many things we have learned and experienced and how enthralling the voyage of discovery has been. No spider has ever been as extensively studied as *Cupiennius*. It not only showed us the path toward an understanding of many peculiarities of a spider's world, but also served as a kind of model animal, again and again raising questions of more general biological interest.

Two crucial questions underlie every page of this book: how do the various sensory systems of spiders work, and what role do they play in behavior? Four hundred million years of evolutionary history, extending back into the Devonian period, have equipped the spiders with a formidable array of biosensors. The first thing that fascinates physiologists like me is their "technical" perfection and complexity; but the subject matter here is much broader. Perfection and adaptation of sensory systems are not isolated phenomena; they are integrated into species-specific behavior in a species-specific habitat. The selection pressures operating in evolution are not exerted on individual sensory cells, neurons or ion channels, but on the phenotype of the animal in its entirety. What behaves and survives is the whole organism. The overarching context is thus not technological, but biological. Here the fit between environment and sense organs and the selectivity of the senses, as an interface between environment and behavior, play a dominant role. I hope that the twenty-five chapters of the book will make that clear.

The attempt to do neuro*biology* in this sense is also the reason why this book is not only the product of laboratory experiments on sensory physiology and neuroethology. Field observations and a quantitative assessment of the stimuli the animals encounter in their Central American habitat proved highly relevant to an approach that may well be termed sensory ecology. Even questions about the taxonomic position and the geographic distribution of the various species of the genus *Cupiennius*, or about their daily activity rhythms, call for answers if the mechanisms of reproductive isolation, for example, and how they involve sensation are to be understood. In addition, woven throughout the fabric of the book is a simple conviction that Vogel and Wainwright (1969) once expressed in these words: "Structure without function is a corpse; function without structure is a ghost."

Some of the following 25 chapters, I fear, will be a little harder to assimilate than the reader in a hurry will find comfortable; but before becoming impatient, please remember that *details* are sometimes just as important as the principles they are meant to elucidate. Often detail is like wine: it must be savored at leisure if it is to be truly enjoyed. Moreover, I have tried not only to allow readers to share the experience of specialized research but also to encourage them time and again by providing a broader vista.

After 30 years, someone who has been involved all along begins to acquire a *comprehensive view* that is very difficult to piece together by brief or intermittent surveys of the many odds and ends in the specialized literature. In this book I am trying to convey something of this comprehensive view, in the hope of giving the next generation of sensory biologists and arachnophilic zoologists a useful foundation for further questioning and investigation. We live in a time of continual headlong progress in biology and also of a rapid shifting of interest; and we live in a time when analytical work seems all-important. To pause occasionally for synthesis and taking stock is not a luxury but a necessity.

Through all these many years, of course, we have been enthusiastic hunters of experimentally confirmable facts and causal relationships. That is of the essence of the natural sciences. The more facts became available, the more often it seemed to me as though the story we were deciphering took, for the initiated, a genuinely poetic turn. Struck by wonder, we enjoyed ever more moments in which time seemed to stand still, in which research was not simply being done but celebrated, and in which research trips were rightly dignified by calling them expeditions. I should be especially happy if readers of my book were now and then to perceive even a little of this special charm of the work of a biologist, who does not pursue an ordinary profession but makes his profession a way of life.

Incompleteness and *prematurity* are two inescapable attendants of even scientific work. I have tried at many points to highlight this fact. Sometimes we are deeply troubled by them, but then we should bear in mind that they are important forces, driving us to keep on pushing forward and wanting to know more. We are well advised to agree at least a little with Ed Ricketts, the biologist of Monterey, whose wisdom and way of life John Steinbeck's book *The Log from the Sea of Cortez* anchored in the world's literature: "Number one and first in importance we must have as much fun as we can with what we have." Knowing our goal only in part, and being only somewhat familiar with the route, we make many detours and wrong turns. Sometimes we must fall back on the insight of the cameleer: "He who goes into the desert cannot turn back. When there is no turning back, we must find the best way possible to go forward" (Paulo Coelho in *Der Alchimist*, Diogenes 1996). Anyone who has had as many enthusiastic companions as I on this long journey must count himself fortunate.

Without them, the "expedition" into the sensory world of the spiders would have been impossible. The reader will find names of my fellow wanderers on nearly every page.

The German (DFG) and later the Austrian (FWF) Science Foundation generously supported our research through many years. I thank Johannes Halbritter and Heidemarie Grillitsch for their help with the preparation of many of the figures and Maria Wieser for secretarial help. I am obliged to Ann Biederman-Thorson for her competent and careful translation from the German and to the Austrian Academy of Sciences for a grant making this translation possible. Collaborating with Dieter Czeschlik, Ursula Gramm, and Karl-Heinz Winter of Springer-Verlag was a pleasure and I much appreciate their effort to turn my manuscript into a handsome book.

My most profound thanks go to my wife for her understanding and tolerance.

Lofer and Vienna, Spring 2001 FRIEDRICH G. BARTH

Gracián B (1637) The Art of Worldly Wisdom. Shambhala Publications Inc., Boston 1993

Steinbeck J (1951) The Log from the Sea of Cortez. Mandarin Classic, Reed International Books Ltd., London 1955

Vogel S, Wainwright SA (1969) A Functional Bestiary. Addison – Wesley, Reading, Mass USA

Contents

The General Biology of *Cupiennius*

I How it All Began

As far as research is concerned, *Cupiennius salei* (Keyserling 1877) first came to light in Munich's large indoor marketplace. Among the culinary delights on offer there were imported tropical fruits, and these proved quite regularly to harbor zoological rarities. More out of fear of poisonous animals and pests than interest in zoology, an official site was established for registration and examination of this exotic living flotsam. For obvious reasons, the Zoological Institute of the University of Munich maintained good relations with this office. And so it happened that around 1960 there came into the hands of Mechthild Melchers three female spiders, and a few years later some males, of a respectable size, with bodies about 3 cm long and a leg span of over 10 cm. They had arrived in the heart of Bavaria with a shipment of bananas – no one is quite sure from where, but presumably Central or South America – and had survived the trip, which must have lasted weeks, in excellent condition. The young doctoral student was fascinated by these animals. She found a place for them in a tiny dark cellar at the Zoological Institute (where large colonies of cockroaches and flour beetles were also housed), and succeeded in breeding them in considerable numbers. In 1963, finally, she earned her degree with a very nice and thorough dissertation entitled (in translation) "On the biology and the behavior of *Cupiennius salei*, an American Ctenidae" (Melchers 1963b). Her professor was Alfred Kästner, whose name has long been well known in arachnological circles but whose interest was not limited to spiders. Alfred Kästner was one of those zoologists, now unfortunately on the verge of extinction, whose gaze took in the whole animal kingdom in all its brilliant diversity, to a truly marvellous degree. He was able to keep students almost infinitely enthralled with details as well as countless anecdotes, which revealed not only an understanding of the subject but also a priceless level of personal involvement and persuasiveness.

So that is how *Cupiennieus salei* was snatched from anonymity among the mass of more than 30,000 known spider species. Although the genus *Cupiennius* was established by Simon (1891) over 100 years ago, by 1963 even taxonomic questions were still wide open. About the natural habitat and the many ways the spiders had adapted to it, practically nothing reliable was known. The senses of *Cupiennius* were largely a mystery, as was the case for spiders in general, to say nothing of its neuroethology or sensory ecology. At that time spiders had hardly made their way into any physiological laboratory and, as large and important a group of arthropods as they are, the experimental zoologists severely neglected them in favor of insects and crustaceans (they still do, although much has changed).

My own relationship with *Cupiennius* began in 1963. I had just returned to Germany from Ted Bullock's laboratory at UCLA, full of new ideas about neurons, sense organs and behavior, when I started doing electrophysiological experiments on the lyriform organs of *Cupiennius salei* with Hansjochem Autrum in Munich. The work was written up in 1967, as the third doctoral thesis on this spider. The second dissertation was that of R. Keller (1961); it turned out not to be entirely accurate, and interpreted the lyriform organs erroneously as olfactory organs. Therefore Professor Autrum thought another doctoral student should follow up this problem, and turned to me. Since then *Cupiennius salei* in my team alone, over the years, has provided the material for 21 Ph.D. theses, 41 theses for lesser degrees and more than 150 scientific publications. Papers from other laboratories that have also employed *Cupiennius* successfully as an experimental animal are concerned not only with the subjects treated in this book but also with embryology and development (Seitz 1966, 1967, 1970, 1971), the fine structure of the hemocytes and the Malpighian vessels (Seitz 1975, 1976), the biochemistry of hemocyanin and muscular metabolism (Loewe et al. 1970; Loewe and Linzen 1975; Linzen and Gallowitz 1975; Linzen et al. 1977, 1985; Markl 1980) and the functional morphology of the lungs, gas transport in the hemolymph, circulatory physiology and the fine structure of the heart and the hemocytes (Fincke and Paul 1989; Paul 1991; Paul et al. 1989 a, b; Paul et al. 1994 a–c; Seitz 1972). Finally, the fine structure of the cuticle of *Cupiennius salei* and its biomechanical implications have been the subject of extensive investigation (Barth 1969, 1970, 1973). And very recently *Cupiennius* has returned to Munich, where Diethard Tautz (now in Cologne) and his coworkers have been studying the Hox genes of this genus and questions related to segmentation (Damen and Tautz 1998; Damen et al. 1998).

Surely no other spider has been anything like as extensively investigated. Apart from Vienna and Munich, to my knowledge work on *Cupiennius* is currently underway in Frankfurt am Main (E.-A. Seyfarth), in Jena (R. Blickhan), Bern (W. Nentwig), Cologne (D. Tautz), and Edmonton and Halifax, Canada (A. French, J. A. Meinertzhagen, and E.-A. Seyfarth).

II The Relatives – Who Is Who?

The genus *Cupiennius* as it is now understood was established in 1891 by the French arachnologist Eugène Simon (Simon 1891, 1897, 1898).

Eugène Simon (Fig. 1) was born on April 30, 1848 in Paris, the same year in which the Republican movement gained the upper hand in the February Revolution and the man later to become the Emperor Napoleon III was first elected president. Simon died in the city of his birth in 1924, on November 17. Pierre Bonnet says of him in his Bibliographia Araneorum (1945): "c'est certainement le plus grand arachnologue que nous ayons eu jusqu'ici". Indeed, he was one of the really great heroes of zoology in his time; he worked not only on spiders but also on crustaceans, insects, birds and even mushrooms! The first edition of his "Histoire naturelle des Araignées" was published when he was only 16. At the end of his productive scientific life, he had written 319 articles about spiders, in addition to his two main works in book form, "Les Arachnides de France" and the second edition of "L'Histoire naturelle des Araignées".

Fig. 1. Eugène Simon, 1848–1924. (Bonnet 1945)

Eugène Simon occupied himself with spiders from all over the world that various researchers had brought back from their long journeys, and he himself collected specimens from many places including Corsica, Sicily, Spain, Morocco, Algeria, Venezuela, the Philippines and South Africa. A tremendous amount of travelling for that time! As might be expected, then, the name Simon is not only associated with the genus *Cupiennius*. Many other genera and species of spiders are recognized because of his taxonomic work (Bonnet 1945).

93 years later, we have revised the genus (Lachmuth et al. 1984) with the expert help of M. Grasshoff of the Senckenberg Museum in Frankfurt am Main (a neighbor of my own former institution), and we have observed large numbers of *Cupiennius* in the field and raised them in laboratories in Frankfurt and Vienna (Barth, Seyfarth et al. 1988). Finally, we have also carried out DNA sequence analyses (Huber et al. 1993; Felber 1994). In a very recent paper we described a second (after *Cupiennius panamensis*) new species, *C. remedius* (Barth and Cordes 1998). Of the last two authors, the physiologist was especially pleased with this achievement – like a ski jumper who has managed a successful slalom run. The same publication includes a revised key (see below) with which to identify the nine *Cupiennius* species that exist according to our present knowledge. The picture is as follows.

1 The Genus

From a morphological viewpoint – that is, the classical viewpoint of the taxonomist – the genus *Cupiennius* can readily be distinguished by a few characters. For the field biologist, who is already familiar with the habitat of *Cupiennius* and its general body appearance, the eight eyes are especially valuable and easy identifiers. As with other spider genera, the first step is to check their size and relative positions (Fig. 2a). But eye position alone leaves room for doubt, since the eye arrangement of *Cupiennius* resembles that in other genera of the family Ctenidae and also of the Thalassiinae, a subfamily of the Pisauridae. However, in no spider of another genus or family with a similar eye arrangement are all eight eyes circular in shape, as they are in *Cupiennius.* In the other spiders, the anterior lateral eyes have an oval to kidney-shaped outline.

The second easily checked character of the genus is the arrangement of spines on the legs (Fig. 2a). A final important distinguishing character is the third claw on the tarsus together with claw tufts in which the hairs point apically to diagonally downward, but never diagonally upward as in other genera (Fig. 2a). Concerning the leg length: in *Cupiennius* leg III is always the shortest, legs I, II and IV being all about equally long. Hence the "leg formula" is $1 \triangleq 2 \triangleq 4, 3$.

It is harder to understand the copulatory organs, which Melchers (1963b) was the first to describe in detail for *Cupiennius salei.* Their complex anatomy arouses shock and confusion in most physiologists (including the author) at first sight. The figure (Fig. 2b) shows the components and the scientific terms used for them (Comstock 1913). Here we have to go into a little general arachnology, in order to make clear the features characteristic of *Cupiennius.*

First the *Male*. The terminal segment of its pedipalp, the pedipalp tarsus, is expanded and is called the cymbium. In all adult spiders this feature clearly distinguishes the males from the females. The wall of the tarsus forms an ellipsoidal to pear-shaped genital bulb, which the male fills with his sperm for indirect insemination. This process is unique among the arthropods: the male tugs his sperm out of the sperm-duct opening onto a web spun specially for this purpose (Fig. 3, Chapter X), and transfers it from there into the palp. In *Cupiennius* the bulb is situated in the two basal thirds of the cymbium. Its wall consists of stiff cuticular sclerites and zones of soft membrane, the hematodochae. Thanks to these membranous zones the whole bulb swells up under the pressure of the hemolymph during copulation. Then the hard parts of the bulbs – the sclerites and their processes, some of which have complex shapes – are brought into their functional positions. In *Cupiennius* the most conspicuous of these structures is the genus-specific, hatchet-shaped median apophysis, with its two shovel- or spoon-shaped processes (Melchers 1963b; Lachmuth et al. 1984; Barth and Cordes 1998).

The *Females*. Eugène Simon, who established the genus *Cupiennius* in 1891 by segregating certain forms previously included in the genus *Ctenus* Walckenaer 1805, called spiders with a complicated sexual apparatus "entelegynes". The females of *Cupiennius*, like the males, leave no doubt that the genus belongs in this category. Here the complication lies in a special copulatory organ, the epigynum, which is located just outside the actual genital pore, in the anterior ventral region of the abdomen (which in spiders is called the "opisthosoma"). The epigynum bulges slightly outward and in its interior forms sperm ducts and seminal receptacles. Details of its structure in *Cupiennius* are shown in Fig. 2b. On its exterior it consists of a central lamella and two side plates, symmetrically arranged with respect to the lamella and partially inserted under it. The laterally overlapping central lamella is a genus-specific character. At the front end of the epigynum are two epigynum pockets, right and left, each of which leads to the opening of its own seminal receptacle by way of a copulatory duct and a sperm duct. Thus in *Cupiennius* there are two such receptacles on each side of the body. During copulation the male's bulb is pushed into the epigynum, distending it, and becomes anchored there by means of its

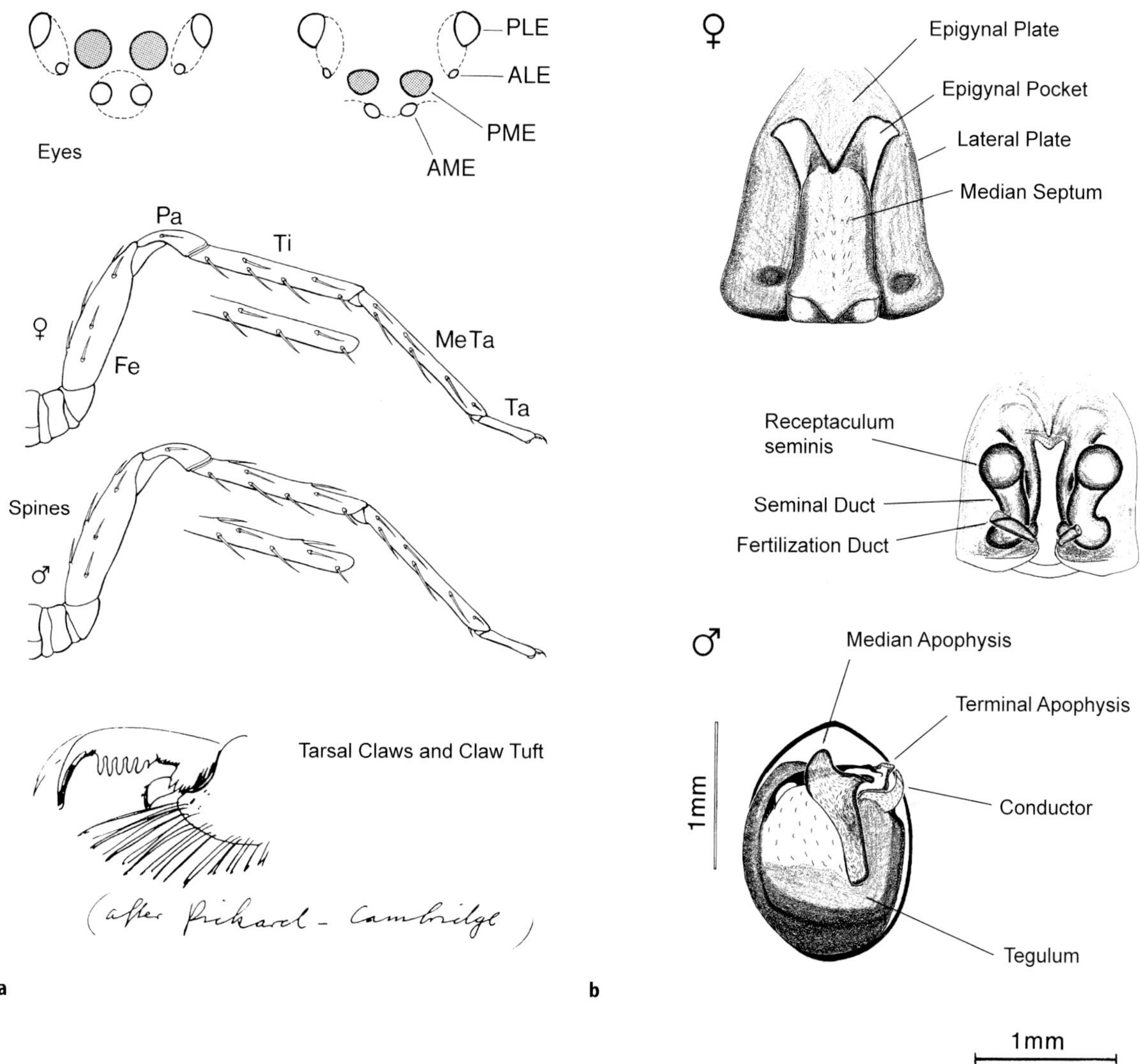

Fig. 2a, b. Characteristics of the genus *Cupiennius*. **a** *Top* The position of the eyes in frontal (*left*) and dorsal view (*right*). *PME* Posterior median eyes; *PLE* posterior lateral eyes; *AME* anterior median (principal) eyes; *PME* posterior median eyes. *Middle* Bristles on the walking leg; female: legs I and II, tibia of legs III and IV; male: legs I and II, tibia legs III and IV. *Fe* Femur; *Pa* patella; *Ti* tibia; *MeTa* metatarsus, *Ta* tarsus. *Bottom* Third claw and claw tufts (Drawing M. Melchers). **b** The copulatory organs, taking *Cupiennius getazi* as the example. *Female* Epigynum in ventral view (*top*) and dorsal view (*below*) after dissecting it free. *Male* Left bulb in ventral view. (After Lachmuth et al. 1984)

median apophysis. The tip of the bulb, the embolus, is inserted into one of the sperm ducts and the sperm are released. They may be stored there for months before fertilization occurs, as the eggs do not become fertilized until they are being laid. Then the sperm cells migrate through a fertilization duct to a structure called the "external uterus", where they unite with the egg cells.

As we shall soon see, the morphology of the copulatory organs also plays an important role in identifying the species of *Cupiennius*. Therefore it is essential to include at least this cursory description.

All other characters in the literature used to distinguish the genus are of lesser importance, so there is no need to go into them here. The interested reader will find more details in the publications by Lachmuth et al. (1984) and Barth and Cordes (1998). The geographical distribution of the genus is treated in Chapter III.

2 The Species

The revision of the genus *Cupiennius* (Lachmuth et al. 1984) was based on the examination of 533 adult animals from the arachnological collections of museums throughout the world, as well as many specimens captured in their Central American habitat by the author and his students and associates. This major cleaning-up exercise reduced the nominal 21 *Cupiennius* species to seven: the three large-bodied species (Plates 1–3) *Cupiennius getazi* (Simon 1891) (type species), *Cupiennius coccineus* (F. Pickard-Cambridge 1901) and *Cupiennius salei* (Keyserling 1877), with leg spans up to 10 cm or more; and the four smaller species *Cupiennius granadensis* (Keyserling 1877), *Cupiennius foliatus* (F. Pickard-Cambridge 1901), *Cupiennius cubae* (Strand 1910) (Plate 6) and the previously overlooked new species *Cupiennius panamensis* (Lachmuth et al. 1984) (Plate 8), with leg spans up to about 4 cm (Fig. 3). *Cupiennius salei* (Plate 1), with which our spider research began and which is still its focal point, was assigned by E. Simon to the genus *Cupiennius* in 1897, having previously been known as *Ctenus Saléi* Keyserling 1877. The holotype of *C. salei* has been lost, but Keyserling, a nobleman living at Glogau in Silesia, described it so precisely in the 1877 Proceedings of the Zoological-Botanical Society of Vienna, that there can be no doubt of its identity (Lachmuth et al. 1984). The Count had presented his findings the preceding year at a meeting on October 4, 1876 in Vienna, the same city where *C. salei* has now found such a permanent residence, so far from its native homeland. What strange paths history sometimes follows!

Count Eugen Wilhelm Theodor von Keyserling was born on March 22, 1832 in Pokroy, Lithuania, and died on April 14, 1889 in Ernsdorf, at his estate in Silesia. His premature demise was due to an illness, cerebral tuberculosis.

After extensive journeys between 1856 and 1864 to the Empire of the Russian Tsars, to the Caucasus, and to Armenia, Persia, England, Algeria and France, Count Keyserling planned to visit South America. But it came to nothing, because by then he had met his future wife in Switzerland; abandoning his great travel plan, he instead acquired considerable tracts of land in Silesia and devoted himself to agriculture. There, in his free time, he studied the many spiders his friends had sent him from America. One of these friends was George Marx, who was born in the Grand-Duchy of Hessen in 1838 and emigrated to the USA at age 22. Marx was later to play a particularly important role, in that he undertook to complete and publish Count Keyserling's great multi-volume work on "The Spiders of America" after the latter's death (Bonnet 1945).

His son Robert wrote about Eugen von Keyserling as follows: "The fourth Keyserling in this generation who should

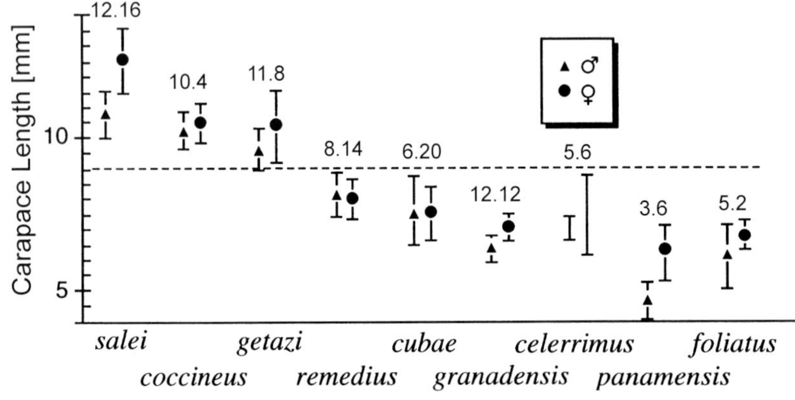

Fig. 3. Size distribution of the nine known species of *Cupiennius*, using carapace length as the measure. *Numbers above symbols* give the number of measured individuals, and *bars* the standard deviation (highest and lowest values in cases with n<7). (Barth and Cordes 1998 after Lachmuth et al. 1984 and Brescovit and von Eickstedt 1995)

be mentioned is the older brother of the above-mentioned Hugo, Eugen Keyserling, whose quiet scholarly life ran its course on the sidelines, noticed only by the closest professional associates ... The area he most made his own was arachnid research. The major works of the "spider-keyserling", as he was known to his kinsmen, ... include many fine drawings from his own hand. His spider collection, containing more than 10,000 species, was purchased by the Museum of Natural Sciences in London after his son had offered them in vain to the Berlin museum." (Thanks to Harald C. von Keyserlingk for providing me with information on E. von Keyserling.)

Another new species, one of the smaller kind, was discovered in Guatemala in 1992. We named it *Cupiennius remedius* (Barth and Cordes 1998) after the place where it was found, the Finca Remedios in the mountains of Alta Verapaz, where *Cupiennius salei* also lives.

We chose this name also in honor of the owners of the farm, Alfred and Inge Schleehauf and their family, who in the most generous and friendly way gave us access to this beautiful part of the earth (Fig. 4). Strictly speaking, the farm is called "Nuestra Señora De Los Remedios" (in translation: "Our Lady of the Remedies") and hence is dedicated to medicine. However, the crops there are not medicinal herbs but coffee, cardamom and sweet peppers. The estate was founded long ago, in 1832. Nowadays about 150 Quekchi Mayas work there; counting the families, about 1000 people live on the farm. For the biologist the nearby Sierra de las Nubes is a very special experience. In this cloud forest lives the quetzal, splendid bird of the gods and symbol of freedom for Guatemala. This treasure, now in danger of dying out, is thought to inhabit this particular piece of land in greater density than all the other remnants of Central American cloud forests. The Proyecto Ecologico Quetzal has been concerned for some years with preserving this paradise.

Cupiennius remedius (Plate 8) has a striking spotted coloration, which distinguishes it unequivocally from all other species of the genus. We have recently summarized details of the species-specific characters of the male bulbs and the female epigyna for all *Cupiennius* species (Barth and Cordes 1998). The interested reader may wish to consult the identification key and the associated figures in the Appendix. Surely still more unknown species must be awaiting discovery. In fact, the list of species has very recently been extended by another name: *Cupiennius celerrimus* (Simon 1891). Because the holotype for this species could not be located and the species had then never been seen again, in

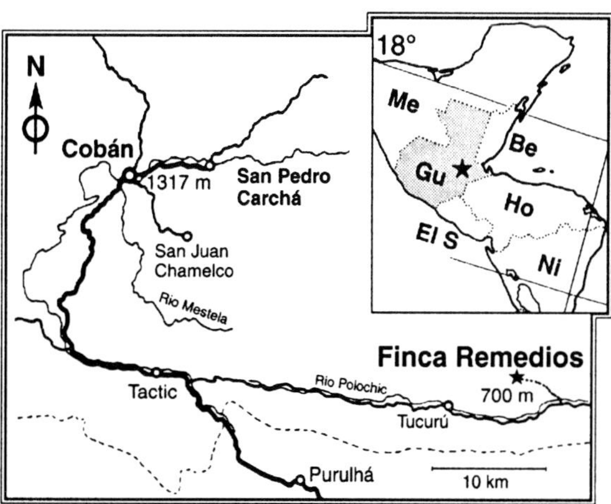

Fig. 4. The location of the Finca Remedios in Guatemala (Alta Verapaz), where the new species *Cupiennius remedius* was found. (Barth and Cordes 1998)

1984 we (Lachmuth et al. 1984) eliminated it from the genus, particularly because Brazil seemed an unlikely habitat for such a spider. Now, however, *Cupiennius celerrimus* is back: Brescovit and von Eickstedt (1995) found it again in Brazil and described it anew, more than a century after it was first described by Simon (Simon 1891).

Almost all *Cupiennius* species can now be categorized not merely as morphospecies, for which the configuration of the sex organs is a particularly important distinguishing character (as is the case for other spider genera as well). Having bred these spiders in the laboratory, we can be sure that *Cupiennius salei, C. coccineus, C. getazi, C. foliatus, C. panamensis, C. cubae* and *C. remedius* are also biospecies. The key in its new extended form in the Appendix (see also Barth and Cordes 1998) is an additional guide to their identification.

In association with questions about species isolation and vibratory communication during courtship (see Chapter XX) we tried to find out how the various forms within the genus are related to one another, and to see whether certain hypotheses we had made about their kinship on the basis of behavior would have to be rejected or not. For this purpose we used the polymerase chain reaction (PCR) method to examine the se-

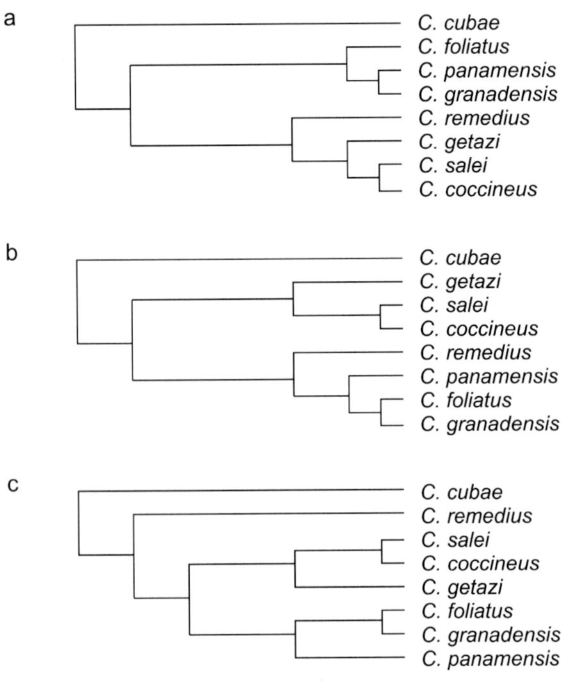

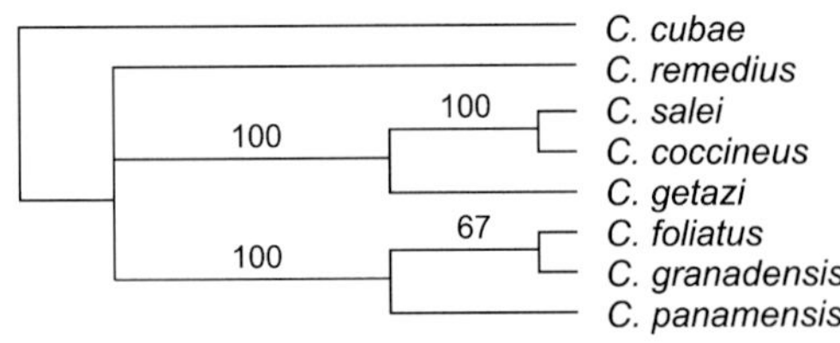

Fig. 5. Cladogram of the genus *Cupiennius* based on the sequential analysis of ca. 684 nucleotides of mitochondrial 12S + 16S rDNA and constructed by means of the programs PHYLIP and PAUP. *Top* Three equivalent "most parsimonious trees". *Below* The majority-rule consensus cladogram derived from the three trees shown above. The clades formed by the three large species *C. getazi* – *C. salei* – *C. coccineus* and by *C. salei* – *C. coccineus* seem to be as well established as the group formed by the smaller species *C. foliatus* – *C. granadensis* – *C. panamensis*. These results are consistent with those obtained by the neighbor-joining method. The numbers in the *lowermost cladogram* give the frequency of the respective group. (Felber 1994)

quence of base pairs from fragments of mitochondrial DNA (16S and 12S rDNA) obtained from femur and prosoma musculature (Felber 1994). About 684 nucleotides were sequenced from each species and by means of "parsimony" and the "neighbor-joining" method cladistic trees (cladograms) were constructed (Fig. 5). All these evaluations indicate a close kinship within two sister groups, one of which comprises the three large-bodied species (*C. salei, C. coccineus, C. getazi*) and the other, the three relatively small species we tested (*C. granadensis, C. foliatus, C. panamensis*). Among the large species, *C. salei* and *C. coccineus* seem to be particularly close to one another phylogenetically, a point to which we shall return later (see Chapter XX, Courtship). Of the smaller species, *C. granadensis* appears especially similar to *C. panamensis*. *C. remedius* occupies an intermediate position between the large and the smaller species. *C. cubae*, the insular species, is the most primitive according to the DNA analysis and hence is at the base of the cladogram; it is presumably more closely related to the large than to the smaller species. Analysis including an outgroup (*Pardosa* sp., *Pisaura* sp.; Huber et al. 1993) gives strong indications that the genus *Cupiennius* is monophyletic.

These kinship hypotheses weigh all the more heavily in that the most important conclusions are corroborated by data of quite different kinds. Foremost among these are the number and frequency of the various behavioral elements involved in the courtship of the different *Cupiennius* species. As will be detailed in Chapter XX (courtship), such comparisons also indicate a close relationship between the large species *C. salei, C. getazi* and *C. coccineus*, single out *C. cubae* as the most primitive species, and place *C. remedius* in a phylogenetic position between *C. cubae* and the large species.

The particularly close phylogenetic relationship between *C. salei* and *C. coccineus* seems also to be reflected in the similarity of their cuticular carbohydrates, which Manfred Kaib of the University of Bayreuth has investigated for us in *C. salei, C. coccineus, C. getazi* and *C. cubae*.

3 The Family

The genus *Cupiennius* has been assigned, by different authors, to four different families: the Ctenidae (comb spiders), Clubionidae (sack spiders), Pisauridae (fishing spiders) and Lycosidae (wolf spiders) (Barth and Seyfarth 1979). According to Homann (1961), the Lycosidae, Ctenidae and Pisauridae cannot be separated from one another as families. Instead, they should be considered subfamilies (Lycosinae, Cteninae, Pisaurinae) of the Lycosidae, in particular because they all have the same retinal structure (see Chapter XI), which differs from that of the Clubionidae. Recent cladistic analyses also support the close phylogenetic relationships of these three families (Coddington and Levi 1991). However, all this has little to do with the position of the genus *Cupiennius*. A comparison of its genus-specific characters with the distinguishing characters of the three families reveals substantial differences in eye position with respect to the Clubionidae and the Lycosidae. Among the pisaurids, only the Thalassiinae have eyes arranged like the Ctenidae. *Cupiennius* differs from the pisaurids, though, in having claw tufts; these are lacking in the Pisauridae and Lycosidae but are important characters of the Ctenidae and Clubionidae. A third claw is present in the Lycosidae and Pisauridae but not in the Clubionidae and most Ctenidae. The net result, then, is that the character combination "third claw, claw tuft, eye position and retinal structure" clearly places the genus *Cupiennius* in the Ctenidae.

The base sequences of mitochondrial DNA of nine species of the above-mentioned four families corroborate the conclusions drawn from morphological taxonomy. However, they also indicate a distinct possibility that the family Ctenidae, established in 1877 by Keyserling, is polyphyletic (Fig. 6). The family to which *Cupiennius* belongs thus remains debatable. Ctenids of the genus *Phoneutria* (notorious as the genus that includes the most poisonous spiders of South America) (Plate 11) show especially pronounced differences in their DNA from *Cupiennius*, greater than the representatives of the Lycosidae and Pisauridae that have been investigated (Huber et al. 1993). An analysis of the same 16S rDNA from the femur muscle of *Ancylometes vulpes, Phoneutria boliviensis, Ph. nigriventer, Ph. keyserlingi* and *Ctenus coxalis* revealed over 94% matching sequences in all five species (Barth and Schlick, unpublished). This result implies close kinship between these species, whereas the other representatives of the Ctenidae that have been tested are considerably less closely related (Fig. 6). We may conclude that eventually a new family will be created, either for *Cupiennius* or for *Phoneutria, Ancylometes* and *Ctenus*, which certainly belong together.

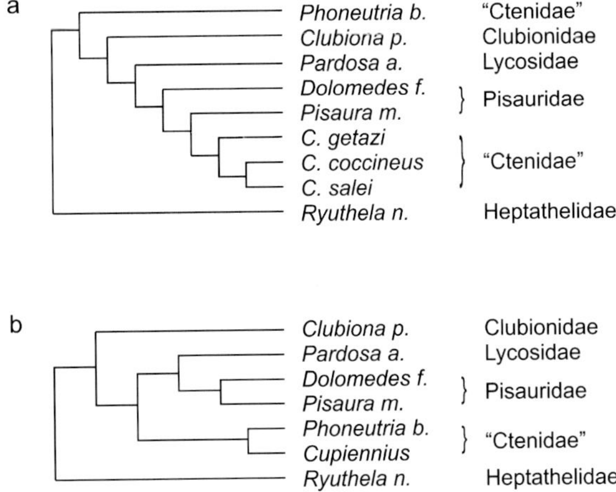

Fig. 6a, b. Cladograms regarding the family assignment of *Cupiennius*. **a** Most parsimonious tree (PHYLIP 3.4) based on the comparison of mitochondrial 16S rDNA. *Phoneutria* and *Cupiennius* are the species most distant from each other. There are serious doubts concerning the monophyly of the Ctenidae. **b** The present view (based on morphological cladistic analysis) of the phylogenetic relationships of the taxa in **a**. (Huber et al. 1993; **b** after Coddington and Levi 1991)

III The Habitat

1 Geographic Distribution

Both the relevant literature and the labeling of the museum material we used to revise the genus indicate that *Cupiennius* is a Central American genus (Fig. 1). However, the term "Central American" is used in a rather broad sense here, including the Caribbean, with Cuba, Haiti and Jamaica, as well as parts of Colombia north of the Andes. The present state of our knowledge is also based on extensive contacts with South American arachnologists and searches that I have made in the Atlantic rain forest of Brazil. Until recently, all the evidence was against the possibility of *Cupiennius* living in South America. The only exception was northern Colombia, where male *Cupiennius cubae* had been found twice (in Bogotá) and for

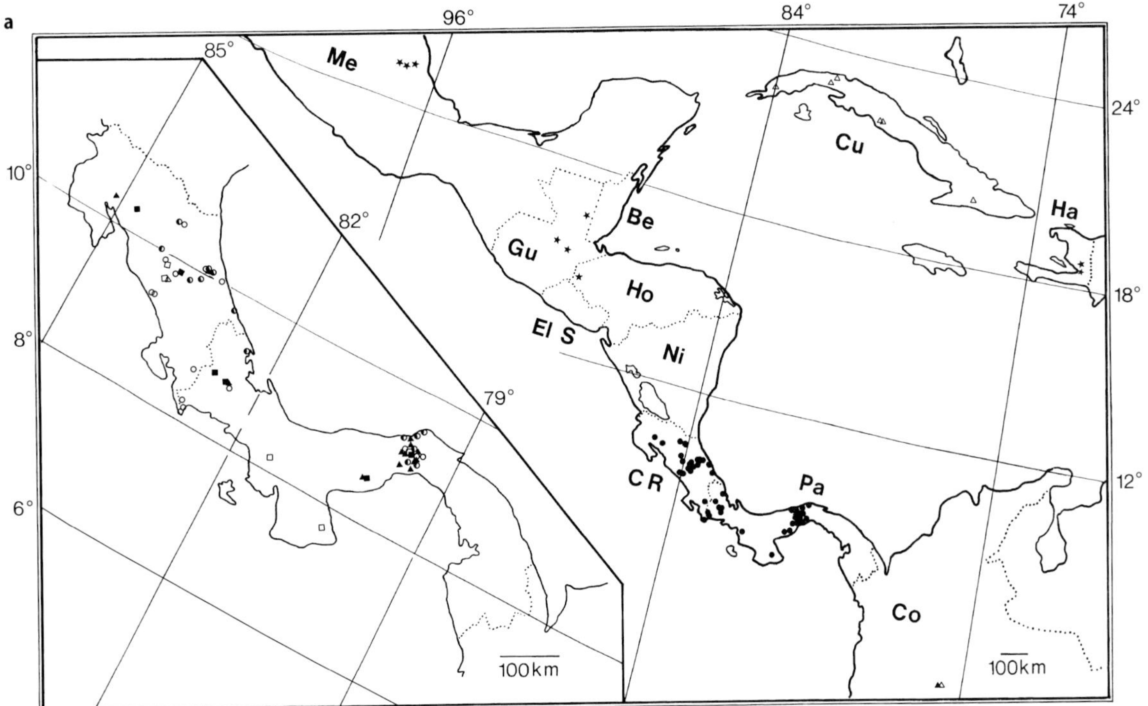

Fig. 1 a. Geographical distribution of seven out of the nine known species of the genus *Cupiennius*. *C. remedius* n.sp. was only recently described and is only known from Guatemala (see Chap. II). *C. celerrimus* was redescribed when found in Tefé (Amazonas) and, according to Brescovit and von Eickstedt, occurs in the north and northeast of Brazil and Venezuela. ○ *C. coccineus*; △ *C. cubae*; □ *C. foliatus*; ◐ *C. getazi*; ▲ *C. granadensis*; ■ *C. panamensis*; * *C. salei* (continued on page 14)

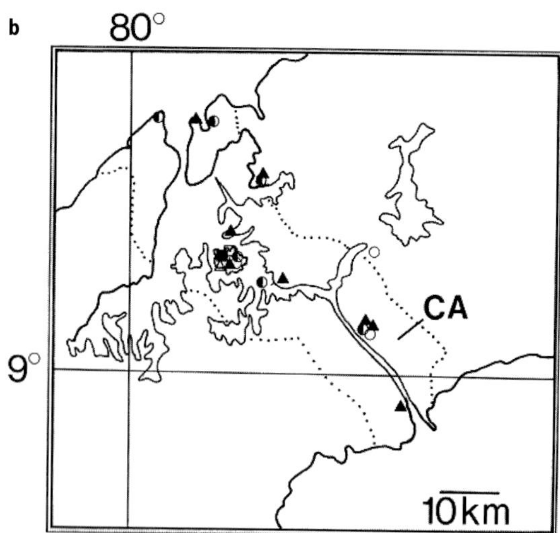

Fig. 1 (cont.). Central America and the Caribbean with an enlarged view of Costa Rica and Panama; in the overview map symbols do not identify species in Costa Rica and Panama. **b** The Canal Area of Panama. *Be* Belize; *CA* Canal Area; *Co* Colombia; *CR* Costa Rica; *Cu* Cuba; *El S* El Salvador; *Gu* Guatemala; *Ha* Haiti; *Ho* Honduras; *Me* Mexico; *Ni* Nicaragua. (Barth, Seyfarth et al. 1988)

which one finding of *Cupiennius granadensis* had been recorded (British Museum, Natural History, London) in 1890 (Lachmuth et al. 1984). But now things have become more complicated, because *Cupiennius celerrimus* has once again been found in South America (near Tefé, Amazonas, Brazil); from comparisons with material in various Brazilian collections, it has been concluded that this species is present in the northern and northeastern parts of Brazil and also in Venezuela (Brescovit and von Eickstedt 1995).

Since 1977 some of my coworkers and I have taken a number of research trips to Central America. For the problems that interested us, in the areas of neuroethology and sensory physiology, it became increasingly important to make a personal acquaintance with the natural habitat, to observe the animals in the field, and to get a better idea of the range of distribution and the population structure of individual species. We also needed to refresh our spider colony, every two years or so, with new animals caught in the wild.

After a long, fruitless correspondence with arachnologists in presumed *Cupiennius* countries, in June of 1975 the first clear – in fact, living – proof of the existence of *C. salei* in Central America reached us in Frankfurt. It came from Dr. Gerda Kramer, who was then carrying out ecological studies in Guatemala (the Chipati/San Pedro Garchá area of Alta Verapaz). I recently ran across the letters we then exchanged, the important and exciting prelude to our long years of field studies, which began in Guatemala. The actual information about where the spiders were found, however, was less than encouraging.

Frankfurt, June 24, 1975

Dear Miss Kramer,

I was happy to take a living spider out of your package! To judge from the first inspection, it is either really C. salei or at least a species so closely related as to be just as interesting to us. Now I have two questions: 1. Where did you find the animal? At what time of day or night? How far from the research station? 2. ... [This question concerned the plans for our first trip into the home of C. salei.]

For the already successful spider search, many thanks indeed. You have done us a really great favor.

With best greetings,
Yours, Friedrich G. Barth

The answer to this missive arrived soon afterward.

Chipati, July 7, 1975

Dear Mr. Barth,

Thank you very much for your letter of June 24, 1975. I am glad to hear that we were lucky with the spider. The search was actually quite simple, as you will be able to see from the site where it was found.
Site: farmhouse privy, 10 m away from the house. The spider was hanging on a side wall.
Time: June 4, 1975, 6 p.m. Altitude: ca. 800 m.

... If we find any more spiders, we'll send them to you.

Many greetings,
Gerda Kramer

According to our subsequent investigations in the years 1977 to 2000, *Cupiennius* lives in a region extending from the Mexican state Veracruz in the north to Panama in the south. The altitudes range from sea level to ca. 1500 m for all species; *Cupiennius salei*, according to the available data (including recent findings in Chiapas, Mexico), is restricted to altitudes between ca. 200 m and 1250 m. *Cupiennius cubae* is the species with the northernmost distribution. *Cupiennius salei*, the species most thoroughly investigated, ranges as far north as Veracruz state in Mexico but in the south is known only as far as Honduras. Count Keyserling in 1877 described the distribution of *Cupiennius salei* (which he was still calling *Ctenus Saléi* n.sp.) as follows: "South America. In Mexico collected by Mr. Salé at Veracruz and Cordoba." A hundred years later we found *C. salei* still right there, in particular near Fortin de las Flores (altitude 1006 m), only a few kilometers to the west of Cordoba (Barth et al. 1988a).

The density of *Cupiennius* finds is by far the highest in the Canal Area in Panama. This presumably reflects the collecting activity of a particularly large number of zoologists, based at the Smithsonian Research Institute on Barro Colorado Island (Fig. 1b). Maps like the one shown here are necessarily incomplete. They are based on reports of sites where the animals were caught and/or observed – places, therefore, where *Cupiennius* actually exists. It would be equally interesting to know where people have searched for *Cupiennius* in vain. It could also be that the absence in places where it could in principle exist is simply due to the fact that no one has looked there, or not thoroughly enough. Perhaps this is the reason for the large gap in El Salvador and Nicaragua. And, possibly, *Cupiennius* has not been detected in many areas because it is living in inaccessible places, such as the epiphytes in the forest canopy.

Occasionally two or more species are found living as close neighbors in the same habitat. One clear case of such sympatry is presented by *Cupiennius coccineus* and *Cupiennius getazi*. We shall return to it with regard to questions of vibratory communication during courtship (see Chapter XX). The Canal Area in Panama is inhabited by five *Cupiennius* species (the ones not known there are *Cupiennius salei* and *Cupiennius cubae*, the two species with the northernmost distribution, and the new species *Cupiennius remedius* as well as the species recently rediscovered in Brazil, *Cupiennius celerrimus*). Overlapping habitats are particularly likely in Panama (Barth, Seyfarth et al. 1988).

2 Plants and Retreats

In order to understand the way of life and the sensory environment of *Cupiennius*, one must be aware of its close relationship to particular plants. *Cupiennius* belongs to the category of wandering or hunting spiders. None of the *Cupiennius* species builds webs in order to catch prey. Instead, *Cupiennius* typically spends the day hidden in retreats on the plants on which by night it hunts, courts a partner or molts. Someone searching for *Cupiennius* in the field is well advised to use these plants as signposts. Table 1 summarizes all the plants on which we have found various species of *Cupiennius* in Mexico,

Guatemala, Costa Rica and Panama. Most of these plants are characterized by mechanically strong, large leaves, which at their bases provide the spiders with shelter in the form of small spaces that are effectively open only upward. A human observer can detect these hideaways most easily in the bromeliads, agaves and plants of the banana family (Fig. 2, Plate 5).

Impatiens and *Tradescantia* are conspicuous exceptions: these plants offer no such hiding places. Nevertheless, we have often found *Cupiennius panamensis* on them in the highlands of Costa Rica, below the cloud forest of Monteverde. An explanation may be that *Cupiennius panamensis* is one of the smallest of all *Cupiennius* species. Only about 1.3 cm in body length,

Table 1

Spider	Plant	
	Species	Family
Cupiennius coccineus	Dieffenbachia spec.	Araceae
	Geonoma cuncuta	Arecaceae
	Aechmea mexicana	Bromeliaceae
	Neoregalia spec.	Bromeliaceae
	Cyclanthus bipartatus	Cyclanthaceae
	Sansevieria spec.	Liliaceae
	Musa sapientium	Musaceae
	Heliconia imbricata	Musaceae
	not identified	Araceae
Cupiennius cubae	Sansevieria spec.	Liliaceae
	diverse trees, not identified	
Cupiennius foliatus	Ananas comosus	Bromeliaceae
Cupiennius getazi	Ananas comosus	Bromeliaceae
	Bromelia pinguin	Bromeliaceae
	Guzmania spec.	Bromeliaceae
	Heliconia imbricata	Musaceae
	Musa spec.	Musaceae
	Cynerium sagittatum	Gramineae
	not identified	Araceae
Cupiennius panamensis	Impatiens balsamina	Balsaminaceae
	Campelia spec.	Commelinaceae
	Tradescantia spec.	Commelinaceae
	Musa sapientum	Musaceae
	Zingiber spec.	Zingiberaceae
	not identified	Araceae
	not identified	Arecaceae
	not identified	Filicatae
Cupiennius remedius	Musa spec.	Musaceae
Cupiennius salei	Furcrea melanodonta	Amaryllidaceae
	Xanthosoma sagittifolium	Araceae
	Aechmea bractea	Bromeliaceae
	Aechmea lueddemanniana	Bromeliaceae
	Aechmea mexicana	Bromeliaceae
	Sansevieria spec.	Liliaceae
	Musa sapientium	Musaceae

Fig. 2. Examples of typical retreats of *Cupiennius. Upper left* Ginger (*Zingiber* sp.) with leaves spun together (*arrow*) by *C. panamensis. Lower left* Bromeliad (*Aechmea bractea*) with retreat between the bases of two neighboring leaves; females carrying an egg sac cover this space with a sheet of silk (*arrow*). *Right* Pseudostem of a banana plant (*Musa sapientium*); the sheath of one of the outer leaves forms the shelter together with the leaf below (*arrow*). (After Barth, Seyfarth et al. 1988)

it is small and light enough to sit under a leaf and be completely covered (Plate 8).

The aquatic uva grass (*Gynerium*), ginger plants (genus *Zingiber*) and the Araceae are intermediate between these extremes with respect to hiding places. Although their leaves are large, they do not provide "prefabricated" refuges. We have often observed that the spiders (*Cupiennius salei, C. coccineus, C. getazi, C. panamensis*) choosing these plants build a shelter themselves, either closing open spaces by spinning together parts of the plant such as adjacent leaves, or constructing protective tunnels by bending and rolling up leaves and fastening them at the ends with silk (Barth, Seyfarth et al. 1988, Fig. 2). The females behave similarly once they have copulated, or when they are already carrying an egg sac around with them. Then they spin their retreat shut with a silk cover (Fig. 2; Plates 10, 11). When such females occupy bromeliads and agaves, they often sit in the central funnel of the plant instead of at the periphery between the leaves.

3 Choice Behavior

Although there is a long list of plant names in Table 1, spiders are very particular in choosing which one to make their home. In virgin forest with its many plant species this is especially obvious. *Cupiennius* seeks out those that offer a hiding place, and is very rarely to be found on any of the multitude of other plants in its habitat. The exception mentioned above, *Cupiennius panamensis*, confirms the rule insofar as these small spiders surely have little difficulty finding a secluded place even on plants that do not have large, mechanically stable and unbranched leaves like those of the typical *Cupiennius* plants (bromeliads, banana plants).

The botanically well-informed reader will have noticed that most of the *Cupiennius* plants are monocotyledons. One reason may be that plants in this group not only offer hiding places but also are good conductors of the vibrations that are so important in a spider's life (see Chapter VIII). It is quite conceivable that differences in the arrangement of the vascular tissue lead to significant differences in the way vibration is propagated in monocotyledons and dicotyledons (Barth 1985, Barth et al. 1988b). However, this possibility has not yet been examined in detail.

The choice of the right habitat is a process of fundamental significance for all animals, as it is one of the main factors determining their probability of survival. The performance of sensory systems, then, must also be evaluated in this context.

Given that the life of *Cupiennius* is so closely bound to the plants and the retreats they provide, we wanted to know what criteria the spiders use to make their choice. Is the attractiveness of a plant determined only by the availability of a retreat, or do the presence of prey and other attributes also contribute? Erich Mitter

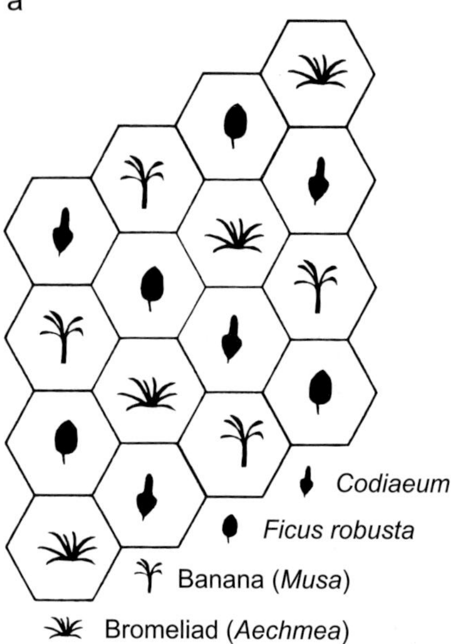

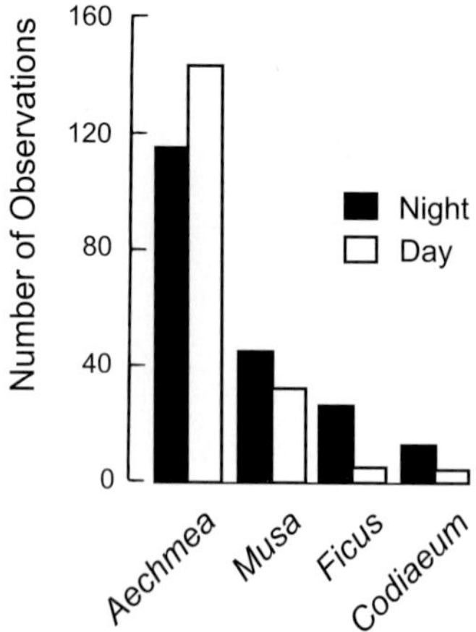

Fig. 3a. The choice of the retreat. **a** Monocots (*Musa, Aechmea*) versus dicots (*Codiaeum, Ficus*). Arrangement of the 16 plants offered to the spiders in the laboratory; 96 spiders released on the plants were observed for a total of 384 times on the plants during a period of 72 h. The *lower panel* shows the number of observations made during the day and at night (continued on opposite page)

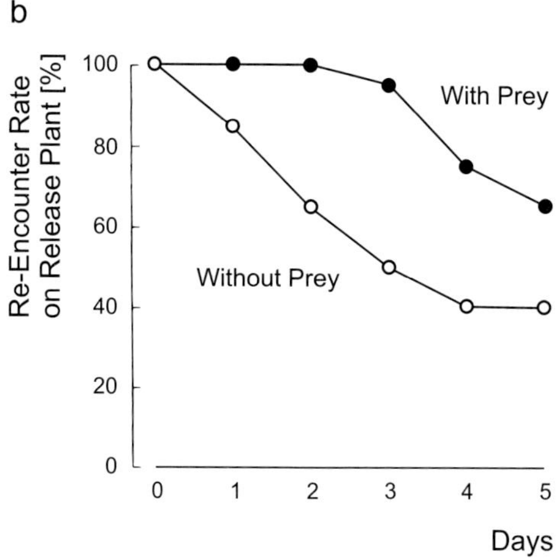

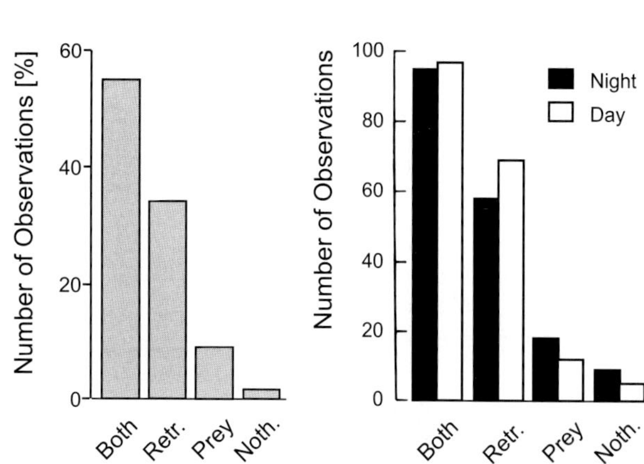

Fig. 3b,c. b Plants with prey and plants without. In this field experiment 20 spiders were released on bromeliads with prey animals and 20 on those without. After 5 days the preference for the plants with prey is significant ($p<0.05$). **c** Multifactorial choice: 96 spiders were released on a total of 16 bromeliads which were divided into four categories: (1) with prey and retreat; (2) with prey but without retreat; (3) without prey but with retreat; (4) without prey and without retreat. For 72 h it was observed where the spiders went (363 observations). The histograms show the relative distribution (in % of all observations) of the animals during the day and at night on the four plant categories (*left*) and in addition the absolute number of the respective observations. (Mitter 1994)

(1994) prepared a dissertation on this subject, for which he spent several months in Costa Rica; in brief, this is what he learned (Mitter 1994).

In choice experiments in the laboratory, *Cupiennius getazi* and *Cupiennius coccineus* still prefer monocots to dicots (Fig. 4a). Furthermore, they show a highly significant ($p<0.01$) preference for monocots with hiding places, over those in which the hiding places were blocked up for the experiment. Plants on which prey was present were preferred to those without prey (Fig. 4b). Probably the most interesting finding was that the spiders optimize their choice of microhabitat in the sense that they clearly choose plants with both prey and shelter rather than plants with either one of these advantages but not the other (Fig. 4c). That is, they do not orient themselves by a single environmental parameter, automatically acquiring a whole set of factors that are associated with that parameter. Instead, unlike other spiders (Riechert 1985), in one of the most important decisions of their lives they perceive and take into account several factors at once. Evidently, though, the hiding place is more important to the spider than the presence of prey. The role of the increased relative humidity in a hiding place, as compared to that in the surroundings, is considered in the next section.

4 The Climate in the Habitat

Macroclimate

The habitats of spiders of the genus *Cupiennius* are located in two climatic regions, which in technical terms (Walter and Lieth 1967) are defined as follows:
(i) "equatorial, always humid zone or with two rainy seasons, with no frost, temperatures mostly above 20 °C and little change in temperature over the year",
(ii) "tropical and subtropical zones of summer rain and a cooler dry season."

The climate diagrams of figure 4 (Barth, Seyfarth et al. 1988) show the monthly average val-

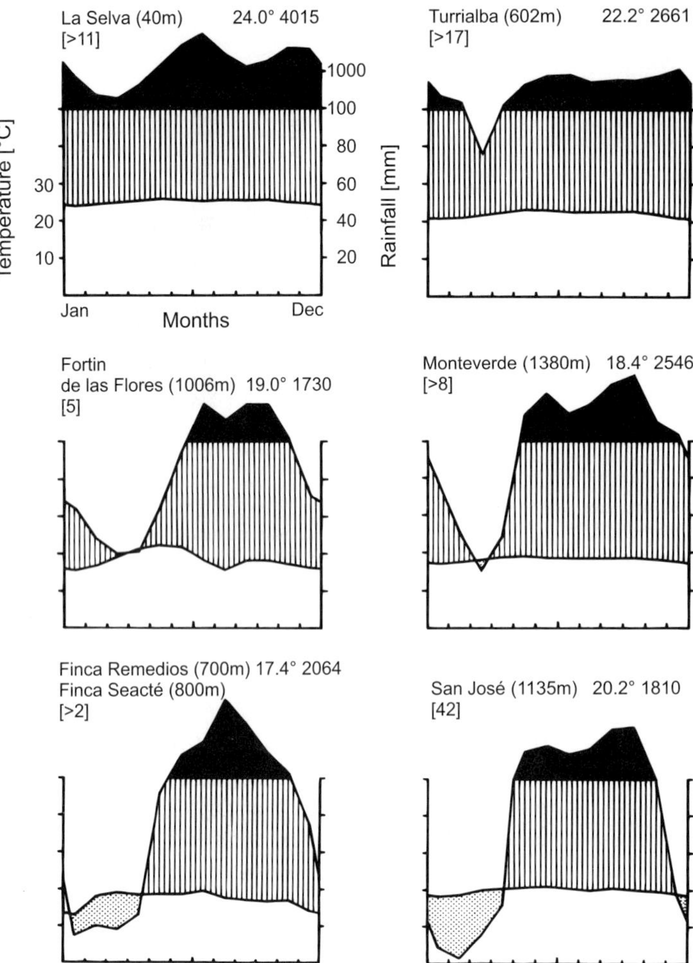

Fig. 4. Climate diagrams for the habitats of *Cupiennius* in Central America, showing average monthly temperatures and rainfall. Dry seasons (*dotted area*) occur only in some of the locations. The number of years for which measurements were taken is found below the name of the site *in square brackets*. The two numbers at the *upper right* represent the average annual temperature (°C) and the average annual rainfall (mm). *Vertical shading* indicates the wet season, whereas the *black areas* indicate monthly rainfall exceeding 100 mm. Note the reduction of calibration in the corresponding upper part of ordinate to 1/10. Sources: La Selva and San José, Costa Rica: Instituto Meteorologico Nacional San José and Walter and Lieth (1967); Turrialba, Costa Rica: CATIE; Fortin de las Flores, Mexico: Dr. P. Böhm; Monteverde, Costa Rica: JH Campbell and Instituto Meteorologico Nacional, San José; Finca Remedios, Guatemala: A. Schleehauf. (Barth, Seyfarth et al. 1988)

ues of temperature and precipitation for 6 habitats of *Cupiennius* in Central America, where we have found five different species: lowland (La Selva, 37 m, *Cupiennius getazi* and *Cupiennius coccineus*) and medium altitude (Turrialba, 602 m, *Cupiennius getazi)* in Costa Rica, and highland in Costa Rica (San José, 1135 m, *Cupiennius coccineus*; and Monteverde, 1380 m, *Cupiennius panamensis*), Mexico (Fortin de las Flores, 1006 m, *Cupiennius salei*) and Guatemala (Fincas Remedios and Seacté, ca. 800 m, *Cupiennius salei* and *Cupiennius remedius* n. sp.).

As the graphs show, the range of climate values extends from habitats with no dry period in the lowland (La Selva) to those with a clearly distinguishable dry period in the highland (San José, Fincas Remedios and Seacté). At the high altitudes the annual average temperature is below 20 °C. The climate data in the Canal Area of Panama, where no less than five *Cupiennius* species live, are similar to those for San José, Costa Rica.

Microclimate

The microclimate in the immediate vicinity of an animal can differ considerably from the average values. Some animals are masters at seeking out comfortable niches in otherwise hostile surroundings. And, of course, many construct their own shelters: think of a fox's den, a bird's nest, a bumblebee's hole in the ground, a termite mound or countless other microhabitats made by the animals themselves. Where does *Cupiennius* stand in this respect?

Like spiders in general, *Cupiennius* is not endothermic. Not only is its metabolic rate too low for it to generate its own heat, it also lacks musculature suitable for this purpose, such as a bumblebee has in the form of its flight muscles (Anderson 1970). Therefore its temperature regulation can only be ectothermic. Does *Cupiennius* do anything to keep its body temperature and its water content, which is closely related to temperature, within the required limits? For a

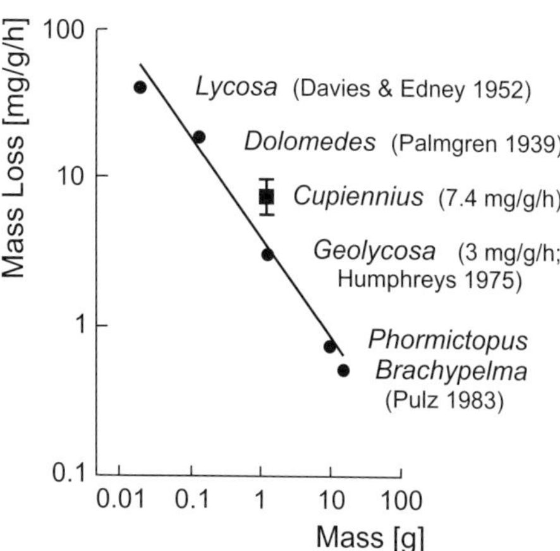

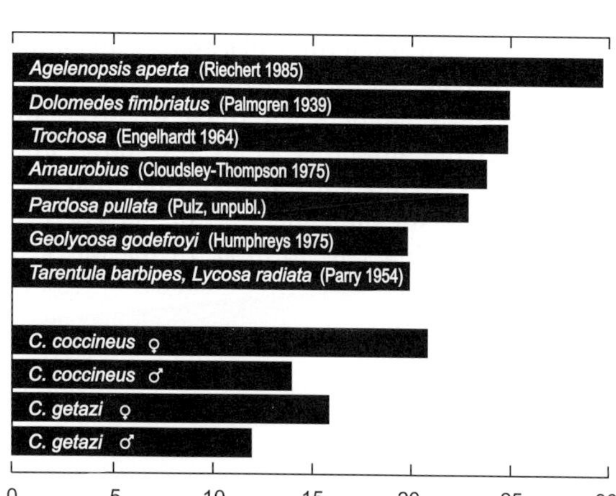

Fig. 6 a, b. Water balance in *Cupiennius*. **a** Relationship between the evaporation rate and body mass in different spiders under roughly similar environmental conditions. *Lycosa amentata* air current 5 cm s^{-1}, "dry air"; *Dolomedes fimbriatus*: "weak air current", 10% RH; *Geolycosa godeffroyi*: 0.7 cm/s, 10% RH Humphreys 1975); *Phormictopus, Brachypelma*: 1.7 cm/s, 8% RH (Pulz 1986); *Cupiennius* (female): 0.2 cm/s, 60% RH. **b** Lethal water loss for different spider species in percent of the initial body mass. In the case of *Cupiennius* each value from measurements of six animals. *Tarentula* is now called *Alopecosa*. (Parry 1954; Cloudsley-Thompson 1970; Pulz 1987; Mitter 1994)

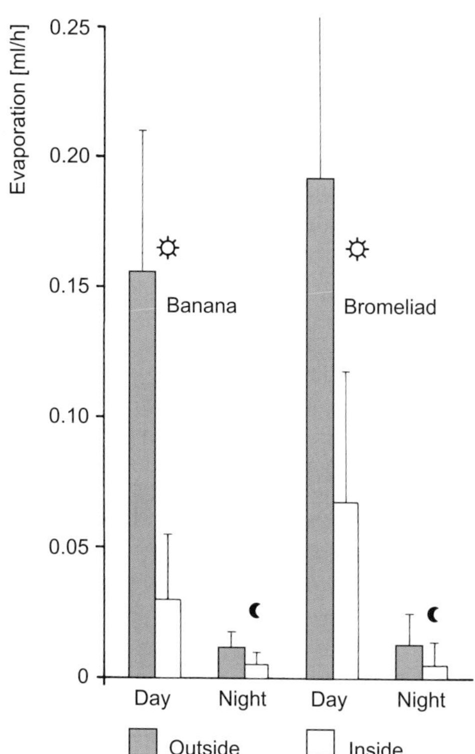

Fig. 5. Average water evaporation per h inside and outside typical retreats of *Cupiennius* in a banana plant (*Musa sapientium*) and a bromeliad (*Aechmea bractea*). Measurements were taken in the field during 15 consecutive days and nights. (Barth, Seyfarth et al. 1988)

terrestrial arthropod with a large surface-to-volume ratio, this is potentially a serious problem (*re* spiders see Pulz 1987).

Cupiennius has a marked day-night activity rhythm (Barth and Seyfarth 1979; Seyfarth 1980; Schmitt et al. 1990) (see Chapter IV). It spends the light, daytime hours in its natural or fabricated hiding place on the plant. Apart from the mechanical protection and concealment from predators that this behavior provides, it is clearly advantageous in shielding the spider from the sun and preventing heat desiccation. Figure 6 shows microclimatic details of two typical retreats of *Cupiennius salei*, one on a banana plant and the other on a bromeliad. During the day the average rate of water evaporation inside the retreat is considerably lower than just outside. By night, on the other hand, when the spider has left its retreat and is sitting or walking around in the open, the conditions inside and outside the retreat are substantially the same.

The measurements of relative humidity we made in Mexico for hiding places of *Cupiennius salei* on banana plants point in the same direction. Where the spiders withdraw to spend the day, the relative humidity is always above 90%. This value is significantly higher than that at a corresponding site "outside" the plant, where at noon the humidity falls to 67%. In the Mexican midday heat, it is quite impressive to come across a shelter in which the spider is still completely covered with dewdrops from the night before (Plate 5).

That the hiding place is a refuge for *Cupiennius* in this climatic sense as well, and helps the spider to conserve water, has been demonstrated clearly in laboratory experiments (Mitter 1994). When the relative humidity is set at 60%, the evaporative water loss of *Cupiennius* is about twice as high (about 0.6% of the initial weight per hour) than at 80% relative humidity. The water loss due to evaporation in *Cupiennius getazi* and *Cupiennius coccineus* is considerably higher, as a percentage of body weight, than in other spiders under the same conditions (Fig. 6a). Furthermore, even moderate water loss (ca. 12 to 21% of the original body weight) is lethal (Fig. 6b). All this underlines the significance of humidity and hence of the hiding place for the well-being and survival of *Cupiennius*. The daily activity rhythm (see Chapter IV) can also be interpreted in this light: all activities are scheduled for the climatically favorable twilight and dark hours. Humidity is even more critical for males than for females, as their rate of evaporation is about 1.7 times as high. There is no evidence, so far, of any physiological means of compensating water loss in *Cupiennius*.

IV Daily Activity Rhythm

Anyone looking for *Cupiennius* in its natural habitat by day will need to have a good eye for its typical plants and to be familiar with its particular hideaways. On our very first trip into the highlands of Alta Verapaz in Guatemala's *tierra templada*, we soon realized that the locomotor activity of *Cupiennius salei* has a well-developed day-night rhythm (Barth and Seyfarth 1979). After sundown, when the light intensity has fallen to about 15 lux, the spider turns around and leaves its retreat. It stays nearby for about half an hour, sitting motionless with the prosoma pointing upward until to the human eye it has become completely dark (light intensity below 0.1 lux). Then *Cupiennius* walks slowly onto a leaf and lies in wait for prey (Fig. 1). This sequence of turning around, waiting and walking is stereotyped and can also be observed in the laboratory. The timing of the return to the retreat varies widely. The spider may come back after three or four hours or later on. Having captured prey, the spider eats it on the spot, rather than immediately carrying it back to the protection of the retreat.

These initial observations of *Cupiennius salei* in March 1977, on the grounds of the Finca Seacté and Finca Remedios, were simultaneously exciting and encouraging for us, because until then it had been so difficult to find *Cupiennius salei* in its homeland at all, and also because we had already found out that in order to do behavioral experiments, for instance on courtship (see Chapter XX), in the laboratory it might be very important to choose the right time of day.

Day-Night Rhythm

Since then, two laboratory studies have investigated spider daily rhythms in some detail. One involved *Cupiennius salei* and was published by my long-term coworker and colleague Ernst-August Seyfarth in 1980, when we were both at the University of Frankfurt am Main. Here, an "actograph" was used to record the times during which the spiders were in motion, under controlled conditions with 12 h light, 12 h darkness (LD 12:12); the light phase was arranged to coincide either with the daytime outdoors or with the night. In either case, the animals began to

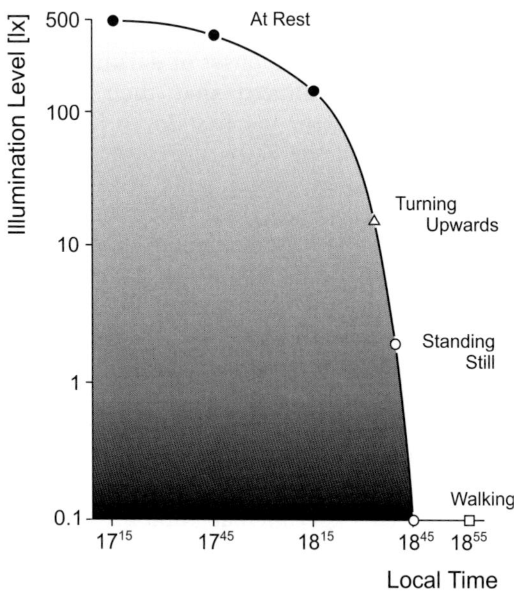

Fig. 1. Sequence of behaviors shown by *Cupiennius salei* as the illumination level decreases at dusk; field observations in Guatemala. (Seyfarth 1980)

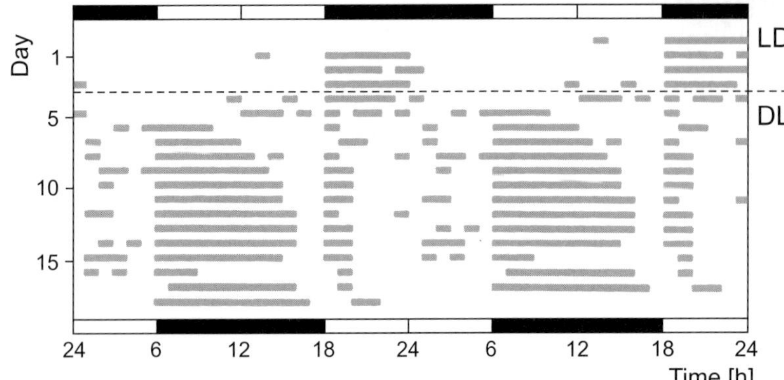

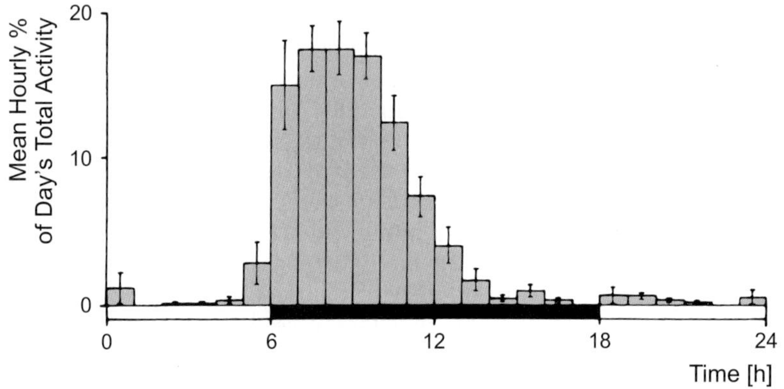

Fig. 2. Entrained activity rhythm of *Cupiennius salei* (adult females) after reversal of LD cycle (12:12 light-dark) to DL cycle. *Top* This actogram shows high activity of the animal when the light is switched on at 06:00 h (initial LD cycle); within 48 h, however, the spider is synchronized with the new DL cycle. *Bottom* Distribution of hourly activity (±SD) during 21 days. (Seyfarth 1980)

walk shortly after the onset of darkness. The experiments also showed that their activity is usually maximal in the first three hours of the dark phase and the activity period usually lasts 6 to 7 hours. In experiments with permanent darkness (DD) the activity rhythm was shown to be endogenous and circadian; that is, its period is not exactly 24 hours but 24.9 h (standard error ±0.31 h). In permanently light conditions (LL; 26 lux) the rhythmicity of walking activity breaks down after about 3 days. Furthermore, when the light-dark alternation was suddenly reversed (LD → DL), it took only 48 hours for the animals to become synchronized with the new DL rhythm. This last finding had major implications for us experimental zoologists: It means that under laboratory conditions it is easy to turn our daytime into nighttime for *Cupiennius*, so that we can have active animals available for observation at a time when we ourselves are normally active (Figs. 2 and 3).

The importance of studying *Cupiennius salei* at the right time of day, even when we were interested in supposedly simple and stereotyped forms of behavior, had been clear to us since we had found that certain leg-muscle reflexes can be triggered much more easily during twilight or in a darkened laboratory than in the light phase of the day (Seyfarth 1978a). In fact, fifteen years earlier Melchers (1963b) had noticed that certain activities such as molting, courting and copulation, as well as spinning – of the sperm web by the male and of an egg sac by the female – occur exclusively in the dark phase of the day. The endogenous circadian rhythm, demonstrated at least for locomotor behavior, and the strict limitation of activity to the night are not specialties of *Cupiennius* alone. Most of the other arachnids that have been investigated in this respect behave in just the same way (Cloudsley-Thompson 1978, 1987). One special feature of *Cupiennius salei* is the ca. 24.9-h peri-

od of the free-running (in DD) rhythm (Fig. 3). Scorpions (Wuttke 1966, Cloudsley-Thompson 1968, 1973, 1975), a bird spider (Cloudsley-Thompson 1968) and a whip spider (Beck 1972) all were found to have a circadian period shorter than or equal to 24 h. *Cupiennius*, in contrast to these spiders but like other arthropods, does not follow "Aschoff's rule", which says that the period in DD should be shorter than 24 h and also shorter than the period in permanent light (LL) (Pittendrigh 1960; Aschoff 1979). In LL the locomotor activity of *Cupiennius* becomes arrhythmic after only a few cycles.

Because we were less interested in the rhythms *per se* than in their significance for the normal behavior of *Cupiennius*, we asked, "Of what benefit is a nocturnal mode of life to this spider?" One factor that immediately sprang to mind was that most of the animals we know to be its prey (Barth and Seyfarth 1979 and subsequent field observations) – that is, cockroaches, crickets, earwigs, moths and small frogs – are also night-active. On the other hand, predators such as birds, spider-wasps (Pompilidae: the females paralyze spiders with their venomous sting and leave them as the sole food for their larvae) and certain reptiles are day-active. Presumably the most important advantage *Cupiennius* derives from its nocturnal activity is avoiding the heat and low relative humidity of the daytime. As described in Chapter III, at temperatures above about 40°, which are easily reached in direct sunlight, spiders lose considerable amounts of water (Davies and Edney 1952; Mitter 1994). When they are molting, a process that lasts several hours and is probably the most dangerous situation in a spider's life (it is repeated 11 times during the lifetime of *Cupiennius*), the exuvial fluid would evaporate very rapidly at such temperatures and loss of its lubricant action could cause severe problems. The hydraulic mechanism for extending the legs is also likely to be affected (see Chapter XXIV).

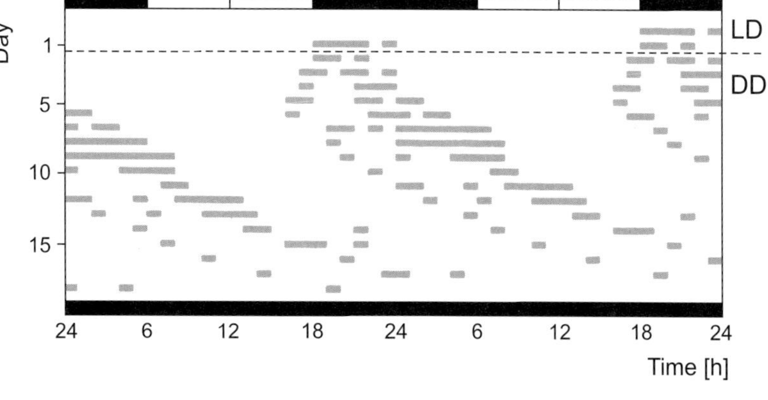

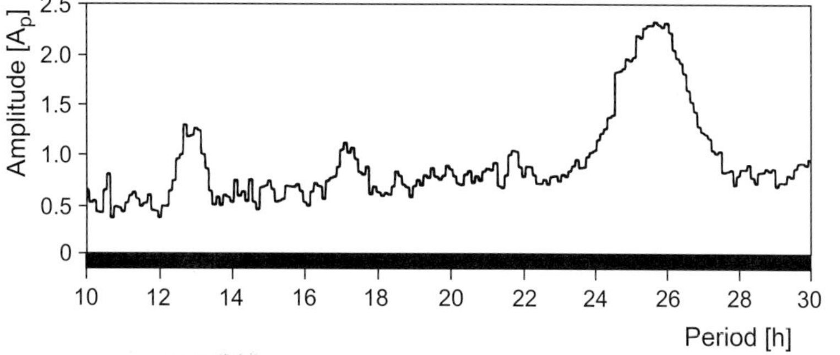

Fig. 3. Free-running activity rhythm of *Cupiennius salei* (subadult females) in DD. *Top* Typical actogram, which shows splitting during a period of 3 to 4 days with an apparent dissociation of the evening activity (one period of activity slightly shorter, the other slightly longer than 24 h; see days 3 to 6). *Bottom* Periodogram of data shown above. A_p is a measure for the standard deviation of the average activity during 0.1-h intervals. The main peak is at 25.7 h and indicates the period of the endogenous rhythm. The average of seven animals is 24.9 ± 0.31 h. (Seyfarth 1980)

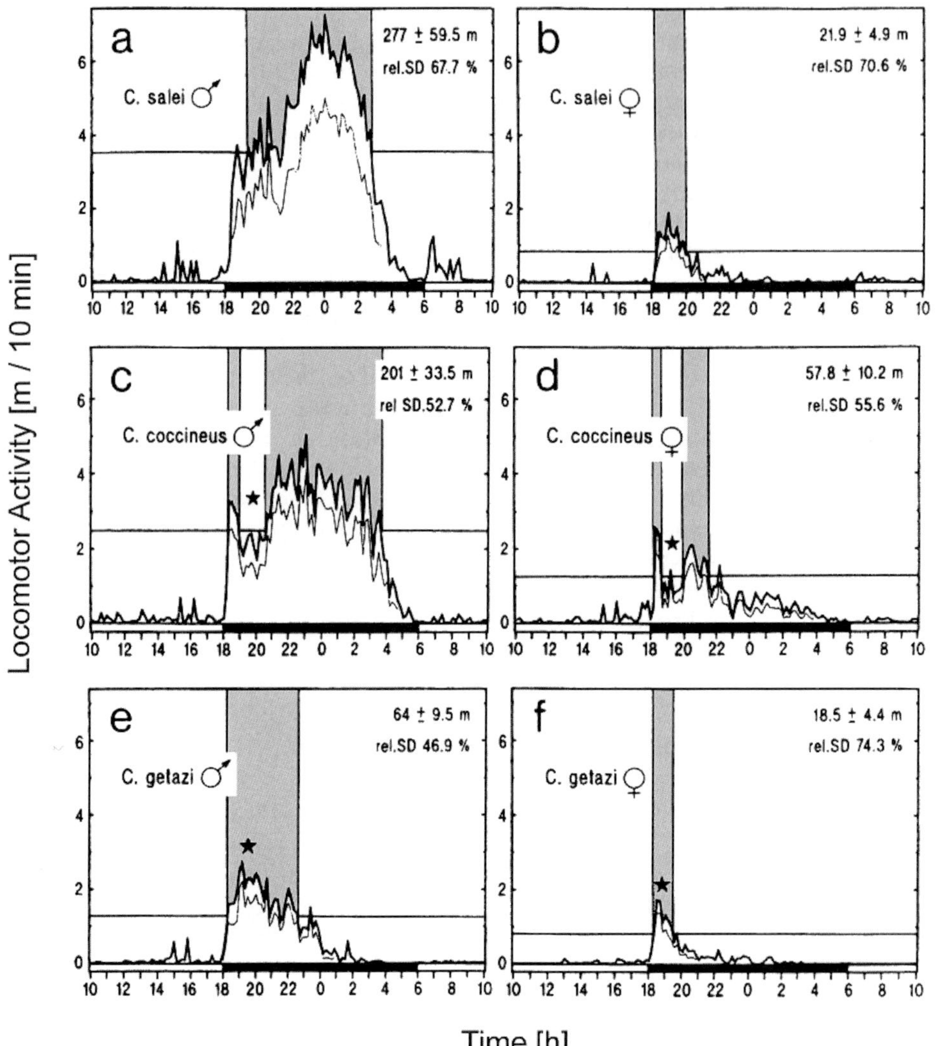

Fig. 4a–f. Daily locomotor activity patterns of adult males and females of *Cupiennius salei*, *C. coccineus*, and *C. getazi* (in all cases n = 10); mean (*thick line*) and standard error (*thin line*, only lower limits shown). *Horizontal lines* indicate 50% values of activity and the *shaded areas* the time periods of maximum activity. The *asterisks* mark time of maximum activity in *C. getazi* and of relative activity minimum in *C. coccineus*. *Black area* on time axis represents the dark period (18:00 to 06:00). The numbers on the *upper right* give the total amount of activity (m) (mean, standard error, relative SD). (Schmitt et al. 1990)

Comparison among species

And this is not yet the entire story. In another actograph experiment, we returned to the question of activity distribution over the day in the entirely different context of courtship and the reproductive isolation of species from one another (Schmitt et al. 1990). This involved a comparison between males and females of the three large *Cupiennius* species (Plates 1–3).

The main findings are shown in figure 4. Like *Cupiennius salei*, *Cupiennius getazi* and *Cupiennius coccineus* are strictly night-active. The males of all three species carry out only about

4% of their total activity in the light phase; the corresponding figure for females is 8.7%. Again, lights-out proved to be a very effective "zeitgeber" (timing signal): the active time began as soon as the light was turned off. Twenty minutes after the onset of darkness, the activity levels of all species had already exceeded 50% of the maximum. On average the males were 3.5 times (*Cupiennius coccineus* and *Cupiennius getazi*) to 12.7 times (*Cupiennius salei*) as active as the females. When we use the term "wandering spiders", then, we should bear in mind that the males in particular deserve this name. Their strong tendency to move about presumably has a mainly sexual motivation and is associated with searching for females (see Chapter XX, courtship).

The graph shows something else interesting, in view of the fact that *Cupiennius getazi* and *Cupiennius coccineus* are sympatric with one another whereas they and *Cupiennius salei*, the third of these three large *Cupiennius* species, are allopatric – that is, *C. salei* so far was never found together with the other two (see Fig. 1, Chapter III). As will be described in Chapter XX, vibratory communication plays an especially important role in reproductive isolation of the species. Nevertheless, it is likely that differences in the daily activity patterns also contribute by isolating the species temporally from one another, producing a partial "time sharing". *Cupiennius coccineus* in particular (males and females) has a relative activity minimum at the time when activity of *Cupiennius getazi* (males and females) is maximal. Furthermore, *Cupiennius coccineus* is most active when the activity of *Cupiennius getazi* is beginning to decline. And finally, the activity period of both sexes is distinctly longer in *Cupiennius coccineus* than in *Cupiennius getazi*.

V No Spider Without Poison

Whatever ingenious forms of behavior a spider may use for hunting, and however acute the senses used to detect prey, in the end it is poison that settles the matter. Killing prey by injecting poison is typical of spiders (with very few exceptions: the Uloboridae and *Heptathela*, Liphistiidae; Foelix 1992), and this characteristic has contributed much to the almost universal attitude of antipathy towards these wonderful animals. However, fear of a spider bite is almost always unfounded. The species dangerous to humans mainly belong to only five genera. Even though the spiders' prey is subdued by their venom, an injection into a person – very unlikely to occur in the first place – would give reason for sorrow in only a small percentage of cases. This should not come as a surprise, because we are by nature neither prey nor predators of spiders. In Europe, at least, I think it is more probable that a person will be struck by lightning than be hurt by a spider bite. The large, hairy bird spiders, which appear so threatening to the layman, have relatively small poison glands and hence are ordinarily not dangerous to us, although a bite by their massive chelicerae may present an appreciable risk of infection.

Among the vast majority of spiders harmless to humans is *Cupiennius*. Some information is available about its poison, but before summarizing this I should like to introduce briefly the most important and best known of the truly dangerous spiders. There are about 30 species of these. Only two are likely to be encountered at our northern latitudes: the water spider (*Argyroneta aquatica*, Agelenidae) and *Chiracanthium punctorium* (Clubionidae, called in German *Dornfingerspinne* or "thorn-finger" spider). Their bite may be unpleasant, and can later cause piercing, burning pain, a shivering fit and swelling of the reddened site of the bite; but after three days these symptoms disappear. According to Habermehl (1976) the species of five genera can be categorized as really dangerous to humans. They are responsible for several thousand spider-bite incidents each year, of which a few hundred end in death. This horror list comprises the following genera:

- *Latrodectus* (Theridiidae), the black widows; their poison is neurotoxic and they live in North and South America, in Africa and also in the Mediterranean region, sometimes in large numbers.
- *Atrax* (Dipluridae), which is limited to Australia.
- *Loxosceles* (Sicariidae), infamous in the USA as the "brown spider" (*L. reclusa*) but particularly widespread in South America; its poison has a cytotoxic and hemolytic action.
- *Phoneutria* (Ctenidae), of Central and South America, the toxin of which is a particularly effective nerve poison that causes fatal paralysis (Plate 11).
- *Harpactirella* (Barychelidae, Mygalomorphae), called "baboon spiders" by the locals; a dangerous South African species belongs to this genus.

In the publications by Maretić (1987) and by Bettini (1978) the interested reader will find more about the pharmacology and clinical aspects of spider poisons and bites. About the consequences of being bitten by *Phoneutria nigriventer*, which currently, like *Cupiennius*, is still being counted as a member of the Ctenidae (but see Chapter II), Zvonimir Mareti writes: "The pain due to the bite is so severe that it may cause shock if not combated immediately. It irradiates to the whole affected limb extending to the trunk. ... Since with the bite only small doses of this potent venom are injected, death rarely occurs." A summary devoted specifically to the poisons of *Phoneutria*, a spider known and notorious in Brazil as "aranha armadeira" because of the aggressive posture it adopts when threatened, has been published by Schenberg and Pereira Lima (1978).

Although most spiders, unlike the ones referred to above, certainly are better than their reputation, my essentially positive view needs an addendum. According to recent studies there may be representatives even of species so far considered harmless whose bite may cause long-lasting necrotic arachnidism (*Tegenaria agrestis*, Vest 1987, 1993; Anonymous 1996).

The Harmlessness of *Cupiennius*

So as not to inspire yet more fear of spiders, or to depart too far from the main subject, let us now return to *Cupiennius*. To the worried reader I can say at the outset that I have survived being bitten by *C. salei, C. getazi, C. coccineus* and *C. panamensis* with no problems at all. All these bites were sustained in situations to which a reckless biologist exposed himself out of curiosity and interest in the object of his research, but which a – shall we say – normal person would hardly ever encounter: for instance, catching spiders in the thickets of a tropical forest at night. Nevertheless, the "first bite" can be nerve-racking. I still remember very well. It was in Costa Rica, at the OTS station Las Cruces in the south of the country, during spring of 1985. The perpetrator: a large female *C. coccineus*; the victim: Ortrun, my wife. While being fed, the spider had escaped from its home container, and without deliberation or inhibition she lifted it off the wall of the room. What sensible animal would not defend itself with all available means in such desperate straits? The bite was a vigorous one to a finger; the pain was about like that of a bee sting. What would happen next? We watched the clock and, benevolently guessing that the spider would turn out to be innocuous, let things take their course. The first violent pain was followed a few minutes later by a feeling of numbness around the bitten place, after 20 minutes the pain had almost died away, and after 30 minutes it was all over and the result of the experiment was clear: harmless! Because we knew how closely *Cupiennius* is related to *Phoneutria*, the experience was not without a certain tension.

The happy outcome of this involuntary but still welcome experiment with *Cupiennius coccineus* reminds me of a letter dated September 5, 1957. I was given it long ago by Mechthild Melchers, who introduced *C. salei* to science as a laboratory animal. Sender: Dr. J. Vellard, Institut Francais d'Études Andines, Lima, Peru. Translated from French, the letter reads as follows:

"Sir and dear colleague: On returning from a trip I find your letter of July 27 and am responding immediately. In 1936 I published with Masson of Paris, in the series of monographs of the Institut Pasteur, a book on: THE VENOM OF THE SPIDERS. In this work, which you can obtain from the editor or find in university libraries, I am concerned with the genus *Cupiennius* on page 191 to 193, and more particularly with *Cupiennius salei*; it is a species not very dangerous to man and to mammals, mainly causing local accidents; but this venom is extremely active for amphibians ..."

Monsieur Vellard was evidently unaware that Mechthild is a woman's first name.

The Poisons

Wolfgang Nentwig and his colleagues at the Zoological Inistitute of the University of Bern in Switzerland, in particular Thomas Friedel, who later came to my laboratory in Vienna, subjected the poison of *Cupiennius salei* to a variety of analyses. For such tests, too, this species is very useful because of its size and the ease with which it can be bred and kept in the laboratory.

The two *venom glands* are of the "endocephalous" type according to the old classification by Millot (1931); this type is also characteristic of the Pisauridae, Agelenidae, Linyphiidae, Lycosidae and Pholcidae. Figure 1 shows the position and the impressive dimensions of the venom glands of *C. salei*. They extend far into the prosoma, within which is situated the cylindrically shaped main gland with a slight constriction in its middle region. From the front end of the gland itself a duct passes forward through the basal segment of the chelicera to its opening at the end of the movable fang. Before reaching the

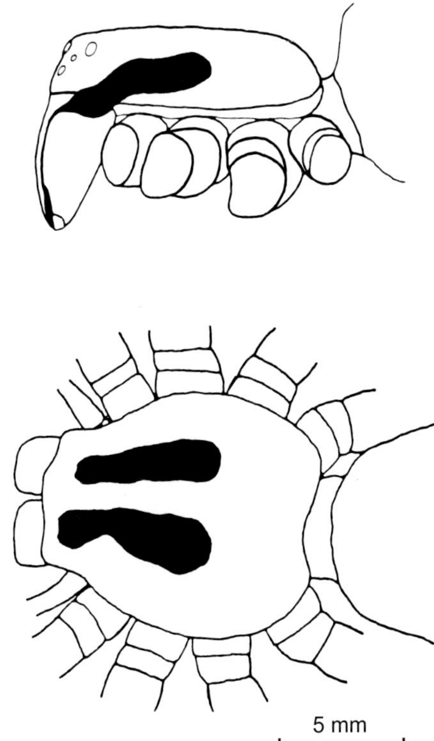

Fig. 1. Position and size of the venom glands in the prosoma of an adult female of *Cupiennius salei* in lateral (*top*) and dorsal (*bottom*) view. (Malli et al. 1993)

opening, it expands to form a so-called ampulla, which in the opinion of some arachnologists serves as a reservoir for the poison that is ready to be injected. The volume of the venom gland approximately doubles from one molt to the next. The gland of the male is smaller than that of the female. In adult animals it can hold around 8.6 μl or 12.1 μl, respectively. The amount of poison that can be extracted from the gland is about 3.0 μl for males and 5.6 μl for females (Malli et al. 1993).

The clear, colorless venom of *C. salei* is a natural insecticide, which acts by causing a very rapid, dose-dependent paralysis of the prey insect. If a large enough amount of poison is injected, the victim dies (Friedel and Nentwig 1989). This occurrence makes little difference biologically, because *C. salei* usually eats its prey as soon as it has been caught. The poison of adult females is evidently more toxic than that of the male (Malli et al. 1993).

When the strength of the poison is compared among various spider species in terms of the *lethal dose* for 50% of test animals (LD_{50}), differences of 2 to 3 orders of magnitude are found (Friedel and Nentwig 1989; Nentwig et al. 1992). The same applies to the lower, and presumably more biologically relevant, dose that causes immobilization of the insect (ID_{50}). When the poisons of 18 spider species belonging to 15 families were tested on the cockroach *Blatta*

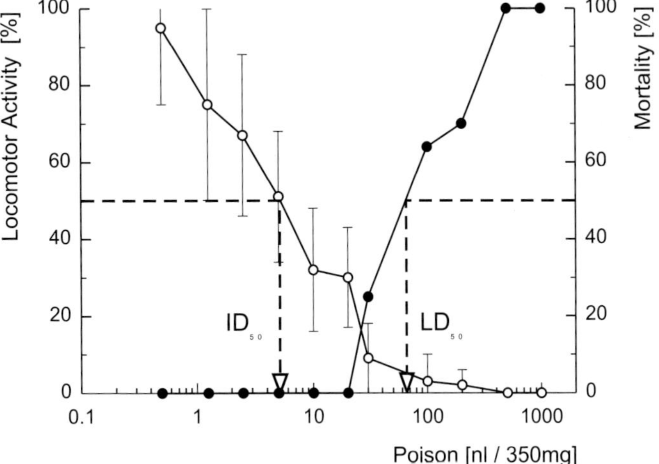

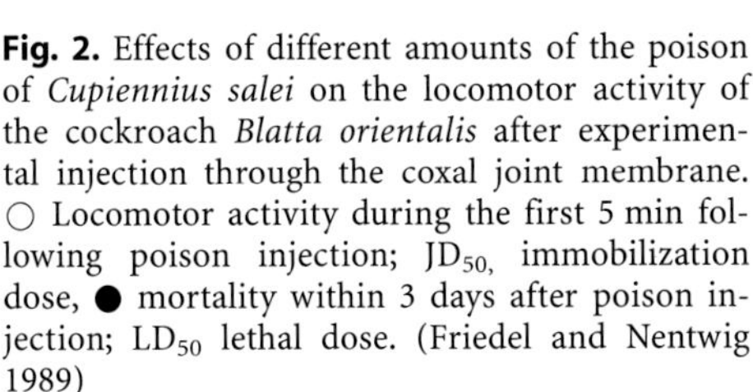

Fig. 2. Effects of different amounts of the poison of *Cupiennius salei* on the locomotor activity of the cockroach *Blatta orientalis* after experimental injection through the coxal joint membrane. ○ Locomotor activity during the first 5 min following poison injection; JD_{50}, immobilization dose, ● mortality within 3 days after poison injection; LD_{50} lethal dose. (Friedel and Nentwig 1989)

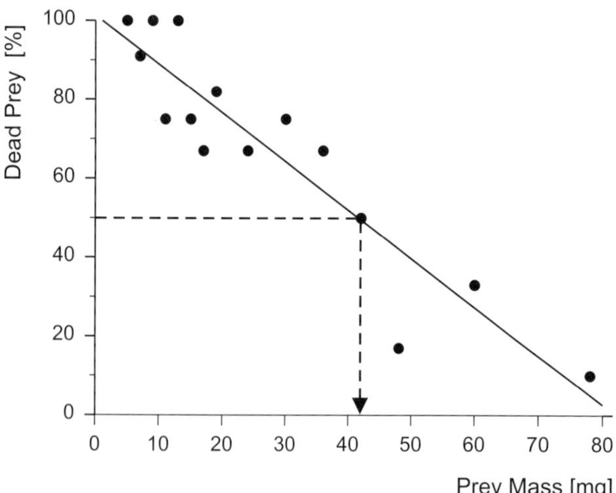

Fig. 3. The lethal effect of the poison of *Cupiennius salei* on crickets (*Acheta domesticus*) 24 h after being bitten (and subsequent separation from the spider) as a function of body mass. More than 50% of the crickets with a body mass <40 mg are found dead 24 h after the bite. (Boevé 1991)

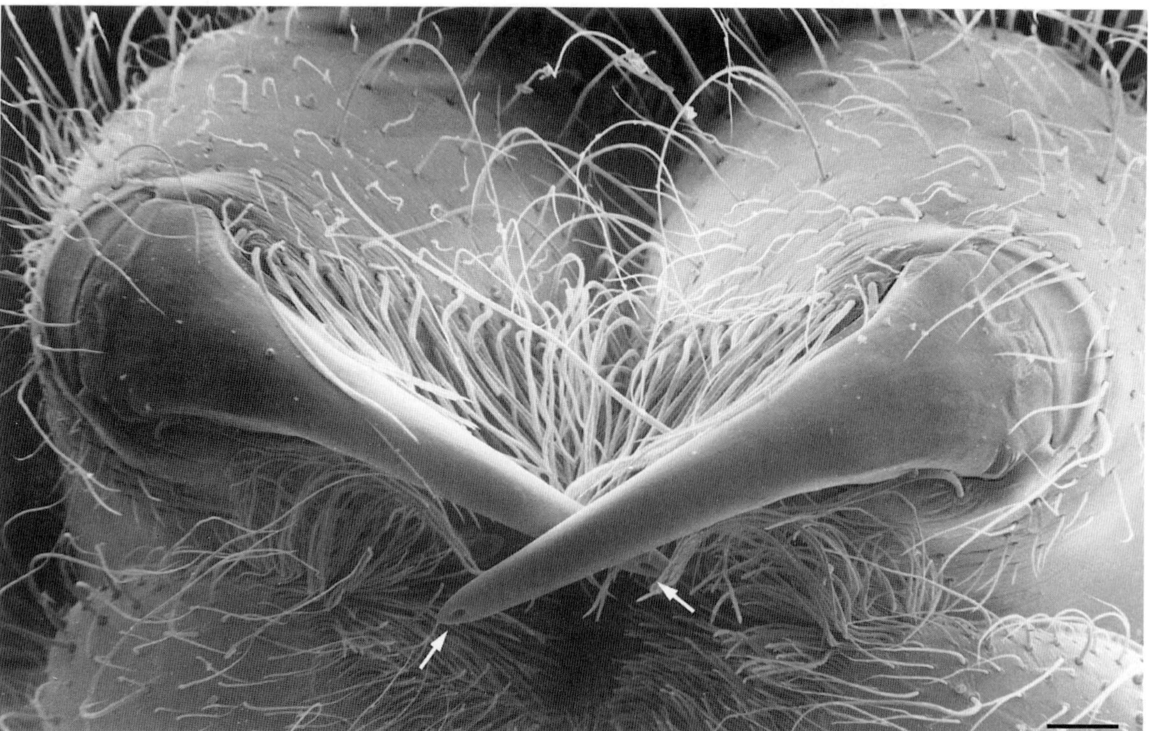

Fig. 4. The chelicerae of *Cupiennius salei*. *Arrows* point to openings of venom gland duct at the tip of the fangs. Bar 100 μm

orientalis, *Cupiennius salei* was roughly in the middle of the range of toxicity measured as LD_{50} (Fig. 2).

From figure 3 it can be seen that when the test animal is taken away from the spider as soon as it has been bitten, after 24 hours prey weighing less than 40 mg have a mortality $\geq 50\%$ (Boevé 1994). The ID_{50} varies between about 2.2 and 150 nl among the 18 spider species tested. *Cupiennius salei* belongs to the group with poison that most quickly causes immobility. In addition to its rapid onset, the immobility produced by injection of *Cupiennius* poison is largely irreversible (Nentwig et al. 1992), which is not true of the poison of many other spiders. An interesting hypothesis, supported by the same analyses, states that the spiders with the most rapidly acting poisons are those most in need of them – spiders that, like *Cupiennius*, hunt in vegetation and are likely to lose a prey animal for good if it struggles too long and falls to the ground. As would be predicted, most of the spiders that hunt on the ground or catch prey in webs have higher ID_{50} values, even spiders as poisonous as the black widow (*Latrodectus hesperus*).

Very recently *molecular biologists* have also become interested in spider poisons. The reason: toxins that cause paralysis act on channel proteins in cell membranes in such a specific way that they are excellent tools for characterizing ion channels. For instance, there are neurotoxic spider poisons that block voltage-dependent presynaptic calcium channels and in this way prevent the release of transmitter at neuromuscular synapses (Bowers et al. 1987; Branton et al. 1987). Other poisons also act presynaptically but oppositely, causing the transmitter at neuromuscular synapses of insects to be released in massive quantities (Cull-Candy et al. 1973). Still others act postsynaptically on glutamate-receptor channels (Usherwood et al. 1984; Adams et al. 1987, 1989). Adams et al. (1989) were able to show that the venom of *Agelenopsis* contains two toxins that have different effects on the neuromuscular synapse of flies: one acts postsynaptically to cause rapid, reversible paralysis, while the other acts presynaptically and paralyzes slowly but irreversibly. With such complex properties, spider venoms have aroused equal interest in analysts who want to isolate and discover the chemical characteristics of their toxic components. The *Cupiennius* venom, like that of other spiders, contains proteins and peptides with enzymatic and/or neurotoxic actions. Kuhn-Nentwig et al. (1994) have isolated 13 such components and determined the amino acid sequences of the toxin CSTX-1 (CS for *Cupiennius salei*, TX for toxin). CSTX-1 is the most active toxic peptide in the venom of this species, with an LD_{50} of 10.5 ng/mg fly (*Protophormia* sp.). The raw venom of *C. salei* is much more effective, however, with an LD_{50} of 1.43 ng/mg fly.

Apart from several disulfide-rich peptides like the neurotoxic CSTX1 the venom of *Cupiennius salei* also contains peptides of smaller molecular mass which show antibacterial activity (Kuhn-Nentwig et al 1998).

B Sensory Systems

What is the biological significance of sense organs? This question is central to an understanding of the spider's world, toward which we have striven all these long years. Sense organs are the interfaces between the environment and the central nervous system, and hence between environment and behavior. These organs and the neural structures that transmit their signals are not simply windows through which the animal views its surroundings. The picture of the environment generated by the brain on the basis of information from the sense organs has been thoroughly filtered, giving a species-specific view of the world that may differ dramatically in different animals. Perceptions are by no means comprehensive in a physical sense. Instead, they provide a limited and distorted image of the objectively measurable reality. Pablo Picasso once said, so we are told, that art is the lie that helps us to recognize the truth (Wilson 1984). In this sense, art and perception are close kin.

We have become much more aware of the technical perfection of many sense organs, and their associated information-processing structures, thanks to a number of recent experimental innovations. To a great extent, the fascination that motivates sensory physiologists' research lies in the excellent performance of sense organs when they are examined in the same way as technical measurement devices. Evidently there has been a strong evolutionary selection pressure on the sensory systems. This should come as no surprise, given their role as mediators between environment and behavior: evolution gave the advantage to those organisms that created for themselves a more useful image of their habitat than did their conspecifics. Continual feedback by way of experience steadily corrected and refined this image. *A priori*, then, we would expect significant properties of this environment, the species-specific habitat, to be reflected in the properties of the sense organs. Although an animal can never know everything about its surroundings, the things it does know are of particular biological importance. And we can distinguish genetic information about the characteristics of the habitats of former generations from the sensory information that an individual – one of our spiders, for instance – obtains from its sense organs about the habitat in which it is actually living.

Here we see the roots of the young discipline of sensory ecology. This is by no means a new concept, but it is becoming more "fashionable". The idea had already been formulated at the beginning of this century, when in particular Jakob von Uexküll (1909, 1920) emphatically pointed out both the subjectivity and the predetermined nature of the relations between an organism and its environment. He regarded the things in the environment as carriers of meanings, which play quite specific roles in animals' lives. Accordingly, von Uexküll ascribed considerable importance to their subjective experience, at a time when animals were considered more as reflex machines and there was not much room in biology for the older ideas of Immanuel Kant about subjective time and subjective space.

Quite clearly, then, the technical refinement of these biological measurement devices, as such, is only part of the story; the way they are matched to the specific features of the habitat is the other. Wherever possible, therefore, we should think about their performance in the context of the whole organism and its environment, and avoid viewing their "technical" perfection in isolation. David B. Dusenbery (1992) recently published an important book entitled "Sensory Ecology". It presents an initial, comprehensive attempt to describe the basic principles governing the acquisition of information from the environment and how that information is used. The diversity of problems to be dealt with when we consider the constraints and opportunities associated with the uptake and use of stimuli representing different forms of energy is also the subject of a recent book on the "Ecology of Sensing" (Barth and Schmid 2001).

Konrad Lorenz made the monumental assertion that life is a process of obtaining knowledge (Lorenz 1943, 1973). Viewed in this light, comprehension of the surroundings is a step toward emancipation from the compulsions of nature. Obviously, in this process it must have been an advantage for an organism to be so adapted that its perceptions match the real world. A false reconstruction of the real world by sense organs and brain is eliminated, because it reduces fitness. At the same time, though, we should not forget that our experience of the world, and that of all other animals, is constrained by the biological limitations of the "filter properties" mentioned above. The perceived reality is a species-specific reality, and it looks quite different to a spider than to a fly or a bird.

I shall return to such thoughts in due course, after presenting some concrete examples of what a spider's senses can achieve, the purposes they serve and why they are necessary. But first let us briefly consider another quite general problem: how many senses are there?

This question occupied Aristotle over 2000 years ago. What he wrote in his treatise *De anima* remains the classical point of departure for further debates: there are five senses, namely sight, hearing, taste, smell and touch. How about the sense of temperature, though? And feelings of pain, dizziness, hunger and satiety? Scholars today are still concerned with what lies beyond the five standard senses and their definition and classification. Before we proceed to the spider senses, at least a few explanatory words on this topic may be helpful.

First: the classical subdivision of the senses and its modifications derive from medicine and refer to humans. A zoologist cannot be content with this, because the senses of animals are often organized quite differently from human senses, and within the animal kingdom there are a number of means of detecting stimuli imperceptible to humans. Both of these aspects apply especially to the invertebrates, including our spiders; for example, spiders have a highly developed "wind sense" and finely tuned skeletal strain sensors. Such abilities are particularly appealing to zoologists because of their alien qualities: the detection of polarization patterns in sky light by bees and ants, the hearing of ultrasound in grasshoppers and bats, the infrared sense of some snakes, the magnetic-field sensitivity of bees, pigeons and migratory birds, and not least the electric sense of certain fishes.

The message here, then, is that the classical subdivision is of only limited use, not merely because it is based on humans but because it refers to human sensations. Given that we can never know for sure everything that a fellow human is feeling, even one with whom we are most intimately acquainted, it must be utterly impossible to comprehend the sensations mediated by a spider brain when, for example, it is exposed to the concerted activity of several thousand sensors signalling the mechanical state of the skeleton. It follows that the senses should be classified not on the basis of sensations but according to the physical (or chemical) properties of the relevant stimulus, the form of energy involved. That it can be difficult to specify this "adequate" (to use the technical term) stimulus will be impressively demonstrated by the spiders.

VI The Special Significance of Mechanical Senses

Of all the senses of spiders, the mechanical senses are especially well developed. Spiders have a multitude of differently specialized receptors that together respond to a broad spectrum of mechanical stimuli: the most delicate air currents and vibrations as well as slight deformations of the exoskeleton and gentle touches. The role of mechanoreception in the behavior of most spiders is correspondingly diverse. The most familiar example may be the responses of spiders to the vibrations of the web when a prey animal is trapped. As will become evident in Part D of the book, however, this is only one of many ethological situations in which the mechanical sense dominates. To placate my "visual" colleagues, I should mention that there are other spiders in which vision is a similarly dominant sense, in particular among the jumping spiders (see also Chapters XI and XXII).

The Exoskeleton

The mechanosensory systems that we have studied in our spiders – with patience bordering on the pernickety, because it was their expected technical perfection that was initially most fascinating to us – are intimately related to the cuticular exoskeleton. As in other arthropods, this exoskeleton makes two crucial contributions to sensation: first as a base on which the sense organs are seated, and second as an accessory structure through which stimuli are propagated, and within which mechanical stresses (some caused by the activity of the animal's own muscles, as during locomotion) are detected by specialized biological sensors. This second aspect is manifest only in arthropods, which therefore possess a unique skeletal sense.

In spiders this skeletal sense is more elaborate than in other arthropods, and from a technical point of view it amounts to a particularly refined sensory system. For those who agree with me that enjoyment is enhanced by immersion in detail: please be patient, as Chapter VII will give you a chance to participate in a long story of research.

Meanwhile, we may say simply that the abundance of cuticular sensilla, different from one another in many respects, is easily visible on the outer surface of a spider exoskeleton. Although so diverse, they can be classified as two major groups, the many innervated "hairs" and the innervated "holes". This classification is justified not only by their gross configuration, but by the physical events associated with stimulation. The following descriptions of the individual sensory systems will show, among other things, what functional possibilities reside in the two structural categories "hair" and "hole" and to what a great extent the structure of a sensillum reflects the physical properties of its particular "adequate" stimulus (that is, the stimulus especially effective under natural conditions). In this way I hope to convey some idea of the selection pressures and evolutionary constraints that arose from the natural stimulus situation and resulted in the present state of adaptedness.

Remarkable Sensitivities

Before we consider the details of hair and hole sensilla, a few impressive numbers will show what these organs can do. It should be kept in mind that sensory systems are closely interlinked with behavior, and with the biotic and abiotic factors in the habitat. Therefore, as is the case with other organs as well, the biologically optimal performance of sense organs is usually a compromise, a response to many selection pressures. The most spectacular achievements are thus not always the best, by which I mean the most economical. But peak performances in the technical sense are the best indicators of the potential hidden in the structural plan of a sensillum.

In the category of *hair sensilla*, the technically most perfect version is represented by the trichobothria of the arachnids (see Chapter IX) and the analogous filiform hairs of insects. Well-founded estimates of absolute sensitivity are available for the filiform hairs on the cricket cercus (Thurm 1982). Here the energy consumption (work) during maximal effective stimulation was found to be about 10^{-16} Ws (1 Ws equals 1 joule). Such a stimulus imposes a force of only 1 to 5×10^{-8} N, and the diameter of the dendrites of the sensory cell receiving the stimulus changes by only about 10 to 20 nm. At the threshold, the corresponding value was 10^{-19} Ws. This is equivalent to less than the energy contained in one photon of green light. The change in dendrite diameter at threshold is a mere 0.05 nm. Tateo Shimozawa and his co-workers at the Hokkaido University in Sapporo have recently declared the filiform hairs of the cricket *Gryllus bimaculatus* to be the most sensitive of all receptors so far described (Shimozawa et al. 1998). As a value for the mechanical energy needed to trigger an action potential (more precisely, the work to be performed) they cite 4×10^{-21} Ws. It follows that the threshold sensitivity of these cells is one hundred times greater than that of photoreceptor cells responsive to a single quantum of light. No doubt about it: these filiform hairs operate at the limits of the physically possible. Their sensitivity is by no means inferior even to that of the human ear or eye (Autrum 1984; Gitter and Klinke 1989). All this serves as a reminder that the exchange of information in general is characterized by very small amounts of energy, as compared for example with the energy involved in trophic interactions.

As far as the *hole sensilla* are concerned (see Chapter VII), much the same applies, although in this case it is very difficult to obtain the reliable measurements needed for energy calculations, which are not yet as well established as we would wish. For a composite hole sensillum of a bird spider (the lyriform organ) the force threshold for detection of slow deflection of the metatarsus is 8 to 10 mN, which puts the organ under a strain of 60 $\mu\varepsilon$ (this corresponds to compression of the cuticle by 60 millionths of its initial length). During slow locomotion one slit of this organ is compressed at most by less than 3 nm (Blickhan and Barth 1985). In the case of "our" spider *Cupiennius salei*, the smallest measured deflection of the metatarsus that was an effective stimulus to the same organ was also tiny, 0.006°. This movement applies to the joint a force of only 40 μN (Bohnenberger 1979).

We felt that an effort to understand better the function of these finely tuned biological measuring instruments was bound to be rewarded, and we have been engaged in that effort for many years. The following chapters describe what happened.

VII The Measurement of Strain in the Exoskeleton

How, then, is strain measured in the spider skeleton? Why can we claim that this strain measurement provides the spider with a highly detailed picture of mechanical events in the cuticle, and why is that picture so hard for us to understand?

Before we can answer such questions, we must tackle some problems in mechanics. We shall also have to consider some details if we are to understand the functional and evolutionary potential contained in the Bauplan of a strain sensor in the arthropod exoskeleton. First we shall see where the strain sensors are located in the skeleton. Then a few basic terms used in mechanics have to be clarified. The functional morphology of the sensors and receptor mechanisms will be treated next. Finally, the characteristic formation of groups of slits and the origin and effect of natural stimuli will be discussed.

1 Occurrence and Topography of Slit Sensilla

Slit sense organs are a feature unique to the arachnids. They were first mentioned in print over 100 years ago, by Bertkau (1878). There followed detailed reports of their presence in various spiders by McIndoo (1911), Vogel (1923) and Kaston (1935). After a 20-year pause, Pringle (1955) took up the subject again and, in a pioneering work, showed that these sensilla function as mechanoreceptors. The spiders still remain the best-investigated group of arachnids. The different species resemble one another closely in the kinds of slit sense organs they possess and their topography.

Philip Bertkau, who first described slit sensilla, was born in Cologne on January 11, 1849 (barely a year after Eugène Simon). Those were turbulent times, with the March revolution of 1848 barely over. After leaving school he moved to Bonn, where he died on October 22, 1894. Bertkau obtained his doctorate at the age of 23 in Bonn, and began teaching at the local Zoological Institute in 1874. Eight years later he became a professor at the Agricultural Academy in the Bonn suburb of Poppelsdorf, and in 1890 the position of Curator was specially created for him at the Institute for Zoology and Comparative Anatomy. It was a nervous disorder that caused his death at just 44 years of age. In Bertkau's research on spiders, anatomy and physiology were his chief interests. He also worked on the olfactory sense of butterflies and the structure of hermaphroditic arthropods. Pierre Bonnet (1945), in whose Bibliographia Araneorum I found these details, writes quite fondly about Bertkau's human qualities: "He was an amiable man, with an affable character, esteemed by all who approached him."

In the exoskeleton of *Cupiennius salei* there are around 3300 slits (Barth and Libera 1970) (Fig. 1). The great majority (86%) are situated on the legs and pedipalps, embedded in hard, sclerotized exocuticle. The opisthosoma (without the petiolus) is equipped with 96 slits, which here are surrounded by soft mesocuticle.

As in other arachnids, the slit sense organs of the spiders can be assigned to three types on the basis of the slit arrangement (Fig. 2): (i) isolated single slits, (ii) "loose" groups of several slits, and (iii) composite or lyriform organs, made up of as many as 30 slits close together in parallel (Plate 16). In *Cupiennius salei* about half of all the slits are either isolated or part of a "loose" group; the other half form a total of 144 lyriform organs, all situated on the extremities: 134 on the walking legs and pedipalps, 10 on the spinnerets and chelicerae.

Most lyriform organs are very close to joints, whereas the single slits are typically some distance away from a joint. Single slits are ar-

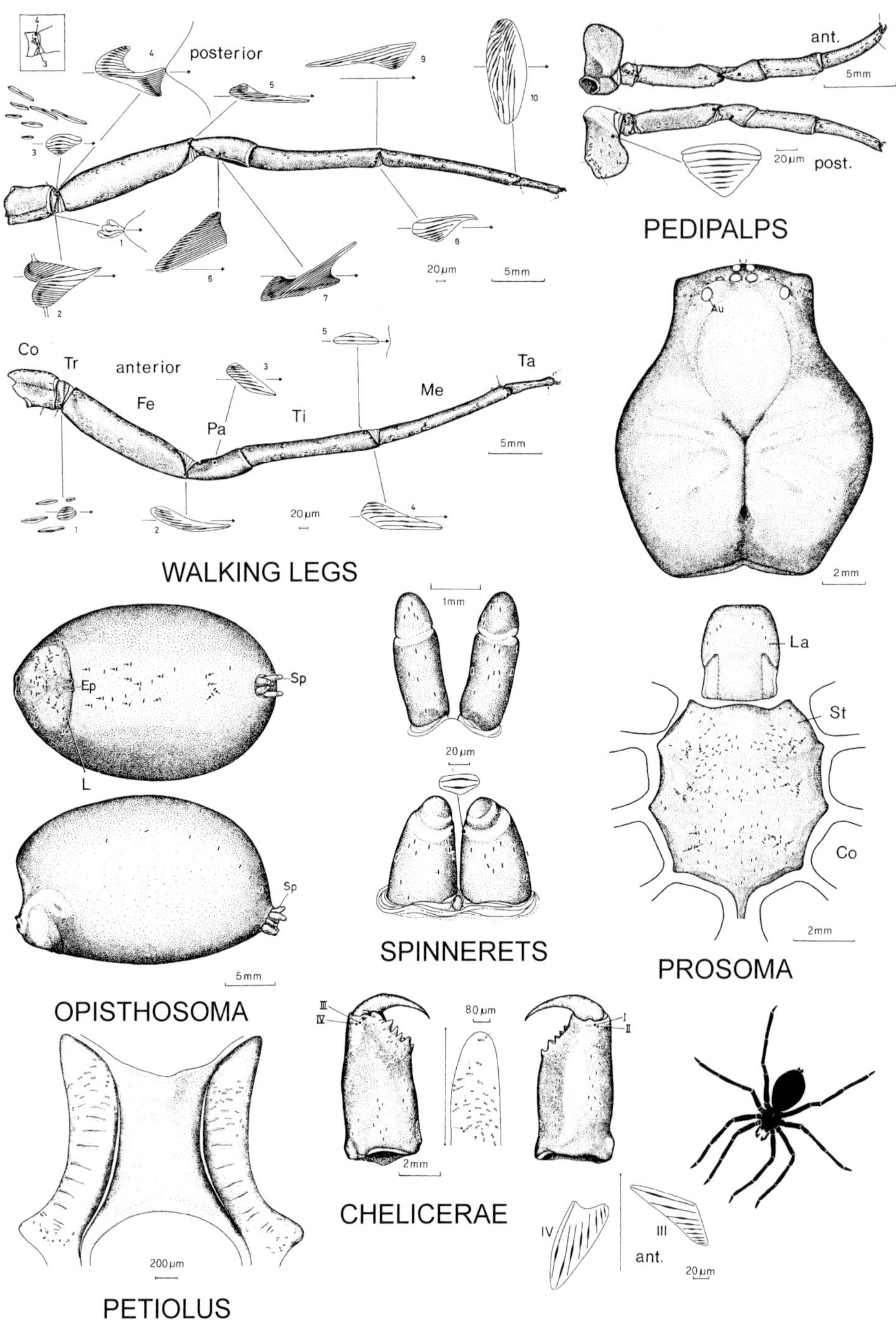

Fig. 1. Distribution of the slit sensilla in the exoskeleton of *Cupiennius salei*. *Small arrows* point to individual single slits longer than 30 µm; *long arrows in insets* indicate the orientation of the long axis of the respective body part and point distally. *Co* Coxa; *Tr* trochanter; *Fe* femur; *Pa* patella; *Ti* tibia; *Me* metatarsus; *Ta* tarsus; *Sp* spinnerets; *L* lung slit; *Ep* epigynum; *La* labium; *St* sternum. (Barth 1985b)

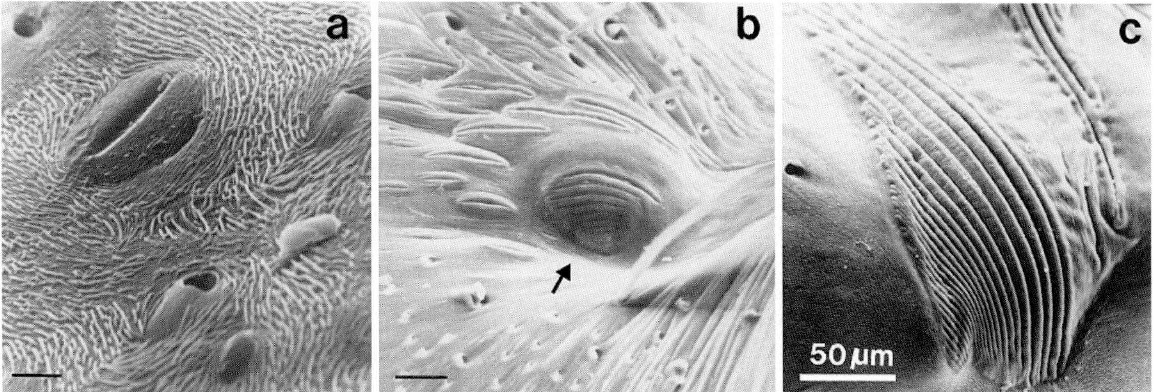

Fig. 2a–c. Types of slit sensilla. Slit sensilla of arachnids occur as single isolated slits (**a** on opisthosoma), as relatively loosely arranged groups (**b** trochanter of walking leg; *arrow* points to lyriform organ), and as lyriform or compound organs with a close parallel arrangement of several slits (**c** trochanter of walking leg). All examples are taken from *Cupiennius salei*; *bar* in all cases 50 µm. (Barth 1981)

ranged in rows on the front and back surfaces of the legs and pedipalps, and some of them lie conspicuously close to the sites of muscle attachment. The majority of both the single slits and the lyriform organs are found laterally on the limbs, and in both cases the slits are oriented approximately parallel to the long axis of the limb. Loose groups of typical structure are found on the trochanter of the walking legs, on the chelicerae and on the petiolus.

Peters and Pfreundt (1986) tried to correlate differences in the slit sense organs possessed by various spiders with differences in their modes of life. The result is not very impressive, because the similarity among spiders is considerably more striking than the difference. It did turn out that hunting spiders have slightly fewer slits in their lyriform organs than the web-building spiders that had previously been examined. However, no far-reaching interpretations of this finding are possible, especially since these authors did not include single slits and loose groups in their analysis.

How does the inventory of slit sense organs in other arachnids compare? *Cupiennius* has been found to differ widely from a scorpion (Barth and Wadepuhl 1975) as well as from a whip spider, a whip scorpion and a daddy-long-legs (Barth and Stagl 1976), both in the total number of slits and in the proportions of single slits, groups and lyriform organs. Whereas in the spiders there are 15 lyriform organs on each leg, in the other arachnids there are none, or only one or two. Where spiders have lyriform organs, the scorpion and whip scorpion have groups of single slits. The whip spider and daddy-long-legs are the least well equipped, with only 58 or 45 slits per walking leg, respectively, in comparison with 352 on the leg of *Cupiennius salei* (Fig. 3). Some details of this comparison can be interpreted in terms of functional morphology, but first it is necessary to explain how a slit functions.

2 Terminology of Mechanics

Slit sensilla measure the effects of mechanical stresses in the cuticle, such as are produced by muscular activity, hemolymph pressure, gravity and substrate vibrations. What happens in a piece of material exposed to such external forces? To answer this we must distinguish several quantities in the science of mechanics: stress, strain, Young's modulus (of elasticity) and displacement.

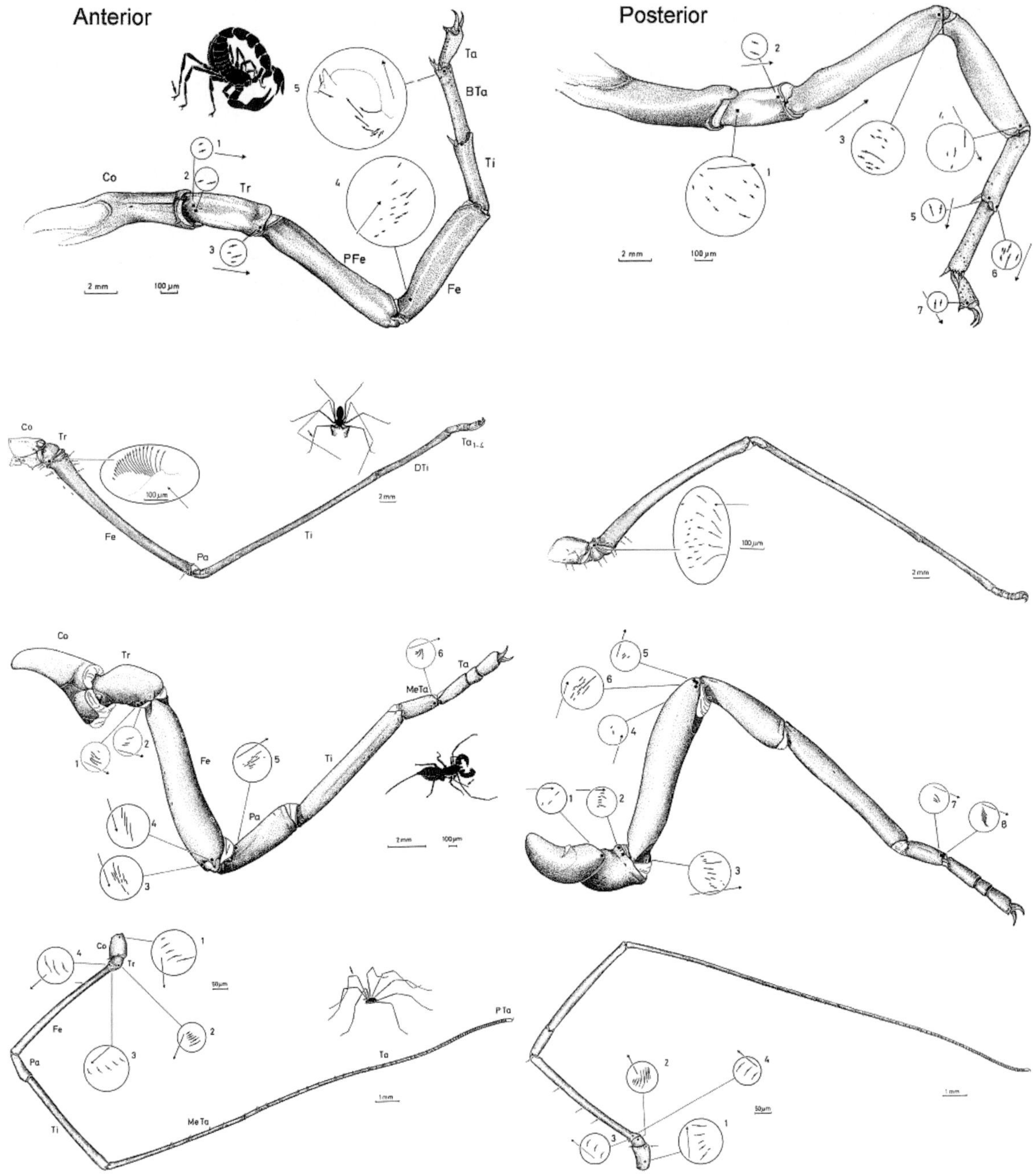

Fig. 3. Comparison of topography of slit sensilla on the walking legs of different species of arachnids. *From top to bottom* Scorpion *Androctonus australis* (Scorpiones); whip spider *Heterophrynus sp.* (*Admetus pumilio*) (Amblypygi); whip scorpion *Mastigoproctus brasilianus* (Uropygi); daddy longlegs *Amilenus aurantiacus* (Opiliones). (Barth 1978)

Stress δ

This is the force (P) per unit area (A) that is produced in the material by the action of an external force, and opposes the latter force. The stress perpendicular to a plane is given by

$$\delta = \frac{P}{A}[N/m^2 \text{ or Pa}].$$

Strictly speaking, a distinction must be made between the stress acting on the material and the stress that results. The acting stress is equal to the load imposed per unit area and has the dimensions of pressure.

Strain ε

Stress and strain are closely related to one another. Stress inevitably leads to a deformation of the material, however tiny this may be. Strain is defined as the relative change in length of a unit volume in a specific direction:

$$\varepsilon = \frac{\Delta l}{l_0}.$$

Here l_0 denotes the initial length and Δl is the change in length. Hence strain is a dimensionless quantity. Strain is negative when the length is made less than l_0; that is, when the material is compressed.

Modulus of Elasticity (Young's Modulus) E

Up to a limiting value (Hooke's region) stress and strain are often proportional to one another. In the simplest case – say, a bar upon which a pulling force is imposed – this situation is expressed by

$$E = \frac{\delta}{\varepsilon}[Pa].$$

In addition to the modulus of elasticity for tensile stress (E, or Young's modulus), there is a modulus of elasticity for shear stress; this gives the ratio between the tangentially acting force per unit area and the deformation in degrees of arc.

Displacement ω

It is essential to distinguish between the strain in a unit of volume and the displacement of a particular point in a piece of material under stress. This is illustrated by the simple case of a bar with stress imposed on a single axis (Fig. 4a). The displacement ω at the point s is equal to the sum of the strains of all volume units up to this point:

$$\omega = \int_0^s \varepsilon ds.$$

As is evident in the figure, the distributions of both strain and stress along the bar are different from that of displacement. Whereas strain and stress are constant over the entire long axis of the bar, the displacement increases with its length. Strain situations involving more than one axis are more complicated than is indicated by the present simple example.

A different simple example (Fig. 4b) illustrates another important aspect: the sites of greatest strain do not coincide with the sites of greatest displacement. A bar fixed at one end with a force imposed at its free end will exhibit the greatest stress where the resistance to displacement is greatest, namely at the point where it is fixed. Conversely, the largest displacement will occur at the free end, where the strain (like the stress) is smallest.

Effects at a Surface

A piece of material can be put under stress in a variety of ways: by tensile, compressive and shear forces, by buckling, bending and by twisting (Fig. 4c). Expressed in biological terms: by muscular activity, the pressure of the hemolymph, or gravity. Despite all this diversity, what

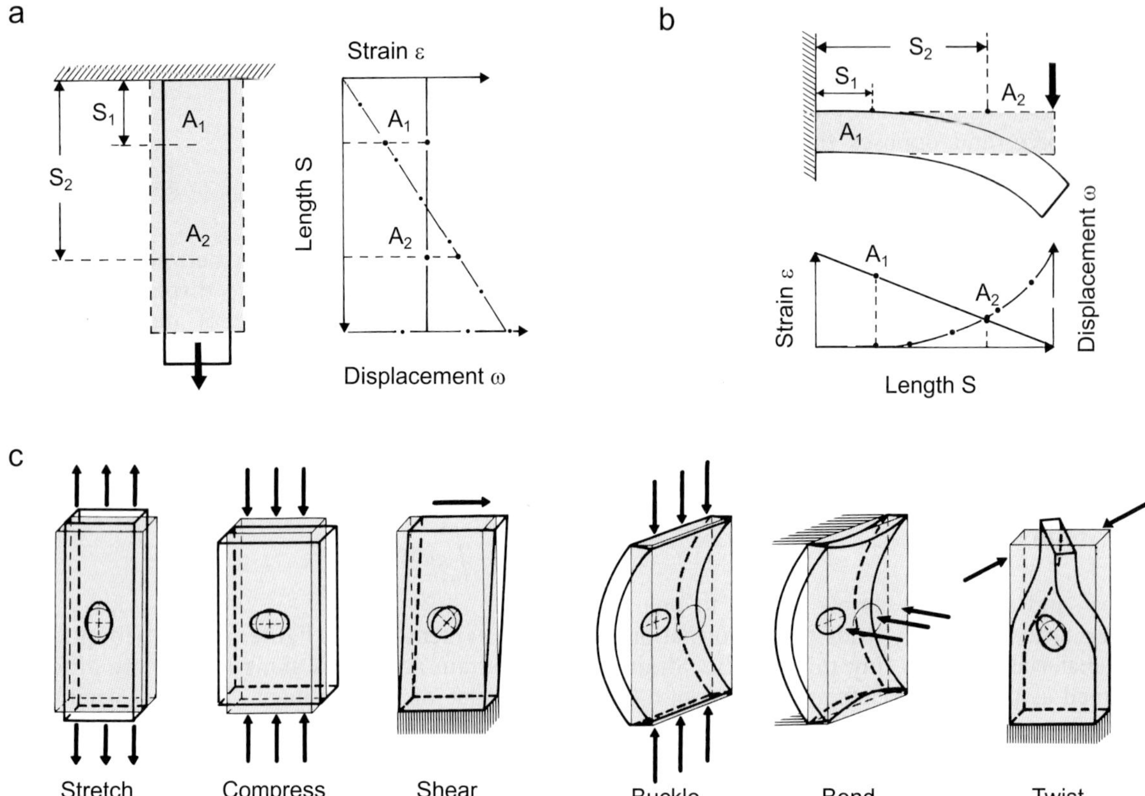

Fig. 4a–c. Some basic mechanical terms and principles. **a** A bar under tension (*arrow*) shows that displacement has to be distinguished from strain. **b** As seen from a single-ended cantilever under load (*arrow*), the sites of maximal stress or strain need not coincide with those of maximal displacement. **c** A piece of material (*shaded*) is loaded (*arrows*) in different ways and thereby deformed (*unshaded*). Basically, the same phenomenon occurs in all cases at the surface: a *circle* drawn onto it deforms into an *ellipse*, indicating strains in two directions perpendicular to each other and with opposite sign. (Barth 1985b)

happens at the surface of the material is always the same: at a given site, there is always extension (+ε) in one direction and compression (–ε) in the perpendicular direction. At the surface there exist only stresses and strains parallel to the surface, assuming that the external force is not acting directly on the surface. All these considerations are crucial in the present context, because the dendritic endings of the slit sensilla are attached to the surface of the cuticle. It is considerably simpler to interpret the mechanical consequences of various forms of load in this situation than it would be if the endings were embedded in a mass of cuticle.

3 Functional Morphology

Structure

When examining the outer surface of a spider's cuticle, after some practice it is possible to see the slit sensilla even under low magnification (for instance, with a dissection microscope). They look like elongated, delicate skeletal structures, though in fact they are holes or clefts in the skeleton, sealed off from the exterior by a largely epicuticular membrane 0.25 μm thick, and bounded by two cuticular ridges that look

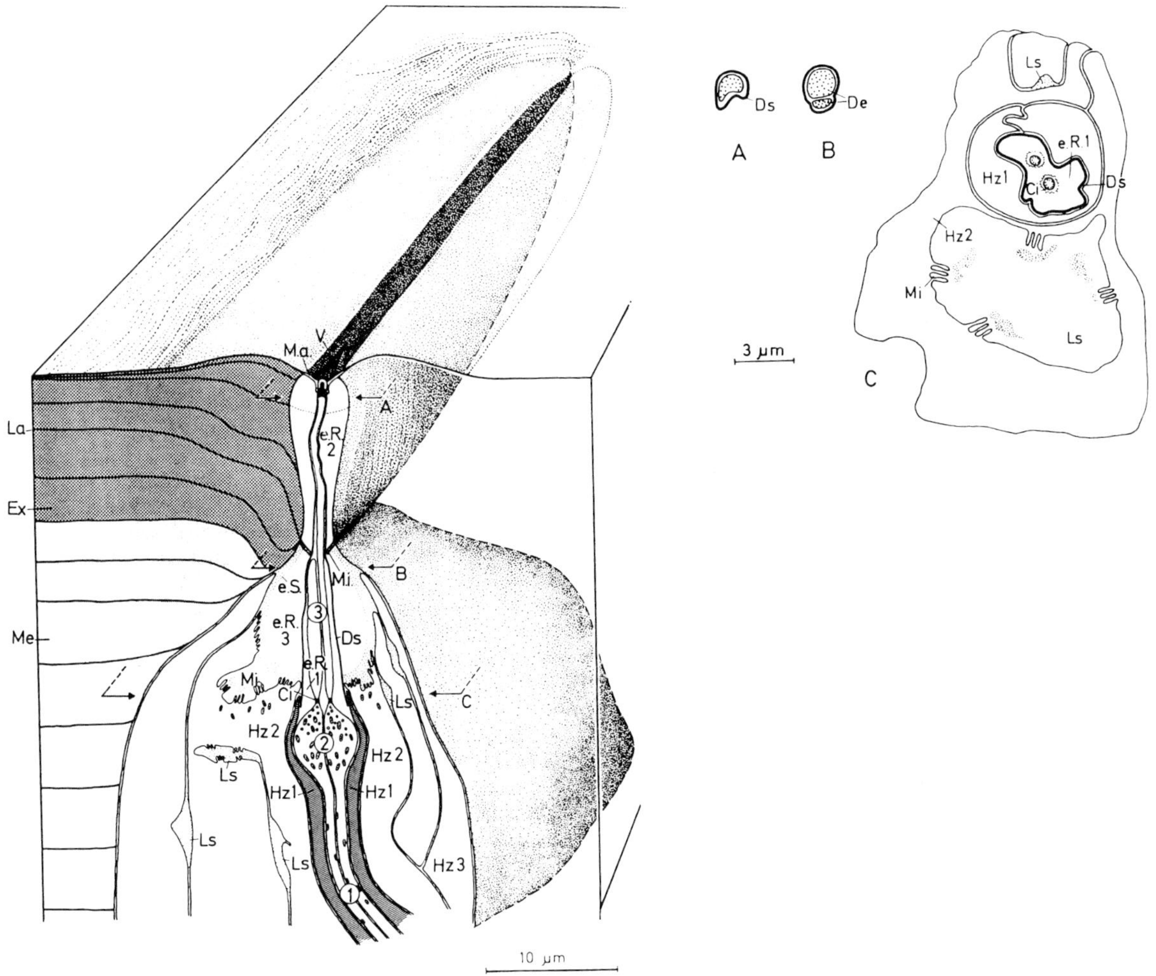

Fig. 5. The fine structure of slit sensilla, taking a tarsal single slit of *Cupiennius salei* as an example. *Left* Cross-section through the middle part of the slit. *Ex* Exocuticle; *Me* mesocuticle; *La* course taken by the cuticular lamellae (their number is much higher in the exocuticle than indicated by those shown in the drawing); *Ma* outer membrane; *Mi* inner membrane. The dendrite of only one of the two sensory cells reaches *Ma*, whereas the other ends close to *Mi*. Both dendrites are subdivided into three sections: (1) section close to soma with tubuli and a few mitochondria; (2) mitochondrion-rich swelling; (3) distal section rich in tubuli but lacking mitochondria. *Ci* Ciliary segment; *Ds* dendritic sheath; *Hz* sheath cells; *Mi* microvilli, *es* extracellular substance (also in the lacunar system *Ls*); eR_{1-3} extracellular spaces. *Right* Schematic drawing of horizontal sections through the slit at different levels as indicated by *A*, *B*, and *C* in *left figure*; *De* dendrite. (Barth 1971 a)

like lips. Such slits are between 8 and 200 µm long and between 1 and 2 µm wide. The dendrite of a bipolar sensory cell ends in a specialized structure of the covering membrane, which we call the coupling cylinder (Figs. 5, 6). The coupling cylinder is about 1 µm deep and about 0.5 µm in diameter. Within the dendrite terminal is a so-called tubular body, consisting of densely packed tubules in electron-dense material. This is one of the few modality-specific structures of the cuticular mechanoreceptors of arthropods; that is, it is diagnostic of their function, just as light-sensing organs can be identified by the enlargements of the cell membrane

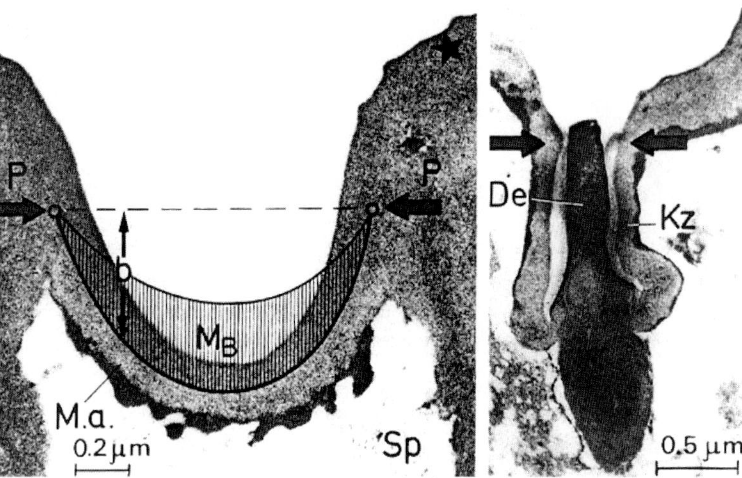

Fig. 6. Stimulus transformation at the slit sensillum: the covering membrane (*Ma, left*) and the dendritic end (*De*), which is coupled to a coupling cylinder (*Kz*). M_B Bending moment due to force *P*, which leads to slit compression. *Arrows on the right* electron micrograph indicate deformation of dendrite by monoaxial compression forces. (Barth 1976)

that contain the photopigment. As the site of sensory transduction, the tubular body is particularly interesting. Unfortunately, very little is known about what it actually does (see below).

The hole in the cuticle, the slit itself, on closer inspection consists of two parts (Fig. 5). The outer part is in the exocuticle and is shaped like a groove that becomes shallower at each end; the dendrite, together with its extracellular dendritic sheath, passes through a small hole in the bottom of this groove to reach the covering membrane. Apart from the dendrite, this outer part contains no cellular components, but is filled with the so-called receptor lymph.

Figure 7 shows a nice detail of the arrangement of the exocuticular lamellae around the outer part of the slit. The orientation of these layers is just like that of the stress lines that would be expected near a notch in any industrial material subjected to tensile force or pressure (Peterson 1966): ordinarily the lamellae lie parallel to the surface but here, like the lines of greatest principal stress, they curve inward and at increasing depths conform ever more closely to the contours of the slit. What does this mean? The direction in which the load-bearing capacity of the cuticle is largest is the same as the direction of the greatest principal stress. If the lamellae were to be oriented parallel to the surface in the vicinity of the slit, there would be a risk that fractures would develop parallel to the surface; this is known to be especially likely when a notch has a high length:width ratio and the stress concentration reaches values above 20 (Peterson 1966). In the meso- and endocuticle (inner part of the slit) the lamellae hardly change their orientation near the slit. This is also expected on mechanical grounds, because these cuticular zones contribute very little to the resistance of the whole cuticle to pressure and tensile forces.

Some readers may now be asking what exactly these cuticular lamellae are. In technical terms, the cuticle of arthropods in general, including the spiders, is a composite material with fiber-reinforced laminations. Microfibers, which consist basically of high-molecular-weight chitin polymers resembling the cellulose of plants, are disposed within a protein matrix so as to form layers parallel to the surface. If this matrix is extensively cross-linked (sclerotized), as it is in the cuticle of the leg skeleton, then the composite material as a whole resists bending. If there is less cross-linking, the material is more flexible, like the joint membranes. The remarkable mechanical adaptability of the cuticle is enhanced by the fact that various aspects of its composition can easily be modified: the proportions of fibers and matrix, the water content and the orientations of the fibers.

Within a given layer parallel to the surface, as a rule all the fibers are parallel to one another. However, their direction can vary from one layer to the next, and if it shifts in a regular and continuous progression, a new property emerges. Not only does the composite material exhibit isotropy (the same mechanical properties in all directions parallel to the surface), but in addition the lamellar structure becomes visible as an optical artefact (Fig. 8). On the other hand, if the cuticular fibers tend to be aligned in only one or a few directions, the resutling cuticle will be mechanically anisotropic. If the preferred fiber directions correspond to the directions of the mechanical stresses that exist in the exoskeleton under natural conditions, this matching to "stress trajectories" (trajectorial design) allows the system to operate with a minimum of material.

More information about the mechanical properties of the material of which the exoskeleton of *Cupiennius* is made can be found in previous publications (Barth 1969, 1970, 1973).

The *inner part* of the cuticular hole is bell-shaped and fits under the outer part. Unlike the latter, it is filled with cells, of the kinds typically found in arthropod cuticular sensilla (Fig. 5). Three cells form successive layers around the dendrite. The innermost cell presumably secretes the dendritic sheath (it extends to the proximal end of the sheath, with which it is in close morphological contact). The two outer enveloping cells are probably equivalent to the trichogen and tormogen cells of the cuticular sensilla of insects; in insect receptors the trichogen cell gives rise to the actual hair (the hair shaft) and the tormogen cell forms its base. A striking feature of the 2nd enveloping cell (Hz2) is its large invagination on the side toward the inner membrane (Mi) of the outer slit. The cell surface in this invagination is covered with irregularly distributed microvilli. The outer enveloping cells (Hz1 and Hz2) together enclose the receptor-lymph space in the interior of the cuticular slit. Their significance for the primary processes will be considered later.

We cannot yet be sure whether these enveloping cells are really homologous to those of insects, because their ontogeny has not yet been studied in spiders in sufficient detail (Harris 1977; Haupt 1982).

Let us now return to the dendrite. It is subdivided into three sections, from proximal to distal: the section next to the cell body, a swollen section full of mitochondria, and finally the outer section, which begins with a ciliated structure and contains many tubules (Fig. 5). A remarkable feature of the slit sense organs is the presence of a similarly subdivided second dendrite (Barth 1971a), part of a second sensory cell (Seyfarth and Pflüger 1984; Seyfarth et al. 1994). Unlike the first, this second dendrite ends at the inner membrane of the outer part of the slit and its end lacks the electron-dense substance between the many tubules that is typical of the tubular body of the long dendrite; furthermore, the second dendrite is not coupled to the cuticle

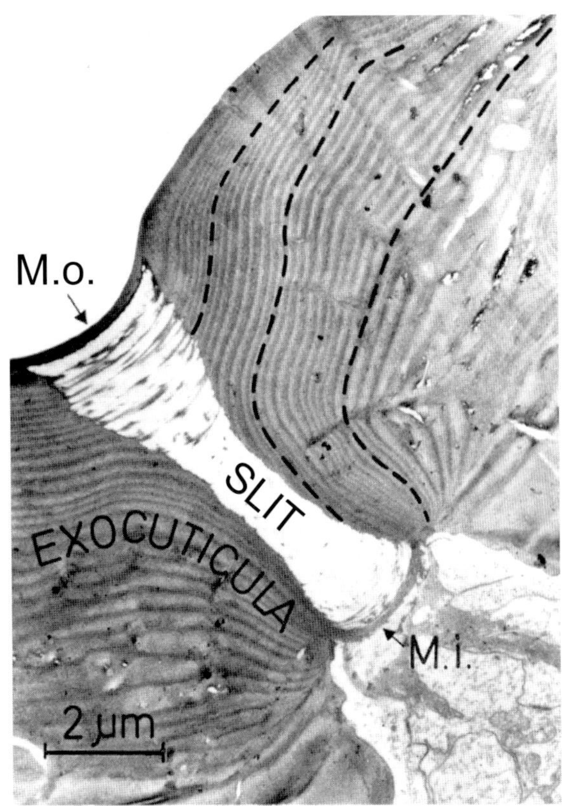

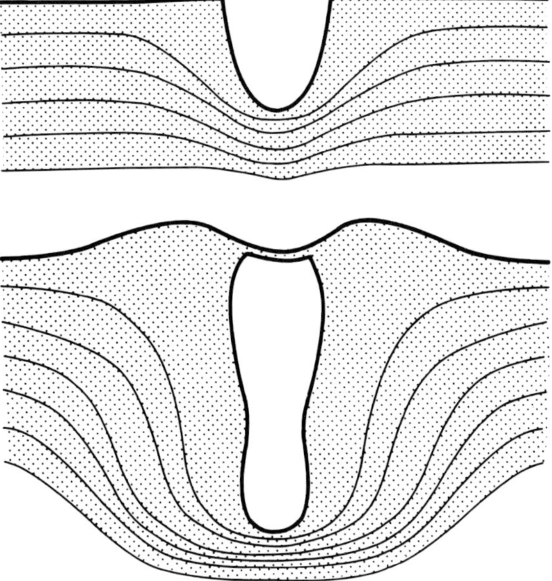

Fig. 7. Slits and cracks. *Top* Electron micrograph of cross-section through a single slit on the tarsus of *Cupiennius salei*. *Ma* Covering membrane; *Mi* inner membrane. *Middle* Stress trajectories close to a crack under compressional or tension load (Peterson 1966). *Bottom* Course taken by the lamellae in the exocuticle surrounding the middle region of a single slit. See text. (Barth 1972a)

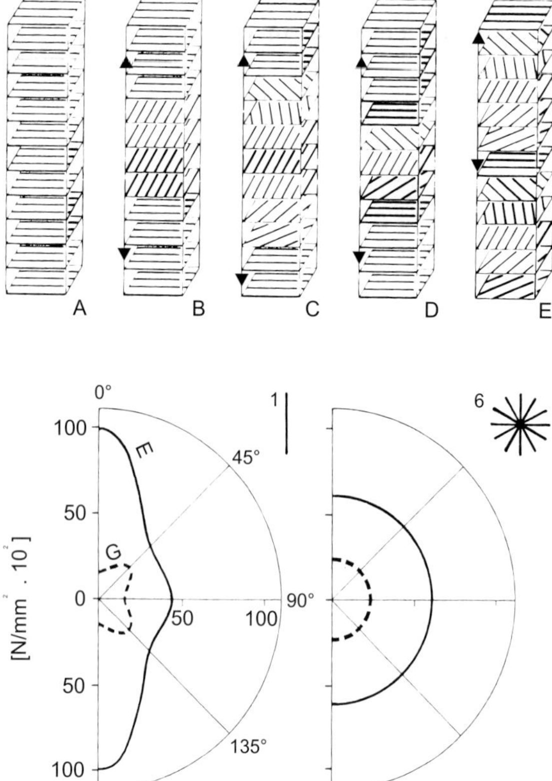

Fig. 8. Fine structure and mechanical properties of the cuticle. *Above* Variations in the basic pattern of microfiber architecture; *A* unidirectional type of arrangement; *E* multidirectional type with the fiber direction rotating regularly and continuously. Intermediates between these extreme cases: *B* bidirectional reinforcement with the fiber orientation changing abruptly by 90° between zones of unidirectional reinforcement; *C* and *D* different combinations of properties of *A*, *B*, and *E*. *Arrows* delimit one lamella, defined by the rotation of the fiber long axis by 180°. *Below* Directionality of the mechanical properties of composite laminated material. The proportion of fibers with a particular orientation greatly influences the mechanical isotropy and anisotropy. The pronounced anisotropy typical of material with only one fiber orientation disappears when the number of fiber orientations rises to as little as six. *E* Young's modulus; *G* shear modulus. (Barth 1973)

of the exoskeleton. The function of this second dendrite is not yet clear, but there are several indications that it is not involved in signalling slit deformation. No second type of action potential can be seen in electrophysiological recordings (Barth 1972a, b). However, more recent intracellular recordings from sensory cells in a lyriform organ on the patella (Seyfarth and French 1994) have revealed that the second dendrite is active only very briefly, so that its response may simply have been overlooked (see Sect. 4a, "Potentials and currents").

The phenomenon of such "supernumerary" dendrites with unknown or uncertain function is typical of the cuticular sensilla of spiders, and is one of the open questions deserving future study (Barth 1981, 1985b).

The Technical and the Biological Version

Holes in the cuticle – whether they are the slit sense organs of spiders or the campaniform sensilla of insects – change shape minimally when the exoskeleton is deformed. Such deformation is caused by forces produced, for instance, when the musculature contracts or the hemolymph pressure is increased in order to extend the leg hydraulically. Then the tiny deformations of the sensory hole in turn deform the end of the dendrite, where it is attached to the covering membrane, locally changing the resistance of the cell membrane. The end result is a flow of current through the membrane and excitation of the sensory cell.

The technical analog of a slit sense organ is a strain gauge. In the simplest case this is a piece of wire fixed to the object to be measured, in such a way that any strain in the object necessarily also appears in the wire, changing its diameter and hence its electrical resistance. This resistance change can be measured very accurately with a bridge circuit and serves as a good indicator of the length change. Without strain gauges modern technology would be inconceivable, especially experimental stress analysis, in which strain-gauge measurements allow the mechanical response to loading of a component to be measured over a wide range, with no damage to the component. This is crucial, because a construction engineer must be certain that all components will be able to bear the expected loads: the wings of an airplane, the chassis of a car or the structural elements of cranes, fuel tanks or pipelines.

The Misunderstood Invention

1938 is a special year in the history of measurement technology, because it marks the birth of the strain gauge. In 1827 Georg Simon Ohm had discovered the relationships between current, voltage and resistance in an electrical circuit, and in 1843 Wheatstone designed a bridge circuit to measure small resistances. Building on these developments, by 1928 Nernst had already used the change in the resistance of freely suspended wires to measure the pressure in the cylinder of an internal-combustion engine.

But it took another ten years until Arthur C. Ruge and his assistant at the Massachusetts Institute of Technology, while studying the behavior of water towers during earthquakes, had the seminal idea of gluing fine wires from a potentiometer directly to the wall of the tank. In a later refinement, they glued the wires to a supporting device and attached that to the tank.

By now this technology has become highly developed, in particular by the introduction of foil and semiconductor designs. Strain gauges are commercially available in enormous variety and are now essential for the construction of industrial machines, buildings, ground and air vehicles as well as in measurement and control-systems technology, not least because many mechanical problems in these areas are too complex to be solved by precise calculation. It is one of the ironies of history that in 1939, when Ruge had submitted the standard description of his new device to the patent committee at MIT, they replied, "while this development is interesting, the Committee does not feel that the commercial use is likely to be of major importance ... This leaves you free to treat the invention entirely as a personal matter." He did, and established a successful private business. (Keil St. 1988, 33 DMS-Produktion in Darmstadt – Eine historische Rückschau aus Anlass des 50-jährigen Jubiläums des Dehnungsmessstreifens. Messtechnische Briefe MTB 24,2, 43-52, Hottinger-Baldwin-Messtechnik GmbH Darmstadt).

The story would probably have developed differently if people had known more about the inventions nature had created millions of years before, in order to measure just such mechanical variables precisely and with high sensitivity. Although the slit sense organs of spiders work quite differently from strain gauges, they are entirely comparable in terms of the measurement to be achieved. In both cases there is a mechanical stress that causes deformation, the magnitude of which is a measure of the imposed force. As we shall see, the range of applications of the slit sense organs is also large. The measured parameters – strain, stress, pressure, force or vibration – depend on the particular way they are "installed" in the skeleton.

How, exactly, do the spiders do it?

Stimulus Transmission (Mechanical Primary Processes) and the Deformation of the Dendrite

Many details of the cuticular structures in the slit sensillum and of their arrangement can be understood in the context of their stimulus-transmitting function. We have analyzed this function both by electrophysiological recording and in mechanical experiments on models (Barth 1972a,b, 1976, 1981, 1985b; Barth and Blickhan 1984). As was said above, these sensory holes are built to detect the tiniest deformations of the skeleton, which are caused by forces that in turn are generated by loads originating either inside or outside the spider. The structural components of the stimulus-transmitting apparatus of the slit sensilla are so formed and disposed that the initially quite diffuse stimulus, from cuticle of the surrounding exoskeleton, is focussed onto an area only about 1 μm^2 at the end of the dendrite, with high mechanical sensitivity.

The sequence of events is as follows.

1. The "adequate" input to the sensillum is compression of the slit (Fig. 9); that is, only this action, and not slit dilation, triggers action potentials (Barth 1972a,b). Elongated structures such as the slit exhibit the following physiologically significant mechanical properties: both the absolute amount of slit deformation and its dependence on the direction of the load increase as the absolute length and the length:width ratio increase. The load direction that most effectively deforms the slit is perpendicular to its long axis.
2. Because it is only 0.25 μm thick and is curved like a roof gutter, the covering membrane is mechanically labile, especially with respect to compression of the slit (that is, laterally imposed stress). Its curvature increases its bending moment M_B ($M_B = F \times b$, where F is the imposed force and b is the depth of bending; see Fig. 6a). The gutter-like shape simultaneously increases the stiffness of the covering membrane in the direction of the long axis of the slit.

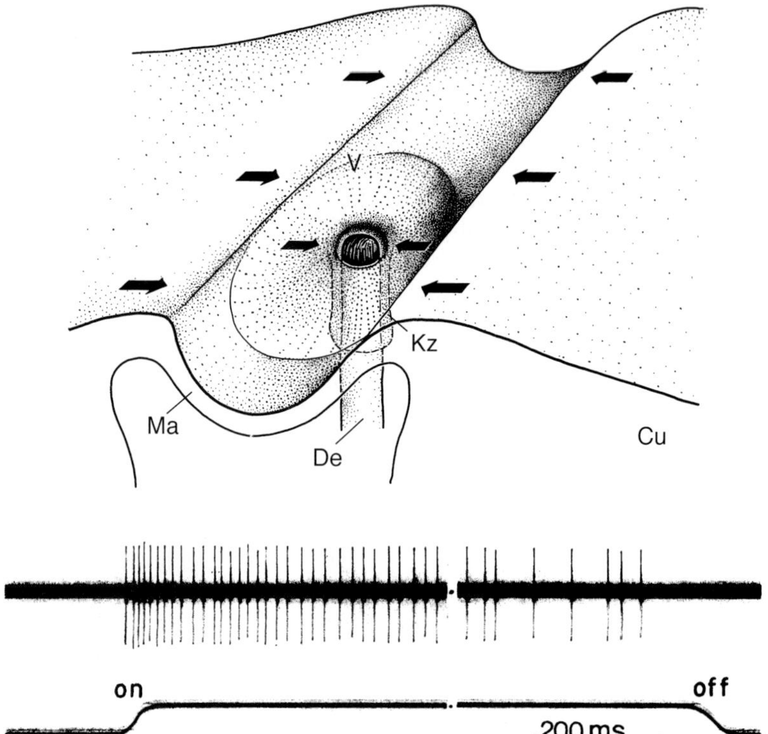

Fig. 9. Adequate deformation of the slit results from compression by forces (*arrows*) at a right angle to the slit long axis. This elicits action potentials as illustrated below by the response of a slit of a lyriform organ on the tibia. The *lower trace* represents the compressional stimulus. *Ma* Covering membrane; *V* depression in *Ma*; *De* dendrite; *Kz* coupling cylinder; *Cu* cuticle. (Barth 1976)

3. The coupling cylinder is the next station the stimulus passes on the way to the dendrite. It couples the end of the dendrite to the "floor" of the covering membrane, where the bending moment is greatest when the slit is compressed. As the covering membrane is additionally bent by the stimulus, the outer region of the coupling cylinder is compressed mainly in a direction perpendicular to the long axis of the slit (Fig. 9).
4. At the end of the stimulus-transformation process, there is a slight deformation of the dendrite by monoaxial compressive forces acting perpendicular to the long axis of the slit. This nonuniform pressure causes a nonuniform deformation of the dendrite, which very probably is associated with negative strain (compression) perpendicular to the dendrite's long axis and positive strain parallel to it.

The cuticular structures on the route from original stimulus to dendrite deformation are evidently adapted to being deformed during the transmission and transformation of the stimulus. The particular configuration of the slit itself causes its deformation during this process to be reduced rather than enhanced. Nevertheless, the slit sensillum is very sensitive because the stimulus-transmitting structures can easily be deformed. This deformability is especially interesting in view of the properties of the material in which the slit is embedded: the modulus of elasticity of the cuticle is about 18 GPa (Blickhan and Barth 1979), about equal to that of bone (Yamada 1970). Under these conditions ordinary stimuli would be expected to deform the skeleton only very slightly, particularly because the cylindrical shape (large geometrical moments of inertia) of the leg segments endows them with high mechanical stability.

The morphological parameters just described are also the main basis for the adaptive radiation of the slit sensilla. A comparison of the slit sensilla of various arachnids (Figs. 1, 3) (Barth

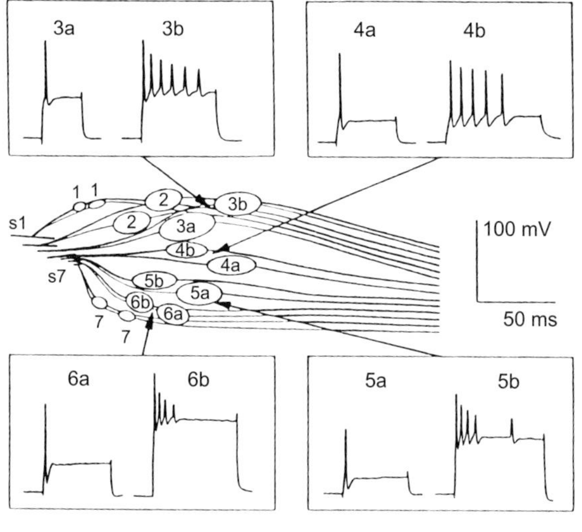

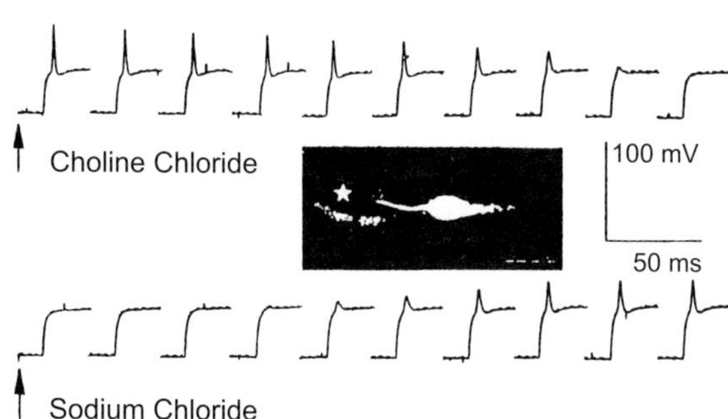

Fig. 10. Intracellular recordings of the activity of individual sensory cells of lyriform organ VS-3 on the walking leg patella of *Cupiennius salei*. *Above* Responses of the two cells innervating each of the seven slits (S1–S7) to electrical stimulation. Whereas the a-type cell generates only one action potential even with strong stimulation, the b-type cell adapts much more slowly (Seyfarth et al. 1994). *Below* Demonstration of the sodium dependence of the action potentials in a neuron of the a-type. The response to step current injection disappears after replacement of NaCl by choline chloride in the bath solution. It reappears when the cell is again exposed to solution containing NaCl. The time interval between the recordings was 10 s. The *inset* shows the sensory cell after injection of Lucifer Yellow. (Seyfarth and French 1994)

and Libera 1970; Barth and Wadepuhl 1975; Barth and Stagl 1976), has shown that they all vary, and with them the mechanical sensitivity of the slits.

Once again, the most important morphological findings are as follows.

- Slit length: varies from a few micrometers to about 200 μm. Both the absolute sensitivity (deformation) and the directional sensitivity increase with length of the slit (Barth 1972a, b, 1981; Barth and Pickelmann 1975).
- Position of dendrite in slit: the deformation of a single slit is always greatest in the middle of its longitudinal extent. This becomes obvious when the edges of the slit are viewed as two flexible bars fixed in position at both ends. To maximize sensitivity, therefore, the midregion along the slit is the best position for the end of the dendrite, and it is indeed the typical position in single slits. In the case of composite (lyriform) organs, the situation is more complicated (see below).
- The degree of curvature of the covering membrane in the unstimulated state also varies, and because this influences the bending moment, it is one of the factors affecting membrane deformability. In the short slits of lyriform organs the trough formed by the covering membrane is shallower than in the long slits, which presumably contributes to their lower sensitivity (Barth 1972b).
- Finally, the list should include a parameter that derives from the pronounced directional differences in deformability of the slit: the slit's orientation in the skeleton with respect to the direction in which the imposed pressures act. A single slit is most sensitive when its long axis is perpendicular to the direction of imposed force. The situation in the lyriform organs will be discussed later. Here it suffices to say that the orientation of the slits in the skeleton (for example, relative to the long axis of the leg) does in fact vary, and that it is worth considering the significance of this variation.

4 Receptor Mechanisms

The non-neuronal accessory structures are of fundamental importance in receiving, transmitting and transforming a stimulus and hence affect the way a sense organ functions as well as its specificity. They are largely responsible for the variability among sense organs. The sensory cells themselves tend to be relatively uniform and conservative. Nevertheless, they are the heart of the matter, for it is here that the mechanical stimulus is translated into the language of the nervous system. In this regard the sensory physiologist wants to know two things. What specific events occur during the process of mechano-electrical transduction? And how are the stimulus and the resulting excitation formally related to one another?

Potentials and Currents

The primary processes at the dendritic membrane of slit sensilla are still poorly understood. As also applies to many other mechanoreceptors, it is very difficult to gain access to the crucial terminal section of the dendrite for experimental purposes, because this section is so intimately coupled to the stimulus-transmitting structure. Recently, however, a preparation has been developed in which the activity of sensory cells can be recorded intracellularly, and which can be expected to yield much more information (Seyfarth and French 1994).

Some important findings are already available. It is especially interesting that clear differences have been found between the slit sensilla of the spiders and campaniform sensilla, the "hole sensilla" of insects, which are structurally quite similar to the slit sense organs (Barth 1981; Thurm 1982; Thurm and Küppers 1980).

In his studies of the campaniform sensilla Thurm (1974, 1982) found many indications that the end of the dendrite amounts to a variable resistor, the resistance of which decreases when the appropriate stimulus is applied. As a result, more current flows through the sensory cell, driven to a considerable extent by a non-neural battery. In the campaniform sensilla, this battery is situated in the apical microvilli of the tormogen cell and is an electrogenic pump that transports potassium ions into the receptor-lymph space. A "standing potential" is generated; that is, the receptor lymph is positively charged with respect to the hemolymph, by 10–100 mV. The parallel to the human inner ear is astonishing (Davis 1965): the stria vascularis is comparable to the tormogen cell and the endolymph space, to the receptor-lymph space. In both cases the standing potential is largely the result of a K^+ current.

In spiders, curiously, no standing potential has been detected in the slit sense organs; furthermore, extracellular recording also fails to detect a receptor potential. The lack of a standing potential has to do with the chemical composition of the receptor lymph: whereas in insects this fluid is rich in K^+ (Küppers 1974; Kaissling and Thorson 1980), in the slit sensilla of *Cupiennius* it is Na^+-rich and K^+-poor. Correspondingly, electron-microscopic examination of the apical membrane of the tormogen cell has revealed no evidence of a K^+ pump there (Barth 1971a; Rick et al. 1976).

Because no receptor potential can be recorded from the cuticular surface, we have so far assumed that the action potentials are generated in the dendrite and that the soma was as little involved in this process as in transduction itself. This would be a reasonable assumption, given that the soma is far away from the slit, as far as 200 μm (Seyfarth et al. 1982). The high Na^+ content of the receptor lymph is interpreted in the sense of a conventional electrical excitability.

More recent intracellular recordings from cells of a lyriform organ on the patella support the idea that the action potentials depend on an inward current of Na^+. When the Na^+ in the physiological saline was replaced by choline ions (which cannot pass through the cell membrane), or when the sodium channels were blocked by tetrodotoxin, electrical stimulation of the cells no longer elicited action potentials (Seyfarth and French 1994). It has since been shown that these manipulations abolish the response to mechanical stimulation of the slit as well (Juusola et al. 1994); these experiments also confirm that the receptor current is carried mainly by Na^+ ions (Fig. 10).

In another recent development, Ulli Höger and Ernst-August Seyfarth have collaborated with Päivi Torkkeli and Andrew French at Dalhousie University in Halifax to measure receptor currents in the sensory cells of the patellar lyriform organ VS3 with the voltage-clamp technique (Höger et al. 1997). Although they changed the membrane potential over a wide range (–200 mV to +200 mV), the receptor current always flowed in one direction: into the cell. This inability to reverse the current direction provides excellent support for the earlier findings and hypotheses that the receptor current is carried substantially by Na^+. This result implies that the permeability of the membrane to Na^+ during mechanical stimulation is much greater than the K^+ permeability. The behavior of the mechanosensitive channels in the sensory cells of our spiders thus really is different from that of insect mechanoreceptors and the hair cells of vertebrates, and these channels are also unlike the mechanically activated channels in the membranes of non-sensory cells. In all these other cases the mechanical stimulus increases conductivity, either for univalent cations in general or specifically for K^+.

It also seems that the puzzle of the second cell has been solved – that is, its apparent failure to produce action potentials (Barth 1967; Bohnenberger 1981; Seyfarth et al. 1982). When they stimulated the cells of the patellar organ electrically, Seyfarth and French (1994) found that one of the two cells innervating each slit gave an extremely phasic response, with only one or two action potentials. In extracellular recordings this response was presumably masked by the activity of the other cell. However, it is not yet known how this cell behaves during mechanical stimulation, or the processes that cause it to be stimulated under natural conditions.

Transfer Properties

The formal relationship between the rate at which a slit sensillum fires impulses and a change in the mechanical stimulus eliciting that response can be described by a simple power function:

$$y(t) = a \cdot d \cdot t^{-k}.$$

Here y is the response of the sensory cell, t is time, a is a constant representing the amplification, d is stimulus amplitude and k is a receptor constant that describes how quickly the response of the receptor to a maintained stimulus declines, and how it responds to stimuli presented at different frequencies (Fig. 11).

The absolute values of k found for the slit sensilla tested so far lie between 0.2 and 0.7. What does that mean? For a pure displacement sensor (the response of which is independent of frequency) k would equal 0, whereas for a pure velocity sensor (a differentiator of first order) k = 1, at least as considered ideally in the frequency domain. In other words, the response behavior of the slit sensilla is somewhere between those of a displacement and of a velocity

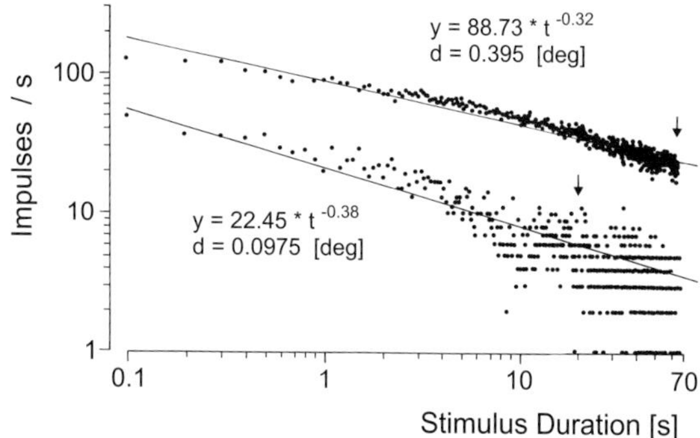

Fig. 11. The step response of slit sensilla follows a power-law function. The example shows the response of slit 2 of lyriform organ HS-8 to two step stimuli differing in amplitude (given as lateral displacement d of the metatarsus in degrees) by a factor of 4. Note the corresponding change in the equation describing the response, which implies practically constant gain (see text). *Continuous lines* are fitted to measured points to the *left of the arrows* and correspond to the power-law functions given. (Barth 1985b)

sensor. These sensilla adapt slowly and respond over a large frequency range, between ca. 5 mHz and at least 500 Hz – that is, over at least five decades of frequency (Bohnenberger 1979).

All these values are very similar to those for the campaniform sensilla of insects (Chapman et al. 1979). Furthermore, the equation given above describes the typical response behavior of many other receptors (Thorson and Biederman-Thorson 1974).

The response characteristics of the slit sensilla are based neither exclusively on the mechanical filter properties of the stimulus-conducting structures nor exclusively on the properties of the impulse-generating mechanism. The arguments that led us to this conclusion are as follows.

- The cuticle is a highly polymerized material. The dynamic mechanical behavior of such materials is usually described by power functions (Bohnenberger 1981; Wainwright et al. 1976). The covering membrane, the coupling cylinder and the receptor-lymph space are all very likely to have viscosities that could at least partially explain the adaptation of the response to a maintained stimulus.
- We have investigated one lyriform organ (HS-8) specifically with respect to the possibility that mechanical viscosities are involved in the response behavior. The organ is situated on the distal tibia, ventral to the joint with the metatarsus, and can best be stimulated by the sideways movement of the metatarsus. After a stepwise deflection of the metatarsus, the force acting on the metatarsus and the strain in the cuticle near the organ were found to show only a slight degree of "adaptation" to this stimulus (Bohnenberger 1981; Blickhan et al. 1984). The viscoelastic properties of the joint are consistent with this result, in that the k value for the decline in force (which follows a power function) is only 0.04; that is, the joint is almost perfectly elastic. Hence the high-pass characteristics of the slit are only very slightly influenced by the mechanics of the cuticle and joint.

Expressed the other way around: the power function reflects mainly the non-mechanical processes of transduction. This view is corroborated by the power-law behavior also found in chemo- and photoreceptors, where there is no mechanical filter (Biederman-Thorson and Thorson 1971; Thorson and Biederman-Thorson 1974). Furthermore, Mann and Chapman (1975) showed for the campaniform sensilla that the time course of adaptation is based primarily on the coding process (generation of action potentials) and not on the mechanical coupling or on mechano-electrical transduction.

5 Groups of Slits and Lyriform Organs

One of the most conspicuous properties of slit sensilla is to assemble into groups and "lyriform organs". I think this is also their most interesting property, because it illustrates so clearly how many mechanical tricks and opportunities for adaptation are hidden in the simple structural plan of the slit sense organs.

First, a definition: the term "composite" or *lyriform organ* refers to an aggregation of 2 to 29 (the largest known number) of slits arranged close together in parallel. In contrast, a "group" is a less strictly arranged aggregation of slits, which are appreciably further apart from one another (Figs. 2, 12) (Barth and Libera 1970; Barth and Stagl 1976). In quantitative terms, an isolated single slit is one that is at least 100 μm away from the next slit. By definition, a group comprises at least one slit that is at least 30 μm long plus at least one other slit, separated from the first one by less than 100 μm. Lyriform organs are unmistakable because the (at least two) slits of which they are composed are parallel to one another and separated by a very small distance (usually 5 μm and never more than 10 μm).

With these definitions, almost all slits can readily be assigned to one of the three categories. More important, however, are the mechanical consequences of these purely morphological differences. The deformability of a slit depends to a great extent on the presence or absence of a nearby slit. Slits cannot influence their neighbors when the distance between them exceeds a critical value, about 1.5 times the slit length. This value has been derived from experiments with models in which three slits were placed at various distances from one another, and it is consistent with technological guidelines (Barth et al. 1984). When considering "groups", therefore, one should bear in mind the mechanical significance of the spacing within the group.

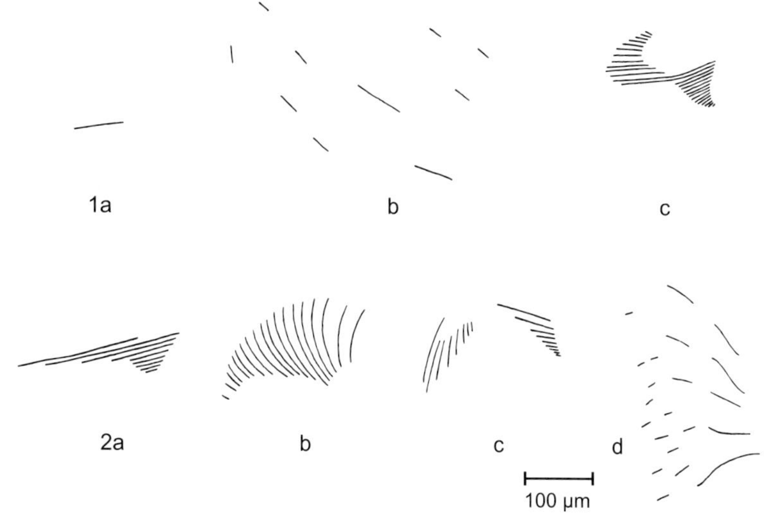

Fig. 12. Types of configurations of slit sense organs. *Above* Degree of aggregation; *1a* large single slit; *1b* group of single slits (trochanter, scorpion); *1c* lyriform organ (trochanter, spider). *Below* Orientation of slits within lyriform organs and groups; *2a* lyriform organ (tibia, spider); *2b* lyriform organ (trochanter, whip spider); *2c* lyriform organs (chelicera, spider); *2d* group of slits (trochanter, whip spider). (Barth 1976)

Lyriform Organs

So far, our research has concentrated on the lyriform organs, the extreme case of aggregation with an almost farcical range of morphological variability. The best studied of these is the organ called HS-8, one of which is located on the back of each walking-leg tibia. No other organ of its kind has been studied in so many ways: with the help of mechanical models and morphological analyses to explain the most important mechanical implications of the aggregation of slits, and by electrophysiological recording from individual slits to examine their physiological characteristics directly (Barth and Pickelmann 1975; Barth and Bohnenberger 1978; Bohnenberger 1981; Barth et al. 1984).

What can lyriform organs do better, or more, than single slits? What functional differences are there between the slits of which they are composed?

Thresholds. The initial experiments with Plexiglas models, together with microscopic observation of the original organs (by interference contrast with reflected light), had shown that the slits at the periphery of the "lyre" absorb considerably more of an imposed load, and hence are more severely deformed, than the slits in between. The deformation of a slit flanked by parallel slits is much less than that of an isolated single slit of the same length (Barth and Pickelmann 1975). The prediction that there would be corresponding differences in the physiological threshold sensitivities of the individual slits of the lyriform organ was then fully confirmed by electrophysiological experiments. HS-8 is stimulated by deflection of the metatarsus; it comprises seven slits, and it was possible to obtain reliable recordings of the responses of all but the most and the least sensitive of them. The remaining five slits varied in their threshold levels of metatarsal deflection, over a range of about 40 dB (Barth and Bohnenberger 1978; Bohnenberger 1981). Apart from the parallel arrangement of the slits, differences in length also contribute to the differences in threshold (see Fig. 1, posterior 8).

Working Ranges. The stimulus amplitudes to which the individual slits respond overlap to a great extent. That is, apart from differences in threshold there is no physiological range fractionation. A more sensitive slit responds over a range that includes the response range of a less sensitive one (Barth and Bohnenberger 1978). However, in his doctoral dissertation Johannes Bohnenberger (1978) was able to show that the regions in which individual slits give linear responses are largely separate, with very little

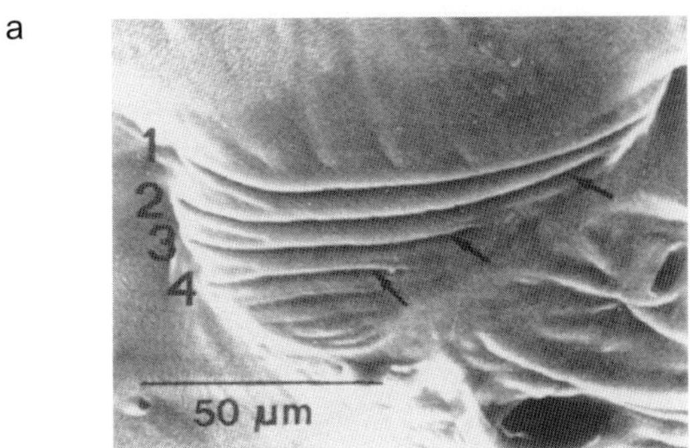

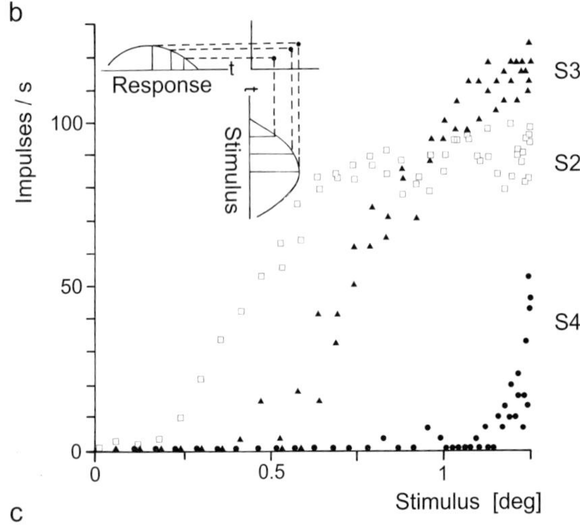

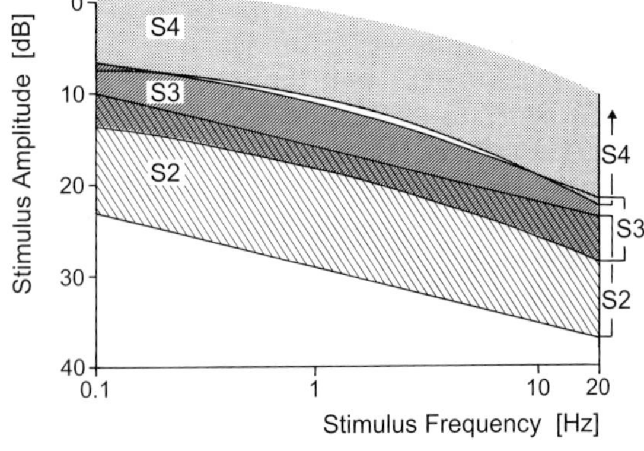

◀ **Fig. 13 a–c.** Functional organization of a lyriform organ (HS-8, walking leg tibia, see Fig. 1) of *Cupiennius salei*. **a** Scanning micrograph of the organ on the posterior side of the tibia. Four (*1–4*) of the eight slits of the organ were studied electrophysiologically; *arrows* point to slits also studied for experiment shown in **b** and **c. b** Responses of slits S2, S3, and S4 to sinusoidal mechanical stimulation of the organ (0.01 Hz). The instantaneous response (frequency of action potentials) is plotted against stimulus amplitude. **c** Linear response ranges of slits S2, S3, and S4 calculated using the values presented in **b.** (Barth 1985b)

overlap. This is an important finding: the linear range of the organ as a whole is considerably greater than that of the individual slits, and so is the range of high absolute difference sensitivity (Fig. 13). An individual slit can differentiate stimulus intensities considerably less well when they reach 10 dB above threshold, or even less, whereas the whole organ does much better. This expansion of the range of high difference sensitivity seems quite useful, as does the ability to respond to stimulus frequencies down to 1 Hz or lower. The lyriform organ HS-8 functions as a proprioreceptor in kinesthetic orientation and in a synergic leg reflex (Barth and Seyfarth 1971; Seyfarth and Barth 1972; Seyfarth 1978a,b, 1985; see Chapters X and XXI) and is presumably part of a feedback system in which its role is to signal even very small departures from the set point. In comparison with exteroreceptors such as our eye or ear, the working range of the lyriform organ HS-8 is small. This, too, is consistent with its proprioreceptive function.

Tuning. The individual slits of the lyriform organ HS-8 respond to deformations at frequencies in the range from 0.005 Hz to 1 kHz. They behave as high-pass filters, in that the threshold is high at low frequencies (up to about 10 Hz) and falls as the frequency increases. The length differences between the slits are not correlated with their tuning to different narrow frequency ranges (Bohnenberger 1981). In this sense, the term "lyriform" is misleading: there is much evidence that the slits, unlike the strings of a lyre, do not function on the basis of differences in resonant frequencies; their frequency discrimination evidently has another mechanism (Fig. 13) (Bohnenberger 1981). Very similar results have been obtained for the metatarsal lyriform organ (Barth and Geethabali 1982; Bleckmann and Barth 1984), which functions as a vibration receptor (see Chapter VIII). Even without such resonances, of course, the patterns of excitation over the whole organ can vary depending on stimulus frequency (this pattern results from the differences in absolute sensitivity of the slits at a given frequency), so that it would be possible for the central nervous system to make use of the differences in overall pattern for frequency discrimination.

Slit 1. One of the slits of the lyriform organ HS-8 differs from all the others: slit 1, the longest slit, situated at the dorsal edge of the organ, exhibits spontaneous activity, gives purely phasic responses, and has a distinctly lower threshold, which makes it much more difficult to use in experiments.

Patterns of Arrangement

A lyriform organ can thus accomplish much more than can a single slit. But this is not yet the end of the story. The arrangement of the slits that form a lyriform organ can vary widely and sometimes seems quite peculiar (Figs. 1,12). Does this variability have any functional significance?

It is practically impossible to make theoretical predictions concerning the detailed mechanical consequences of the various arrangements, given the methodology currently available, and experiments also present considerable technical problems. Furthermore, it is not yet possible to apply tangential stresses to the organs in situ cleanly in different directions, while at the same time measuring the complex slit deformations with the necessary spatial resolution and/or recording the activity of an individual slit electrophysiologically. So, again we resorted to model experiments. On several occasions these have helped us identify the most important mechanical effects and pointed the way to hypotheses that would not otherwise have come to mind.

Models. Figure 14 gives an idea of the apparatus used to apply stress to disks containing model lyriform organs and also shows the transducer developed to measure the slit deformation.

Of course, the disk model of a lyriform organ cannot copy the original in all structural details, so we cannot claim that it represents all details of the mechanical responses that occur in the spider. However, we believe that the most important effects we have observed with the model are also valid for the living organ (Barth 1972a,b; Barth and Pickelmann 1975; Barth et al. 1984). Neither the material used for the models (carefully annealed Araldite B or Plexiglas) nor the size of the models presents a problem in this regard (Barth 1972a). It can also be taken for granted that the conclusions drawn from static loading in the biologically relevant frequency range also apply qualitatively to dynamic loading, because viscosities do not affect the nature of the deformation at frequencies below several kHz (Barth and Pickelmann 1975).

The many variations in detail on the general lyriform arrangement can be classified as three basic patterns, distinguished from one another by the extent of lateral offset of adjacent slits from one another and the nature of the distribution of slit lengths within the group. All models contained just 5 slits each; we did not try to duplicate all possible slit numbers, so as not to mask the effects of differences in slit arrangement and length distribution. The three categories were named according to the outline of the slit pattern: type A, "slanted bar" (all slits equally long and each offset from the next by the same amount); type B, "triangle" (graded lengths of the five slits and various lateral offsets, depending on subgroup); and type C, "heart" (the longest slit in the middle and shorter ones on each side of it). The resemblance to the originals can be seen in figure 15.

Mechanical Directionality. To find out the stimulus patterns to which the dendrites in various slits are likely to be exposed, we loaded these disk models from various directions within the plane of the model (which corresponds to the application of force to the spider leg in a direction parallel to the surface) and measured the resulting deformation. The following effects were observed.

- For all slit arrangements slit compression (the effective stimulus for the original organ) reaches values three to five times higher than slit dilation.

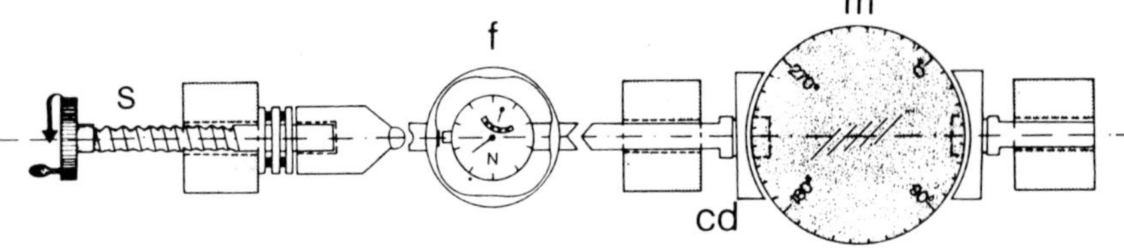

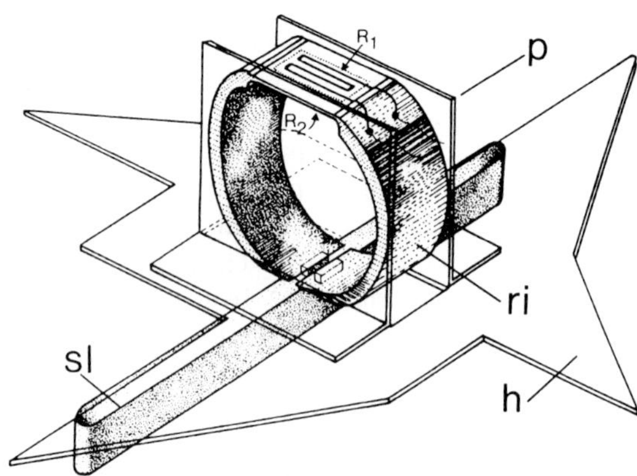

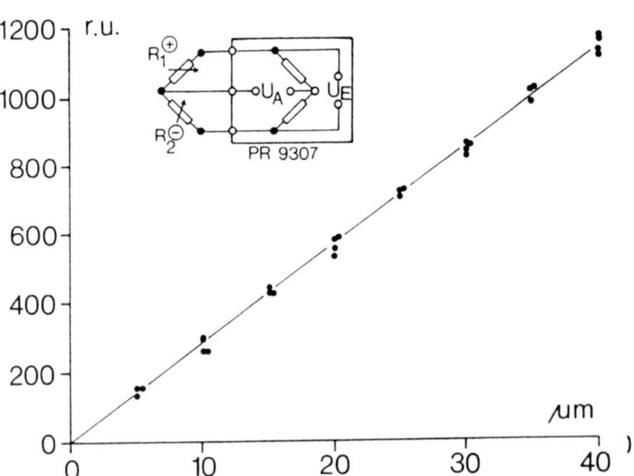

Fig. 14. Model experiments designed to identify the mechanical consequences of slit aggregation in lyriform organs. *Top* Apparatus used to load the models. Both the model (*m*) and the force gauge (*f*) are held by a clamping device (*cd*) mounted on a V-way. Force can be applied by turning the screw (*s*). Markings on the model allow precise adjustment of the load angle α by rotating the model disk into its desired position before loading. *Middle* Device to measure the deformation of a slit under load. The metal torus *ri*, with strain gauges R_1 and R_2 attached to both surfaces of its thin and flat upper part, is inserted into the model slit *sl* with the small ridges bordering the cut at its lower part. Deformation of the slit changes the width of the cut in the torus and, as a consequence, the resistance of the strain gauges. Transparent devices perpendicular (*p*) and horizontal (*h*) to the model disk serve to align the torus precisely with respect to the slit axis and the point along its length to be measured. *Bottom* Calibration of the torus shown above. The inset shows the position of strain gauges R_1 and R_2 in the Wheatstone bridge circuit. U_A, output voltage plotted as relative units (*r.u.*) as a function of the change in the width (µm) of the cut in the torus. (Barth 1984)

- The range of load angles leading to compression or dilation of the slits differed considerably among the various arrangements. In model A load imposed at angles between about 30° and 150° (0° being parallel to the long axis of the slit) caused compression of all slits; in models C and B (see Fig. 15) the corresponding range was about 75°–105°. The values for the two other "triangle" models (B_2, B_3) were intermediate.
- In all models the highest compression values were obtained with the load angle 90°.
- In model A all slits behaved the same for all load directions. Only the sites of greatest deformation along the slit varied. In all other models there were distinct differences in slit

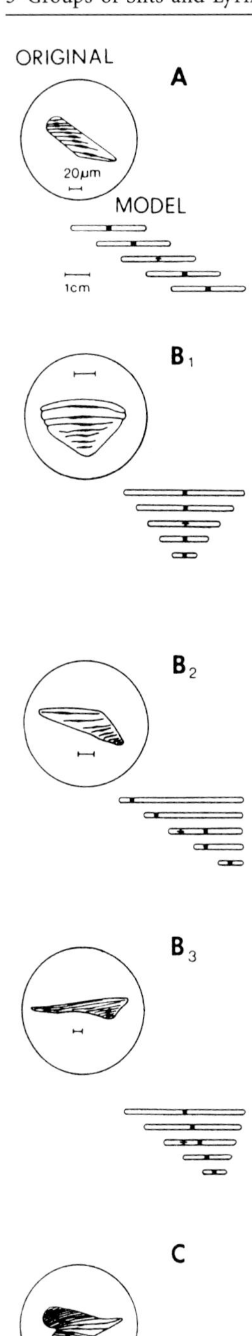

deformation; some were compressed and at the same time others were dilated. A striking feature of the configuration "triangle" was that the longest slit, at the edge, was always considerably more strongly deformed than any of the others.

- The transition from compression to dilation occurs at load angles between 15° and 30° (or 180° minus these angles). Whereas in model A this transition was simultaneous in all slits, the situation was more complicated in models B and C. In the "heart" configuration (C) the longest slit, the one in the middle, was dilated at load angles between 15° and 60° (compressed at 120°–165°), while two neighboring slits on the same side were compressed (dilated). In addition, the deformation of the two neighboring slits changed sides at angles in certain ranges. In the configuration B, furthermore, the individual slits behaved non-uniformly in themselves, so that often one part of a slit was compressed while the rest was dilated.
- What part of a slit is maximally deformed? In the case of an isolated single slit, the maximum is always in the middle of its longitudinal extent (Barth and Pickelmann 1975). For slits in lyriform arrangements it depends on the position of the slit within the group, on the load direction and on the configuration of the slits. A change in load angle between ca. 75° and 105° hardly shifts the position of the maximum in the slit at all, but changes in other angular ranges can have a marked effect.

Location of Dendrite. To the hurried reader it may seem overly meticulous to study such details. But this information is indispensable if we want to truly understand the functional morphology of a slit. Answers to questions about the stimulus pattern to be expected in a lyri-

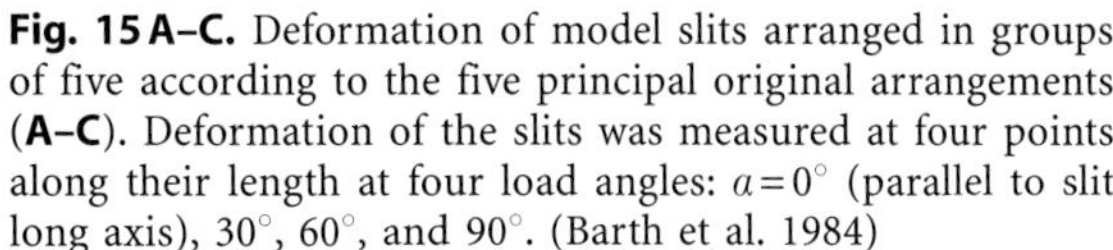

Fig. 15 A–C. Deformation of model slits arranged in groups of five according to the five principal original arrangements (**A–C**). Deformation of the slits was measured at four points along their length at four load angles: $\alpha = 0°$ (parallel to slit long axis), 30°, 60°, and 90°. (Barth et al. 1984)

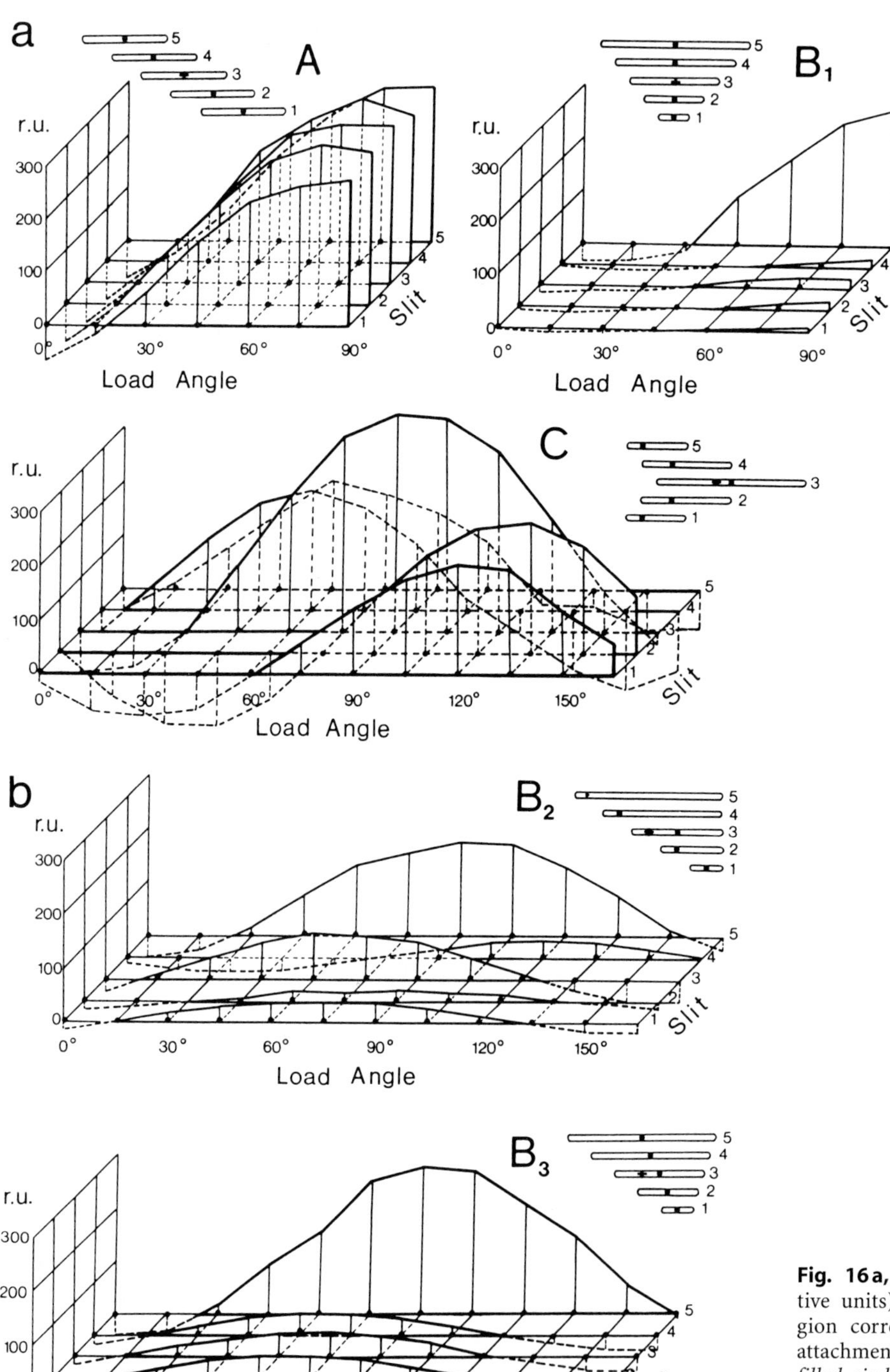

Fig. 16a, b. Deformation (*r. u.* relative units) of model slits in the region corresponding to the dendrite attachment area in the original (see *filled circles in insets*). Load angle α was changed in steps of 15°. Compression values *above x-axis*, dilation values *below* it. (Barth et al. 1984)

form organ are just as important as those to the question whether the end of the dendrite is positioned at the "right" place along the slit. Therefore it is good to know that in the original lyriform organs (Barth and Libera 1970; Barth and Stagl 1976) the dendrite ends are situated precisely where the largest compression values were found in the models. From this we conclude that the arrangement of the dendrite end is such as to maximize sensitivity.

Figure 16 shows what happens in the regions of the model corresponding to the dendrite attachment sites (where in the original the response of the sensory cell is initiated). Group A is distinguished by the similarity of the deformation in all slits and by its large amplitude for load angles around 90°. Here, too, compression is produced over a broad range of load angles (145°). The striking aspect of configuration B is that the long edge slit is much more deformed than any of the other slits. Even slight lateral displacements of the slits have clear consequences: for instance, enlargement of the range of load angles in which compression is produced, and the increase in deformation of the slits 1–4 (see Fig. 16). Configuration C, finally, is notable for the fact that it is the only one in which stress is effective over almost the entire range of possible directions (180°); always at least one of the slits is compressed.

Hypotheses Concerning Function. One more thing should be said about the functional differences between the various lyriform organs inferred from these experiments. From the behavior of these models, we have generated working hypotheses for the next generation of experiments (and for sensory physiologists interested in mechanics): that the lyriform organs are specifically adapted to the conditions that apply to the measurement of the strains that exist in their particular locations. Among these conditions are, for instance, a particularly great "angular working range", a large range of absolute sensitivities and/or a mechanical directional characteristic that permits analysis of strain directions. Let us take one last look at the three types of models in this context.

- *Type A, "slanted bar"* (examples include organ VS-2 on the femur, VS-3 on the patella, VS-4 on the tibia; femur of daddy-long-legs and scorpion, Figs. 1, 3). From a mechanical point of view no appreciable differences in the absolute sensitivity of these slits would be expected, nor should the working range (the range of angles over which stress causes compression) of the whole organ be greater than those of the individual slits (no range fractionation). Configuration A also seems unsuitable to signal stress direction. Because the pattern of excitation within the group of slits is the same for all load angles, it ought to be very difficult, at the least, to discriminate the effects of stimulus amplitude from those of stimulus direction. In this case, what advantages has the configuration A to offer? (i) A particularly broad "field of view"; that is, high sensitivity to strains over a particularly large range of directions. It may be that a broad scatter of stimulus directions (strain trajectories) is typical of the sites where these lyriform organs are located. (ii) The similarity of the mechanical behavior of the various slits suggests that convergence may be involved here; that is, the neural signals from the individual slits may be combined in the CNS in order to increase absolute sensitivity by increasing the signal-to-noise ratio.
- *Type B, "triangle"* (examples: HS-8 and HS-9 on the tibia, VS-1 and HS-3 on the trochanter, HS-6 on the patella, the lyriform organs of the chelicerae and the coxa of the pedipalps; trochanter of the daddy-long-legs). Here, unlike model A, there are major differences in the degree of deformation of the slits; therefore, it seems reasonable to postulate that the range of stimulus amplitudes is fractionated, in that the responses of the slits are distributed. For the lyriform organ HS-8 we already have physiological evidence to corroborate this idea (Barth and Bohnenberger 1978; Bohnenberger 1981). As a corollary, it seems likely that the stimulus amplitude will ordinarily be extremely variable at the sites where lyriform organs with approximately triangular outline are located.

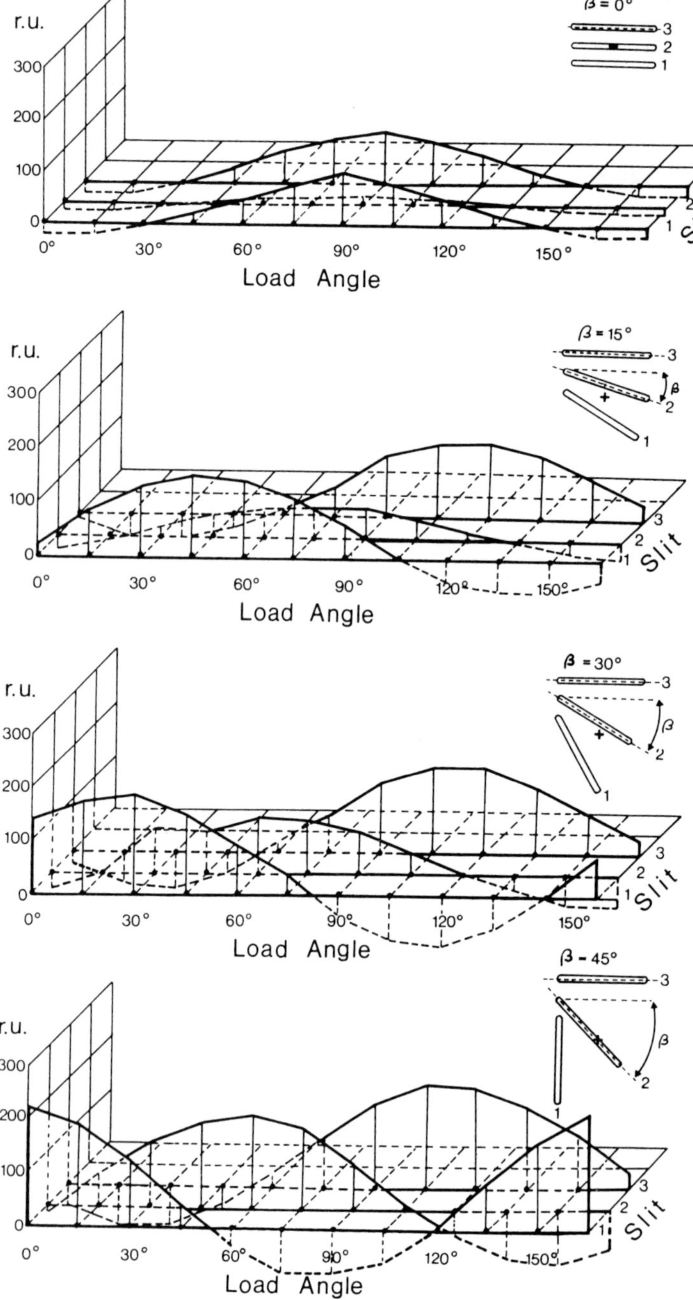

Fig. 17. Slit deformation (*r.u.* relative units) as a function of both the load angle α and angle β between neighboring slits in a group of three. *From top to bottom* β varies between 0° and 45° (see *insets*). All deformation values refer to the center of the slit and are plotted as in Fig. 16. (Barth et al. 1984)

- *Type C, "heart"* (example: HS-2 on the trochanter). Although the parallel arrangement of the slits may not at first suggest such a thing, from a mechanical viewpoint the configuration C would be suitable for analyzing the stress direction. Therefore one would expect that where such organs are located, the range of stimulus directions would tend to be large. The high deformability of all the slits and the marked differences in pattern of excitation within the group, depending on stimulus direction, are presumably adaptations to this situation.

Divergence of Slits. The model results have still more to tell us. The impatient reader may be consoled by the thought that the time spent in reading is vanishingly small in comparison with the time it took to carry out the research reported here.

It will be evident by now that the deformation of individual slits depends strongly on the stress direction and is greatest for angles between 75° and 105°. The model experiments just described produced the less obvious result that even parallel arrangements of slits can be used to gain information about stimulus direction, if the central nervous system is able to evaluate the changing patterns of excitation within a group of slits. The apparently simpler means to the same end is also represented: in some organs different directional sensitivities are combined at the same place by assembling slits with differently oriented long axes. Within a lyriform organ on the trochanter of the whip spider, the slit axes diverge by almost 90° (see Figs. 3, 12). There are similar examples in the daddy-long-legs (coxa) and the whip scorpion (metatarsus) (Barth and Stagl 1976). Next to one another on the spider chelicera are two lyriform organs, the slits of which are perpendicular to one another (Fig. 1) (Barth and Libera 1970). The model experiments show the effect of this pairing quite clearly (see Fig. 17) (Barth et al. 1984). As the angle β between the slit axes increases, so does their deformation, while the difference between the maximal deformations of the slits simultaneously decreases. The difference between the load angles most effective for the slits also increases with increasing β.

The relationships illustrated in figure 17 suggest comparison with the spectral sensitivity curves of the various types of photoreceptors in the eye. The basis of color vision in both arthropods and vertebrates is that the visible spectrum is subdivided into several components (three, for instance, in the trichromatic color vision of honeybees and humans) by the specific sensitivities of different receptor cells. Might it not be that a similar multichannel analysis is being carried out here to determine stress direction?

For a more formal approach to understanding the complex mechanics underlying the deformation of lyriform slit groups, see the original literature (Barth et al. 1984).

6 The Positions of Slit Sensilla on the Body and the Natural Stimuli

What are the functional causes and consequences of the typical, specific positions of the various slit sensilla in the exoskeleton? This question cannot be answered until we know the answer to another: where do the effective stimuli come from?

This, again, is a complex area of research, not least because of the difficulties that must be overcome in order to measure the actual strains within the spider skeleton.

Here we proceed in four steps. (i) The first is simply to make a few important general observations about the organs' position on the body. (ii) Not quite so simply, we then clarify the stimuli. (iii) In the third step, the mechanical sensitivity of the exoskeleton is specified by means of a concrete representative example, and (iv) finally we proceed to the in vivo measurement of biologically relevant strain stimuli in the spider skeleton.

Position on the Body

It is important to remember here that many kinds of load are potentially effective stimuli for slit sensilla (see Fig. 4c); bending, buckling,

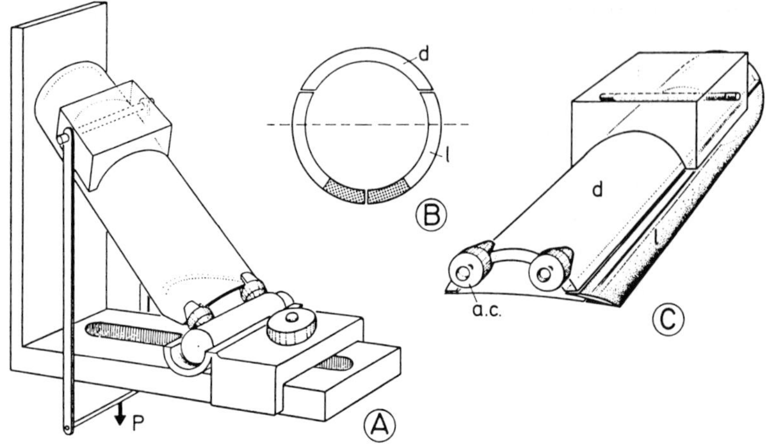

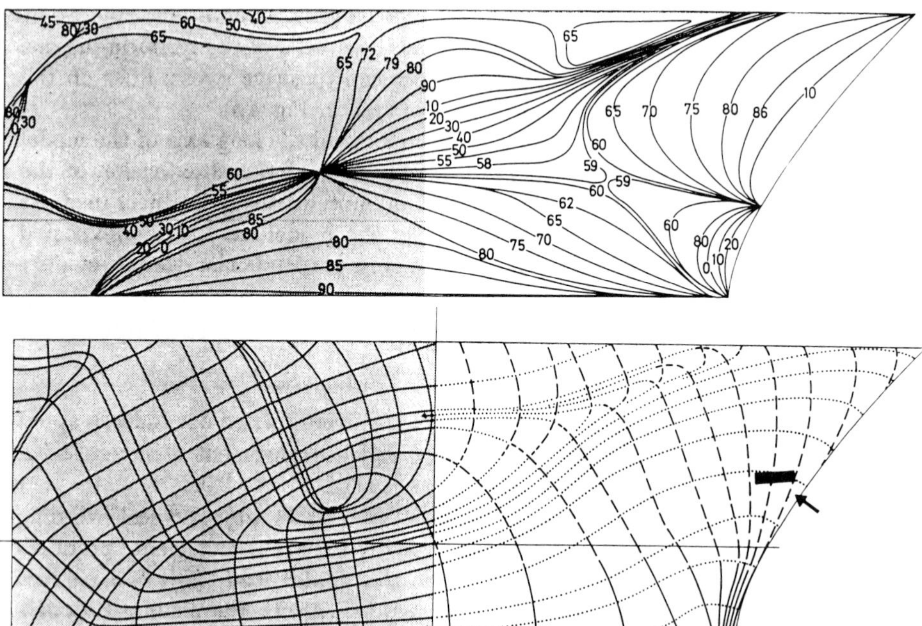

Fig. 18. Photoelastic experiment designed to determine the tension and pressure lines in a model tibia under quasi-natural load. *Top* Model tibia with two articular condyli (*a.c.*) in the device (*A*) used for loading. By tilting the tibia through various angles the direction of the load vector *P* can be varied. After freezing to preserve the stresses the models are cut into a dorsal and two ventrolateral pieces (*d* and *l* in *B*) for the analysis with polarized light. *Middle* and *bottom* Result of an experiment in which the angle between the load vector and the long axis of the model tibia was 60°. Analysis of the ventrolateral areas (*l*) where most of the lyriform organs are located in the original spider leg. The *upper drawing* shows the isoclinic lines (lines connecting all points where the principal stresses are oriented in the same direction or perpendicular to it) as the angle of linear polarization of the transmitted light. The *lower drawing* shows the lines of principal stresses deduced from the isoclinics. The sign of the stresses was determined for the surface of the model: - - - - pressure, —— tension, ... no stress. The *shaded areas* are doubtful in significance for mechanical reasons. *Arrow* points to site of lyriform organ HS-8 in the original leg. (Barth and Pickelmann 1975)

twisting, compression, stretching, shearing or combinations of any of the above! This may be one of the reasons for the broad spectrum of topographic peculiarities among the slit sense organs. However, comparison of the walking legs (where most of the slits are situated) of representatives of five orders of arachnids revealed a few general rules (Fig. 3) (Barth and Stagl 1976).

- In all cases the slits are most numerous at the proximal end of the leg. This is associated with the concentration of musculature in the proximal leg section (Frank 1957; Parry 1957; Millot and Vachon 1949; Snodgrass 1965) and also with the fact that the load-bearing phase (stance phase) of the walking movement is produced here. The activity of these muscles undoubtedly causes strains in the skeleton.
- All lyriform organs are located directly at or near joints, where forces are transmitted from one leg segment to the next by way of structures with areas that are small relative to the cross-sectional area of the leg: joint tubercles or hinges. In these regions a concentration of stresses would be expected, which is a good reason for the sensilla to be positioned next to surfaces thus loaded. However, some of these organs are not in immediate proximity to a joint, although they are not far away (see Figs. 1, 3). It is entirely possible for a stress-induced displacement to increase up to a certain distance away from the site of force transmission. The degree of cuticular bending due to forces acting on the joint can also be greater some distance away than at the joint itself (which is characterized by an especially thick, strongly sclerotized cuticle as well as by a large geometrical moment of inertia).
- Unlike the lyriform organs, the single slits are often situated far away from joints, on the legs and the opisthosoma, but they are frequently close to the sites of muscle attachment, where bending forces can also originate.
- On the legs, most of the lyriform organs are not only near joints but lateral to them and approximately parallel to the long axis of the leg. It occurred to us that at this particular position compression (negative and hence excitatory strain) is produced by the forces of the muscles and, in addition, is especially effective because its direction is approximately perpendicular to the long axes of the slits. This hypothesis was tested first with a model for the organ HS-8 and later by direct measurements on a spider. The model was a mock tibia made of Araldite, upon which stress was imposed in just the way that we thought would occur during activity of the flexor muscles in the real leg. The courses of the lines of compression and extension (positive strain) were determined by photoelastic measurements (Fig. 18). The result fully confirmed our hypothesis (Barth and Pickelmann 1975). The position and orientation of HS-8 are such as to maximize the mechanical sensitivity of this organ, given the nature and direction of the force transmission in the skeleton during muscle activity and the organ's own directional sensitivity.

These general rules account for the specific positions of at least a large number of slits. But they are certainly not exhaustive. In many cases it is necessary to examine closely the particular area of the leg or body. One example is the vibration-sensitive lyriform organ of the metatarsus, treated in Chapter VIII. We shall also see (in Chapter XXIV) that what matters is not merely the magnitude and direction of strains in a particular skeletal area; the time course of these strains can also be related to the presence or absence of a sensor at a particular place on the exoskeleton.

Forces and Moments

Natural scientists measure, and if something cannot yet be measured they make it measurable. All the models, hypotheses and plausible interpretations cannot take the place of direct measurements. Therefore, in years of meticulous work we have developed miniaturized instruments to detect strain, force and pressure in objects with the tiny dimensions of a spider leg. I shouldn't really say "we": the main role here was played by my doctoral student Reinhard Blickhan, who has since become a professor at the Institute for Sport Sciences of the University of Jena. As a physicist, he brought with him not only the necessary know-how but also a practically infinite patience with the solution of methodological problems, which biologists often lack. By the time he had finished, strain, force and hemolymph pressure could be measured not only *in vivo* but even simultaneously. Of course, for technical reasons some of the places of interest remained inaccessible. We concentrated on the region of the distal tibia and the joint between that and the metatarsus. There were several strong arguments in favor of this choice. First, no less than four lyriform organs are present here, among them HS-8, by far the most thoroughly studied of all. Second, there are no extensor muscles at this joint; their function is performed by the hemolymph pressure, which extends the leg hydraulically. As a result, here the effects of muscle activity can be distinguished from those of hemolymph pressure. Blickhan and Barth (1985) give details about this highly refined measurement technique.

The measurements showed us how strains are induced by natural loads and how these loads are transformed into deformation at the site of the sense organs (Blickhan and Barth 1985). First, however, the natural loading parameters should be explained.

The exoskeleton of the arthropod leg consists of thin-walled, tubular components connected together by joints. The joint surface area is considerably smaller than the cross-sectional area of the leg. In joints that rotate through large angles, the flexible joint membrane is often folded like a bellows. Spiders can move their legs rapidly and with large amplitudes. We expect and indeed find long, slender legs with musculature located primarily in the proximal section so as to keep the forces of inertia from being too large. Similarly, the joints with several degrees of freedom, such as ball joints (just like our hip joint), are positioned close to the body, whereas hinge joints such as the tibia-metatarsus joint (like our knee) are usually in the middle of a leg, because they do not need muscles to stabilize them in the direction perpendicular to the main movement plane. The most distal segment, finally, is often especially short and contains no muscles at all; it is moved by means of long tendons. The last joint is often a ball joint, because errors in the position of this segment have only a slight influence on the position of the whole animal.

The two load parameters that bring about the active flexion and extension of the tibia-metatarsus joint (and are also present in a spider that is sitting on a substrate and bearing its own weight) are the forces developed by the muscles and the hemolymph pressure, respectively (Fig. 19) (Blickhan and Barth 1985).

(i) Muscle Forces. The arrangement of the flexors in the tibia is shown in figure 20. The moment generated by their contraction is $\vec{M}_{\mathrm{m}}$ and equals the product of the magnitude and direction of the muscle force vector $\vec{F}_{\mathrm{m}}$ and the position of the center of gravity of the muscle attachment area $\vec{r}_{\mathrm{c}}$ with respect to the axis of rotation of the joint:

$$\vec{M}_{\mathrm{m}} = \vec{r}_{\mathrm{c}} \times \vec{F}_{\mathrm{m}} \ .$$

(ii) Hemolymph Pressure. The moment $\vec{M}_{\mathrm{h}}$ depends on four variables: the pressure amplitude P, the effective area A of the surface (on which the pressure acts), the orientation n of this surface and the position of its center of gravity $\vec{r}_{\mathrm{s}}$ with reference to the joint axis:

$$\vec{M}_{\mathrm{h}} = \mathrm{P} \times \mathrm{A}(\vec{r}_{\mathrm{s}} \times \vec{n}_{\mathrm{A}}) \ .$$

A and $\vec{r}_{\mathrm{s}}$ and hence the moment associated with a given pressure can be reduced by the activation of muscles that pass through the effective area.

The hemolymph pressure acts antagonistically to the flexor muscles. Although spiders have an open circulatory system, the blood flows between the muscles along well defined, fixed pathways (Fig. 21). Of particular importance in the present context are the dorsoventral channels, which end ventral to the tibia-metatarsus joint and thus at the greatest possible distance from the dorsal joint axis. This produces a large moment $\vec{M}_{\mathrm{h}}$.

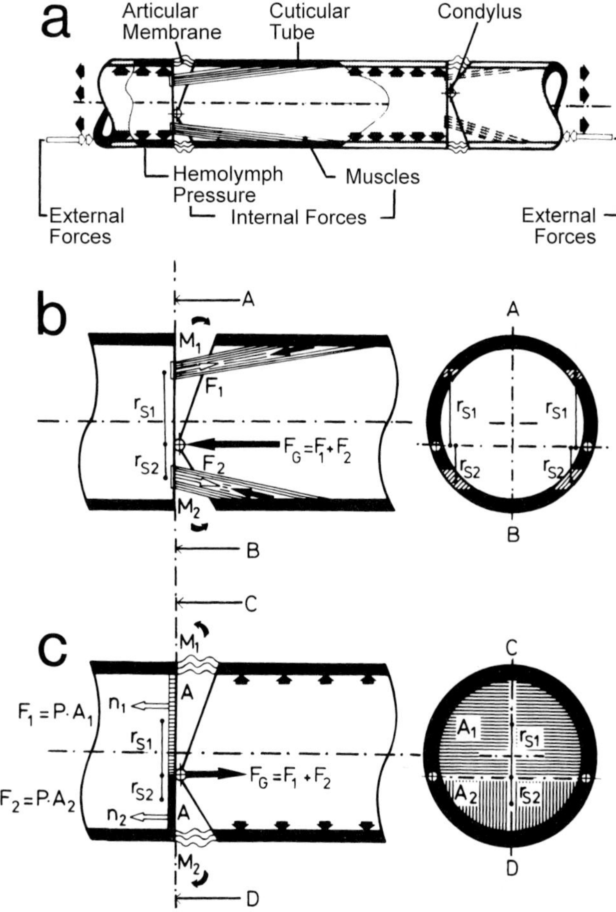

Fig. 19 a–c. Mechanical "Bauplan" of arthropod legs. **a** Arthropod legs consist of tube-like segments, which are connected by joints with contacting surfaces much smaller than the segment's cross-section. External (⇔) and internal (→) forces (hemolymph pressure and muscle activity) determine the loading condition of a segment. **b** Torques developed by muscle forces. The vectors of the muscle forces (F_1, F_2; ⇒) apply at the center of gravity (r_{s1} and r_{s2}) of the attaching surfaces developing the torques M_1 and M_2. **c** Torques developed by hemolymph pressure (P). P acts on the effective areas A and induces the torques M_1 and M_2. → forces developed at the segment under study; F_G, resulting joint force. (Blickhan and Barth 1985)

The force that dominates at the joint, however, is determined not only by the above moments $\vec{M}_m$ and $\vec{M}_h$. A third moment $\vec{M}_G$ also contributes, the resultant of the ground reaction force $\vec{F}_G$ and its moment arms of force vector $\vec{r}_A$:

$$\vec{M}_G = \vec{r}_A \times \vec{F}_G \,.$$

$\vec{M}_G$ increases with increasing distance of the leg tip from the joint axis. Unlike the dorsoventral forces (F_y), the lateral (F_z) and axial (F_x) forces cannot be actively adjusted. They are passively absorbed by the joint and its joint membrane. The internal forces (muscle, hemolymph) also act with or against gravity, yet another external force.

Figure 19 summarizes the most important features of the mechanical structural plan of arthropod legs, and of our spider leg joint in particular.

Mechanical Sensitivity

The next question is: how do the various areas of the tibial cuticle in which lyriform organs are situated differ in their responsiveness to the various components of imposed load? The mechanical sensitivity of interest here can be defined as

$$S = \frac{\text{Strain}}{\text{Force}} \,[\varepsilon/\text{N}]$$

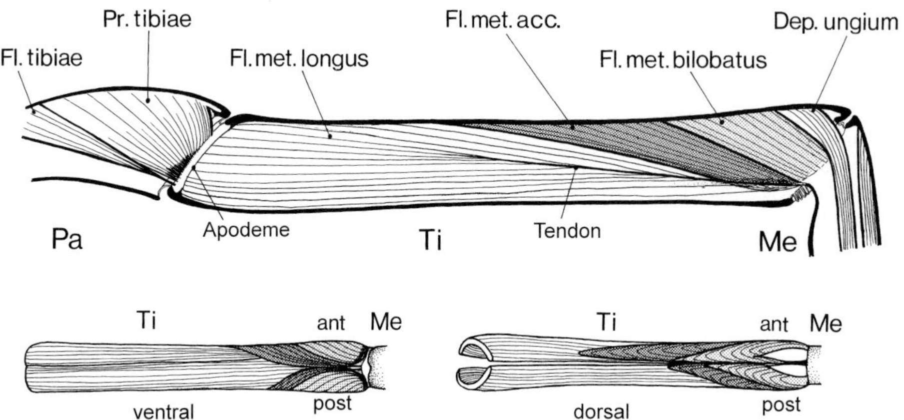

Fig. 20. Arrangement of muscles (flexors) in the walking-leg tibia (*Ti*) of *Cupiennius salei*. *Pa* Patella, *Me* metatarsus. *Above* Longitudinal section; *below* ventral and dorsal aspects. Muscles *from the left* (proximal) *to the right* (distal) are: Flexor tibiae, Promotor (remotor) tibiae, F. metatarsi longus, F. metatarsi accessorius, F. metatarsi bilobatus promotor (remotor), and Depressor ungium. (Blickhan and Barth 1985)

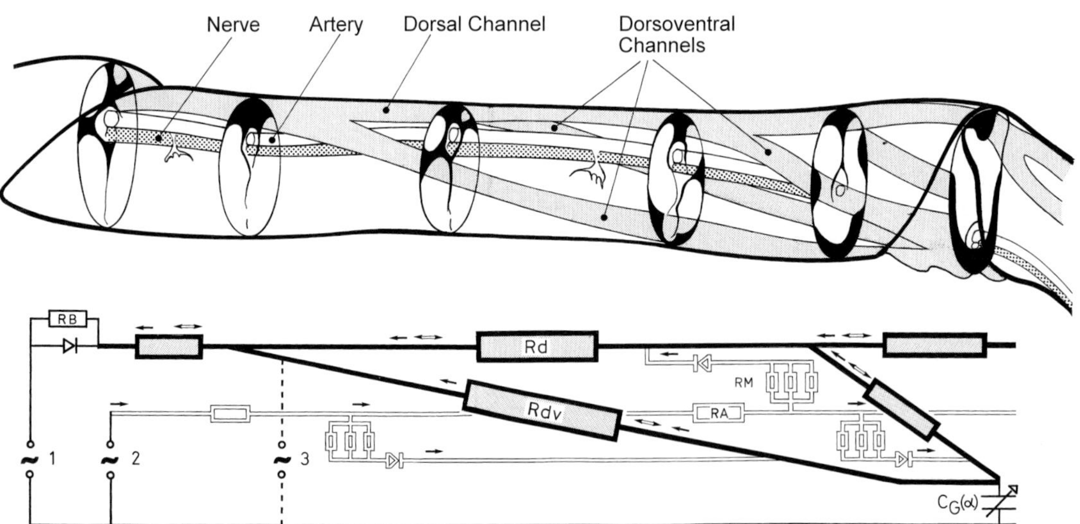

Fig. 21. Lacunae in the tibia of the spider walking leg form channels for the fast transport of the hemolymph during active and passive joint movements. Pressure measurements have to be done in these channels because the transducer displaces an appreciable volume. *Above* The dorsal channel is embedded between muscle attachments and provides for a quick supply to the adjacent segments. The dorsoventral channels are closed distally, however. The course taken by the channels is simplified here (*black* original cross-sections). *Below* Circuit explaining fluid mechanics in the tibia. — supply by arterial hemolymph; —▭— fluid resistances of the arteries (R_A) and the muscular capillaries (R_M); ⊳⊢ valves prohibiting backflow; — lacunae; -▭- fluid resistances of the dorsal (R_d) and dorsoventral (R_{dv}) hemolymph channels; ⫳ C_G (α) volume displacement of the articular membrane; → flow direction of the circulation and pumping direction during leg movement; ⊳⊢ valves and resistance (R_B) of the channels in the proximal leg segments; ~ pressure sources: *1* prosoma; *2* heart; *3* muscles. (Blickhan and Barth 1985)

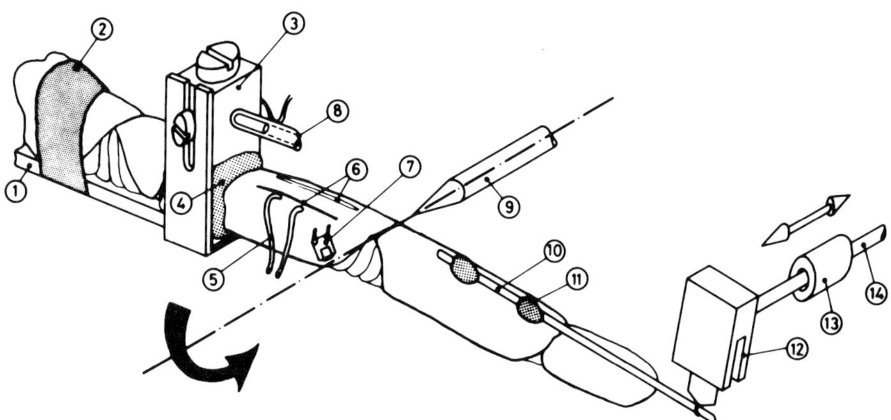

Fig. 22. Experimental arrangement to determine the mechanical sensitivity at the sites of slit sensilla. The animals were tethered with adhesive tape (*2*) on a brass spider dummy (*1*) and the leg to be studied fixed with dental cement (*4*). *3* pressure transducer (*8* reference channel). The position of the animal was adjusted to allow its rotation around the axis of the tibia-metatarsus joint (*9* spine in the axis of rotation). A steel needle (*10*) glued (*11*) onto the metatarsus was attached to a force transducer (*12*) that was moved by an electrodynamic vibrator (*13* displacement transducer, *14* axis of vibrator). Strains induced in the tibia were measured using strain gauges (*7*). At the attachment site of the Flexor metatarsi bilobatus (*6*) electrodes for muscle stimulation and the recording of myograms were implanted (*5*). (Blickhan and Barth 1985)

and as such can be quantitatively determined. For such measurements, miniaturized strain gauges are attached to the places of interest on the lyriform organs HS-8 and HS-9 on the back surface of the tibia, the organ VS-4 on the front surface, and the organ VS-5 on the lower surface, as well as to the cuticle near the site of attachment of the flexor muscle (Flex. met. bil.) on the dorsal surface of the tibia. This is not a simple task. For these experiments we used the largest spider leg available – not one from *Cupiennius*, but the leg of a bird spider (Theraphosidae), with a tibia about 4 mm in diameter. The metatarsus was deflected by applying force to its distal end, in the dorsoventral (F_y), lateral (F_z) and axial (F_x) directions (Fig. 22).

We were pleased to find a good agreement between the result of these tests and the theoretically predicted maximal ε/F values for a tibia-like tube (diameter 4 mm, wall thickness 35 µm, E modulus 18 kPa) (Blickhan and Barth 1985; Szabo 1972, 1975). The following values show which strains actually have to be expected.

- S_y, the mechanical sensitivity to active dorsoventral forces, was found to measure up to 20 µε/mN. In more understandable terms, a piece of cuticle changes its length by a factor of $20 \cdot 10^{-6}$ (that is, by 20 millionths) when a load of 1 mN is imposed. Now something interesting happens: muscular forces induce negative strain – that is, compression – in the region of the organs HS-8, HS-9 and VS-4, but positive strain in the region of VS-5. Hence muscular forces produce a stimulus that is effective for the three laterally situated organs but not for the ventral organ VS-5. The latter is compressed, and thus effectively stimulated, by the action of the hemolymph pressure, which simultaneously dilates the other organs. The strains induced by the hemolymph pressure are considerably larger than those brought about by the flexor muscle, except in the case of HS-8. The site of HS-8 is also distinguished by the fact that strains here are remarkably independent of the joint position.
- S_z, the mechanical sensitivity to lateral forces, is as low as 10 µε/mN in the vicinity of HS-9 and VS-4. It depends strongly on the joint angle α. The organ HS-8 is compressed both when the metatarsus is bent backward, if α <180°, and when it is bent forward, if α >180°.

- S_x, the mechanical sensitivity to axial forces, is the smallest of the three, with values below 0.8 με/mN. Like S_z, it is highly dependent on α. With $\alpha = 170°$ there is hardly any mechanical effect at any of the organs. When pressure is imposed, the organs HS-8, HS-9 and VS-4 are compressed as long as $\alpha < 170°$, whereas traction compresses them when $\alpha > 170°$.

Even this brief description allows us to make a number of hypotheses about the stimulation of the lyriform organs in a freely moving animal. The idea that different organs can be specialized for the measurement of specific stresses because of their different positions has so far been confirmed. To obtain definite proof, however, we need direct measurements of strain in a spider that is walking normally.

In Vivo Measurements

Such an animal is a mobile high-tech installation (Fig. 23). First, its movements – stepping pattern, joint angles, leg kinematics – must be recorded precisely (by video), but at the same time measurements must be obtained of the stresses (by strain gauges) at the sites of the lyriform organs, the hemolymph pressure (by pressure transducer) and the ground reaction force exerted on the ground by the legs (by force platform). We concentrated on the slow walks, with a velocity of 1 to 10 cm/s and a step frequency between 0.3 and 1.5 Hz. In this case inertial forces are negligible. The angle between tibia and metatarsus in all legs varies only between 160° and 180°. Given the dependence of mechanical sensitivity S on angle, it follows that the strains induced in the tibia result mainly from muscular force (up to 200 mN) and the hemolymph pressure (up to 5.3 kPa, or 50 kPa during rapid walking and jumps). Because space on the skeleton is so limited, which presents a problem in spite of the miniaturization of the transducers, for these experiments we again used bird spiders, the legs of which are somewhat larger in diameter than those of *Cupiennius.*

- The four lyriform organs: The strains measured at the site of the various organs on the front and back surfaces of the tibia are very similar, varying only between 13 and 20 με (up to 120 με during fast walking). However, they differ appreciably in their time courses – that is, in the phase relationship of strain to walking pattern (Fig. 24). The organ HS-8 is compressed and stimulated during the stance phase of the step, whereas the single ventrally located organ (VS-5) is stimulated during the swing phase. To put it another way: the organs HS-8, HS-9 and VS-4 signal the muscle force during flexion of the joint, and VS-5 signals the hemolymph pressure during extension. All the other organs are not compressed but instead dilated during the latter phase.
- The lyriform organs are not located at sites characterized by high strain values. For in-

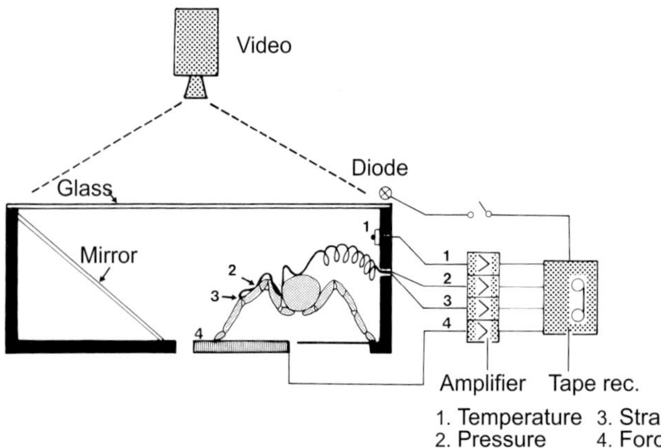

Fig. 23. Apparatus for the simultaneous measurement of leg kinematics, cuticular strains, ground reaction force, and hemolymph pressure. *1* Walking track; *2* mirror; *3* Plexiglas cover; *4* force platform; *5* ground membrane; *6* pressure transducer; *7* wires; *8* temperature sensor; *9* light-emitting diode; *10* amplifier and electronic filter; *11* FM tape recorder; *12* video camera; *13* video recorder; *14* monitor. (Blickhan and Barth 1985)

stance, the strain on the dorsal surface parallel to the long axis of the leg is about 38 με, distinctly higher than in the vicinity of the organs. This underlines the significance of the different time courses.

- The four legs: As was expected, the time course of the strain at the site of a given organ is very similar in all four legs. However, the strain amplitudes clearly increase from the first to the fourth leg. The first leg is special in that its swing phase is shorter and its vertical deflection higher than those of the other legs. During slow locomotion it evidently functions in addition as a "feeler".

Axel Schmid (1997) recently added an interesting facet to this observation. On a walking compensator (a "treadmill") *Cupiennius* proceeds directly toward a visual target only as long as the target can be seen. This stands to reason! In the dark, however, not only does it no longer walk in an oriented way; *Cupiennius* also instantly changes its gait. Now it walks only with the second to fourth pairs of legs and uses the first legs to feel its way. The spider holds these feelers out in front of itself and moves them up, down and sideways, even during the pauses in walking. A neuronal switch activated by the visual input causes the change from one movement pattern to the other (see also Chapter XXIV).

The fourth leg exhibits higher strain values than all the others: for example, up to –80 με at HS-8, which is about four times the corresponding value on the second leg. The main reason is that

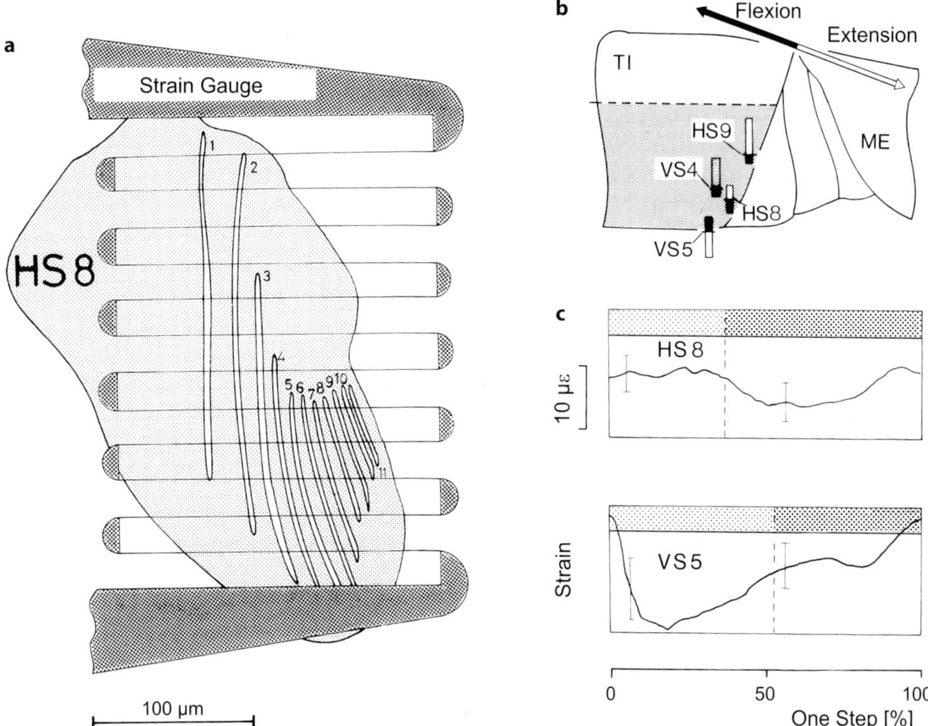

Fig. 24 a–c. Strains at the site of lyriform organs. **a** Miniature strain gauge glued onto the cuticle at the site of organ HS-8 on the tibia for in vivo strain measurements. **b** During flexion of the metatarsus (*Me*) by muscular activity there are negative strains at the sites of the lyriform organs located laterally on the tibia (*VS4, HS8, HS9*), which lead to slit compression and thus effective stimulation. In contrast, the ventral organ *VS5* is exposed to positive strains (dilation). In the figure the site of the respective organ is indicated by a *horizontal line* and negative strains are represented by *bars above this line*, positive strains by *bars below it*. **c** Strains at organs HS8 and VS5 during one step in a freely walking animal. The different relationships to the phases of the stepping cycle are clearly seen. Whereas HS8 is mainly stimulated (*dark shading*) during the stance phase, VS5 is mainly stimulated (*light shading*) during the swing phase. (Barth 1985b)

the fourth leg makes a flatter contact with the substrate, so that the vertical component of the standing force (40 mN) is greater. The hemolymph pressure in leg 4 is also greater than in, for instance, leg 2; furthermore, it rises more abruptly when walking begins, from ca. 2.2 kPa to about 5.3 kPa. From the mechanical sensitivities S it can be concluded that the hemolymph pressure contributes approximately 25% to the strain measured at the site of the organ HS-8.

7 Toward a Definition of the Adequate Stimulus

At the end of all these mechanical ins and outs the attentive reader may well be confused by one question, which it is entirely reasonable to ask: just what is the physical parameter to which a slit sensillum responds? To strain, to displacement, or to force? Should we call the slit sensilla force receptors or strain receptors?

It all depends on the level at which the responses are being considered, so there is no entirely unequivocal answer. At the level of the receptor, the slit measures the relative displacement of two points in the cuticle, and accordingly is a "displacement receiver". If the slit is viewed as part of a larger section of cuticle, then it is justified to compare it with technical strain gauges and call it a "strain receiver", which measures the local deformations of the skeleton. But it does not stop there: slit sensilla can also be seen in the context of whole legs (and other parts of the body), in which case they measure forces that cause deformation of the skeleton. The tension at a joint is a direct function of the force at the tip of the leg, and the slits are detecting this force at the same time as they are measuring the strain. In this context, then, they can rightly be classified as force receptors (Blickhan and Barth 1985; Seyfarth 1978b).

A second and perhaps more important question arises from the answer to the first: what kind of information is of interest to the animal? Certainly a spider would be the least interested in how the edges of the slit, considered in isolation, are displaced with respect to one another. On the other hand, it is presumably highly relevant to know about the deformation of large parts of the skeleton (as signalled by their strain), either in order to avoid fracture of the skeletal material or to deduce from these deformations the forces that are acting on the skeleton, whether internal (muscles, hemolymph pressure) or external.

The many years of occupying myself with biology have convinced me that every animal and every real problem is interesting and worth analyzing. The crucial point is the depth and thoroughness of the analysis. In this respect, our research on the slit sensilla of spiders, and in particular on the complexities of the stimulation events, has always been spurred on by the idea that we were making a broader contribution. The work was meant to add to our understanding of a sense that shows us in an extreme form the implications of the arthropod exoskeleton for sensation in general, revealing how evolution has dealt with the mechanical possibilities offered by the skeleton and, in the process, millions of years ago had already arrived at an efficient analog of the modern technical strain gauge. For an extensive comparison between the slit sensilla of arachnids and the campaniform sensilla of insects see Barth (1981).

VIII The Vibration Sense

What first comes to mind when you think about the special characteristics of spiders? Perhaps that they spin silken threads (the Old English word from which "spider" is derived means "spinner"). But anyone who has observed a spider keeping watch over its web may think of something else: that vibrations play a very important role in their behavior. The world in which spiders live is a world full of vibrations, and their vibration sense is correspondingly well developed. Take a blade of grass, use it to pluck gently on one strand in the web, and the spider will come running, expecting to find a captive meal. The many spiders that do not use webs to catch prey are equally sensitive. In later chapters (XVIII to XX) concerned with the details of prey capture and courtship, we shall see that spiders not only detect vibratory signals but also produce and transmit them as a means of communication.

First, however, some general functional principles of the spiders' vibration sense must be clarified if we want to understand the behavior it controls. This is especially important because we humans, although also surrounded by vibrations, pay little attention to them and hardly respond at all to them on an emotional level. Which, by the way, makes it rather curious that the term "good vibrations" should have become so widely understood as an expression that things are going well (in German, similarly, a feeling of compatibility can be expressed as "having a good wire" to someone, although in this case there appears to be a reference to the undeniable significance of the telephone). For a spider, to speak of good vibrations would be perfectly reasonable. We shall understand this as we delve into their world, strange as it may be to our own experience.

1 The Metatarsal Vibration Sense Organ

The windows through which a spider's nervous system sees its vibratory surroundings are extremely sensitive vibration receptors on the legs. Of particular interest is the so-called metatarsal organ, a lyriform organ so constructed and arranged that it is a very good detector of vibrations of the substrate. Furthermore, the metatarsal organ is the best example of an exteroreceptive slit sense organ. Its sensitivity to vibration has been known since the early publications of Walcott and van der Kloot (1959) and Liesenfeld (1961) on web spiders. We ourselves then took a closer look at the metatarsal organ of *Cupiennius* in 1972 and 1982 and succeeded, for the first time, in obtaining electrophysiological recordings of the activity of single slits. The dedicated work of Geethabali from Bangalore, India, who was visiting our laboratory at the Goethe University in Frankfurt am Main on an Alexander von Humboldt scholarship, greatly assisted this project (Barth 1972a, b; Barth and Geethabali 1982).

Position and Structure

In all spiders, the metatarsal organ is situated behind a cuticular ridge at the distal end of the metatarsus. Two features distinguish it from all other lyriform organs and slit sensilla: its position in the middle of the dorsal surface and the orientation of its slits, which are perpendicular to the long axis of the leg. Both of these contribute substantially to its high sensitivity (Fig. 1). The upward movement of the tarsus produced

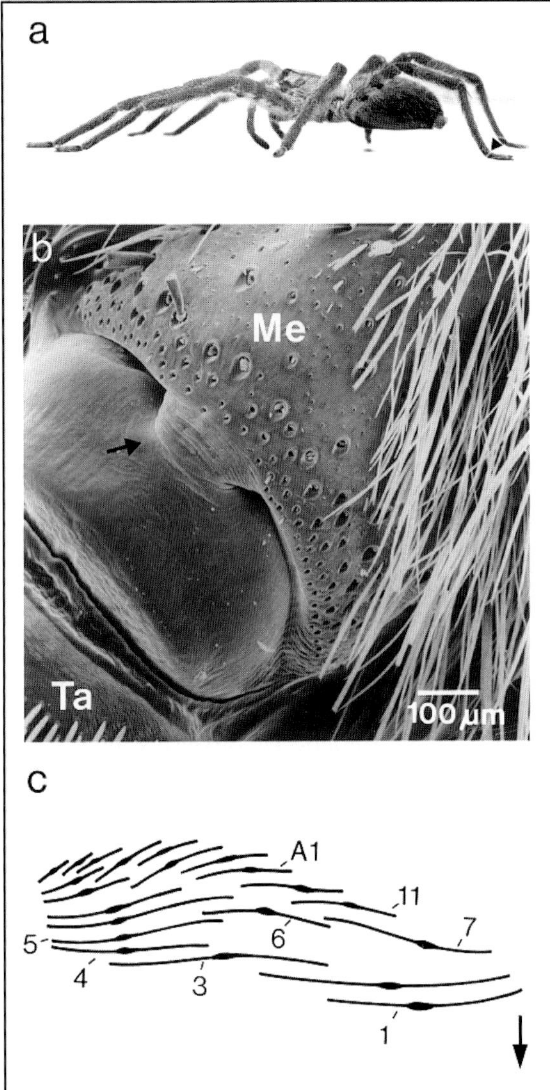

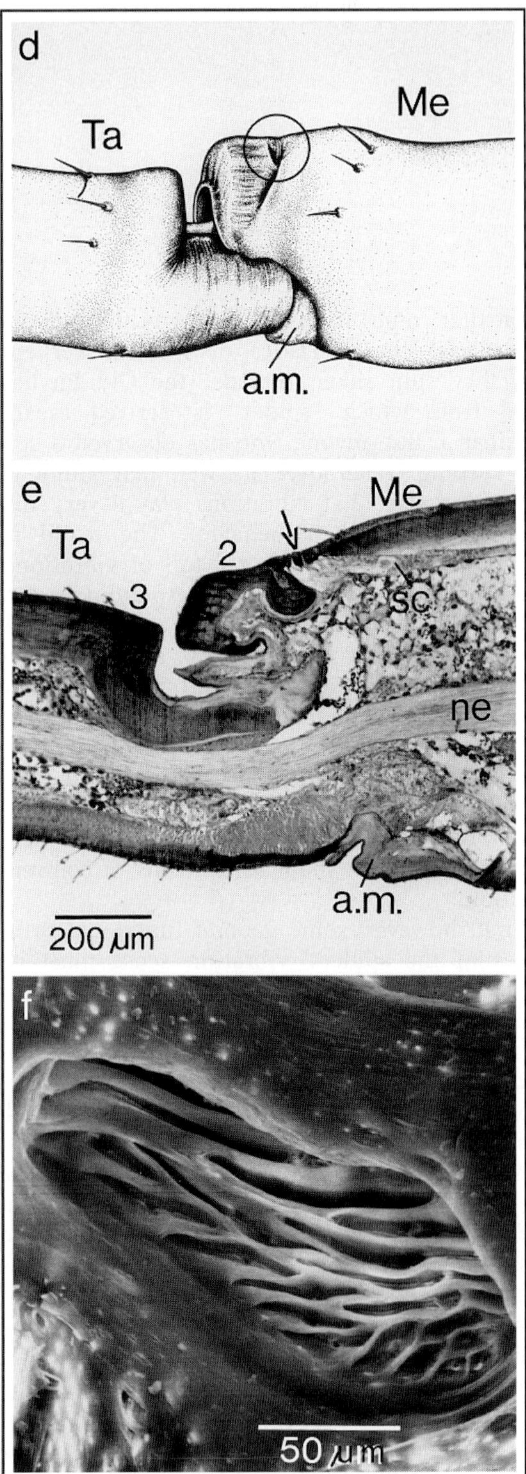

Fig. 1 a–f. The metatarsal lyriform organ of *Cupiennius salei*. **a** Position of the organ distally on the metatarsus of all legs (see *arrowhead*, leg 4). **b** Scanning electron micrograph of metatarsal organ (*arrow*) in dorsal view. *Me* Metatarsus; *Ta* tarsus. **c** Length and arrangement of slits and site of the dendrite attachment to the covering membrane (see *thickening*). *Arrow* points distal toward the tarsus. **d** Lateral view of the joint between metatarsus and tarsus, showing the location of the metatarsal organ (see *circle*); *am* articular membrane. **e** Longitudinal section through the tarsus-metatarsus joint region. *2 (3)* Distal (proximal) end of metatarsus (tarsus); *sc* sensory cells innervating the metatarsal organ, *ne* leg nerve. **f** View of cuticle at the site of a metatarsal organ from inside (exuvia). (Barth and Geethabali 1982; **a** photo E.-A. Seyfarth)

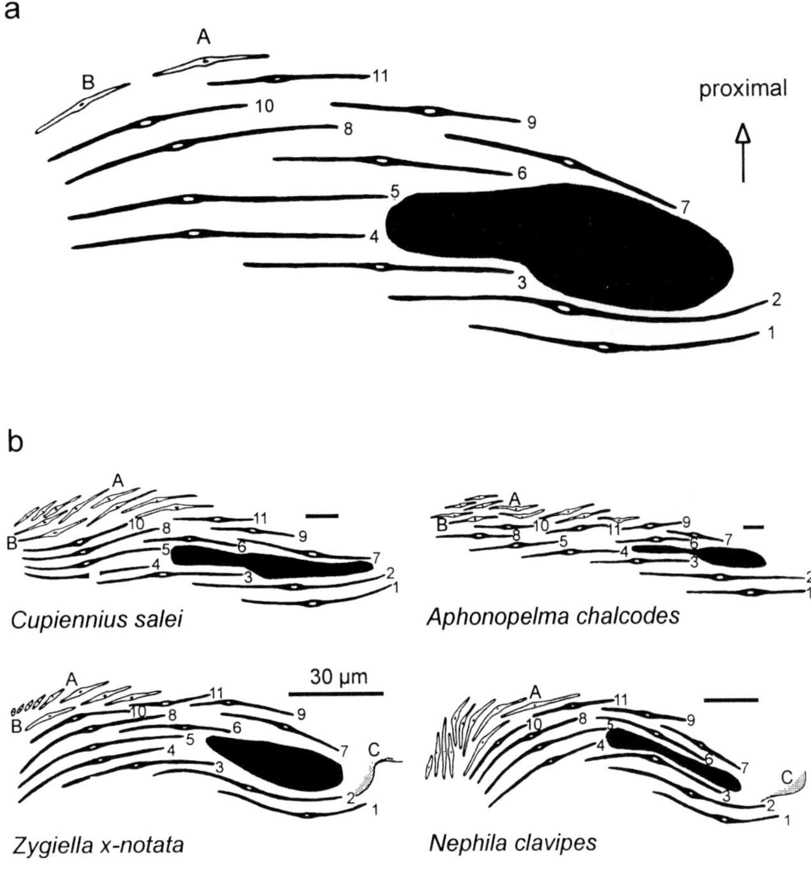

Fig. 2a, b. Slit arrangement in the metatarsal organ. **a** The basic pattern found in all spiders examined, comprising the long slits *1* to *11*, which are individually identifiable, and smaller slits *A* and *B*, which are more variable in number and arrangement. *Black area* without slits. **b** The metatarsal organs of various spider species. (van de Roemer 1980)

by substrate vibrations, as well as its sideways movement, causes compression of the slits as soon as the proximal end of the tarsus presses aginst the distal end of the metatarsus. The topography of the organ ensures that when this occurs, the compression forces to which the slits are exposed are approximately perpendicular to their long axis and hence produce maximal deformation (see Chapter VII, Fig. 9). A deep groove in the cuticle on each side of the organ makes it still more readily deformable; these grooves both increase the mechanical lability of the organ and ensure that the compression forces are concentrated at the site of the metatarsal organ – that is, the organ receives a quasi-punctate stimulus (Barth 1972 a, b).

In *Cupiennius salei* the metatarsal organ comprises 21 slits, between ca. 20 and 120 µm long (Barth 1971 a). Some other spiders have considerably fewer slits. For instance, in the organ of the jumping spider *Salticus scenicus* there are only 11. *Zygiella x-notata*, the "sector spider", and *Nephila clavipes*, the "golden web spider" of neotropical and subtropical regions, each have 20 slits, *Tegenaria larva* has 16 and *Achaearanea tepidariorum* has only 8 to 10.

The arrangement of the slits always conforms to the same basic pattern, regardless of their number. The eleven largest slits are arranged in the same way in every spider and can be individually identified (Figs. 2, 3). In contrast, the joint mechanics vary widely. For example, consider *Cupiennius* and *Nephila*: it is evident at a glance, and has been confirmed by quantitative measurements of forces and deflection angles, that the tarsus of the hunting spider (*Cupien-*

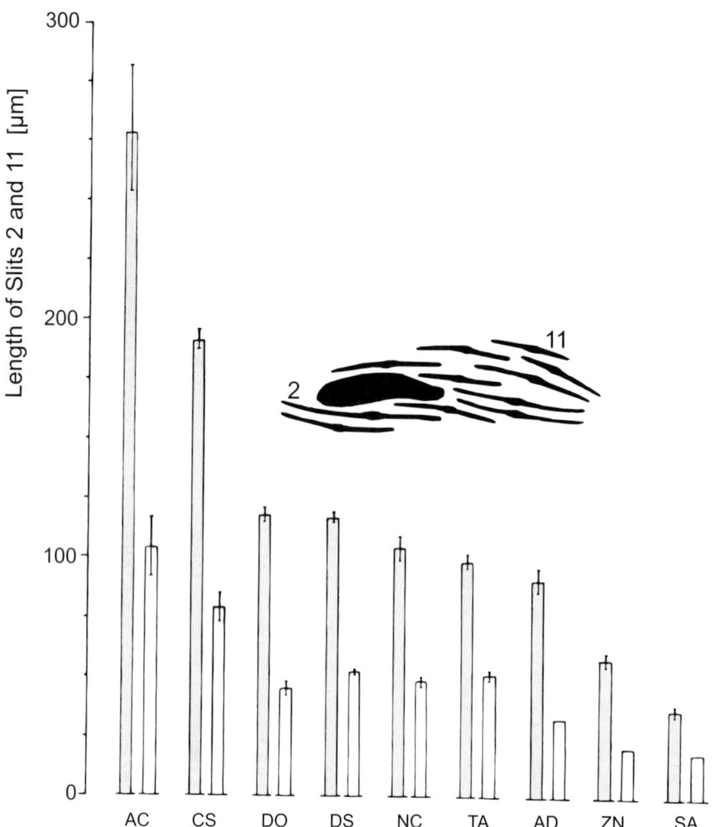

Fig. 3. Length of slits 2 and 11 of the metatarsal organs of various spider species. *AC Aphonopelma chalcodes*; *CS Cupiennius salei*; *DO Dolomedes scriptus*; *DS Dinopis subrufus*; *NC Nephila clavipes*; *TA Tegenaria atrica*; *AD Araneus diadematus*; *ZN Zygiella x-notata*; *SA Salticus scenicus*. (N=3; n=6–12). (van de Roemer 1980)

nius) is considerably more flexibly connected to the metatarsus than is the case in the web spider (*Nephila*). Figure 4 illustrates this by diagrams showing torque as a function of angle of rotation for the tarsus-metatarsus joints of *Nephila clavipes*, *Dolomedes scriptus* and *Cupiennius salei* during dorsoventral and lateral movements of the tarsus. There are clearly large differences in the stiffness of the joint as well as in its working range (defined as the angle through which it can bend before cuticular structures come into contact; in the graph this contact appears as a distinct inflection of the curve, which indicates that considerably greater forces are needed for further bending). The tarsus of *Nephila* can be deflected dorsally by only 1° before an opposing structure is encountered, whereas the corresponding values are 19.5° for *Dolomedes*, 28.5° for *Cupiennius*, and as much as 53° for *Tegenaria*. Hence large stiffness of a joint is associated with a small working range and conversely.

Unlike the large slits, the smaller ones vary in different spiders with respect to their number (0 to 10), positions and curvature.

Thresholds and Frequency Tuning

In one set of experiments the tarsus of a *Cupiennius* leg was coupled to an electrically driven and controlled vibrator and moved with various frequencies and amplitudes, while the action potentials simultaneously generated by the organ were recorded. This allowed us to measure threshold curves and to determine the absolute and the spectral sensitivity of individual slits in the metatarsal organ. We did this for about 10 of the 21 slits with frequencies in the range from 0.1 Hz to 1–3 kHz (Fig. 5) (Barth and Geethabali 1982).

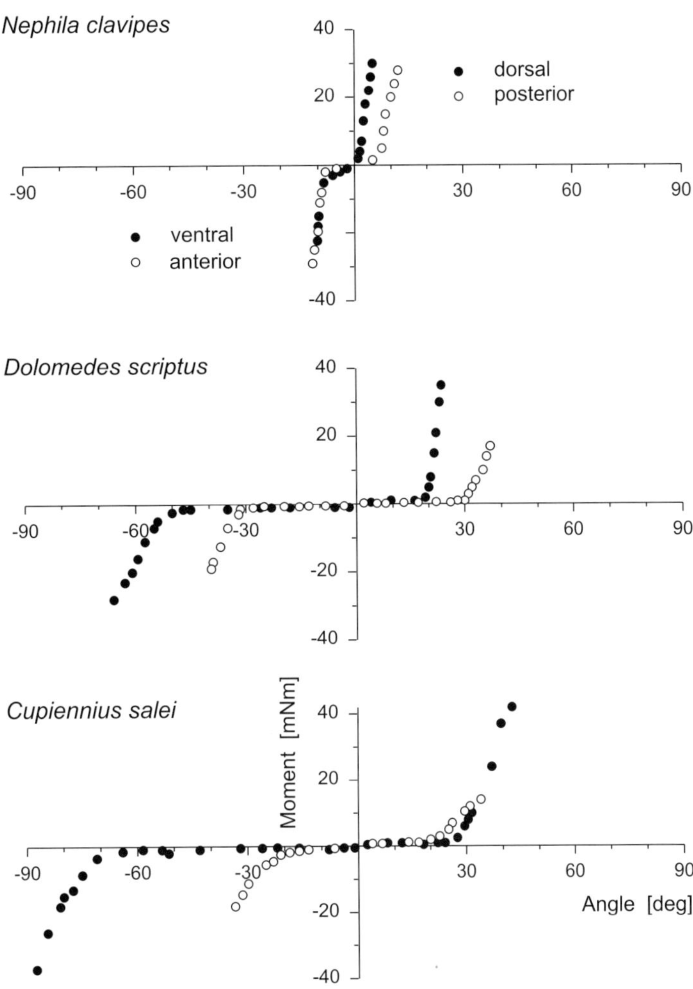

Fig. 4. Mechanics of the tarsus-metatarsaus joint in the web spider *Nephila*, the semiaquatic spider *Dolomedes*, and the hunting spider *Cupiennius*. The graphs show the moments measured when rotating the tarsus against the metatarsus to various degrees in a dorso-ventral and anterior–posterior direction. (van de Roemer 1980)

The result:

1. All the slits behave like high-pass filters (Fig. 6). They are relatively insensitive up to about 40 Hz, but at higher frequencies their sensitivity rapidly increases. In concrete terms, this means that when vibrated at low frequencies the tarsus must be deflected by 10^{-3} to 10^{-2} cm in order to elicit a response, while at higher frequencies this threshold value falls sharply (as much as 40 dB/decade) by 3 to 4 orders of magnitude, reaching 10^{-6} to 10^{-7} cm at 1 kHz. When the stimulus waveform is narrow-band noise (bandwidth 1/3 octave, $Q = 0.35$) instead of pure sinusoidal vibration, the thresholds are as much as 10 dB lower. This is significant with respect to the natural stimulus situation (see Chapter XVIII). Probably most readers will not find the number 10^{-7} cm particularly exciting. So as to visualize better how sensitive the metatarsal organ is, let us scale everything up to more familiar dimensions: if the threshold deflection were 1 mm, then an adult *Cupiennius salei* enlarged by the same factor (10^6) would be a monster with a leg span of 100 km! Still, *Cupiennius* is not the world champion for vibration sensitivity. A comparison in this regard will be given in the last section of this chapter.

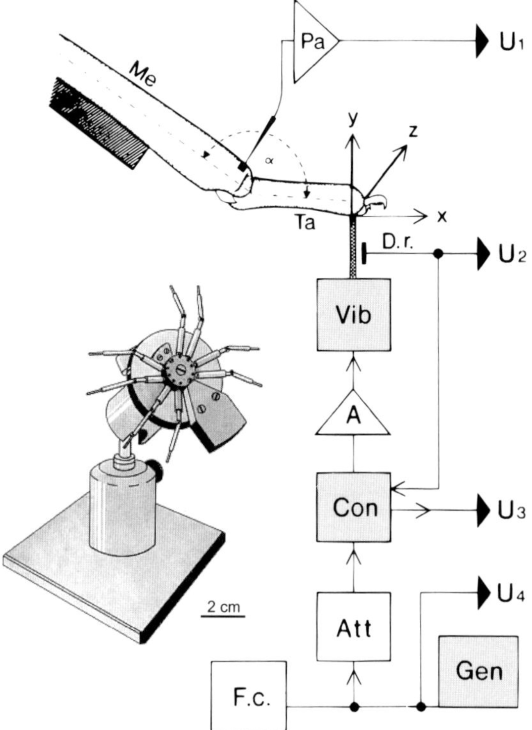

Fig. 5. Setup and arrangement used to analyze the electrophysiological response of single slits of the metatarsal organ to vibratory stimuli. *Me* Metatarsus fastened to distal end (*shaded area*) of metal leg of the spider platform (see *inset*); *Ta* tarsus; *Pa* preamplifier (with tungsten electrode in slit); *y* and *z* directions of tarsal displacement; *Vib* electrodynamic vibrator; *A* power amplifier; *Con* control unit; *Att* attenuator; *Gen* function generator; *F.c.* frequency counter; *D.r.* displacement transducer; U_1 to U_4 recorded voltages; α angle between tarsus and metatarsus; *inset* metal spider; due to the jointed metal "legs" the spider can be glued onto the platform with its legs in a natural position. (Barth and Geethabali 1982)

2. At the lower frequencies the threshold deflection of the tarsus is approximately constant, but beyond the inflection point of the curve at ca. 40 Hz (in some slits at only 1 Hz or 20 Hz), the threshold deflection corresponds to constant acceleration (Fig. 6). The smallest acceleration values are about 0.02 cm/s^2 (1 Hz). That the threshold curve should reflect different stimulus parameters in different frequency ranges seems to make sense. In a general technical context, too, for monitoring very low frequencies it would be preferable to measure the displacement (d), because the accelerations are so small, whereas at high frequencies the acceleration (a) would be measured because it is large even for small displacements ($d = a/4\pi^2 f^2$, where f is frequency).
3. At least some of the slits are also sensitive to lateral movement of the tarsus. Their threshold curves for deflection to the side are quite similar to those for dorsoventral movement. In both cases, the slits do not exhibit any tuning to narrow frequency ranges, even when the stimulus is applied as naturally as possible, with the tarsus resting loosely on the vibrating rod (as it would on a plant) or when the spider is free to move around during the experiment. The lack of genuine "tuning" in the biologically relevant frequency range has also been documented for other slit sensilla. However, it would be premature to conclude that the metatarsal organ is of no use for frequency discrimination. We must bear in mind that suprathreshold stimulation may reveal nonlinearities with a functional role, such as in the analysis of the frequency components of vibratory courtship signals (see Chapter XX).

   Furthermore, the ability to discriminate frequencies could also derive from something other than a differential fine-tuning of the individual slits in an organ. The frequency-response curves of the various slits all look much the same in their general shape, but at certain frequencies the responses may differ by up to 1.5 orders of magnitude. If the central nervous system is capable of analyzing the way the excitation pattern of the whole slit ensemble changes with changes in stimulus frequency, this ability could conceivably be exploited for frequency discrimination (Barth and Geethabali 1982). A number of behavioral experiments have demonstrated that our spiders actually do discriminate signals of different frequencies from one another (Hergenröder and Barth 1983a).
4. Apart from a few slits that give purely phasic responses, the majority adapt slowly when a given tarsus deflection is maintained. It can

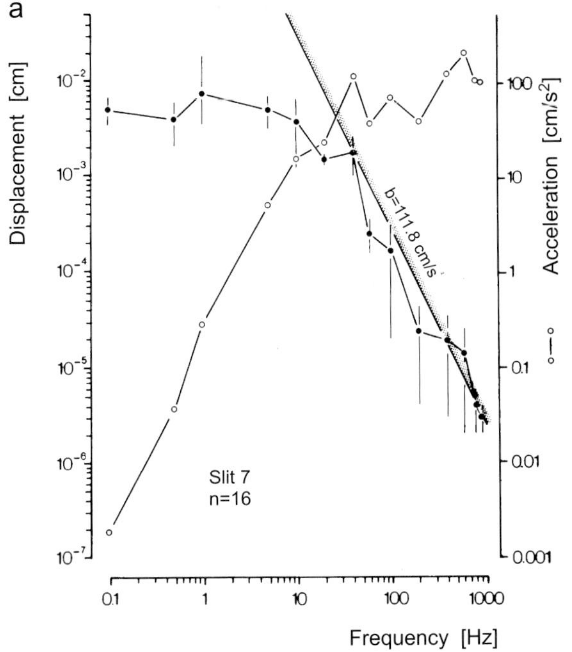

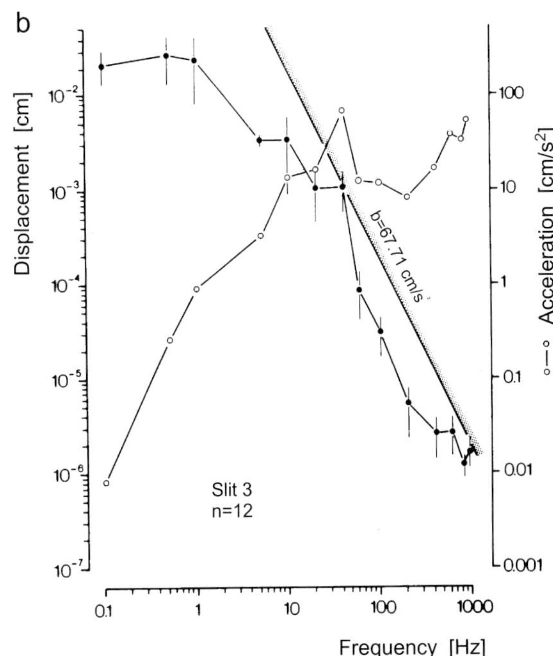

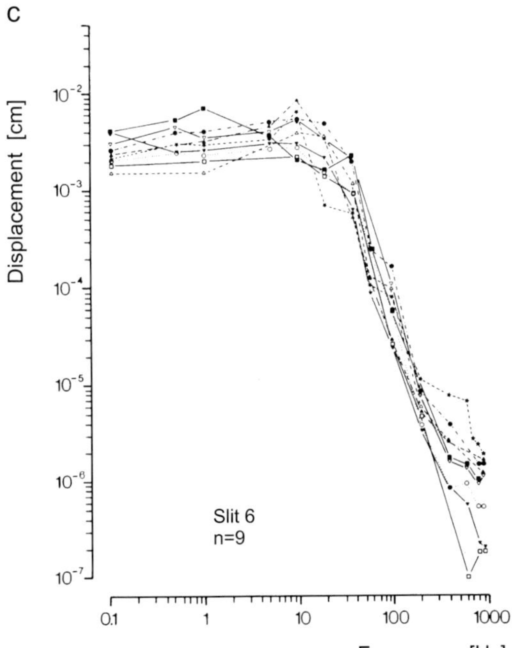

Fig. 6 a–c. Threshold curves for slits of the metatarsal organ of *Cupiennius salei* measured with dorso-ventral vibration of the tarsus. **a** Slit 7. **b** Slit 3. **c** Slit 6. In **a** and **b** both threshold displacement (●) and acceleration (○) are shown. *Shaded bars* are theoretical curves which give the displacements for various frequencies at a constant acceleration; they are fitted to the measured values by eye. **c** shows the threshold curves of slit 6 measured in nine different metatarsal organs. (Barth and Geethabali 1982)

be concluded that the metatarsal organ is not only sensitive to vibration but has an additional proprioceptive function, signalling the position of the tarsus during locomotion.

Looking back at these earlier experiments, I feel that their greatest value was to enable us – much later on – to investigate the responses to natural vibration patterns in an appropriate way and to recognize their special features. There is more on this subject in Chapters XVIII and XX.

Our electrophysiological experiments on the metatarsal organ of the hunting spider *Cupiennius* put us in a position to compare the vibration sensitivity of various ecotypes of spiders. As we shall see, the physical properties of the propagation of vibration signals vary appreciably in such different media as a bromeliad, an orb web, a firmly woven carpet web and the surface of water. We first thought that the threshold curves would differ accordingly, but that turned out to be wrong. Electrophysiologically measured threshold curves are also available for *Zygiella x-notata* and *Tegenaria* spec. (Liesenfeld 1960), as well as for *Dolomedes triton* (Bleckmann and Barth 1984), a semiaquatic spider that catches its prey on the water surface and orients itself by the surface waves generated as the insects move. This collection of species covers the whole range of spider substrates, but the typical high-pass characteristics of the metatarsal organs are the same in all of them. Therefore we must conclude that the physiological properties of the metatarsal organs (at least those that are reflected in the threshold curve) are largely independent not only of the joint mechanics across the most diverse groups of spiders (see above, Fig. 4: in *Dolomedes* the tarsus is about as easily deflected over as large an angle as in *Cupiennius*, whereas the joint of *Zygiella* and other web-weavers such as *Nephila* is considerably stiffer; van de Roemer 1980), but also of the substrate on which the spider is detecting the vibration. It would be interesting to delve further into this question and compare the responses of the organs to suprathreshold natural stimuli.

2 Other Vibration Receptors

At the risk of further complicating the situation, I must mention that the metatarsal organ is not the only vibration-sensitive organ in spiders, although it is the most sensitive and the most obvious in its structure and topography (Fig. 7). *Cupiennius salei*, like other spiders, also has isolated slit sensilla on its pretarsus, about 45 μm long and located one on either side below the two claws; their position alone, at the outermost tip of the leg just behind the claws, makes them good candidates for vibration sensors (Barth and Libera 1970, "claw slits"). Jochen Speck confirmed this speculation in his Master's thesis (Speck and Barth 1982). One adequate stimulus to the two pretarsal slits is active movement of the pretarsus, by an interesting combination of muscle activity and hemolymph pressure (Fig. 8), but they are also effectively stimulated by external vibrations.

The claw slits, again, are not tuned to a particular narrow frequency range (responses have been measured between 0.01 Hz and 1 kHz). Instead, they have a high-pass characteristic very like that of the slits of the metatarsal organ. However, they are less sensitive by about two orders of magnitude: the threshold declines only slowly up to ca. 40 Hz, and considerably more sharply at higher frequencies (ca. 18 dB/decade). The lowest threshold value for displacement (at 1 kHz, the highest frequency tested) was 2×10^{-5} cm, and the lowest for acceleration (at 0.01 Hz) was 0.3×10^{-9} cm/s^2 (Fig. 9). Calculation of the transfer function showed that the receptor properties of the claw slits were intermediate between those of a frequency-independent displacement receptor and those of a velocity receptor ($k=0.39$–0.44, see Chapter VII).

As would be expected, the threshold curve of the claw slit changes as the initial tension in the joint membrane between pretarsus and tarsus changes. It might be that this feature indicates a mechanism for active adjustment of sensitivity (reduction by raising the pretarsus) (Fig. 9).

It should also be mentioned that recordings of the activity of neurons in the central nervous system of *Cupiennius salei* have shown that cer-

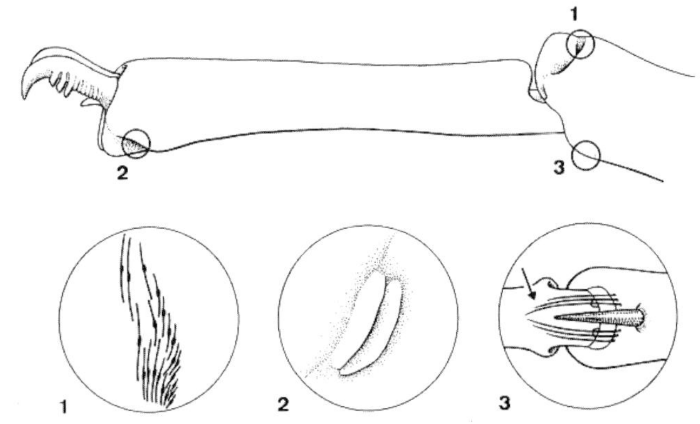

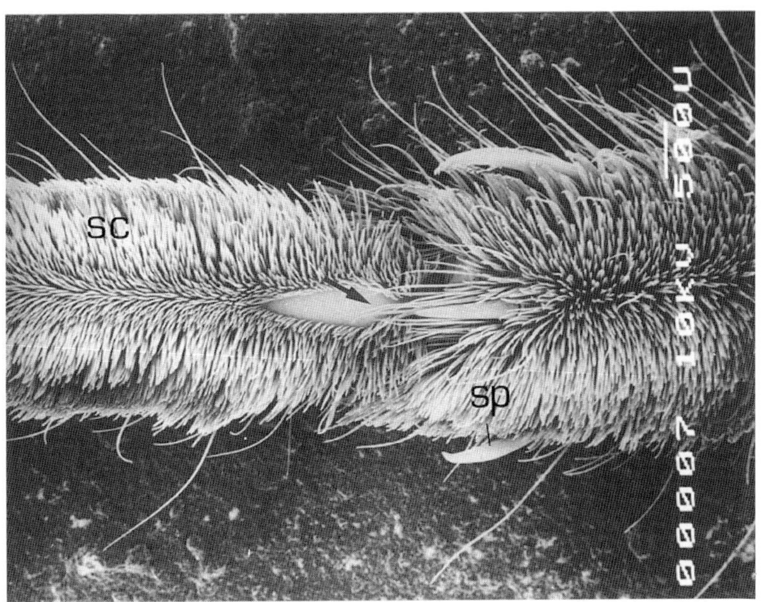

Fig. 7. Three different types of mechanosensitive sensilla distally on the legs, which respond to vibration of the substrate. *Above 1* metatarsal lyriform organ; *2* pretarsal single slit sensillum on each ventrolateral side of the pretarsus behind the claws; *3* metatarsal bridge hairs ventrally at the joint between tarsus and metatarsus (*inset* ventral view of joint). *Below* Ventral aspect of metatarsus (*right*) – tarsus (*left*) joint of *Cupiennius salei* with the vibration-sensitive bridge hairs (*arrow*). *sp* Spines; *sc* scopula hairs. (Speck-Hergenröder and Barth 1988; photo E.-A. Seyfarth)

tain hair sensilla are also vibration-sensitive. We called these hairs metatarsal bridge hairs because, being up to 3 mm long, they form a ventral bridge over the metatarsus-tarsus joint with their tips touching the tarsus, so that when it is deflected, they also move (Fig. 7). At each metatarsus-tarsus joint there are about 16 such hairs. Recordings from central neurons onto which these hairs converge indicate sensitivity peaks at 70 Hz and 150 Hz. As in the case of the pretarsal slits, the absolute threshold values are at least two orders of magnitude higher than those of the metatarsal organ. The sensory cells attached to these hairs have not yet been examined.

So, is that all? It might be that the highly sensitive and extremely phasic slits of lyriform organs in more proximal parts of the spider leg also respond to substrate vibrations (for example, organ HS-8 on the tibia; see Chapter VII and Barth and Bohnenberger 1978; Bohnenberger 1981), especially in view of the fact that vibrations propagate extremely well through the leg (see below). But here, again, no direct experimental evidence is available.

a

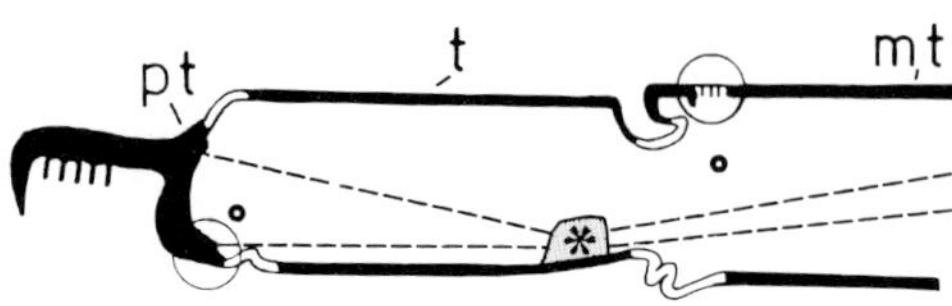

b Dorsal Tendon

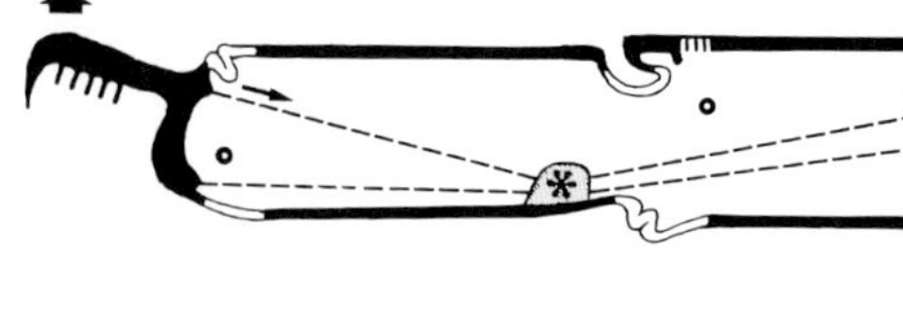

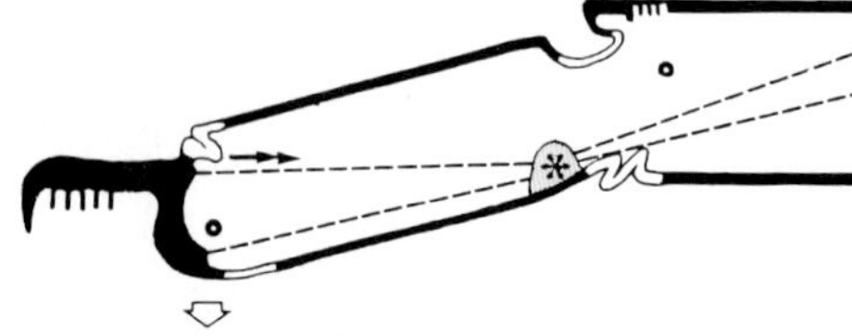

c Ventral Tendon

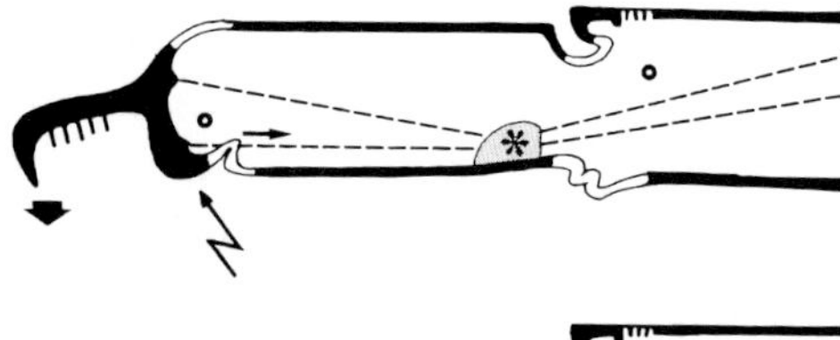

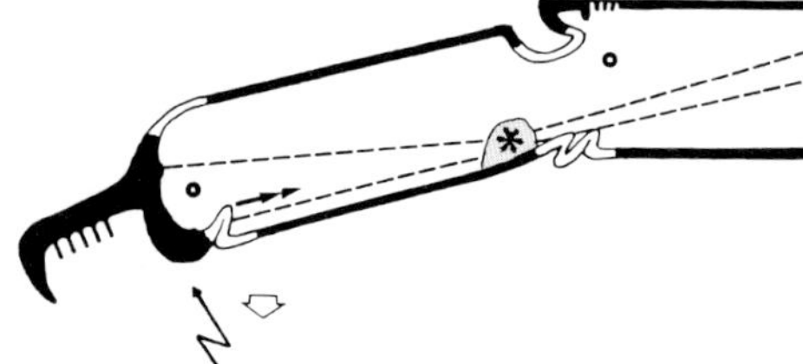

d Hemolymph Pressure

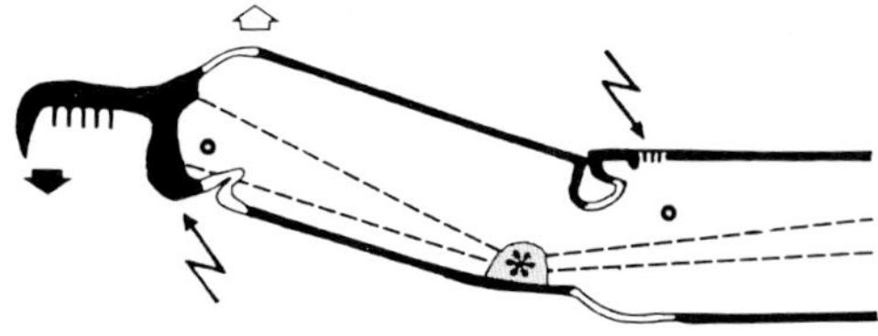

It is more important to conclude by explaining why it ever occurred to us that the metatarsal organ, which proved to be so excellently designed as a vibration sensor, might not be the only source of sensory information about the vibrations transferred into the leg at its tip.

Cupiennius can easily be induced to jump at presumed prey by causing the substrate to vibrate. And as in web-building and semiaquatic spiders, it had been shown that the metatarsal organs participate in the detection of such vibrations (Barth 1982; Hergenröder and Barth 1983a; Bleckmann and Barth 1984; Klärner and Barth 1982). We were quite surprised, then, to find that the effect on this behavior of inactivating the metatarsal organs, though significant, was not dramatic. When the organs of the legs on one side of the body were eliminated and a stimulus was applied under one of these legs, the spider jumped a little too far with respect to both the translational and the rotational component of the movement (Barth 1981). After elimination of the organs the behavioral threshold is significantly higher, but the spider is by no means entirely blind to vibration (Hergenröder and Barth 1983a); a sufficiently strong stimulus is quite capable of inducing prey-capture behavior. It follows that the spider must be using some other vibration-sensitive sensilla. Hence

Fig. 8a–d. The effect of tendon (*broken lines*) pull and hemolymph pressure on the movement of the tarsus (*t*) and pretarsus (*pt*), and on the stimulation of the metatarsal lyriform organ and the pretarsal slits (see circles in **a**). ○ axes of rotation for tarsus and pretarsus; * marks the cuticular collar proximally in the tarsus, which guides the tendons. **a** Neither pull on tendons nor increased hemolymph pressure. **b** Moderate pull on dorsal tendon lifts claw; strong pull on dorsal tendon lifts claws to their uppermost position and flexes the tarsus. **c** Moderate pull on ventral tendon depresses claws and pretarsus and leads to stimulation of pretarsal slits by way of buckling the ventral articular membrane inward; strong pull on ventral tendon not only depresses claws and pretarsus to their lowest position but also flexes the tarsus. **d** Increased hemolymph pressure lifts the tarsus and simultaneously depresses the pretarsus with resulting stimulation of both the metatarsal lyriform organ and pretarsal slits. (Speck and Barth 1982)

our interest in the "B team" of vibration receptors! We shall return to the definitive behavioral experiments in Chapter XVIII.

Our interest in the vibration receptors was always related to our efforts to understand the role of sensations in behavior. Regarding the sense of vibration, some questions of interest are the transmission range of vibratory signals, the bases on which various signal types are distinguished, and how spiders orient by vibration signals. We pursued these by analyzing courtship behavior and prey capture; the answers can be found in Chapters XVIII and XX.

3 A Comparison of Champions

As stated already, spiders are not the world leaders in absolute vibration sensitivity, but they can surely share the podium with those top-class animals. To provide a basis for comparison, here is a summary of some figures for *Cupiennius*: threshold for deflection of the tarsus, as little as 10^{-6} to 10^{-7} cm at frequencies between ca. 400 and 1000 Hz, and 10^{-2} to 10^{-3} cm at low frequencies between 0.1 Hz and ca. 40 Hz; threshold for acceleration of the tarsus down to less than 0.8 cms^{-2} (at 100 Hz, the dominant frequency component of the male courtship signal; this is also the behavioral threshold for the female's response to the male vibration signal, see Chapter XX).

Now consider: the metatarsal organs of a scorpion respond to stepwise deflections of the tarsus by less than 10^{-7} cm (10 Å) (Brownell and Farley 1979); in leaf-cutter ants, too, a deflection amplitude of 10^{-7} cm has been reported as threshold for vibration-sensitive cells (acceleration 2 to 3 cm s^{-2} at 100 Hz) (Markl 1969); grasshoppers and burrower bugs (Cydnidae) respond at the receptor level to accelerations of about 1 cm s^{-2} (Kalmring et al. 1978; Devetak et al. 1978); ghost crabs, which have vibration receptors entirely different from those of spiders and insects, respond between 1 and 3 kHz to accelerations of 0.1 cm s^{-2} (Horch and Salmon 1969).

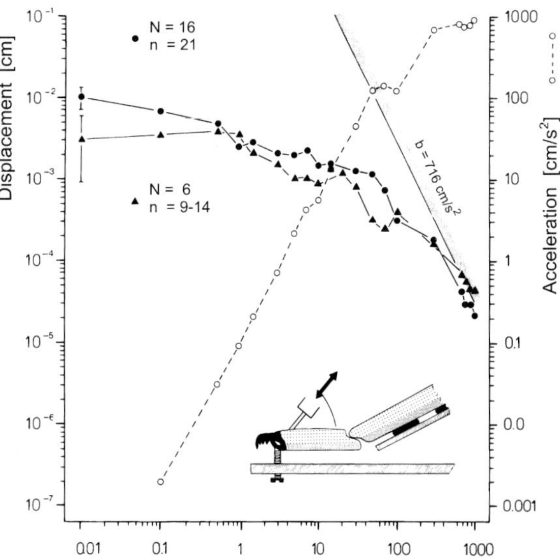

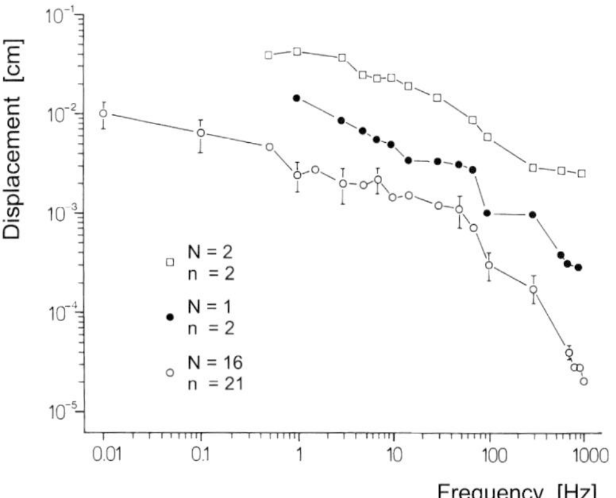

Fig. 9. Electrophysiologically determined threshold curves of pretarsal slits of *Cupiennius salei*. Stimulation by sinusoidally vibrating the tarsus relative to the fixed pretarsus at frequencies between 0.01 Hz and 1 kHz (see *inset*; angle between vibrator stylus and tarsus 50°). *Above* Dorsoventral displacement (pp, *filled symbols*) and acceleration (○) at threshold as a function of vibration frequency; recorded by tungsten electrodes (▲) and glass capillary electrodes (●) without penetration of the cuticle. *Shaded theoretical line* gives displacement at constant acceleration of 716 cm/s^2. Variation given as the standard error of the mean. *Below* Threshold curves as in the upper panel but with different zero settings resulting from changes of the zero position of the vibrator without detaching it from the tarsus (○ normal position; ● shift by 60 µm; □ shift by 100 µm). (Speck and Barth 1982)

According to the classical experiments of Autrum (1941) and later studies by his doctoral student Heidi Schnorbus (1971), the real gold medallists can be found among the Orthoptera. The subgenual organs of the cockroach *Periplaneta americana* responded to substrate displacements as small as 4×10^{-10} cm, or 0.04 Å. This is a sensitivity about 100 times greater than is found in spiders and scorpions. However, this extreme value has recently been called into question and "corrected" upward by two orders of magnitude (Shaw 1994). This would make the subgenual organ of cockroaches roughly as sensitive as that on a cricket hindleg (Dambach 1989). But such sensitivity is still enormous – of the same order as the threshold displacement of the basilar membrane in the mammalian ear or the threshold deflection of the stereovilli of individual auditory hair cells (0.4 nm or 4 Å) (Sellick et al. 1982; Pickles and Corey 1992). These threshold values are at the limits of the physically possible, because Brownian movement comes into play (Hudspeth 1985, 1989; Sivian and White 1933).

Even if we allow the vibration sensitivity of vertebrates into the comparison (see Barth 1998), we are still left with the conclusion that the metatarsal organs of spiders are among the most sensitive vibration receptors in the animal kingdom, and that the lowest thresholds found in the various taxonomic groups of animals are very similar to one another.

The vertebrate most sensitive to vibration, as far as is known at present, is a frog that communicates by vibration signals (*Leptodactylus albilabris*). The substrate acceleration at threshold (for the response of the audio-vestibular nerve and with stimuli having frequency components between 20 and 70 Hz) is only ca. 0.001 cm s^{-2}; at 10 Å the response of individual fibers has already reached saturation (Narins and Lewis 1984; Lewis and Narins 1985; Narins 1995). The corresponding value for displacement is ca. 1×10^{-6} cm (100 Å) when measured at the audio-vestibular nerve, which far exceeds the known sensitivities of warm-blooded vertebrates.

IX Trichobothria – the Measurement of Air Movement

I hope the chapter concerned with slit sensilla made clear that when applied to biological objects, the science of mechanics is not as lifeless as many biologists still suppose. It should also now be apparent that in order to understand mechanosensitive organs, we must first understand the mechanical processes of transmission and transformation as the stimulus passes through intervening tissues to the actual receptor cell. Such details are often crucial, if we are to fully appreciate the fascination of the "design".

So this exercise will now be repeated with reference to the trichobothria – but with a new twist, because from a mechanical viewpoint the trichobothria are almost the opposite of the slit sensilla. They are the second half of the dichotomy first mentioned in Chapter VI; having dealt with one basic design, the "hole", we shall now see the functional possibilities offered by the "hair" design. The trichobothria, in their turn, will demonstrate that sensilla reflect the physics of their effective stimuli and specific surroundings. Further delving into technical matters is therefore inevitable, but don't these excursions into other fields keep biology particularly exciting, and constantly reveal new aspects? One can develop such a mass of clever intuition as to get mired down in it; to avoid this fate, a few years ago we began an extremely fruitful collaboration with JAC Humphrey at the Department of Mechanical Engineering of the University of California, Berkeley. "Pepe" was and remains a good guide to the complexities of fluid mechanics. A delicate little hair, waving in the wind: it is hard to believe what interesting, difficult and important things that hair can tell us. When dressed in such a guise, our biological problems are often hard nuts for even the most qualified technological experts to crack.

First it should be made clear what the trichobothria, the air-movement sensors of spiders, actually are, in terms of their morphology and distribution. Then we shall tackle the difficult question of how air movement causes hair movement, and what the hair in fact does during controlled stimulation in an experiment. The next section of this chapter will be devoted to the response properties of the receptor cells themselves. And finally the connection to the real life of spiders is restored, by considering natural signals and the detection range of the sensory system based on the trichobothria.

1 Topography and Structure of the Sensilla

Number and Topography

The word "trichobothrium" is derived from the Greek *trichos* and the Latin *bothrium* and hence could be translated as "hair cup", which refers to the appearance of the sensillum: a thin, hair-like structure arises from a conspicuous socket in the cuticle. On closer consideration these terms are not entirely sensible. In the first place, hairs are a diagnostic characteristic of mammals; in the case of arthropods, the correct term is "chaetae". A corresponding change of name has been proposed on several occasions, but always without success – probably because "chaeta" actually means "bristle", which is also applied to the thickest mammalian hairs and seems not at all appropriate to fine, flexible structures like the trichobothria. Second, since the hair is certainly more important than the socket, the reverse ordering "bothriotrichos" would seem more suitable.

Trichobothria have been known for well over a century. As early as 1883 Friedrich Dahl wrote in the German journal *Zoologischer Anzeiger* that the trichobothria move when low tones are produced by bowing a violin, and he therefore classified them as auditory hairs. Indeed, it is easy to identify trichobothria under the microscope (or even, after some practice, with the naked eye) in a forest of other hairs, by the fact that the most gentle air movement makes them wave about, even if it was caused merely by the slight trembling of the hands of a non-smoker (Plate 15).

Friedrich Dahl belonged to the same generation of arachnologists as Eugène Simon (see p. 5) and Philip Bertkau (see p. 39). He was born in 1856 in northern Germany (Rosenhofer Brök in the Duchy of Holstein, then and until 1864 under Danish rule). In 1929 he died in a politically much altered Europe, at the age of 73 in Greifswald, district of Rostock. In between he studied at the universities of Leipzig, Freiburg, Berlin and Kiel and made research trips to the Baltic and into the south seas as far as the Bismarck Archipelago, a group of 200 islands that belonged to German New Guinea from 1884 until 1918.

In 1898 Friedrich Dahl was appointed Director of the Section for Arachnids at Berlin's Zoological Museum, where he worked until the end of his days. In the 40 years of his scientific activity Dahl was concerned not only with spiders but also birds, corals, flies and other animals. One of his most significant works was on the wolf spiders (Lycosidae), published in 1908. His wife Maria Dahl, sixteen years younger than him, made his acquaintance at the University of Kiel and raised their four children before she herself turned to spider research in 1920; she prepared the text on some spider families for the multi-volume book "Die Tierwelt Deutschlands" on identification of the fauna of Germany edited by her husband, which appeared in 1925 (Bonnet 1945).

Trichobothria, made so easily movable by their low mass and flexible suspension, are located on the tarsus, the metatarsus and the tibia of the spider leg. Most are on the dorsal surface, presumably because they are most exposed to air movements there. Only in the proximal region of the tibia are trichobothria also found later-

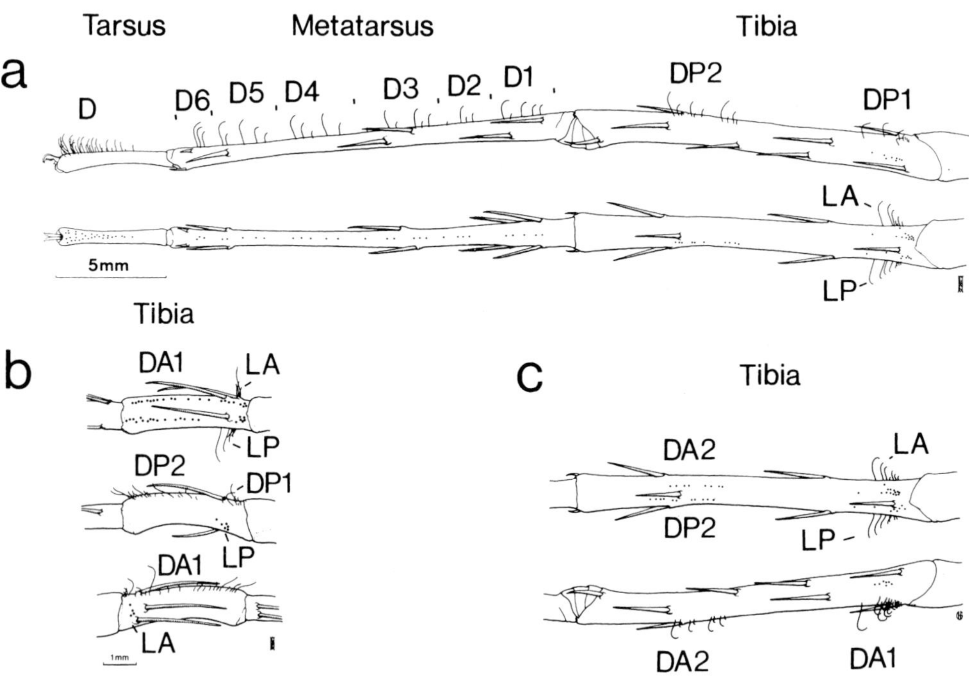

Fig. 1a–c. The arrangement of the trichobothria on the walking legs and pedipalps of an adult *Cupiennius salei*. **a** Dorsal and ventral view of left second walking leg. **b** Left pedipalp in dorsal, postero-lateral, and antero-lateral view (*from top to bottom*). **c** Dorsal and lateral aspect of tibia of left fourth walking leg. *D* Dorsal; *A* anterior; *P* posterior; *L* lateral. (Barth et al. 1993)

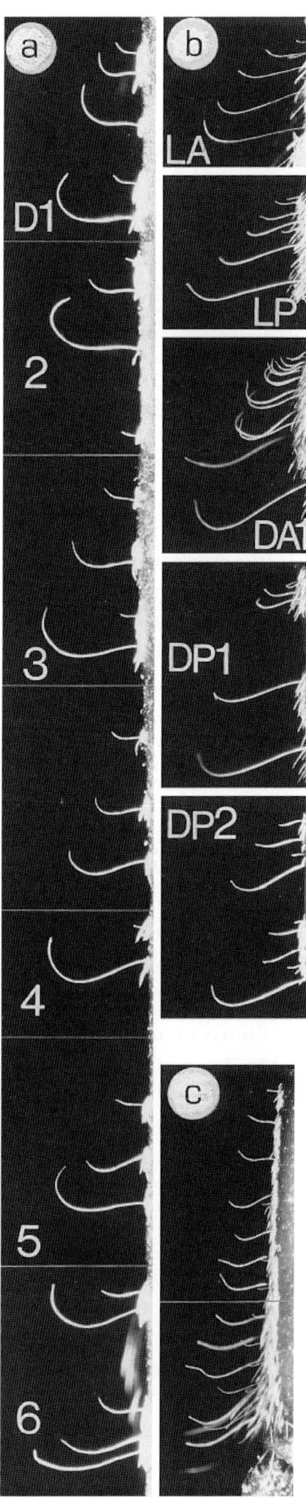

◀ **Fig. 2a–c.** Trichobothria on the walking leg of *Cupiennius salei*. Abbreviations as in Fig. 1. **a** Metatarsus. **b** Tibia. **c** Tarsus. Magnification 16 (Barth et al. 1993)

ally. Figure 1 shows how well equipped *Cupiennius salei* is: to our surprise, our most important experimental animal turned out to have about 100 trichobothria per walking leg, by far the largest number known for spiders. We conclude that the detection of air movements is especially important for "our" species. Altogether, *Cupiennius salei* as an adult has 936 ± 31 (SD; N = 10) trichobothria. Of these, 100 ± 3 are on each leg of the first two leg pairs and 108 ± 3 are on each leg of the back two pairs. The group DA2 (see Fig. 1) makes the difference, being found only on leg pairs 3 and 4. The pedipalps are each provided with 53 ± 3 trichobothria, all of them on the tibia. All these numbers apply to both sexes.

Many trichobothria are assembled into groups of a few (two to three) to 24 sensilla (Fig. 2). This arrangement is particularly visible on the tibia. All the groups are a constant feature from animal to animal and can thus be identified, which facilitates the mechanical and physiological investigation of individual trichobothria (an advantage over, for instance, the filiform hairs on the cricket cercus; Landolfa and Jacobs 1995). In *Cupiennius* the approximately 30 trichobothria on the metatarsus are more uniformly distributed than those on the tibia, forming a straight row. On the tarsus up to 30 trichobothria on the dorsal surface form an arrow-shaped group, the largest of all (Fig. 2, Plate 15).

Morphology

Now we must look more closely at the external structure of the trichobothrium. Later many of its details will turn out to be of functional importance. In *Cupiennius salei* the long trichobothria can easily be seen without a microscope; they are up to 1400 µm long. The shortest trichobothria measure only 100 µm. Groups are distinguished by a gradation in length of the

Fig. 3. 1–6. Surface structure of the hair shaft of a trichobothrium of *Cupiennius salei*. Scanning electron micrographs of hair from its base in the cup (**1**) up to its tip (**6**). Magnification 1900. (Barth et al. 1993)

individual trichobothria (Fig. 2). Typical ranges are 100 to 1200 μm (tibia group DA1), 150 to 1300 μm (tibia group LP), 175 to 850 μm (metatarsus group DI) and 125 to 1125 μm (tarsus).

Hair Shaft. The actual cuticular hair stands nearly perpendicular to the cuticular surface, tilting slightly towards the distal end of the leg. Long hairs are thicker than short ones and in all of them the diameter decreases toward the tip of the hair. At the base long hairs are 10–15 μm thick and short hairs, 5–7 μm, whereas the tip diameters are 5 μm and 1–2 μm, respectively. The most striking feature of the hair shaft of the *Cupiennius* trichobothrium is the feathery-looking surface (Fig. 3); it is covered by twig-like protuberances, with lengths ranging from only about 1 μm at the hair base to about 6 μm at the tip. A second peculiarity of most *Cupiennius* trichobothria is the proximally directed curvature of the shaft in its distal third (Fig. 2). Usually the amount of curvature increases with the overall length of the hair. Remarkably, the plane in which the curved hair lies is inclined at an angle of 5° to 30° with respect to the long axis of the leg. Still more surprising, and as yet completely puzzling, is the asymmetrical arrangement of this slanting, which rotates in a direction around the spider, more or less along with the long axes of the legs. As a consequence, the tips of the trichobothria on the right legs point toward the back leg surface, while those on the left legs point forward.

Socket. The opening of the cup within which the trichobothrium is seated is tilted distalward and has an elliptical outline (ratio of long to short diameter 1.3 ± 0.14 SD, n = 107). The bilateral symmetry of the socket and its orientation on the leg are constant for all trichobothria, even those that differ in their mechanical directional characteristic (see below). Depth and diameter of the sockets increase with the length of the hair (Fig. 4). However, the angle through which the hair can be deflected between the two edges of the socket remains constant at 25°–35°, because the socket of a long hair is not only broader but also deeper (Fig. 4c). In the case of filiform hairs of *Gryllus* the maximal deflection is only 5° (Gnatzy and Tautz 1980; Kämper and Kleindienst 1990), which implies a considerably smaller mechanical working range (for an interpretation in terms of behavioral physiology see Barth et al. 1993, p. 460). Stretched across the bottom of the socket is a cuticular membrane that bears the hair shaft.

Other Spiders

As mentioned above, *Cupiennius salei* has many more trichobothria than any other spider in which they have been counted, about 100 per walking leg. Smaller hunting spiders such as *Philodromus aureolus* and *Pardosa prativaga* have only about 22 or 37, respectively, on each walking leg. And *Agelena labyrinthica*, which sits on a densely woven "carpet" web, has only 25 trichobothria per walking leg (Peters and Pfreundt 1986; Reißland and Görner 1985). However, the trichobothria count does not depend exclusively on the size of the spider; there is a difference between hunting and web-building spiders. Orb-weavers such as *Araneus* (now *Larinioides*) *cornutus* and *Meta reticulata* (now *M. segmentata*) or the sheet-line weaving spider

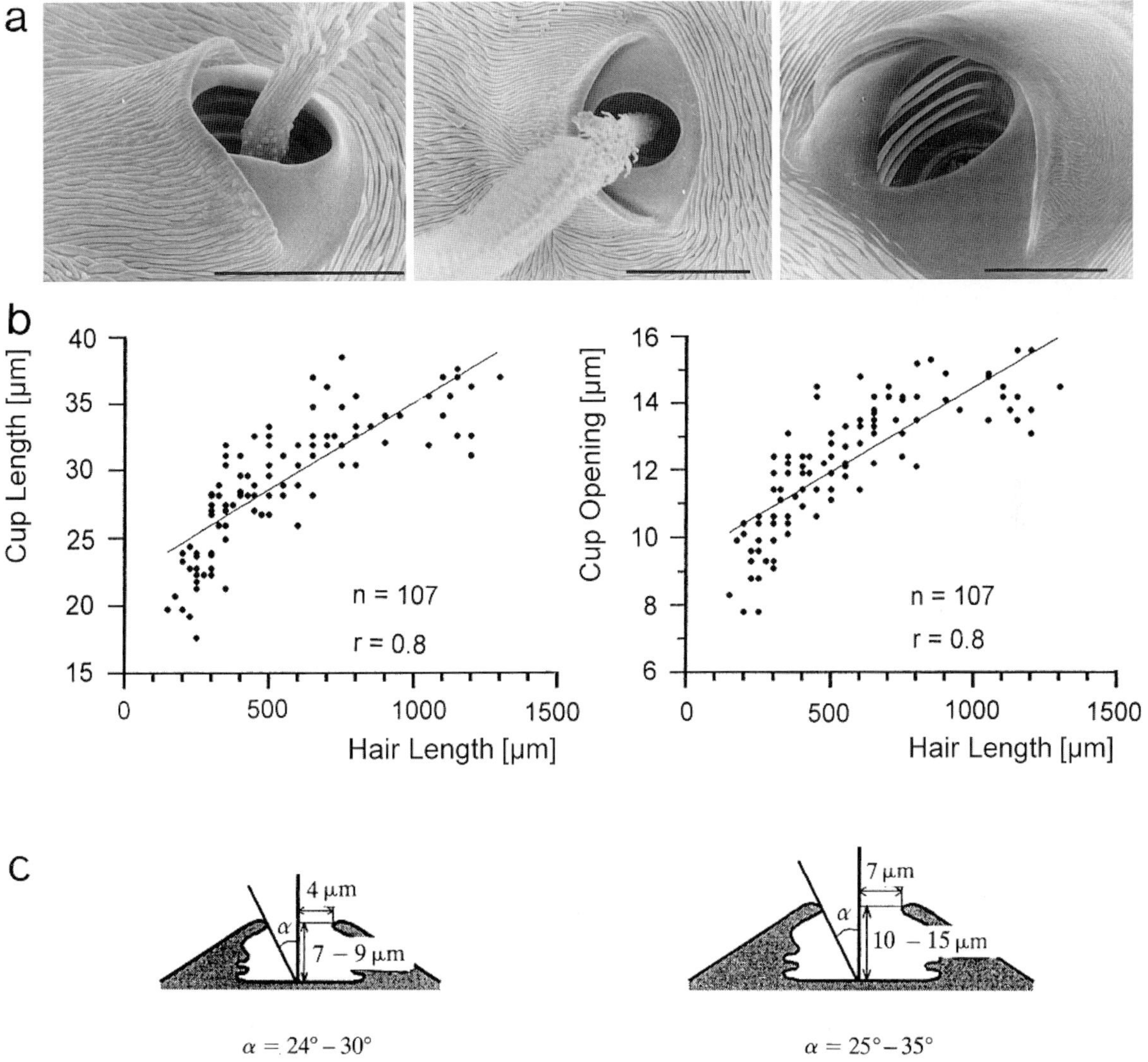

Fig. 4a–c. Structure of the trichobothrial cup. **a** Scanning electron micrographs of cup of a trichobothrium of tibial group TiLP; *bar* 10 µm. **b** Length (*left*) and opening width (*right*) of cup as a function of hair-shaft length. **c** Longitudinal section through the cup of a short (*left*) and a long (*right*) trichobothrium. *α* maximal deflection angle of hair shaft until it contacts the outer rim of the cup. (Barth et al. 1993)

Linyphia triangularis have only 7 to 11 trichobothria per leg. Even *Nephila clavipes*, one of the largest orb-weavers and about the same size as *Cupiennius salei*, has only 40 trichobothria on each of its legs (Fig. 5) (Lehtinen 1980; Peters and Pfreundt 1986; Barth unpubl.).

The two groups also differ with respect to the topography of these organs. Web spiders have none on the tarsi and only one or very few on the metatarsus. The functional significance of these differences is unclear. However, the experiments described below permit a few conclusions. For instance, it can be inferred that the trichobothria on the tarsus of the hunting spiders are situated in a particularly "sensitive" part of the stimulus field, because the velocity of

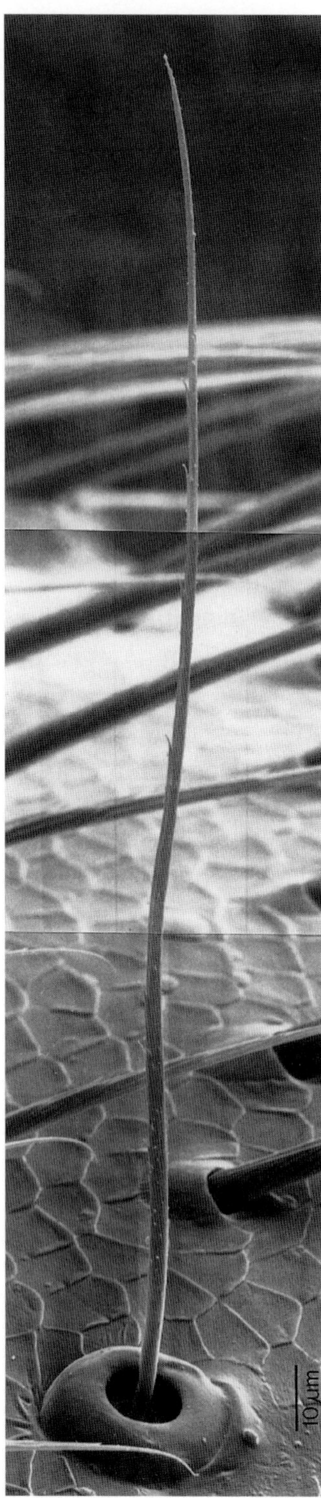

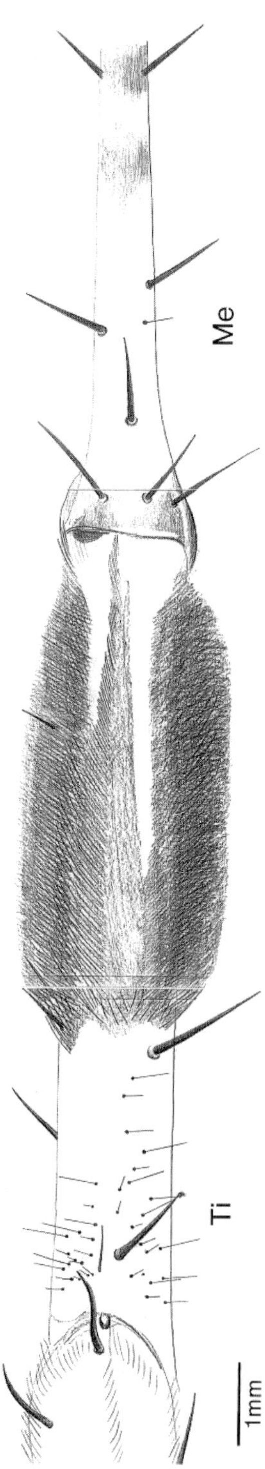

Fig. 5. Trichobothria of the orb-weaver *Nephila clavipes* (second walking leg). Scanning micrograph shows the largely smooth surface of the hair shaft. Length of hair 210 µm, diameter ca. 3 µm. *Me* Metatarsus; *Ti* tibia

air flow is higher over a firm substrate such as the kind of plant on which *Cupiennius* sits (Barth et al. 1993). No such effect would be expected on a web, which is largely "transparent" to air movement. Furthermore, we shall see later (Chapter XIX) that in orb-weaving spiders airborne stimuli elicit not prey-catching behavior but only a defensive raising of the forelegs (Klärner and Barth 1982). This behavior evidently requires relatively little from the sensory apparatus, whereas for successful prey capture spatial analysis of the site of interest is needed, along with precise and rapid grasping movements.

The analysis of fluid mechanics later in this chapter will quantify the effect of individual morphological features of the hair on its deflection by air currents. Then the functional significance of differences in the length and diameter of the hairs, their surface structure and their curvature will also become clear. By the way, the feathering of the hair shaft, which is so conspicuous in *Cupiennius*, does also occur in many other trichobothria and in filiform hairs (Görner 1965; Harris and Mill 1977a), but there is an exception to every rule. In some other spiders (*Nephila clavipes*; Fig. 5) as well as in scorpions (Meßlinger 1987) and insects (Shimozawa and Kanou 1984a) there are filiform hairs with comparatively smooth surfaces. We shall see later why these are also functional.

The marked curvature of the trichobothria of *Cupiennius salei* is found neither in the trichobothria of other spiders and arachnids nor in the insect filiform hairs (Palmgren 1936; Nicklaus 1965; Hoffmann 1967; Görner and Andrews 1969; Harris and Mill 1977; Gnatzy and Tautz 1980; Shimozawa and Kanou 1984a; Peters and Pfreundt 1986; Meßlinger 1987). Only one case is known in insects: the four filiform hairs on the prothorax of the caterpillar of the cabbage moth (*Barathra brassicae*; Tautz 1977).

Fine Structure and Stimulus Transformation

As in other cuticular mechanoreceptors, the site of stimulus transduction is, so to speak, the heart of the trichobothrium. Here the stimulus received and transformed by the stimulus-conducting apparatus is converted into the excitation of a sensory cell. This transduction process changes energy from one form to another: the mechanical energy associated with air movement into the electrochemical energy of the signal produced by the sensory cell.

The functional morphology of the parts of the hair sensilla that are involved in stimulus transduction has therefore attracted considerable interest. In particular, the exemplary electron-microscopic studies by Werner Gnatzy and his colleagues (Gnatzy and Schmidt 1971; Gnatzy and Tautz 1980) on the filiform hairs of the cricket (*Gryllus bimaculatus*) have set a standard. Their results, together with the considerations published by Thurm (1982), allowed the sensitivity of these hairs to be calculated, as mentioned in Chapter VI: at the threshold, the energy that suffices to excite the receptor cell is less than that in a quantum of green light, and the change produced in the diameter of the dendrite is only about 0.05 nm. Figure 6 illustrates the micromechanics underlying this response.

A cricket filiform hair is a two-armed lever, the fulcrum of which is situated slightly below the surface of the cuticle and slightly above the inner end of the hair shaft. Deflection of the long arm, which projects into the stimulus field, presses the short arm against the end of the sensory-cell dendrite, which inserts at the base of the hair. The deflection amplitude is thus scaled down (as in the slit sensilla!). According to the laws of levers, the force exerted on the dendrite is increased by a factor given by the ratio of the two arms. In *Gryllus bimaculatus* the inner end of the hair moves by only 0.03 μm for each angular degree of hair deflection. A small cuticular projection (see Fig. 6) is evidently there to press against the tubular body of the dendrite during stimulation. The osmiophilic filaments or cones, which connect the peripheral tubules of the tubular body to the dendritic cell membrane (Gnatzy and Tautz 1980), are at present thought most likely to contain the receptor molecules (Thurm 1982b, 1996).

So we now see what potential resides in the *Bauplan* "hair". However, we cannot simply apply the details of the cricket filiform hairs to the

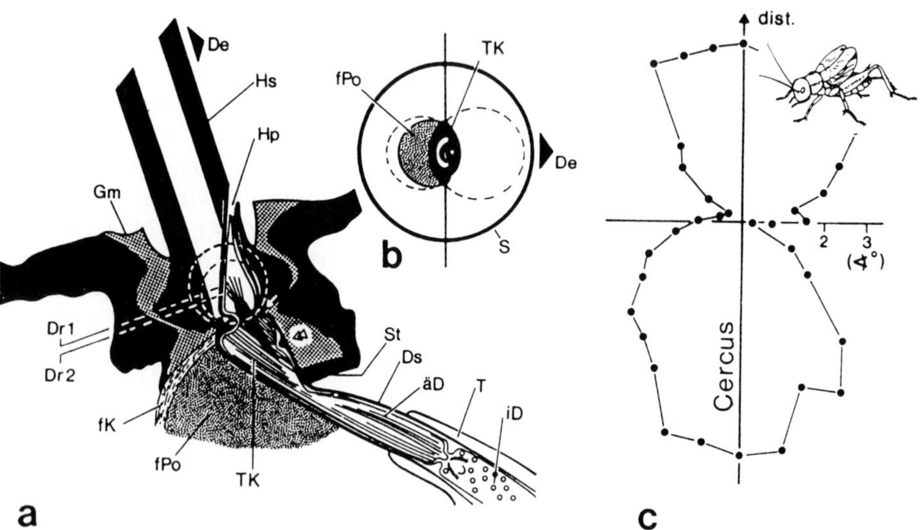

Fig. 6a–c. Functional morphology of stimulus uptake in cricket (*Gryllus bimaculatus*) cercal filiform hairs. **a** Dendrite attachment in the hair base; the hair is a two-armed lever rotating about a point located between *Dr1* and *Dr2*. Displacement of the hair shaft towards *De* leads to a compressional deformation of the dendrite tip at ⇐ and to a depolarization of the sensory cell. *äD* Outer dendritic segment; *Ds* dendritic sheath; *fK* fibrous cap; *fP* fibrous cushion presumably serving as an abutment when the hair is stimulated, *Gm* joint membrane; *Hp* molting pore; *Hs* hair shaft; *iD* inner dendritic segment; *St* cuticular connection; *T* thecogenous cell; *TK* tubular body; *double arrowhead* ribs of the dendritic sheath presumably providing mechanical anchoring of the dendrite. **b** Directional characteristics; *De* direction of displacement leading to depolarization; *S* upper rim of hair socket; *broken lines* indicate the mechanical deflectability of the hair in different directions. **c** Example of mechanical directional characteristics of a cercal filiform hair. (Barth 1986b, after Gnatzy and Tautz 1980)

trichobothria of spiders or other arachnids. The two are not homologous. Instead, we must assume that they have converged onto a similar solution to a given problem, driven in particular by the physical properties of the stimulus. To document this interpretation, a few details (see Fig. 7). In the filiform hair the dendrite ends at the side of the hair shaft, above the fulcrum, in a molting channel that is open to the exterior; in the trichobothria it ends at a helmet-shaped structure at the inner end of the hair. There is one sensory cell for each insect filiform hair, whereas the spider trichobothrium has at least three (Görner 1965; Christian 1971, 1972), and in *Cupiennius salei* there are four (Figs. 8 and 9). And finally, the suspension of the trichobothrium is different: the filiform hair is connected to its base by a joint, but the trichobothrium rests on a membrane at the base of the socket with radially extended fibers. Even within the arachnids the homology of the trichobothria is not quite clear, as can be seen in the relevant literature (Weygoldt and Paulus 1979). The most recent paper on the subject deals with a whip spider (*Typopeltis crucifer*, Uropygi, Arachnida) (Haupt 1996). Its trichobothria, two in number, are near the end of the first pair of legs (at the distal end of the tibia), which are longer than the other six legs and used as feelers. These trichobothria each have 11 (!) sensory cells. However, only six of these cells send dendrites all the way to the dendritic sheath coupled to the base of the hair.

As a consequence of these special features, the ends of the dendrites in the trichobothrium also receive the mechanical stimulus mediated by hair deflection differently than do those in the insect filiform hairs. Our ideas about this situation derive from an early publication by Görner (1965). A central role is played by the "helmet", the interior of which accommodates four dendritic endings (sometimes, as in *Cu-*

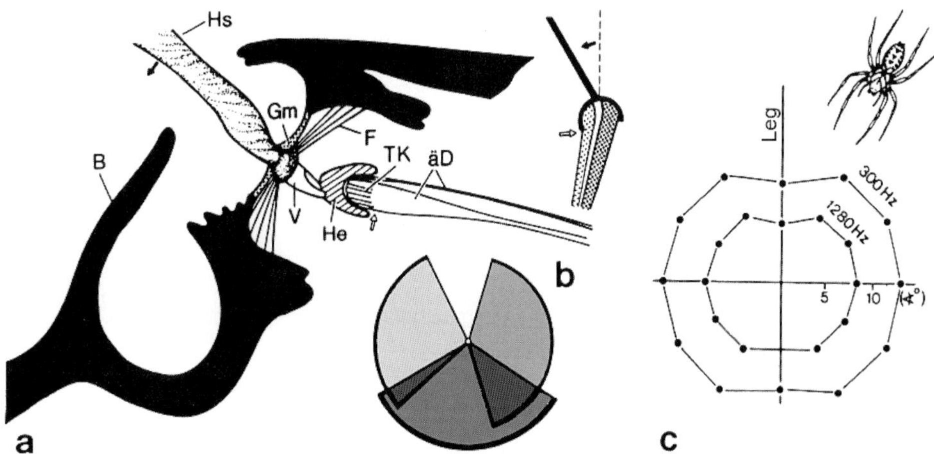

Fig. 7a–c. Functional morphology of stimulus uptake in a trichobothrium of the spider *Tegenaria.* **a** Attachment of dendritic ends to cuticular helmet at the base of the trichobothrium. When the hair shaft is deflected (←) the helmet (*He*) presses preferentially against one of the dendrites (⇒) (see *inset on right*) and thus effectively stimulates it. *äD* Outer dendritic segment; *B* cup; *F* cuticular fibers; *Gm* joint membrane; *Hs* hair shaft; *TK* tubular body; *V* connecting part. **b** Example of the directional ranges of the three sensory cells determined *electrophysiologically.* **c** *Mechanical* directionality of a trichobothrium at two different stimulus frequencies (oscillating air). (Barth 1986b; after Görner 1965; Christian 1971; Reißland and Görner 1978)

piennius salei, one of them has a particularly well developed tubular body; Fig. 8). When the trichobothrium is deflected, the lower edge of the helmet presses against the end of the dendrite that inserts on the side in the direction of deflection (Fig. 7). Different deflection directions stimulate different sensory cells. This has the remarkable effect that a single trichobothrium "looks" in different directions with its three or four sensory cells, so that it can signal the direction of a stimulus to the spider (if the afferents are suitably connected to the central nervous circuitry). To achieve the same thing, a cricket needs several filiform hairs. We shall return to this point in discussing the directional characteristic of the trichobothria.

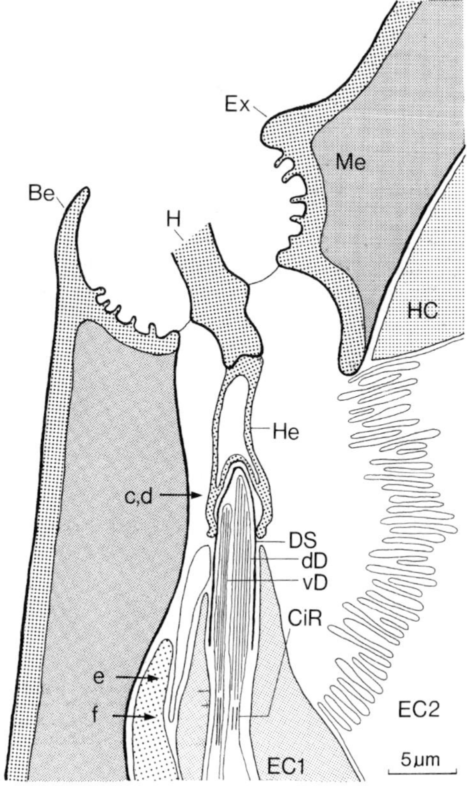

Fig. 8A. Dendrite attachment region of trichobothrium of *Cupiennius salei* according to electron-microscopic analysis. *Be* Cuticular cup; *H* hair shaft; *He* helmet; *DS* dendritic sheath; *dD* dorsally located dendrite; *vD* ventral dendrite; *CiR* ciliary region; *EC1* and *EC2* sheath cells; *HC* hypodermal cell; *Ex* exocuticle; *Me* mesocuticle. *c, d, e,* and *f* indicate levels of cross-sections shown in Fig. 8B. (Anton 1991)

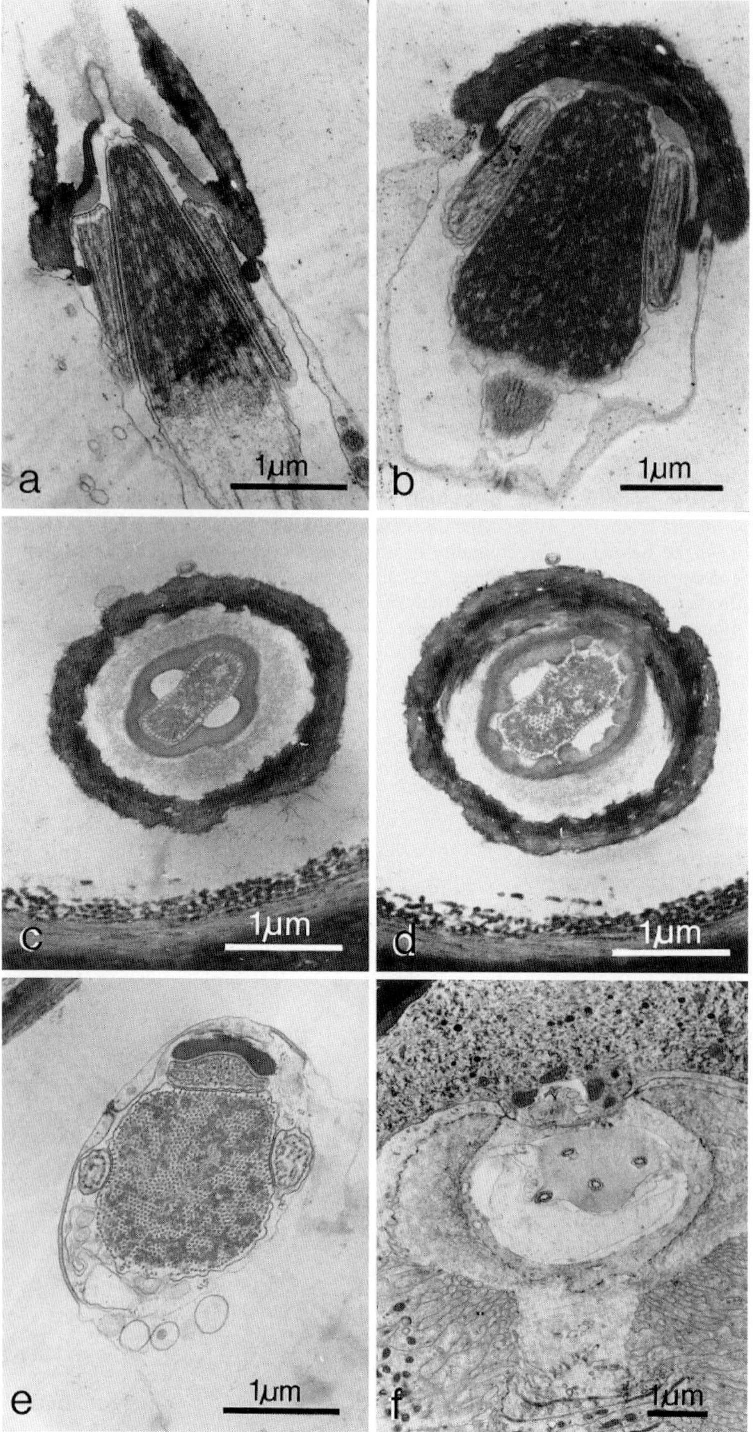

Fig. 8 B a–f. Electron-microscopic sections through dendritic coupling region of trichobothrium of *Cupiennius salei*. **a** and **b** longitudinal sections, **c–f** cross-sections (at the levels indicated in Fig. 8 A). Note three dendrites in **a** and **b**, with the thickest in the middle showing the tubular body most clearly. **c** and **d** show that the thickest dendrite reaches further distally than the other dendrites. It is seen surrounded by the dendritic sheath and the wall of the helmet. In **e** the section is close to the ciliary region, in **f** right in the middle of it with all four dendrites clearly showing the ciliary pattern typical of sensory cells (9 + 0). (Anton 1991)

2 Stimulation by Air Movement – the Interaction Between Air and Hair

It may seem naive to ask how a hair is deflected by an airstream, but serious consideration of the associated physical problems will show that this question challenges even the specialist, who in this case must be an expert in fluid mechanics. Once again we shall be fishing in waters that are not actually biological, but without which we can never reach our goal. As with the slit sensilla, the eminent significance of the receiving and transformation of the stimulus by secondary structures will become clear. Please tolerate the celebration of some details: it is an indispensable part of the exercise.

First, let us go back to the claim that trichobothria and similar hair receptors are just the opposite of slit and other hole sensilla. A severely simplified introductory model designed by Tautz (1979) is shown in figure 9a. Here the interesting properties of a hair sensillum are the mass of its shaft, the stiffness of its connection to the body cuticle, and an internal frictional resistance. If the mass is kept very small and the suspension is very "soft", as is true of a trichobothrium, then the slightest air movement (like that due to the stroking of a violin mentioned by Dahl) suffices to deflect the hair and put it into oscillation (if the air movement is oscillatory). Unlike the slit sensillum, which is usually embedded in (literally) bone-hard cuticle, the hair shaft of the trichobothrium is an extremely movable structure. The same applies to the filiform hairs of insects: Tautz (1977) gives a mass of 3.6×10^{-9} g for the filiform hairs of *Barathra* caterpillars, and according to Shimozawa and Kanou (1984b) the stiffness of the suspension of cricket filiform hairs is of the order of 10^{-12} Nm/rad.

It gets more complicated when we ask how much force acts on the trichobothria when they are stimulated by air. In our Mickey Mouse diagram (Fig. 9a) the force-to-hair coupling is mediated by a damping element. The air into which the trichobothrium projects, and which moves, presents two potential stimulus parameters: pressure and particle movement. The trichobothria respond to the movement of particles, and the driving force is the friction between the air particles and the hair. Friction is proportional to the velocity of air-particle movement, which is why viscosity is included in the diagram, the velocity- and hence frequency-dependent attenuation element.

The force that deflects the hair is not applied at a point but over (almost) the whole length of the hair. It is relevant in this regard that above a surface over which laminar flow is occurring, a "boundary layer" forms, within which the particle velocity is not uniform. Just above the surface the air is not moving at all. Its velocity reaches a maximum – that is, the free-field value – at different distances above the surface, depending on the movement frequency. The thickness of this boundary layer, which acts as a brake, decreases as frequency increases.

It is becoming ever clearer, then, that the question raised at the outset is not trivial at all. It has been a subject of research for almost 20 years now (Fletcher 1978; Tautz 1978, 1979; Reißland and Görner 1978; Shimozawa and Kanou 1984a, b; Kämper and Kleindienst 1990; Shimozawa et al. 1998a, b). N.H. Fletcher (1978), at the University of New England in Australia, made the important connection to an excellent paper, now almost 150 years old (Stokes 1851), that treated the equivalent problem of attenuation of an oscillating pendulum by the viscosity of the medium. Jürgen Tautz (1979), then still at the University of Konstanz, in 1979 published a review summarizing the problems involved in sensory detection of the oscillation of particles in a medium.

A few years ago Pepe Humphrey and I (and of course our students and coworkers) put together our physical-technical and biological know-how and concentrated anew on this medium-movement sense. Our aim was to model the movement of a hair in an oscillating medium mathematically, in the most comprehensive and consistent way possible, without regard to whether the medium is air or water. Although we were attacking the problem on a broad front, in the first instance our attention was focussed on the trichobothrium of *Cupiennius salei* because it was an existing test case. The result was

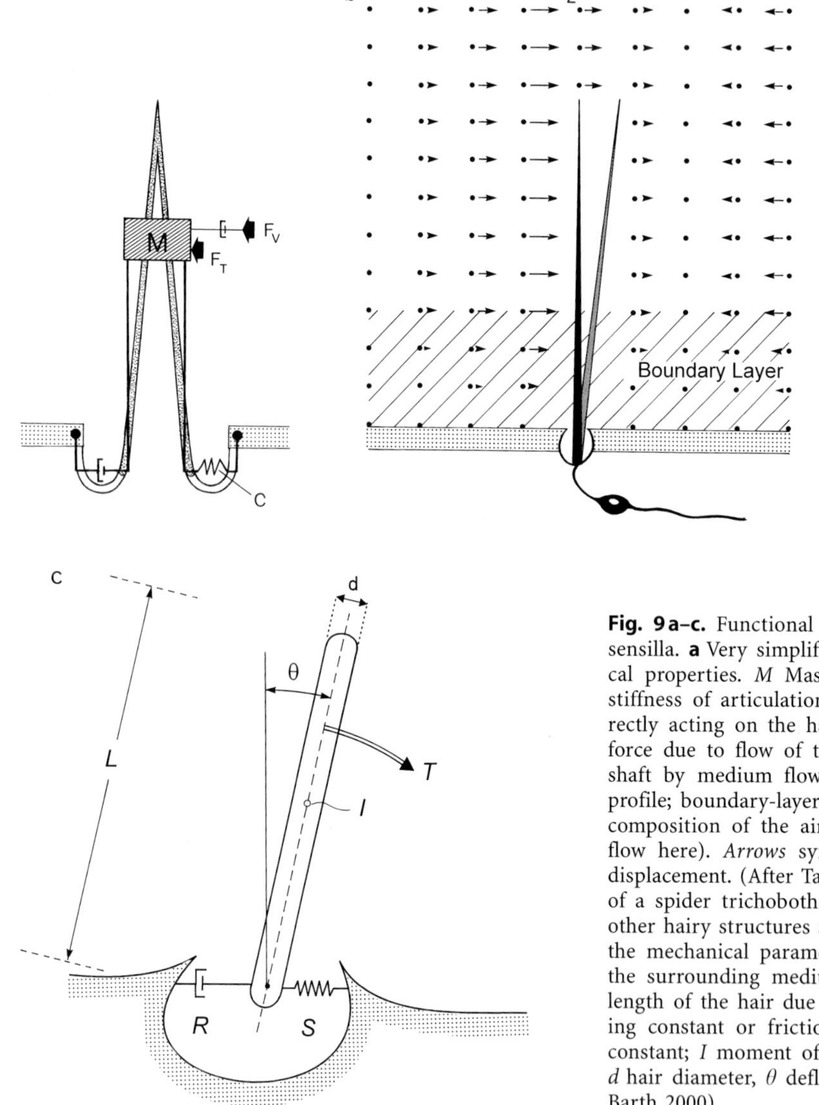

Fig. 9 a–c. Functional properties of mechanoreceptive hair sensilla. **a** Very simplified sketch explaining basic mechanical properties. *M* Mass of hair shaft; *C* elasticity (spring stiffness of articulation); *n* viscous damping; F_T force directly acting on the hair in case of tactile stimulation, F_V force due to flow of the medium. **b** Displacement of hair shaft by medium flow. The hair is exposed to a velocity profile; boundary-layer thickness depends on the frequency composition of the airflow (supposed to be an oscillating flow here). *Arrows* symbolize vectors of medium particle displacement. (After Tautz 1979). **c** Schematic representation of a spider trichobothrium (or insect filiform hair and all other hairy structures sensitive to medium flow) illustrating the mechanical parameters essential for hair deflection by the surrounding medium flow. *T* Torque acting along the length of the hair due to fluid-mechanical effects; *R* damping constant or frictional resistance; *S* torsional restoring constant; *I* moment of inertia of hair shaft; *L* hair length; *d* hair diameter, θ deflection angle. (Humphrey et al. 1993, Barth 2000)

an extension of the fluid-mechanical theory of air and hair movement (Humphrey et al. 1993, 1998) and a comparison of the quantitative predictions of this theory with actual measurements of the trichobothria (Barth et al. 1993). The agreement turned out to be gratifyingly good! Along the way, a few inconsistencies in the literature were corrected. Building on this foundation, we extended the spatial dimension of our model to the whole spider and analyzed the pattern of flow around it, as well as those associated with typical and biologically relevant air stimuli. The most recent step at present is a theoretical comparison of the hair movement in air and – especially with crustaceans in mind – in water.

There is much more to this story, which is detailed in the original papers (Barth et al. 1993,

1995; Humphrey et al. 1993, 1998; Devarakonda et al. 1996). The description that follows is just a summary, although a fairly extensive one.

The Mathematical Model

The Basic Equation. The point of departure was the principle of conservation of angular momentum, applied to a hair with trichobothrial geometry (including the curvature, which is especially pronounced in long hairs). Here two situations are distinguished, both equally relevant to the biology: air-flow oscillations parallel to the long axis of the leg, and those perpendicular to it. For both it holds that

$$I\ddot{\theta} = -R\dot{\theta} - S\theta + T_D + T_{VM}$$

where J (Nm s^2 rad^{-1}) is the angular moment of inertia of the hair with reference to its axis of rotation, R (Nm s rad^{-1}) is the damping constant or frictional resistance, and S (Nm rad^{-1}) is the torsional restoring constant; θ (rad) is the deflection in degrees of arc, $\dot{\theta}$ (rad s^{-1}) is its first derivative over time (that is, the angular velocity), and $\ddot{\theta}$ (rad s^{-2}) is its second derivative, the acceleration of the deflection; finally, T_D represents the torque exerted by the flowing medium and T_{VM}, the torque generated by the mass of medium that must be accelerated along with the hair (added or virtual mass).

This equation is none other than the equation for the enforced oscillation of a simple damped harmonic oscillator. In simpler form it has also been applied to filiform hairs by previous authors, first by Tautz (1977). We learn from this equation that the rate of change of angular momentum depends on four quantities, which together make up the torque that acts on the hair: the frictional force $R\dot{\theta}$ at the pivot point of the hair, the restoring force $S\theta$ at the pivot point of the hair, the force T_D exerted by the medium along the whole hair shaft, and the force T_{VM}, which is exerted by the virtual mass and likewise acts all along the shaft. Whereas $R\dot{\theta}$ and $S\theta$ have a "braking" action (and therefore a negative sign in the equation), T_D and T_{VM} tend to move the hair.

R and S. The values of R and S must be found experimentally, from the actual hair. The restoring constant S was measured by Shimozawa and Kanou (1984b) and Shimozawa et al. (1998) for cricket filiform hairs. The values they found depended on the length of the hair and ranged from about 0.2×10^{-12} Nm rad^{-1} to 8.5×10^{-12} Nm rad^{-1}. They also depend on the direction of deflection: perpendicular to the plane of greatest mobility, the values are 4 to 8 times as large (Kanou et al. 1989). The damping constant R could be calculated from the measurements by Kämper and Kleindienst (1990) of the phase difference between the movements of hair and air for the filiform hairs of crickets. The result: $R = 8.21\times10^{-14}$ (Nm s rad^{-1}) (Humphrey et al. 1993, 1998). Application of the data obtained from cricket filiform hairs to the trichobothria is certainly not without its problems, but the influences of R and S on the hair movement (θ) and its derivatives ($\ddot{\theta}, \dot{\theta}$) can be estimated by systematically comparing the measured and the calculated hair deflection for various pairs of R and S (Barth et al. 1993). In the case of the trichobothria of *Cupiennius salei* in Group D on the metatarsus (Fig. 2), the values thus obtained are $R = 2.20\times10^{-15}$ and $S = 5.77\times10^{-12}$ for a long curved hair, and $R = 0.27\times10^{-15}$ and $S = 0.62\times10^{-12}$ for a short straight one (Fig. 10). The similarity between *Cupiennius* and *Gryllus* with respect to the values for S is remarkable. The values for R are about 40-fold smaller in the trichobothria of *Cupiennius* than those previously calculated for the cercal hairs of *Gryllus* (Humphrey et al. 1993, 1998), indicating higher flexibility of the hair at its articulation due to smaller viscous damping. The changes of S and R with hair length were nicely quantified for cricket cercal hairs in a recent paper by Shimozawa et al. (1998).

T_D and T_{VM}. Now the other two parameters of the equation: T_D and T_{VM}. These are obtained by integration of the "drag" produced by movement of the medium and the force exerted by the virtual mass on the whole length of the hair. If F_D and F_{VM} are the corresponding forces per unit length, we have:

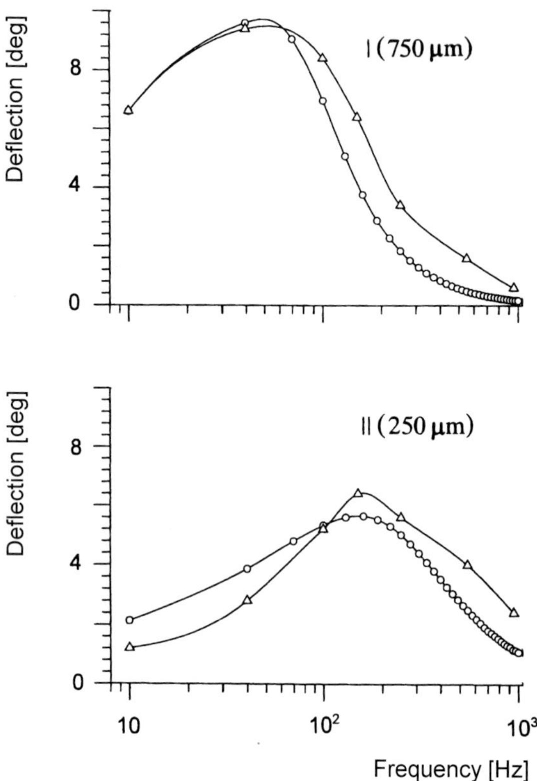

Fig. 10. Determination of the damping constant R and of the torsional restoring constant S by matching calculated (○) with measured (Δ) values of hair deflection over stimulus frequency. Comparison of two hairs differing in length. (Barth et al. 1993)

$$T_D = \int_0^{L_1} F_D y\,dy + \int_0^{L_2} F_D L_1\,dx \quad \text{and}$$

$$T_{VM} = \int_0^{L_1} F_{VM} y\,dy + \int_0^{L_2} F_{VM} L_1\,dx\,.$$

L_1 and L_2 are the lengths of the straight (L_1) and the curved (L_2) sections of the hair, the full length of which is L. Experimentally confirmed theoretical expressions for F_D and F_{VM} are provided by Stokes (1851), at least for medium oscillations perpendicular to the hair. The corresponding equations for oscillation parallel to the "curved" distal hair segment L_2 were derived by Humphrey et al. (1993).

L/d and Re. There are certain restrictions on the validity of the theory of Stokes (1851) for a fluid oscillating perpendicular to a cylinder; the most important for our case are the following two conditions: (i) the ratio of cylinder length L to cylinder diameter d should be considerably greater than 1: $L/d \gg 1$; (ii) the Reynolds number ($Re = V_r d/2\,\rho/\mu$, where V_r is the local relative velocity (that is, the difference between the velocities of medium and hair), ρ is the density of the medium, and μ is the dynamic viscosity) should be considerably smaller than 1: $Re \ll 1$. For a flow of medium parallel to the cylinder only the first restriction applies. Both conditions are met for the trichobothria as well as for the filiform hairs of insects.

There is more to be said about this, but so as not to lose too many readers I shall simply refer those interested in the secrets of the fluid-mechanical theory of trichobothrial deflection to the detailed presentation in Humphrey et al. (1993, 1998), which also extensively surveys the previous literature on the subject.

Before we turn to the actual, experimentally determined mechanical properties of the trichobothria of *Cupiennius salei*, here are two of the most important conclusions drawn from the mathematical model:

1. In calculating the hair deflection it is essential not to ignore the force exerted by the virtual mass (T_{VM}) within the entire range of biologically relevant frequencies, as was done previously (Shimozawa and Kanou 1984b). This applies even more when the medium is water rather than air (see below, Section 4).
2. The direction of the air movement relative to the axis of the substrate – for instance, the cylinder representing a spider leg – has a marked influence on the amplitude of the hair deflection as well as its velocity and acceleration, but not on its resonant frequency.

Flow Direction. The velocity profile over the cylinder surface in the case of parallel orientation of flow differs considerably from that in the case of perpendicular flow. With parallel orientation the method of calculating the velocity profile de-

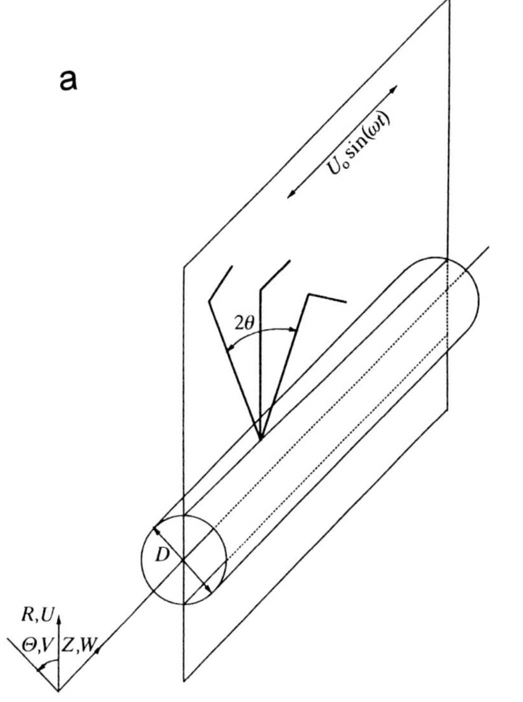

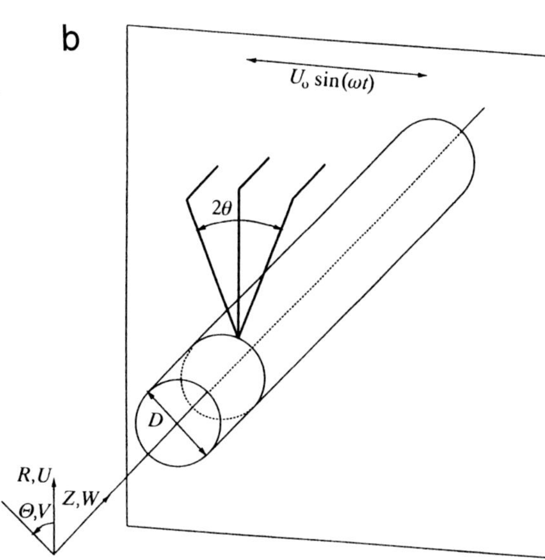

Fig. 11 a, b. Two different stimulus situations. **a** Airflow oscillating parallel to long axis of leg. In the given case the radial (*U*) and the circumferential velocity component (*V*) are zero. **b** Airflow oscillating perpendicular to the longitudinal leg axis. In this case the longitudinal (*W*) velocity component is zero. (Humphrey et al. 1993)

veloped by Stokes (1851) for an infinitely large flat surface is fully applicable, as long as the condition $fD^2/\nu > 20/\pi$ (f, frequency; D, cylinder diameter; ν, kinematic viscosity of the air) is met, which is true of our trichobothria and similar cases. Cylinders in a cross-current need another theory, because of their surface curvature (Wang 1968; Telionis 1981; Humphrey et al. 1993). In this configuration the curvature of the surface plays a substantial role. Upon the oscillating flow there is superimposed another current that is constant on average. The boundary layer is thicker than in the case of parallel flow, and within it the velocities and velocity gradients are distinctly higher (cf. Fig. 18).

The Mechanical Properties of the Trichobothria

The value of theory soon becomes apparent when we use it to compare possible predictions with the actual trichobothria of *Cupiennius salei* and their mechanical properties. An extensive body of data is available for this purpose, from experiments in which we examined the functional significance of the morphological parameters of the trichobothria described above (Barth et al. 1993). To measure the mechanical frequency tuning and directional sensitivity of the trichobothria we needed a controllably variable oscillating flow field. This was produced by two identical loudspeakers connected with opposite polarity and mounted at the open ends of a plastic tube 15 cm in diameter. The flow field in the tube was checked for homogeneity and calibrated with a laser Doppler anemometer. The spider leg with the trichobothria was put into the middle of the tube, and a micromanipulator was used to turn it into the desired orientation relative to the direction of air movement. The displacement of the hair tip was measured by microscopic observation; from that and the length of the hair the deflection angle α was calculated. To measure the boundary layers we again used a laser Doppler anemometer. All other aspects of the method are treated in detail by Barth et al. (1993).

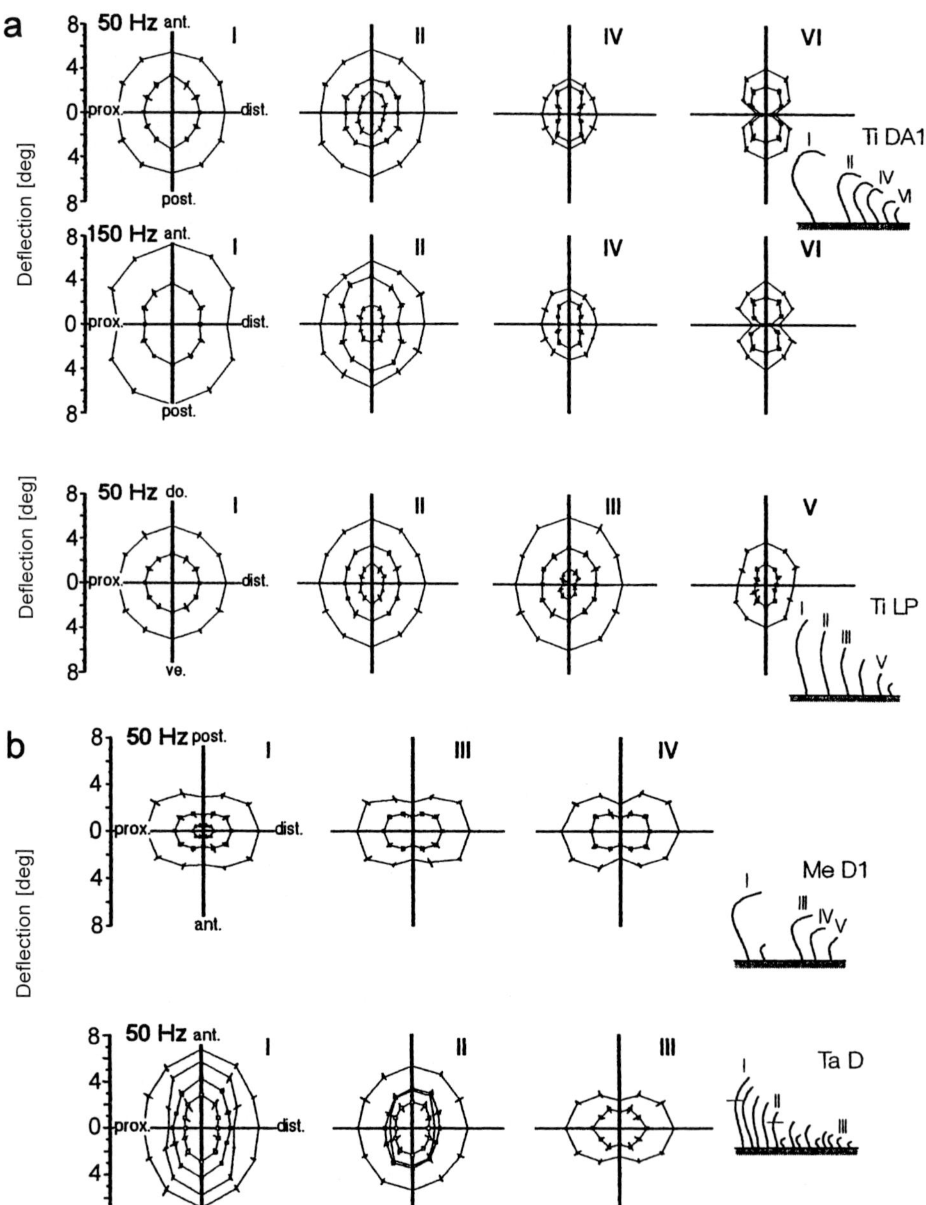

Fig. 12 a, b. The mechanical directional characteristics of the deflection of trichobothria of different groups on the walking leg of *Cupiennius salei* (see Fig. 1) in oscillating airflow. **a** Trichobothria of groups TiDA1 and TiLP stimulated with frequencies 50 and 150 Hz and with different stimulus amplitudes. Group DA1 with hairs I (1250 μm), II (800 μm), IV (550 μm), and VI (350 μm). Air-particle velocities in mm s^{-1} at 50 Hz (150 Hz): hair I – 12 and 29 mm s^{-1} (27 and 59 mm s^{-1}); II – 12, 29, and 105 (8, 24, and 59); IV – 60 and 120 (19 and 51); VI – 69 and 150 (29 and 74). Group LP with hairs I (1120 μm), II (1000 μm), III (700 μm), and V (450 μm). Air-particle velocities at 50 Hz: hair I – 16 and 40 mm s^{-1}; II – 21, 58 and 143; III – 17, 74, and 199; V – 59 and 217. **b** Examples of groups MeD1 and TaD. Group MeD1 with hairs I (750 μm), III (600 μm), and IV (500 μm). Air-particle velocities at 50 Hz: hair I – 5, 16, and 24 mm s^{-1}; III – 12 and 30; IV – 12 and 34. Group TaD with hairs I (900 μm), II (700 μm) and III (300 μm). Air-particle velocities in mm s^{-1}: hair I – 11 and 23; II – 13 and 19; III 47 and 90. The innermost curves for hairs I and II show the practically unaltered directional sensitivity after the bent distal part of the hair has been cut off (see *inset*). (Barth et al. 1993)

Directional Sensitivity. Unlike the filiform hairs of insects, trichobothria were thought to be only slightly or not at all directionally sensitive (Görner 1965; Reißland and Görner 1978). We made measurements in four groups of trichobothria on the tibia, metatarsus and tarsus of *Cupiennius salei* (Fig. 12). Among the long trichobothria in particular, there are some that can be deflected to almost identical degrees from almost every direction parallel to the surface, whereas the short trichobothria show a much clearer preference for the direction either parallel or perpendicular to the long axis of the leg. The large group of trichobothria on the tarsus combines both directional sensitivities: the long trichobothria perpendicular to the leg axis, the short ones parallel to it. The main features of these directional sensitivities are independent of frequency (measured at 10, 50 and 150 Hz). The curved trichobothria retain these properties even when the curved part is cut off (Fig. 13). This is surprising because, in analogy to the filiform hairs of the caterpillar of the cabbage moth *Barathra* (Tautz 1977), we had expected the hair to be deflected further when the particle movement was perpendicular to the plane of the hair shaft. Furthermore, the shortest, straight trichobothria exhibited the most pronounced directional characteristic. Therefore we conclude from this experiment that the origin of the directional sensitivity of the trichobothria should be sought in the suspension of the hair or in the coupling to the dendrite. In the cabbage-moth caterpillar the filiform hair suspension either does not contribute to directional sensitivity or has the same directional sensitivity as is produced by the curvature of the hair. Furthermore, the curvature of the trichobothria seems to be a specialty of *Cupiennius*. Neither the trichobothria of other spiders and arachnids (Palmgren 1936; Hoffmann 1967; Görner and Andrews 1969; Harris and Mill 1977; Peters and Pfreundt 1986; Meßlinger 1987; Haupt 1996; Barth unpubl. re *Nephila clavipes*), nor the filiform hairs of insects (Nicklaus 1965; Gnatzy and Tautz 1980; Shimozawa and Kanou 1984a) – apart from *Barathra* – are curved in such a way. So how are we to interpret the curvature of the *Cupiennius* trichobothria? We speculate that curved hairs situated on the dorsal surface of the leg are considerably more sensitive to air movement from above than straight hairs are. Such air movement is certainly a common natural stimulus situation.

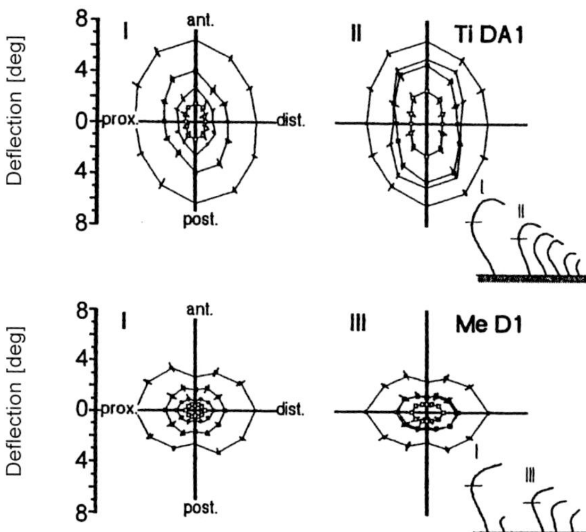

Fig. 13. Mechanical directional characteristics of trichobothria of *Cupiennius salei* before (●) and after (○) cutting their curved distalmost part. Lengths of hairs of tibial group DA1: I – 1000 μm, II – 750 μm; metatarsal group D1: I – 700 μm, III – 600 μm. Oscillation frequency of air 50 Hz; particle velocities TiDA1: hair I – 17 and 31 mm s^{-1}; hair II – 52 and 123 mm s^{-1}; Me D1: hair I – 15 and 26 mm s^{-1}, hair III – 13 and 27 mm s^{-1}. (Barth et al. 1993)

Frequency Tuning. Both long and short trichobothria follow the frequency of oscillation of the medium in a 1:1 ratio over the entire range tested, between 10 Hz and 950 Hz. The deflection angle (constant air-particle velocity) does change with stimulus frequency and also depends on the length of the hair (Fig. 14). The trichobothria of *Cupiennius salei* are tuned to frequencies between 40 and 600 Hz. As hair length increases, the range of best frequencies shifts toward lower frequencies. In addition, the slope of the curve in the low-frequency range increases considerably. If the experiment is done another way, by determining the particle velocity necessary to produce a constant amplitude of hair deflection at the various frequencies, the resulting curves are correspondingly reversed (Fig. 14). In both cases the frequency tuning

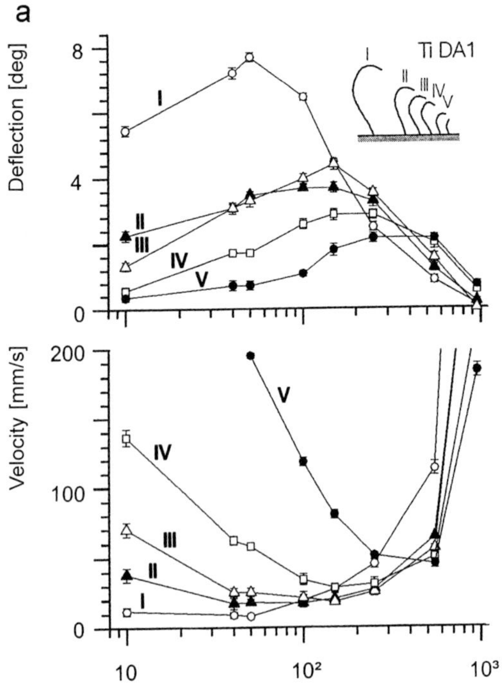

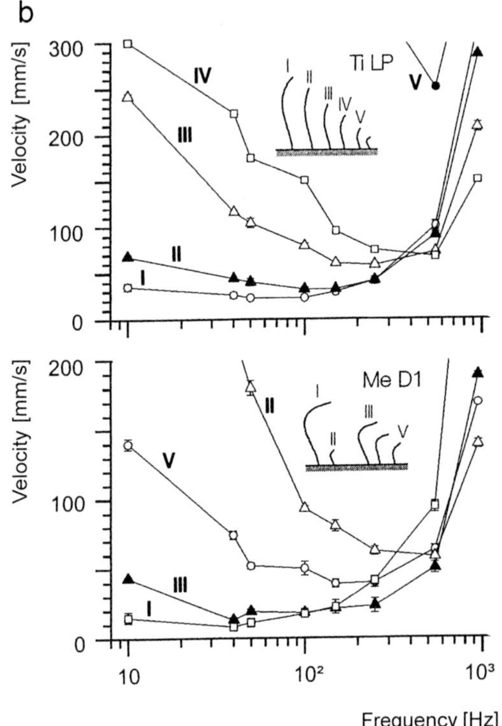

proves to be broad, particularly in the short hairs, and the preferred frequency ranges of the individual hairs overlap. Nevertheless, there is a clear correlation with hair length within a given group of trichobothria (Fig. 15). The situation is complicated by the fact that the best frequency for a particular length can differ from one group of trichobothria to another; for instance, it is higher on the tibia than on the tarsus. Presumably differences in the values of the spring constant S and the damping constant R are responsible for this effect.

Thus the quick, intuitive association of short hairs with high frequencies and long hairs with low turns out to be correct, but that should not mislead us into assuming the wrong reason and thinking only of the resonance of organ pipes, violin strings or the wooden plates of a marimba as analogous. The explanation actually lies in complex relationships of fluid mechanics and has a lot to do with the formation of boundary layers. That the best frequency f of a hair of length L is not given by $L \approx f^{-1/2}$, a relationship to be found in the literature (Fletcher 1978, Tautz 1979), has been demonstrated by both the numerical and the experimental results (Barth et al. 1993).

The Length of the Hair Shaft. Thirty years ago Görner and Andrews (1969) were already able to show that the deflection of the trichobothria of *Agelena labyrinthica* by a constant stimulus increases with the length of the trichobothrium. That this must be the case can be seen from a glance at the basic equations for the hair movement. However, the deflection amplitude does not depend only on the length of the hair shaft, so it is not possible to make generally valid inferences from hair length. A *Cupiennius* tricho-

Fig. 14a,b. Mechanical frequency tuning of trichobothria of *Cupiennius salei*. **a** Tibial group TiDA1; length of hairs I to V – 1150 µm, 700 µm, 650 µm, 500 µm and 400 µm. *Above* Hair deflection at constant air-particle velocity of 50 mm/s. *Below* The air-particle velocity necessary to deflect the hair by 2.5°. **b** Tibial group TiLP and metatarsal group MeD1; length of hairs in TiLP: I – 1150 µm, II – 1050 µm, III – 750 µm, IV – 450 µm, V – 300 µm; in MeD1: I – 850 µm, II – 200 µm, III – 500 µm, V – 300 µm. Deflection angle of hairs always 2.5°. (Barth et al. 1993)

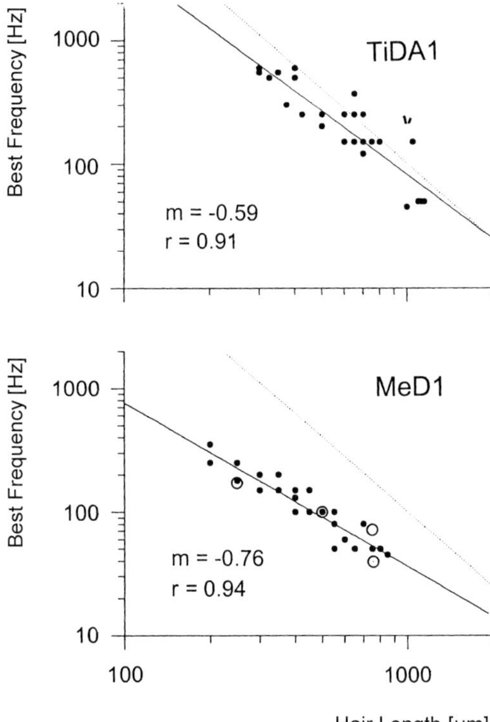

Fig. 15. Mechanical best frequencies as a function of hair length, taking trichobothria of tibial group TiDA1 and metatarsal group MeD1 as examples. *Open circles* in graph for MeD1 represent theoretical values for hair lengths 750 µm (*upper value*: $L_1 + L_2 = 750 + 250 = 1000$ µm; *lower value*: $L_1 = 750$ µm, $L_2 = 0$ µm). *Dotted line* represents calculated boundary-layer thickness for different stimulus frequencies (Stokes flow). *r* Correlation coefficient; *m* reciprocal value of slope of regression line. (Barth et al. 1993)

bothrium of the tibial group DA1 with a length of 1000 µm is deflected by 5° with a particle velocity of 50 mm s^{-1}; the deflection of a trichobothrium only 500 µm long in the metatarsal group D1 or the tarsal group D is significantly larger under the same stimulus conditions. It follows that the hairs of the metatarsus and tarsus are mechanically more sensitive (Fig. 16).

Another expression for sensitivity given in the filiform-hair literature (Fletcher 1978; Tautz 1979) employs the ratio (a) of the maximal hair-tip deflection (x) at the best frequency to the maximal displacement of the oscillating air particles (ξ): $a = x/\xi = \frac{x}{v/2\cdot\pi\cdot f}$ (v, particle velocity; f, best frequency). We measured a for 85 trichobothria and found that out of four groups, only the group LP of the tibia shows a correlation between a and hair length. The highest value found experimentally for a was 1.6, but in 63% of all cases $0.5 < a < 1$, in 22% of cases $1 < a < 2$, and in 15% $a < 0.5$. The calculation of a gave maximal values up to 1.25, which is distinctly below the value of 2 proposed earlier (Fletcher 1978; Tautz 1979) and corresponds better to the experimental finding that in 78% of cases $a < 1$ (Barth et al. 1993).

The best frequencies of a hair calculated for deflection and for the ratio a differ from one

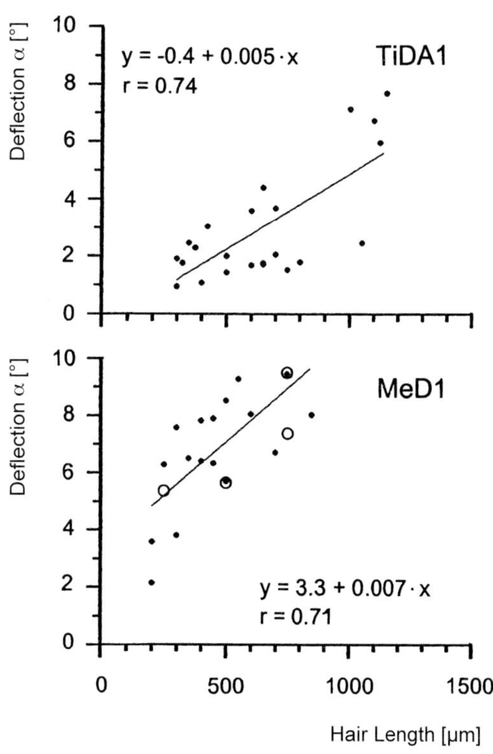

Fig. 16. Deflection angle as a function of hair length, taking trichobothria of tibial group TiDA1 and metatarsal group MeD1 as examples. The trichobothria were each exposed to a particle velocity of 50 mm s^{-1} at their respective best frequencies. *Open circles* in graph for MeD1 represent theoretical values (see Fig. 15). *r* Correlation coefficient. (Barth et al. 1993)

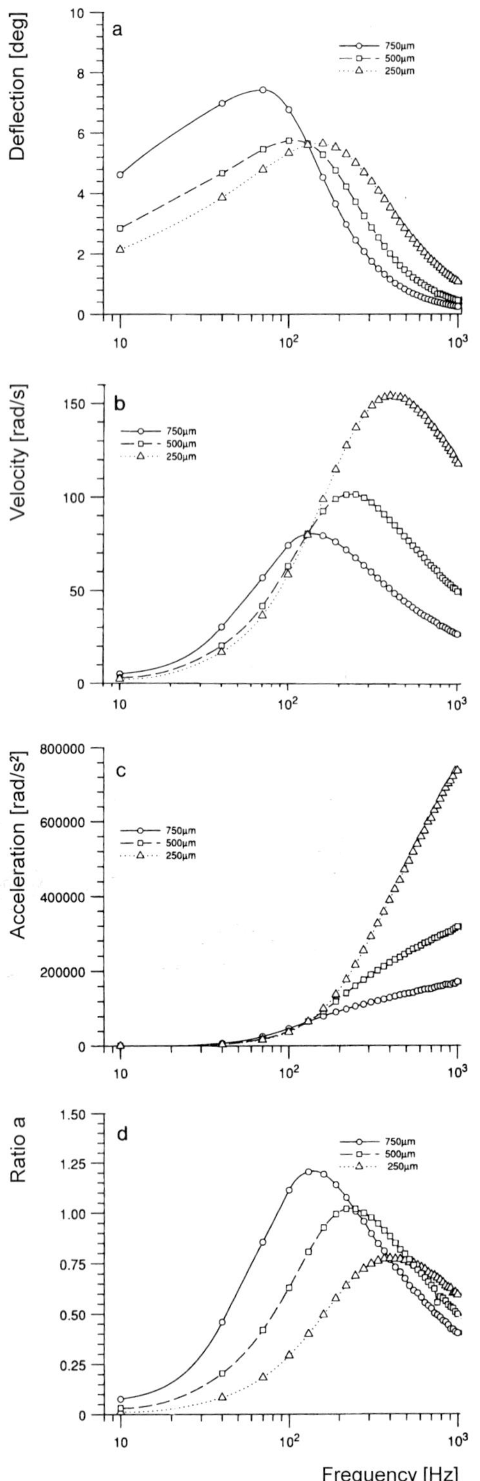

another. The measured values are closer to the first of these (deflection) than the second, as is evident in a comparison between figures 17 and 15: the best frequencies determined by direct measurement of deflection for trichobothria (group MeD1) of lengths 750 µm, 500 µm and 250 µm were ca. 50 Hz, 100 Hz and 300 Hz, respectively.

Groups of hairs with a distinct length gradation are characteristic of *Cupiennius* and are known only for spider trichobothria (see also Palmgren 1936; Harris and Mill 1977; Görner and Andrews 1986; Peters and Pfreundt 1986). One consequence of this arrangement is that in a relatively confined place a broad range of absolute mechanical sensitivities is achieved by assembling the individual ranges of hairs of different lengths. More importantly, however, the range of high sensitivity is expanded, for stimulus frequency and in certain groups (like that on the tarsus) also for stimulus direction. This also answers the question of what a group can do that an individual hair cannot. Presumably the story is still more complicated: there is good evidence that the hairs of a given group oscillate with marked phase shifts, sometimes even in opposite directions, provided their distance from each other is small enough. This gives the group still more functional advantages over a single hair, if the central nervous system can exploit the possibilities; for example, by differential measurement the signal-to-noise ratio and hence the absolute sensitivity could be increased (Humphrey et al. 1993).

From the fluid-mechanical relationships explained briefly at the beginning of this chapter it follows that any hair length "designed" to provide the greatest sensitivity must be related to the thickness of the boundary layer, which changes with frequency: on a spider leg, it var-

Fig. 17 a–d. Calculated dependence of hair deflection (**a**) as well as of its velocity (**b**) and acceleration (**c**) on the oscillation frequency of the airflow. Straight hairs of length 250, 500, and 750 µm and of diameter 5, 6, and 7 µm, respectively. *Graph at bottom* (**d**) shows ratio a (maximal deflection of hair tip/displacement of air particles) as a function of airflow frequency. (Barth et al. 1993)

ies roughly from 2600 µm to 600 µm over the range 10 Hz to 959 Hz (Fig. 18). That is, it is in the same range as the lengths of the trichobothria (100–1400 µm). Both the experimental and the numerical findings indicate that the optimal hair length at the best frequency need not be two to six times greater than the boundary layer thickness (Tautz 1979); in fact, it is not at all greater (Fig. 15). But the principal point remains valid, namely that the trichobothria, like the filiform hairs of insects (Fletcher 1978, Tautz 1979), are mechanical bandpass filters with a special relationship between their low frequency response and the frequency-dependent velocity profiles of the flowing medium (boundary layers). More precisely, the response of a hair depends on the relation between its length and boundary-layer thickness. Thus very short hairs will be insensitive to low stimulus frequencies because they are drowned in a zone of much reduced flow velocities (in a boundary layer large compared to hair length). Whereas this effect is responsible for the low-frequency end of a hair's bandpass properties, the upper limit of its frequency response is determined by the inertia of the hair and of the added mass. Because the moment of inertia increases with the third power of ten of hair length (Shimozawa et al. 1998), the hair becomes increasingly immobile at increasing stimulus frequencies.

A final aspect, closely related to the length of the hair: Shimozawa and Kanou (1984a,b) reported that short hairs are primarily sensitive to acceleration and long hairs, in contrast, to velocity. Measurements of the filiform hairs of the cricket (*Gryllus bimaculatus*) substantially confirmed this view (Kämper and Kleindienst 1990; Landolfa and Miller 1995). Our own calculations, after all factors of the movement equation were taken into account, produced a somewhat more differentiated result (Fig. 17a–c). Short hairs detect velocity as well as or better than long hairs, and at frequencies above ca. 150 Hz short hairs are certainly better acceleration detectors. When the velocity of hair deflection is plotted as a function of frequency, it is evident that hairs of various lengths together form a curve that covers a large frequency range remarkably smoothly. Because the maxima of the different hairs follow one another almost without a gap, there is also a large range of high frequency-discrimination capacity (Fig. 17b).

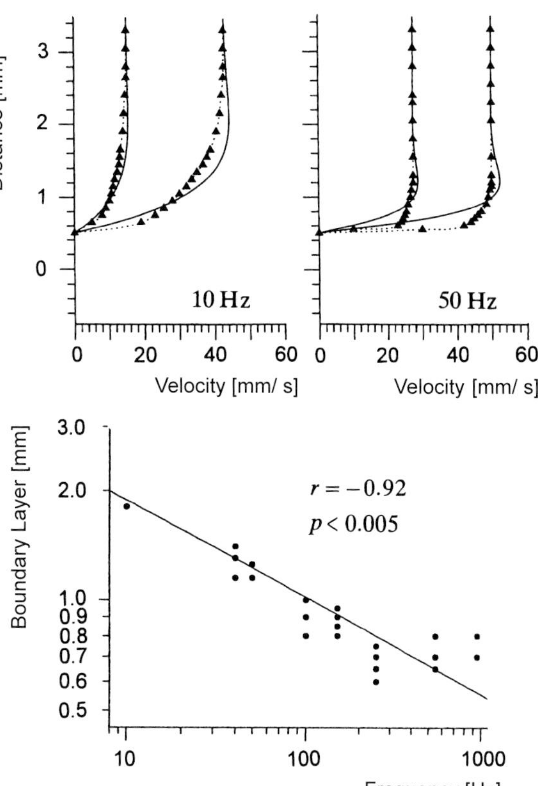

Fig. 18. Velocity profiles and boundary-layer thickness. *Above* Velocity profiles measured dorsally above proximal metatarsus of *Cupiennius salei* at 10 and 50 Hz and at two different free-field velocity (V_∞) values. Orientation of leg parallel to air flow. *Dotted line* Measured values; *continuous line* calculated values (Humphrey et al. 1993). *Below* Measured boundary-layer thickness above metatarsus as a function of stimulus frequency. (Barth et al. 1993)

Surface Structure and Mass. This section concludes with yet another warning against dependence upon intuition when dealing with complex mechanical interrelationships – even if wrong reasoning does lead to the right conclusion.

What functional interpretation can be offered for the striking feathery surface structure of the trichobothria? It points to a form of optimization, for two reasons:

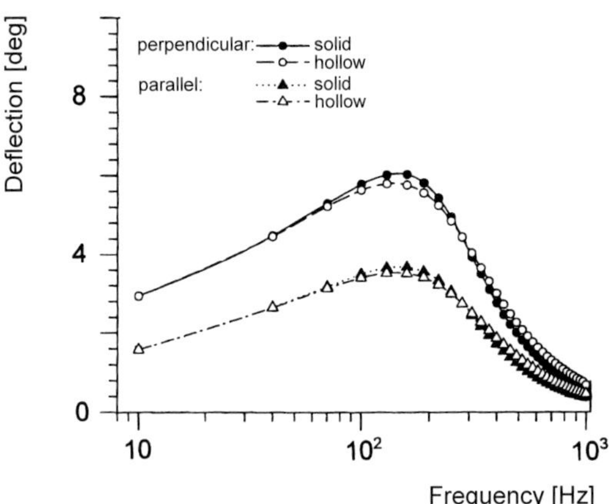

Fig. 19. Independence of deflection of trichobothria from changes of the real mass of the hair shaft, calculated values. The deflections of solid and hollow hairs are almost identical, both in airflows parallel and those perpendicular to the leg axis. The hair modeled here is 500 μm long and straight with a diameter of 7 μm. (Barth et al. 1993)

(a) It increases the drag forces and hence the mechanical sensitivity of the organ, not because the "twigs" projecting from the hair enlarge its overall surface area but because movement of air particles is prevented in the narrow spaces between them and hence the effective hair diameter in the air current is enlarged. The reason is that the current around the trichobothrium is dominated by viscous forces at Reynolds numbers around 10^{-2} and the distances between the individual twigs are extremely small, both absolutely and relative to the thickness of the boundary layer (there are 4 to 14 twigs within a distance of only 10 μm along the long axis of the hair). The viscous forces acting on the air between the twigs keep this air from moving at all (Schlichting 1979).

(b) The second argument for optimization is that the feathery structure reduces the mass of the hair in relation to its effective size. Whereas the diameter of the outer contour of the hair determines the drag forces, the diameter relevant to inertial forces is intermediate between this outer one and the "inner" diameter of the actual hair shaft. However, it has been shown (Fig. 19) that the mass of the hair has no significant influence in the case under discussion here. In any case, calculation of the mechanical frequency response indicated hardly any difference between a hollow and a solid hair. Fletcher (1978) already came to the same conclusion.

3 The Physiology of the Sensory Cells

Mechanical considerations account for a lot, but not everything! In the path leading from air movement to the signal sent to the nervous system, in the form of action potentials, the sensory cells serve as additional filters. The nervous system is not informed about every state or change of state in the spider habitat, and the sensory cells contribute substantially to the selection process.

Electrophysiological recording of the impulse response of individual trichobothria is simplified by the accessibility of the axons, which run just below the cuticle of the leg. Electrodes can be inserted at some distance from a hair, and by deflecting the hairs one by one the sensillum responsible for the recorded activity can be identified.

The most important result is quite straightforward: the sensory cells do not respond to the position of the trichobothrium but to changes in its position. We now know that not only the sensory cells are extremely phasic (rapidly adapting), but also the interneurons linked to the trichobothria (see Chapter XIX). This means that the effective stimuli are changes in velocity of the airstream – irregular, fluctuating stimuli – and not laminar air currents with constant velocity.

The most thorough electrophysiological investigations of spider trichobothria were carried out by Reißland and Görner(1978) with the European house spider *Tegenaria*, a member of the family of funnel-web spiders (Agelenidae). The most important features of the transduction process are as follows (Reißland and Görner 1985):

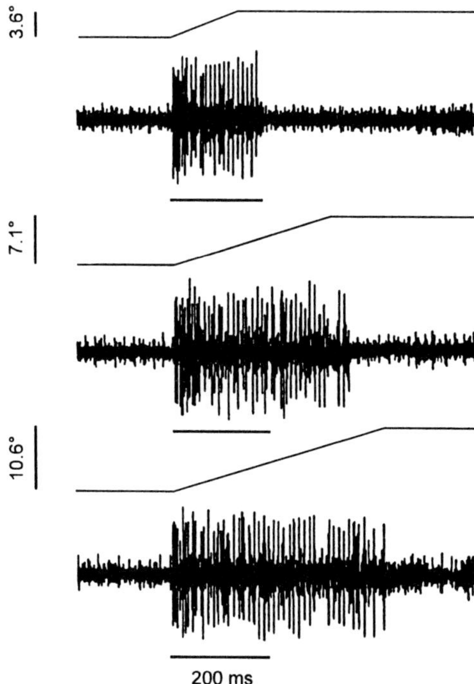

Fig. 20. Impulse response of a trichobothrium of group MeD1 of *Cupiennius salei*. Deflection of hair by a stimulator firmly coupled to its shaft, through increasing angles but at the same deflection velocity. (Barth and Höller 1999)

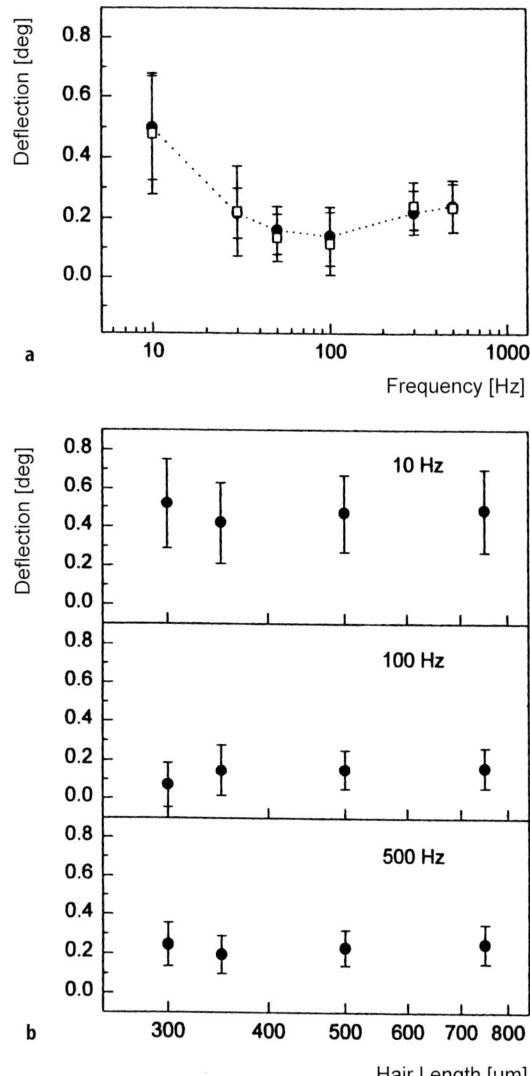

Fig. 21 a, b. a Electrophysiologically measured threshold curves for long (L > 400 μm, *filled circles*) and short (L < 400 μm, *open squares*) trichobothria directly coupled to the vibrator. N = 6, n = 27. **b** Threshold deflection angles in relation to hair length for trichobothria directly coupled to the vibrator, at three different stimulus frequencies. N=6, n=227. (Barth and Höller 1999)

1. Most, if not all sensory cells have a *rectifying action*, inasmuch as they respond only or preferably to deflection out of their resting position but not to movement back into it.
2. *High-pass* characteristics are expressed in the phasic nature of the sensory-cell response; a prolonged deflection is not signalled, for after a stepwise stimulus the impulse discharge soon dies out.
3. The discharge rate of a response in *Tegenaria* is proportional to the logarithm of the stimulus frequency. Stimuli above 200 Hz elicit only one impulse per stimulus cycle.
4. When a sensory cell is stimulated repeatedly, its response adapts: the latency becomes longer and the response duration is reduced.
5. The threshold deflection angle is usually – except for very slow deflections – independent of velocity and amounts to 1°–2°. In a scorpion Hoffmann (1967) found 2°–3°, and for the filiform hairs of the caterpillar of *Barathra* Tautz (1978) gives 2.5°. The threshold angles are thus very similar over the different groups. With regard to deflection velocity, Reißland and Görner (1985) report ca. 1°/s as

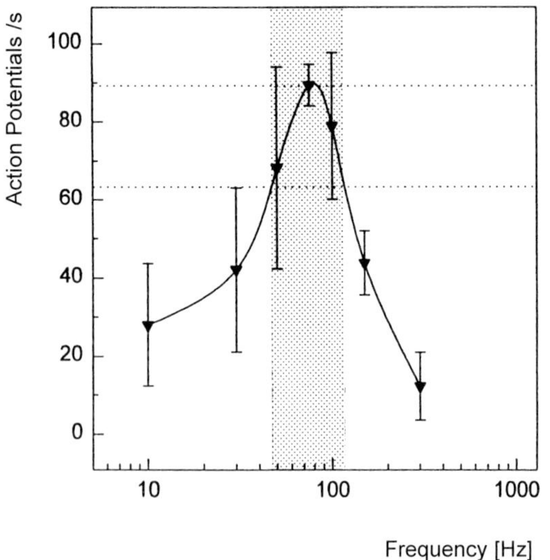

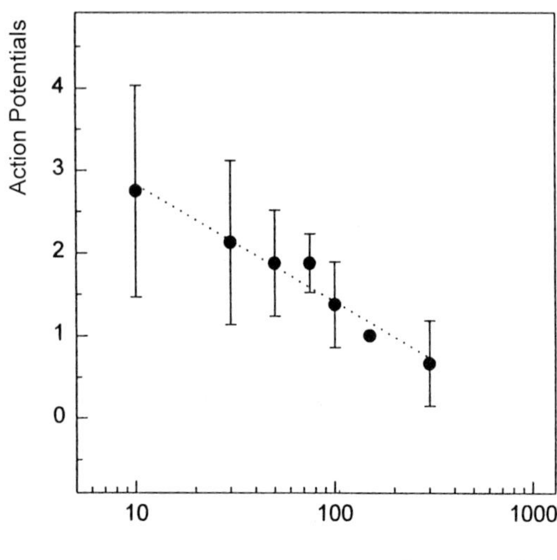

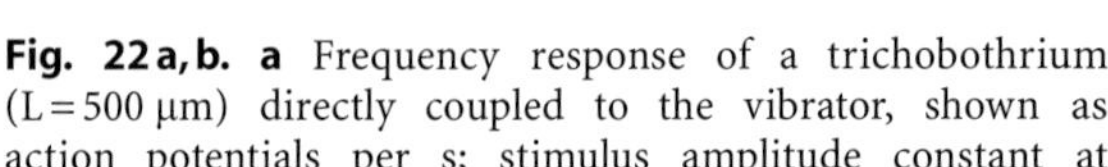

Fig. 22a, b. a Frequency response of a trichobothrium (L = 500 μm) directly coupled to the vibrator, shown as action potentials per s; stimulus amplitude constant at 2 mm/s. The shaded area shows the 3-dB range of maximum response. **b** Number of action potentials per stimulus cycle. N = 4, n = 11. (Barth and Höller 1999)

the lower limit of the working range in *Tegenaria* (response to sinusoidal stimuli partially interrupted).

We have recently also examined single trichobothria of *Cupiennius salei* electrophysiologically (Barth and Höller, 1999).

Here, too, the receptor cells are not spontaneously active and give purely phasic responses to deflection of the hair shaft. This is unsurprising, given that according to their mechanical characteristics, the trichobothria are velocity receptors. Stimulation with ramp deflections revealed that they are also sensitive to acceleration; again, we already knew this (Fig. 20). In the experiments described here, the hair shaft was coupled directly to the stimulator, so that its fluid-mechanical properties were irrelevant. Therefore this mixed sensitivity must reside elsewhere, originating either in the mechanics of the dendrite suspension or in the characteristics of the primary processes at the receptor membrane.

It is interesting that when the trichobothria are directly coupled to a stimulator, their frequency tuning becomes independent of the hair length (Fig. 21). The region of greatest sensitivity is broad and always lies between 50 and 100 Hz. This means that the differences in the frequency tuning of hairs of different lengths observed under natural conditions (no direct coupling to a vibrator) are produced by the fluid mechanics and not by the physiology of the sensory cell. With suprathreshold stimulation (direct coupling to vibrator) the trichobothria are very clearly tuned to the range 50–120 Hz (Fig. 22). These frequencies are important components in the air currents generated by flying insects.

The absolute hair-displacement thresholds for triggering nerve impulses are remarkably low. In *Cupiennius salei* between 10 and 500 Hz stimulus frequency (the tested range) the threshold was less than 1°, and in the most sensitive part of the range, between 50 and 100 Hz, it was only about 0.1° or in some cases even 0.01° (Fig. 21). The trichobothria of *Cupiennius* are thus distinctly more sensitive than those of the spiders and scorpions previously examined. We can now claim that the trichobothria of "our" spider, along with the filiform hairs on the cerci of crickets (Shimozawa and Kanou 1984a, b), are the most sensitive known movement detectors.

When the thresholds are expressed in terms not of deflection angle but rather of the peak velocities reached during this deflection, the trichobothria prove to have pronounced low-pass characteristics, with high sensitivity up to ca. 100 Hz and rapidly rising thresholds beyond that frequency. How the trichobothria handle a natural prey signal and are adapted to detect it will be described in Chapter XIX.

4 Another Medium: Hairs in Water

Hair-shaped movement detectors exist not only in terrestrial arthropods but also in aquatic ones. In fact, the detection of hydrodynamic stimuli by cuticular hairs is widespread and is especially well known in decapod crustaceans (Bleckmann 1994). Our analyses of fluid mechanics (Humphrey et al. 1993) are very general, and can be applied to hairs with various structures and mechanical properties as well as such different media as air and water. Therefore our comparison lets us predict media-dependent differences in the physics of stimulation that would be expected to affect the morphological and mechanical properties of the receptor hairs (Devarakonda et al. 1996).

Briefly, the most important aspects of the comparison are these.

1. The differences in the hair-shaft deflection produced by oscillating movements of the two media are in fact quite large. The most important reason is that the dynamic viscosities of air and water differ, such that in water the virtual or added mass makes the dominant contribution to the effective inertia of the hair.

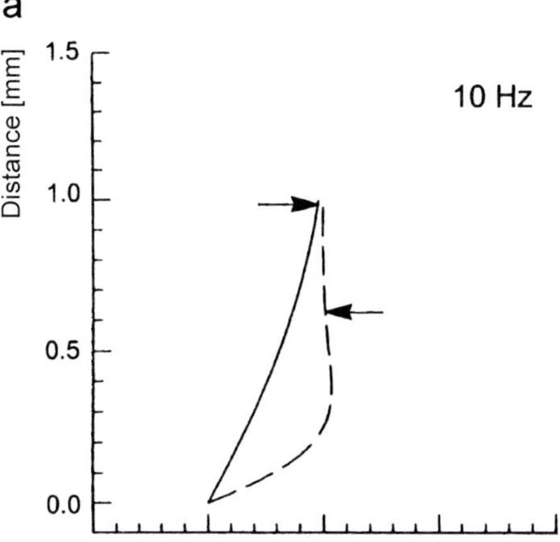

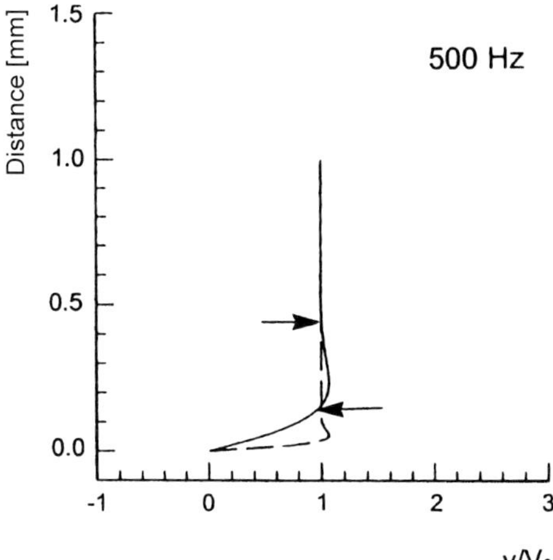

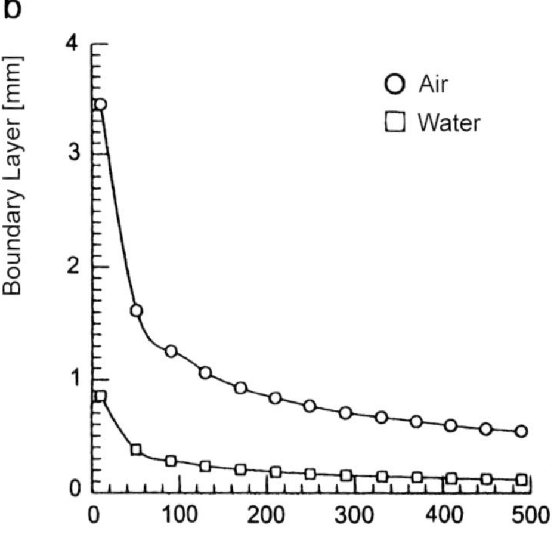

Fig. 23a,b. The effect of the medium on boundary-layer thickness. **a** Velocity profiles above flat surface at time $\omega t = \pi/2$ in air (*continuous line*) and in water (*broken line*) for oscillation frequencies between 10 and 500 Hz. **b** Boundary-layer thickness δ above a flat surface as a function of the frequency of the medium flow in air (*circles*) and in water (*squares*). v/V_0 is the ratio of the given velocity to that of the flow oscillation far from the hair and the substrate. (Devarakonda et al. 1996)

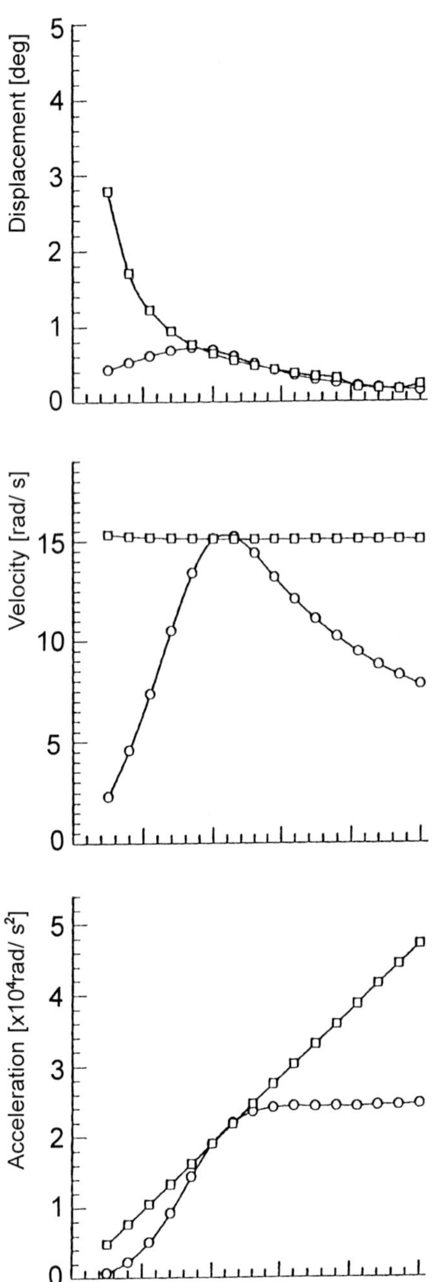

Fig. 24. The effect of the medium on the movement of the trichobothrium in air (*circles*) and in water (*squares*) at different flow oscillation frequencies. Hair parameters: length 500 µm, diameter 7 µm, $S=4\times10^{-12}$ Nm/rad; R=0 Nms/rad. *From above to below* Maximum hair angular displacement, maximum hair angular velocity, maximum hair angular acceleration. (Devarakonda et al. 1996)

2. In water the boundary-layer thickness is smaller than in air, by the factor 0.22 (Fig. 23). The reason: in water the kinematic viscosity (=dynamic viscosity/density) is about 20 times smaller than in air. $\delta = 2.54\,(\nu/f)^{05}$ (δ, boundary-layer thickness; ν, kinematic viscosity; f, frequency). For this reason, and because of the greater drag $D = \rho \cdot S \cdot v^2$ (ρ, density; S, surface area; v, velocity) in water, hairs that function to receive hydrodynamic stimuli would be expected to be shorter than those that operate in air. The hydrodynamic receptors particularly characteristic of crustaceans, called "peg sensilla", are indeed very short (less than 100 µm, mostly 20–30 µm). They remind us of cilia, which detect water movements in many aquatic animals. In contrast, the typical aerodynamic receptors (filiform hairs, trichobothria) are distinctly longer, 100–2000 µm. However, it does not follow that all movement-sensitive hairs in water are so short. There are also long hairs and that is not surprising, because hair length is only one of the sensitivity-determining factors; furthermore, different hairs operate in different sensitivity ranges. What is especially interesting here is that such extremely short flow receptors do occur in water.
3. Hair length has a strong influence on the absolute mechanical sensitivity in both air and water. The hair diameter, in contrast, has hardly any effect in water. Similarly, changes in spring stiffness and damping in water influence the mechanical frequency tuning of a hair considerably less than in air. And according to theory, morphologically and mechanically identical hairs are tuned to substantially lower frequencies, on mechanical grounds, in water than in air. In water they are distinctly low-pass filters, with an operating range between < 1 Hz and ca. 150 Hz (Fig. 24).
4. Our findings lead to the conclusion that the evolutionary pressure to "optimize" morphological and mechanical properties of hair-shaped movement receptors in air must have been considerably stronger than in water. The impressive diversity in form of this particular kind of hair in aquatic crustaceans seems to support this conclusion.

What a precise physical masterpiece a trichobothrium is! So much is already clear, although a great deal remains to be done and discovered, particularly in the realms of cell physiology and central nervous integration.

As in the case of slit sensilla, it is important here to expand our horizons and take into consideration sensory ecology and behavior, for these must have contained important driving forces for the evolution of such remarkable biological measurement devices. In Chapter XIX I shall approach what for most readers will probably be a more lively aspect of air-movement detection: prey capture. Even then, though, the illusion of simplicity will not last long; it will soon become clear that we are only at the threshold of understanding.

X Proprioreception

Although it may seem finicky, I'd like to begin with a small linguistic point. We do not say "ceptor" when referring to receptors, and even when the Latin *recipere* appears in a composite word, we should not truncate it and speak of "proprioception", but rather of proprioreception. The other part of the word also comes from Latin (*proprius*), and the term as a whole means the detection and perception, by way of proprioreceptors, of stimuli that result from an animal's own motor activity (whereas exteroreceptors receive stimuli from outside the animal's body). It should come as no surprise, then, that proprioreceptors are found where such movements occur, especially on the extremities and among them, in turn, especially the legs. On the legs, the proximal part is particularly well equipped with proprioreceptors. Here the number of muscles and the passive mobility of the joints are both large, and active movements in this region must be particularly finely controlled because movements near the body can be amplified in the excursions made by the tip of the leg.

In *Cupiennius* and other spiders there are several extremely diverse kinds of proprioreceptors, which will be considered in sequence below. Much of this description comes from a publication by Seyfarth and Pflüger (1984), who made an exemplary study of the joint between tibia and metatarsus. This is a dorsal hinge joint, in which two dorsolateral joint tubercles confine mobility to substantially dorsoventral extension and flexion movements of the metatarsus, through a range of about 125°. A small third tubercle, positioned asymmetrically between the two large ones, allows the metatarsus to swing sideways through about 10°. Whereas bending of the joint is brought about by two paired muscles (*flexor metatarsi longus* and *flexor metatarsi bilobatus*), extension is based on a hydraulic mechanism (see Chapter XXIV).

1 Slit Sense Organs

Walking Legs

Let's begin with the familiar slit sense organs. At the tibia-metatarsus joint there are no fewer than four, one on the anterior side (VS-4), two on the back side (HS-8 and HS-9) and one on the ventral surface (VS-5) (see Figs. 1, 24 in Chapter VII). As is the case for most lyriform organs, they are situated at the distal end of the proximal segment, the tibia. Altogether they comprise about 25 slits. The smallest organ is the ventral one, with only 2 slits, and the largest is HS-9 on the back side, with 9 slits. Each of the slits in all the organs is innervated by two sensory cells (Fig. 1), one of which is puzzling because no action potentials have ever been detected in extracellular recordings from this cell. Presumably the solution to the puzzle is that its response is so extremely phasic and brief that it is masked by the responses of the other cell. Seyfarth and French (1994) recently showed that such extremely phasic responses do exist, by making the first intracellular recordings from cells of a lyriform organ on the patella. The dendrites of the sensory cells in the organs of interest here are between 36 and 300 μm long, and the spindle-shaped somata (length 20 to 75 μm) are thus as far as 300 μm away from their slits in the cuticular exoskeleton. The axons of the posterior and ventral organs run in a posterior

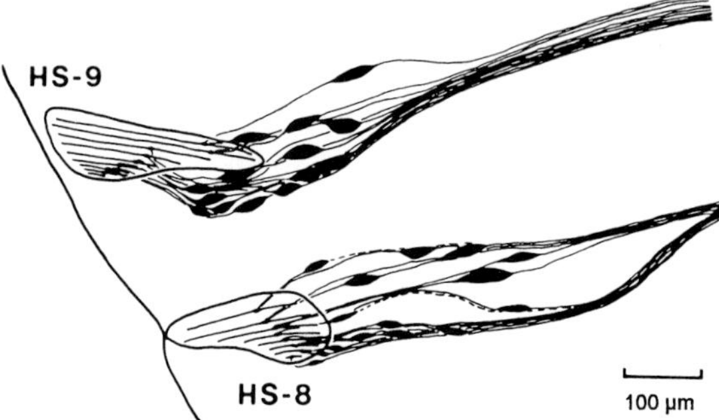

Fig. 1. The sensory cells of lyriform organs *HS-8* and *HS-9* on the walking-leg tibia of *Cupiennius salei* (cobalt stain). Each slit is innervated by two bipolar sensory neurons. In this preparation not all of these cells are stained. (Seyfarth and Pflüger 1984) (with permission of John Wiley & Sons Inc., New York)

sensory nerve, and those of the anterior organ, in the anterior sensory nerve (see Fig. 2, Chapter XV).

We already know from Chapter VII that the lateral lyriform organs HS-8, HS-9 and VS-4 are embedded in a skeletal region where contraction of the joint muscles against a mechanical resistance causes compression of the slits and hence excites the sensory cells. The masterly experiments of Reinhard Blickhan, who was able to measure cuticular strain in freely walking spiders by attaching miniature strain gauges to them, showed us directly that the lateral organs on the posterior surface indeed are stimulated (that is, compressed) during flexion of the joint, which occurs mainly during the support phase

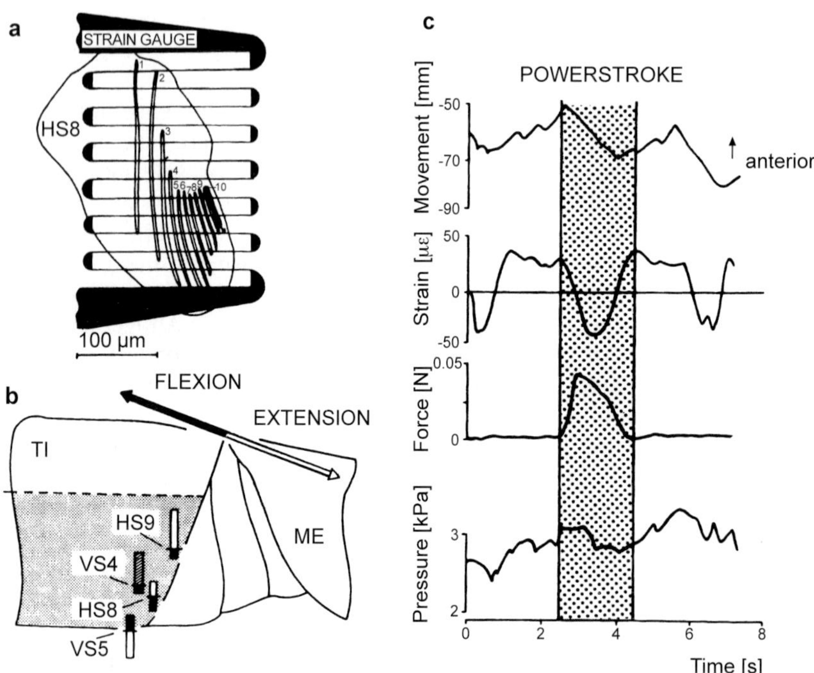

Fig. 2a–c. Cuticular strains at the site of lyriform organs. **a** Miniature strain gauge glued onto the cuticle (here at site of organ *HS-8* on tibia) for in vivo measurements in the freely moving spider. **b** During flexion of the joint between tibia (*TI*) and metatarsus (*ME*) induced by the contraction of the flexor muscles, negative strains leading to slit compression are found at the lyriform organs *VS4* (anterior side) and *HS8* and *HS9* (posterior side of tibia). In contrast, positive strains (slit dilation) are measured at the site of organ VS5 ventrally on the tibia. *Short horizontal lines* represent lyriform organs, the *bars above these* positive strain, the *bars below them* negative strain. During hydraulic joint extension by hemolymph pressure only organ VS5 is compressed. **c** Mechanical parameters measured simultaneously in vivo during slow locomotion of untethered *Cupiennius salei. From top to bottom* Leg movement (leg 4), strain at lyriform organ HS8, ground reaction force (vertical component), and hemolymph pressure. A drop of −80 µε coincides with the power stroke. Hemolymph pressure remains low. (Barth 1985, from Blickhan et al. 1982)

of a step. During leg extension by hemolymph pressure, only the ventral organ (VS-5) is compressed (Fig. 2). We conclude that the presence of several lyriform organs at this joint serves especially to allow the various phases of the stepping cycle to be distinguished from one another, which thus also distinguishes muscular force from hemolymph pressure. The strain amplitudes are similar at all the organs, implying that the strain distribution in the exoskeleton is quite uniform. Under conditions of slow locomotion, therefore, the skeleton should be considered a structure of uniform strength (load-bearing capacity). The organ VS-4 is not compressed during slow locomotion. It may be specialized to signal larger stresses that accompany rapid movement.

These lyriform organs near joints are closely coupled to locomotion, just as are the campaniform sensilla of insects (Zill and Moran 1981a,b; Zill et al. 1981). To our surprise, elimination of lyriform organs nevertheless has little or no effect on normal locomotion. Evidently either the central nervous system plays a highly dominant role in the control of locomotion, or there is a high degree of redundancy in the proprioreceptive equipment at the sensory periphery. As we shall see later (Chapter XXI), lyriform organs do make a major contribution to kinesthetic orientation, which is significantly impaired by the elimination of even a single organ. They are also demonstrably involved in triggering leg reflexes (Seyfarth 1978a,b, 1985; Seyfarth and Pflüger 1984; see Chapter XXIV). The reflex response elicited by excitation of the organs HS-8 and VS-4 is not, as one might imagine, contraction of the intrinsic musculature (which moves the tibia-metatarsus joint), but rather of the pro- or remotor muscle in the patella. This reflex is not one of the ordinary resistance reflexes but a synergistic reflex - an avoidance reflex, to a certain extent - with which the animal withdraws itself from the imposed force (Seyfarth 1978a; see Chapter XXIV).

Spinnerets

No spider without silk! Even in the life of a hunting spider such as *Cupiennius* silk plays a major role, although it is not used to build a web for catching prey.

First consider the dragline. This is a thread that provides a constant link between *Cupiennius* and the substrate, because here and there it is stably connected to the substrate by attachment disks. This situation is reminiscent of a climber who is suspended on a rope and occasionally drives a piton into the rock face. In the case of young spiders the dragline makes possible the "drop-and-swing" behavior (see Chapter XXV). And for an adult *Cupiennius* male the female's dragline points the way to the opposite sex. It carries a pheromone that excites the male during the so-called chemical phase of courtship (see Chapter XX). Finally, it has a remarkable function in another hunting spider, *Dolomedes plantarius*, which stalks insects on the surface of water. It attaches its dragline to the shore and can use it actively for braking - against the inertial forces of its body mass, which tend to drive it forward after it has stopped its active running (Gorb and Barth 1994). The reader interested in the significance of the dragline in the behavior of a web spider can consult a recently published paper on *Nephila* (Gorb et al. 1998).

Cupiennius also uses its silk to construct a cocoon-like egg sac. The first of up to six sacs built after a single copulation is around 2 cm in diameter. Its envelope weighs only about 24 mg and consists of at least 10 km of thread (Melchers 1963b). The female takes good care of her egg sac, carrying it with her at all times, attached to her opisthosoma; it is a valuable cargo, in which as many as 1500–2500 offspring hatch and develop. The mother spider has spun a dense roofing web (see Plate 10) a few days in advance, to close her retreat off from the outside world so that she can proceed to make her egg sacs undisturbed. Spiders transfer sperm indirectly, so that the male must first move his sperm from the opening of the sperm ducts into the receptacles on his palps. To do this, he spins the so-called sperm web. In *Cupiennius* this sperm web is only a few centimeters wide and

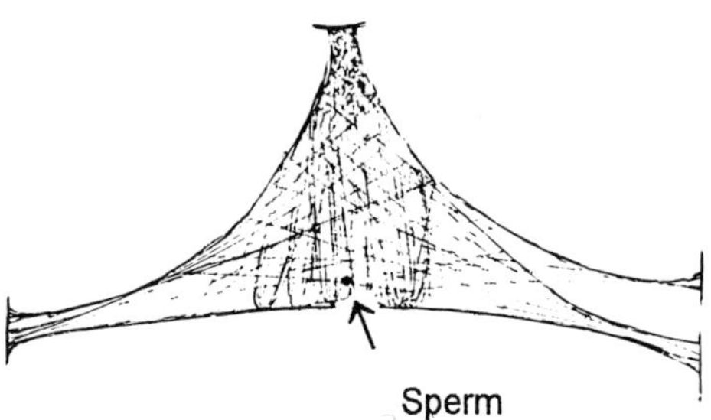

Fig. 3. Sperm web built by a male *Cupiennius salei* (width 4 cm). (Melchers 1963b)

high (Fig. 3). A droplet of sperm about 0.5 mm in diameter is stuck to this web by the male and then taken up by the palps (Melchers 1963b).

Finally, even though *Cupiennius* weaves no prey-trapping web, it does use its silk to immobilize and wrap up animals it has caught, especially if they are large (for instance, cockroaches in contrast to flies). To do this, *Cupiennius* grips its prey with the chelicerae and fastens the dragline to the substrate. Then it swings its body over the prey and attaches the line to the substrate on the other side, pinning the prey down (Plate 7). After this movement has been repeated several times, the victim is criss-crossed by silken threads. Bundles of threads coming out of the posterior spinnerets are not anchored at the attachment disk, but rather stick to the threads already there. Then *Cupiennius* detaches the prey from the substrate with its chelicerae and begins to eat (Melchers 1963, IWF).

Like other spiders, *Cupiennius* will need to check its production and use of silk in many situations, using its sensory systems. We have recently found a complex sense organ that appears to play an important role specifically in the sensory monitoring of the dragline (Gorb and Barth 1996). This composite structure is the product of so-called ampullate glands and is ejected by four large spigots on the anteriormost of the three pairs of spinnerets that *Cupiennius*, like most spiders, possesses. Therefore the dragline usually consists of four threads. Sometimes, though, there are five or even eight threads, which suggests that the median spinnerets occasionally participate in producing it (Fig. 4). The large spigots on the anterior spinnerets are surrounded at the base by a cuticular plate that has a thick edge but is thin and flexible in its central region. In this cuticular plate there are 50–60 sensilla (Fig. 5) that we classify as slit sensilla even though they do not have quite the typical spider structure (see Chapter VII). For one thing, the cuticular lips are lacking, and for another, the ratio of length to width of the slits is so small that they seem more like the campaniform sensilla of insects (usually < 3) than arachnid slit sensilla. However, the innervation by two dendrites of two bipolar sensory cells is typical of spiders (Fig. 6); the campaniform sensilla of insects are supplied by only one bipolar cell, with one dendrite.

The idea that these spigot sensilla on the anterior spinnerets constitute an organ sensitive to strain (or tension), which monitors the movement of the spigots and emission of the dragline, is corroborated by electrophysiological findings: pulling on the dragline initiates activity of these sensory cells (Fig. 7). The large number of individual sensilla and the density of their distribution presumably indicate how important it is for the spider to know exactly what movements the spigots are making and how much tension is being applied to the dragline. Furthermore, the differential arrangement of the spigot sensilla around the bases of the spigots provides a mechanical foundation for analysis of the direction in which the spigots are being deflected.

There is considerable scope for future research here! At present we have made only a very modest beginning, with respect to the motor control of spinning as well as the sensory aspect. Furthermore, other mechanoreceptive sensilla near the spinnerets are presumably involved in the precisely adjusted activities for which silk must be used in various other ways. Because we have no experimental evidence as to their function, these additional sensilla will merely be listed here:

- All three pairs of spinnerets bear numerous hair sensilla on both the proximal and the distal segment (see Fig. 4).
- On the proximal section of each of the anterior spinnerets, and only on their anterior

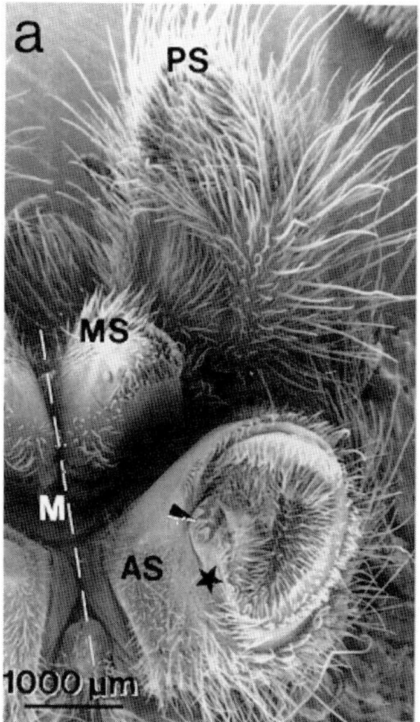

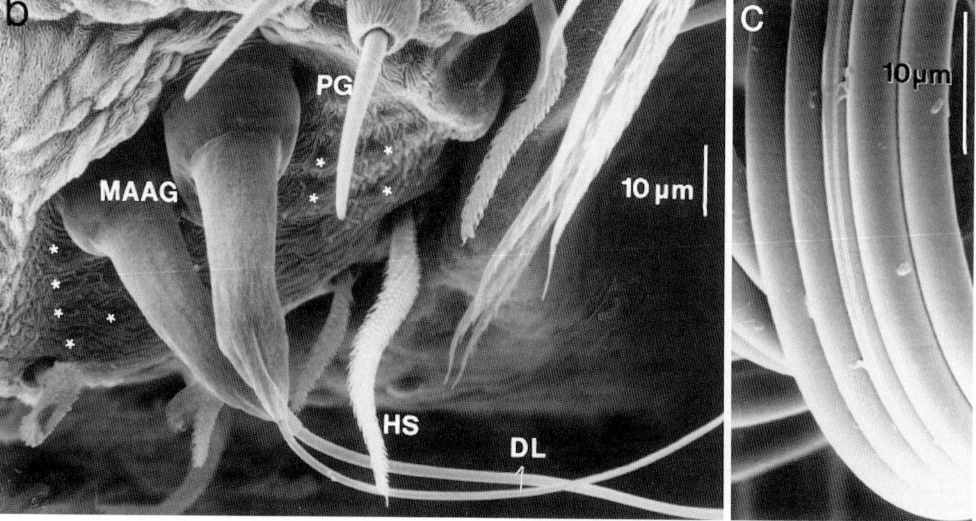

Fig. 4a–c. Spinnerets of an adult female *Cupiennius salei.* **a** The three spinnerets on the right body side. *Arrowhead* points to spigots of the major ampullate glands. *PS, MS, AS* Posterior, median, and anterior spinneret; *asterisk* marks position of a lyriform organ. **b** Spigots of the major ampullate glands (*MAAG*) producing the dragline (*DL*). *PG* Spigot of the piriform gland; *HS* hair sensilla; *small asterisks* mark sites of some slit sensilla. **c** The dragline of an adult female consisting of five threads produced by the spigots of the major (anterior spinneret) and minor (median spinneret) ampullate glands. (Gorb and Barth 1996)

surfaces, there are ca. 10 single slits; on each of the posterior spinnerets, on only the posterior surfaces of both segments together there are ca. 15 single slits (see Fig. 1, Chapter VII).

- The only lyriform organ associated with the spinnerets is at the distal end of the proximal segment of the anterior spinneret (see Fig. 1, Chapter VII). It consists of two slits, each about 40 µm long, which unlike all the single

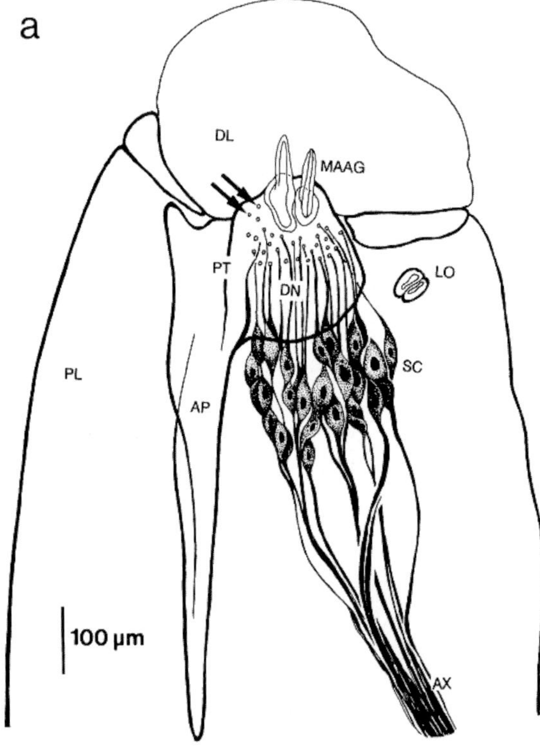

b

slits are oriented not parallel to the long axis of the spinneret but perpendicular to it.

2 Hair Sensilla

A good deal has already been said in this book about cuticular hair sensilla, mostly about the trichobothria, which represent an extreme case with regard to ease of deflection. The legs of *Cupiennius salei* are covered by many thousands of hairs (Fig. 8, Plate 15). The trichobothria are a small minority among the many "tactile" hairs, chemoreceptive hairs and hairs that combine a sensitivity to both of these modalities. Thermo- and hygroreceptive hairs are probably also present but have not yet been investigated. The large bristles or "spines", which are important in the taxonomic classification of *Cupiennius* (see Chapter II), are also innervated. Finally, according to the observations of Foelix (1985a) on jumping spiders it is likely that the scopula hairs on the under surfaces of the tarsi are also equipped with sensory cells; that is, they not only provide mechanical adhesion but also have a sensory function.

Joints

Cupiennius salei has hundreds of hairs that form bridges across the joints and that are deflected and stimulated by joint movement (Fig. 8, Plate 15). Some of these mechanically sensitive hair sensilla at the distal end of the tibia have been investigated by Eckweiler (1983) and Seyfarth and Pflüger (1984). They are stimulated by deflection of the metatarsus; the shaft of the hair

Fig. 5 a, b. Strain detectors on the spigots of the major ampullate glands of the anterior spinnerets (*Cupiennius salei*). **a** The anterior spinneret; *AP* apodeme; *arrows* spigot (*MAAG*) slit sensilla; *AX* axons of sensory cells (*SC*); *DN* dendrites of sensory cells; *DL* distal segment of spinneret; *LO* lyriform organ; *PT* cuticular plate. **b** Strain detectors in dorsal view (scanning electron micrograph). *DN* Dendrite attachment site. (Gorb and Barth 1996)

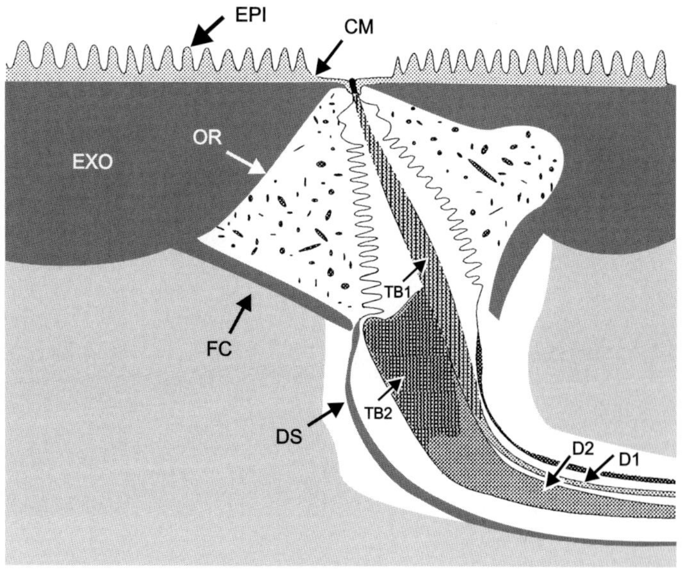

Fig. 6. Schematic representation of single spigot slit sensillum in longitudinal section according to electron microscopic analysis. *CM* Covering membrane; *D1* and *D2* dendrites of sensory cells; *DS* dendritic sheath; *EPI* epicuticle; *EXO* exocuticle; *FC* fibrous cap; *OR* outer receptor lymph space; *TB1* and *TB2* tubular bodies in tip of dendrites. (Gorb and Barth 1996)

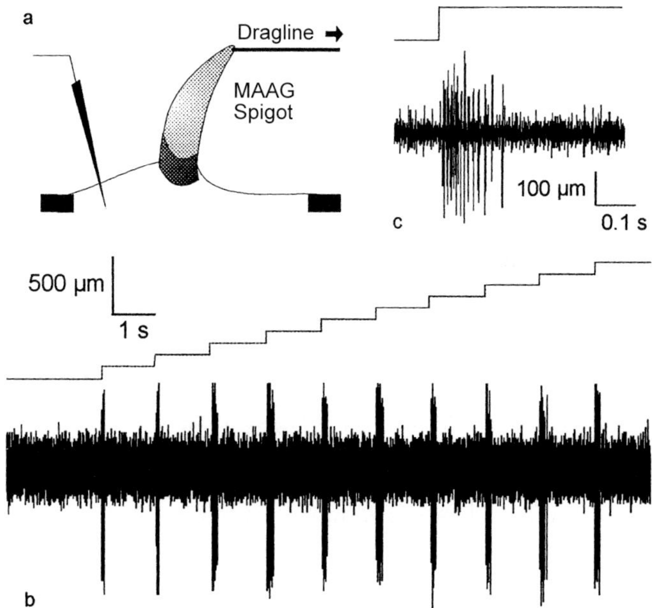

Fig. 7 a–c. Afferent activity of the spigot sensilla in *Cupiennius salei* in response to step-like pulling on the dragline directed posteriorly. **a** Diagram showing electrode position (*left*) and direction of dragline displacement. **b**, **c** Responses of single sensilla to single stimulus and stepwise increasing stimulus, respectively. (Gorb and Barth 1996)

is almost at a right angle to the cuticular surface, which distinguishes these from other sensory hairs. The axons of their sensory cells come together with those of the other joint proprioceptors to form one of the two small, purely sensory lateral nerves (Fig. 8; see also Chapter XV). Like tactile hairs of other spiders (Foelix and Chu-Wang 1973a; Foelix 1985a; Harris and Mill 1977a), they have a threefold innervation; some are innervated by even more than three sensory cells, as has been shown by staining with cobalt sulfide (backfills of sensory nerves).

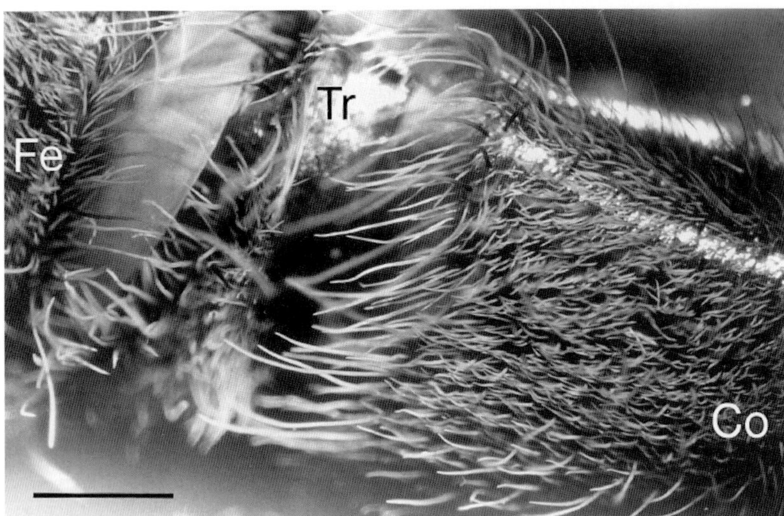

Fig. 8. Mechanoreceptive hair sensilla on the walking leg of *Cupiennius salei;* ventral view. *CO* Coxa; *Tr* trochanter; *Fe* femur. Bar 0.5 mm

Seyfarth and Pflüger (1984) inferred that these bridge hairs were combined mechano- and chemoreceptors, such as had been described for other spiders by Foelix and Chu Wang (1973b) and Harris and Mill (1977b). When they are touched, the spider gives an avoidance response by activating muscles that move the leg away from the stimulus. This reaction always seems to be limited to the affected leg (Seyfarth and Pflüger 1984; Seyfarth 1985; see also Chapter XXII). It is clear that we still do not know nearly enough about the function of the proprioreceptive hairs.

Hair Plates

Another type of proprioreceptor had long been a well-known feature of insects (Pringle 1938, Markl 1962; Pflüger et al. 1981) but was unknown in spiders until Ernst-August Seyfarth (Seyfarth et al. 1985, 1990) discovered it, first in *Cupiennius salei* and then in all the other spiders he examined in this regard. This is the hair plate (Fig. 9). Hair plates are groups of short (25–200 μm) hairs, curved slightly toward the distal end of the leg and located at the proximal end, at the joint between coxa and prosoma. On each leg there are two such fields, one on the anterior side of the coxa (measuring ca. 800×300 μm), and the other on the posterior ventrolateral surface (ca. 700×400 μm). The number of hairs in the anterior group increases from the first leg (ca. 50) to the fourth leg (ca. 70). The number in the posterior group varies in all legs between 27 and 43. These hair plates are stimulated when the pleural membrane "rolls over" the hairs during leg movement and thereby deflects them. In the anterior hair plate this happens when the coxa is moved forward or down, whereas the posterior field is stimulated during backward movement of the coxa and partly when it moves upward.

In contrast to all the other hair sensilla of spiders, the long smooth hairs on the coxa excepted (Fig. 11), each hair of the hair plates has only a single, bipolar sensory cell. Seyfarth, Gnatzy and Hammer (1990) investigated its physiology and fine structure. It is not spontaneously active and responds only when the hair is deflected in the distal direction, with a threshold deflection of 0.5° to 1°. Its response is slowly adapting. A stepwise stimulus elicits a discharge peak, and if the deflection is maintained, the discharge rate slowly declines. As in the slit sensilla (Bohnenberger 1981; Barth 1967) and many other receptors (including some responsive to other than mechanical stimuli) the excitation decreases according to a power function with k values between –0.35 and –0.65 (Fig. 10). These

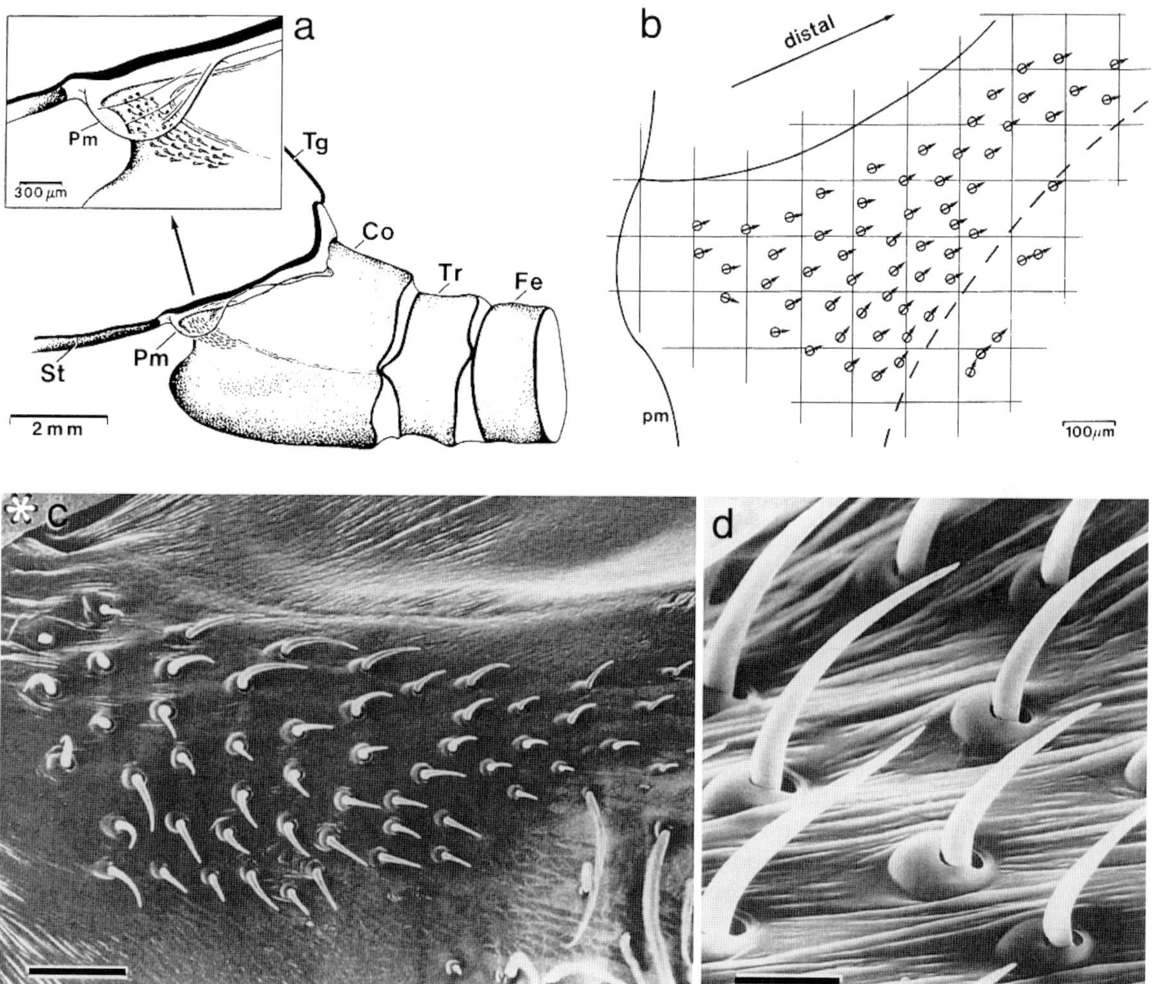

Fig. 9 a–d. Hair plates close to the proximal end of the walking-leg coxa of *Cupiennius salei*. **a** Anterior view of the prosoma – coxa joint (left walking leg 2). *Inset* Enlarged view of hair-plate region. The hairs are sequentially deflected when the soft pleural membrane (*Pm*) rolls over the hair plate during forward movement of the coxa. A similar hair plate is found on the posterior side of the coxa. *Co* coxa; *Tr* trochanter; *Fe* femur; *Tg* tergum; *St* sternum. For clarity, hair shafts and hair arrangement within the plate are drawn much larger than would correspond to the scale bar. **b** Arrangement of hairs and orientation of socket grooves, which also indicate the direction of preferred hair displacement (*arrows*). During forward movement of the coxa the sensilla are stimulated because the pleural membrane rolls over the hair plate from proximal to distal. **c, d** Scanning electron micrographs of a hair plate on the anterior side of a walking-leg coxa. *Asterisk* in **c** marks pleural membrane; *bars* are 100 µm (**c**), and 20 µm (**d**). (Seyfarth et al. 1990)

hair-plate hairs resemble slit sensilla (see Chapter VII) and differ from hair-plate sensilla of insects (Thurm and Küppers 1980; French 1988) in that no standing potential has been found. This lack is yet another indication that the primary processes in the cuticular sensilla of spiders, unlike those in insects, do not depend on such a transepithelial potential.

The many hairs in a hair plate form a composite organ, as do the slits of a lyriform organ. Because the hairs are deflected and stimulated in succession as the joint membrane moves, the

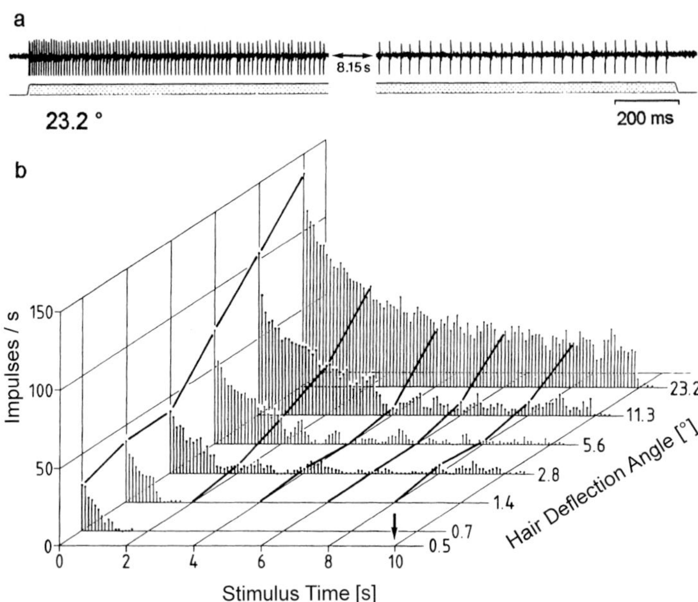

Fig. 10 a, b. Action-potential responses of coxal hair-plate sensilla to steplike deflection (maintained for 10 s). **a** Impulse discharge in response to a deflection of hair shaft by 23.2°. Response only to distalward deflection of hair; note slow adaptation. **b** Mean step responses of hair-plate sensillum to six different stimulus amplitudes; peristimulus-time histograms, bin width 0.1 s; averages of ten consecutive responses. Threshold deflection for the hair studied here was 0.5° from resting position. *Arrow* points to end of stimulus; *bold lines* connect frequency values at equidistant times during stimulation. (Seyfarth et al. 1990)

working range of the hair plate is much larger than that of an individual hair. Each hair is responsible for signalling a particular region of the overall range of possible joint angles.

Insects use their hair plates for diverse forms of behavior: to measure the position of a joint during walking, to monitor body posture, to analyze the configuration of the substrate and to detect gravity (Seyfarth et al. 1990; Markl 1962; Wendler 1964; Wong and Pearson 1976; Bässler 1983; Schmidt and Smith 1987). When the hair plates of *Cupiennius salei* were made nonfunctional by Seyfarth et al. (1990), the behavioral changes they observed were not very dramatic. The normal walking pattern (see Chapter XXIV) persisted unchanged, and the spiders continued to rest in substantially the same, typical posture (prosoma downward when on a vertical surface). It appears that even without functional hair plates, spiders obtain sufficient information from other proprioreceptors. It would be worthwhile to take a closer look at the interplay of all the proprioreceptors.

Coxal Hairs, Yet Again

Now we return to the coxa. This is a prime example of the concentration of proprioreceptors at the proximal leg joints, which are much more movable than the distal joints and therefore are particularly important in locomotion. Here slit sensilla are assembled in unusually large numbers (see Chapter VII), especially at the trochanter, and the hair plates are found only at the prosoma-coxa joint.

As another type of proprioreceptor to monitor movements in the proximal part of the leg, rows of 6 to 11 long hairs are disposed on the anterior and posterior surfaces of the walking-leg coxae, so that they project into the space between adjacent walking legs (Eckweiler et al. 1989) (Fig. 11). These hairs are stimulated when the shaft is deflected by contact with the neighboring coxa, which is completely hairless at the contact site. They can easily be distinguished from all the many other hairs by the fact that they stand nearly perpendicular to the surface of the cuticle, at an angle of 60°–70°. In the many other tactile hairs, this angle is only 15°–30°. The coxae of the pedipalps lack these hairs, nor are there any on the posterior surface of the coxa in the fourth pair of legs.

The length of the hair shafts increases in the proximal or dorsal direction within a group (Fig. 11), from about 40 μm to 1000 μm. In contrast to the many other tactile and chemoreceptive hair sensilla, the hair shafts are characterized by an unusually smooth surface. On one side of the hair socket is a groove, which determines the directional characteristic of deflection of the hair. The orientation of the groove corresponds to the direction in which the hair is deflected at leg contact. As shown in Fig. 12, the smooth, long coxal hairs are arranged in a characteristically different pattern on each coxa.

Like the hairs in the coxal hair plates, the long smooth hairs are each provided with only one bipolar sensory cell, which is atypical for spiders and corresponds to the situation in insects. The sensory cell is not spontaneously active in the unstimulated state. With a stimulus

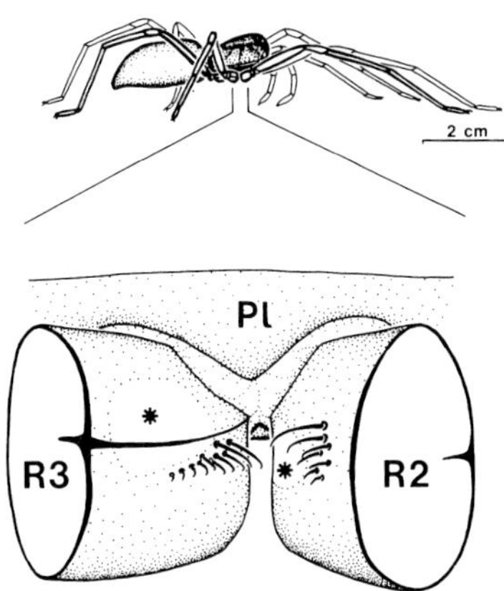

Fig. 11. Position of long smooth hairs on the coxae of two neighboring walking legs (*R2* and *R3*) of *Cupiennius salei.* Hairs on the anterior side of coxa (*R3*) form a line roughly parallel to the long leg axis. In contrast, the row of hairs on the posterior side of the coxa (*R2*) is roughly perpendicular to it. Both groups of hairs lie opposite a region of smooth cuticle without hairs (*) on the adjacent coxa. *Pl* Pleural sclerite of the prosoma. (Eckweiler et al. 1989)

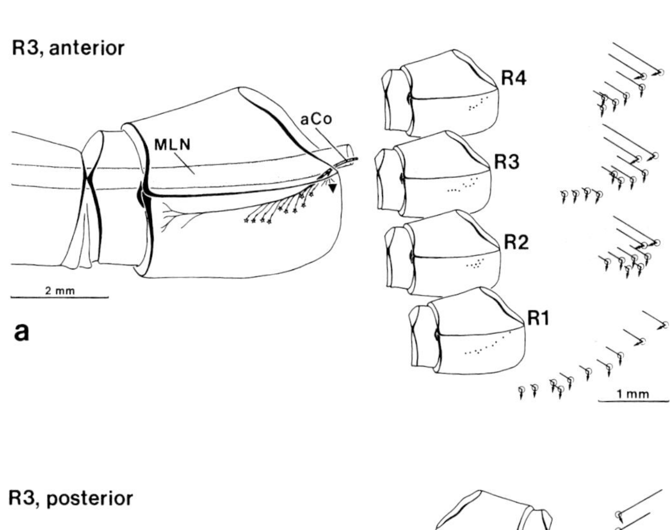

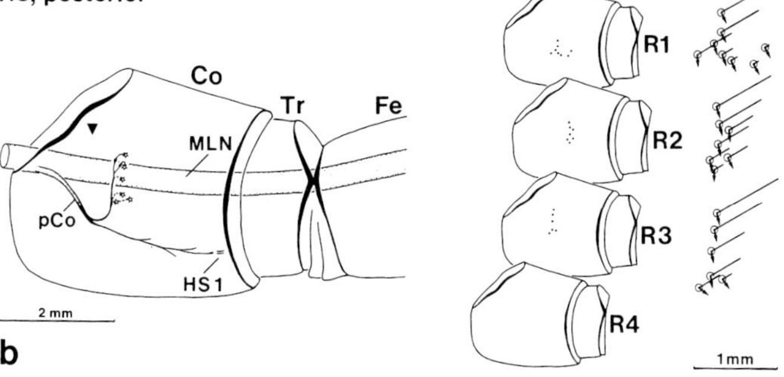

Fig. 12a,b. Topography and innervation of long smooth coxal hairs of *Cupiennius salei.* **a** Anterior side of coxa. **b** Posterior side of coxa. *Left* Innervation of hairs of third leg on right side (*R3*) as seen after cobalt filling. *Asterisks* show location of hair sockets. *Center* Arrangement of long smooth hairs (●) on the coxae of all four walking legs (*R1* to *R4*). *Right* Length distribution of hair shafts in each group; *arrows* indicate direction of hair deflection under natural stimulus conditions. *Inverted triangle* in **a** and **b** (*left*) marks position of coxal hair plates; *Co* coxa; *Tr* trochanter; *Fe* femur; *aCo* and *pCo* anterior and posterior nerve branch in coxa; *MLN* main leg nerve; *HS1* lyriform organ. (Eckweiler et al 1989)

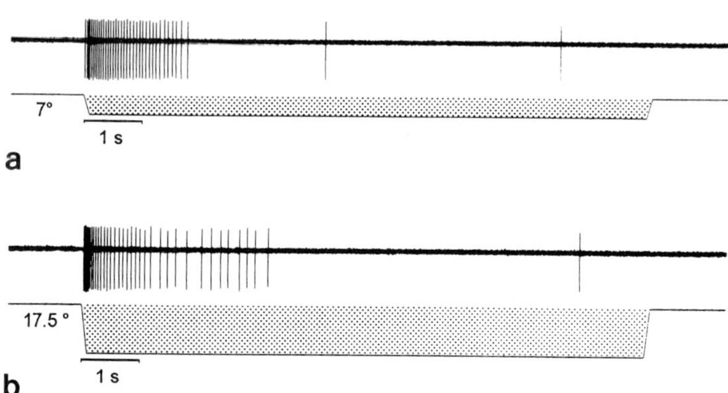

Fig. 13 a, b. Electrophysiologically recorded response of a long smooth hair on the coxa to step-like deflection from its resting position. **a** Deflection 7°, **b** Deflection 17.5°. Only distalward hair deflection (stimulus trace *downward*) elicits action potentials. Response ceases 4 s after stimulus onset. (Eckweiler et al. 1989)

of abrupt onset (rise time 10 ms) the hair must be deflected by 1° to 2° in order to trigger 1 or 2 action potentials. Only deflection in the direction of the groove in the cuticular socket is effective; the movement back into the resting position is not. The response is in the category "slowly adapting", beginning with a phasic peak discharge and declining to zero in a matter of seconds if the deflection is maintained (Fig. 13). The long, smooth coxal hairs thus have a distinctly more phasic behavior than the nearby hair-plate sensilla, the discharge of which lasts longer than 10 s in response to comparable stimulation (Seyfarth et al. 1990).

No special physiological experiments on the proprioreceptive function of the long smooth coxal hairs have been carried out. There is no doubt, however, that they detect contact between the coxae of two adjacent walking legs and signal the relative movements between the neighboring coxae. As the authors (Eckweiler et al. 1989) rightly infer, even during normal, undisturbed locomotion the hairs at the closely spaced coxae are very probably stimulated, particularly since the pro- and retraction of the leg is associated with a considerable rotation of the proximal joints around the long axis of the leg (Kästner 1924; Seyfarth 1985). An involvement of the long smooth coxal hairs in the coordination of walking-leg movements has not yet been directly demonstrated (for example, by elimination experiments). Like the hair plates, the long smooth coxal hairs are no specialty of *Cupiennius salei.* They have also been found, in very similar form, in a number of other spider species including *Nephila clavipes*, *Dolomedes triton*, *Heteropoda juglans*, *Isopoda immanis* and *Araneus* sp. (Eckweiler et al. 1989).

3 Internal Joint Receptors

The absence of chordotonal organs, which are numerous in insects and crustaceans, is a special feature of the spiders. Sensilla with scolopales are known neither in spiders nor in other arachnids. However, spiders do have internal joint receptors. Their number and topography (Rathmayer and Koopmann 1970), fine structure (Foelix and Choms 1979) and electrophysiological properties (Rathmayer 1967) have been described in several publications. According to the evidence available for spiders of quite different taxonomic groups it can be concluded that the picture is generally similar, so that the basic properties of the internal joint receptors of *Cupiennius* can be inferred from those known for other spiders. The receptors inside the tibia-metatarsus joint of *Cupiennius* have been studied by Seyfarth and Pflüger (1984) and are included in the inventory of proprioreceptors at this joint (Fig. 14).

All the joints of both the walking legs and the pedipalps are provided with at least two groups of 3 to 5 multiterminal cells. A given walking leg thus includes 18 groups with a total of about 135 sensory cells (Fig. 15) (Rathmayer and Koopmann 1970). The dendrites of these sensory cells branch in the hypodermis below the joint membrane, and together they form an extensive flat network there (a dendritic fan), which is penetrated by hypodermal cells. Its position in the immediate vicinity of the joint heads, where the forces are transmitted from one segment to the next, at first appears sensible but needs some further discussion. It is not surprising that, once again, the most proximal joints of the leg are particularly well equipped with this type of proprioreceptor as well. The largest number (namely 5) of these internal joint receptors is found in the coxa-trochanter joint, a ball-type joint with particularly great mobility.

The most exciting result of the fine-structural analysis (Foelix and Choms 1979) was the discovery of many synapses on the somata and dendrites of the sensory cells. There is every reason to regard these as efferent. Hence the receptor cells are presumably controlled by the central nervous system, by way of efferent fibers. Foelix and Choms (1979) mention briefly that they also found similar synapses in slit sensilla and hair sensilla. Since then, this finding has been multiply confirmed. Ruth Fabian and Ernst-August Seyfarth (1997) presented strong immunocytochemical evidence that both in the patellar lyriform organ VS-3 and in tactile hairs on the walking leg, acetylcholine and histamine are the transmitters at such synapses. It also turns out that the peripheral synapses are more numerous on the axons of the sensory cells than on the somata (Fabian et al 1997, 1998). However, we still do not know whether and how the sensitivity of the sensory cells is affected by these synapses. It is also not clear in what proportions the synapses are formed by fibers that mediate efferent control, as compared to those coming from other, nearby sensory cells. Both the presence of efferent connectivity so far peripheral and the fine structure of the synapses (Fig. 16) have not been documented in any arthropods but the spiders and other arachnids (Foelix 1985b).

The electron microscope has also revealed that the sensory cells of the joint receptors have neither a ciliary region nor a tubular body. A ciliary region is typical of cuticular sensory cells, including those with functions other than mechanoreception. We have already encountered

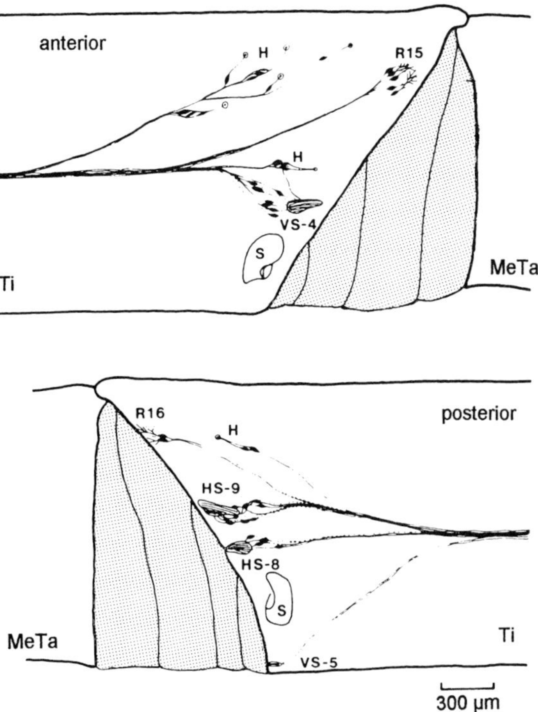

Fig. 14. Innervation of various types of mechanoreceptors at the tibia-metatarsal joint of a walking leg of *Cupiennius salei* (*top* anterior side; *bottom* posterior side) as seen after retrograde cobalt-filling of the fibers in the lateral sensory nerve. *Ti* Tibia; *MeTa* metatarsus. Lyriform organs: *HS-8*, *HS-9*, *VS-4*, *VS-5*; internal joint receptors: *R15*, *R16*. Axons of several hair sensilla are stained as well (*H*); *S* socket of large bristle. On each side of tibia all afferent fibers gather in one bundle ascending to the central nervous system. Not all sensory cells are stained in typical preparations like these. (Seyfarth and Pflüger 1984) (with permission of John Wiley & Sons Inc., New York)

tubular bodies in the discussions of slit sensilla and trichobothria: they are the most important modality-specific structure, and in arthropods their presence indicates a mechanoreceptive function just as reliably as a border of microvilli indicates the function of photoreceptors. In their lack of a tubular body the sensory cells of the internal joint receptors approach the category of multipolar neurons that in other arthropods have been described as proprioreceptors (Pringle 1963). Unlike the multipolar cells, however, these sensory cells have only one dendrite, which arborizes distally.

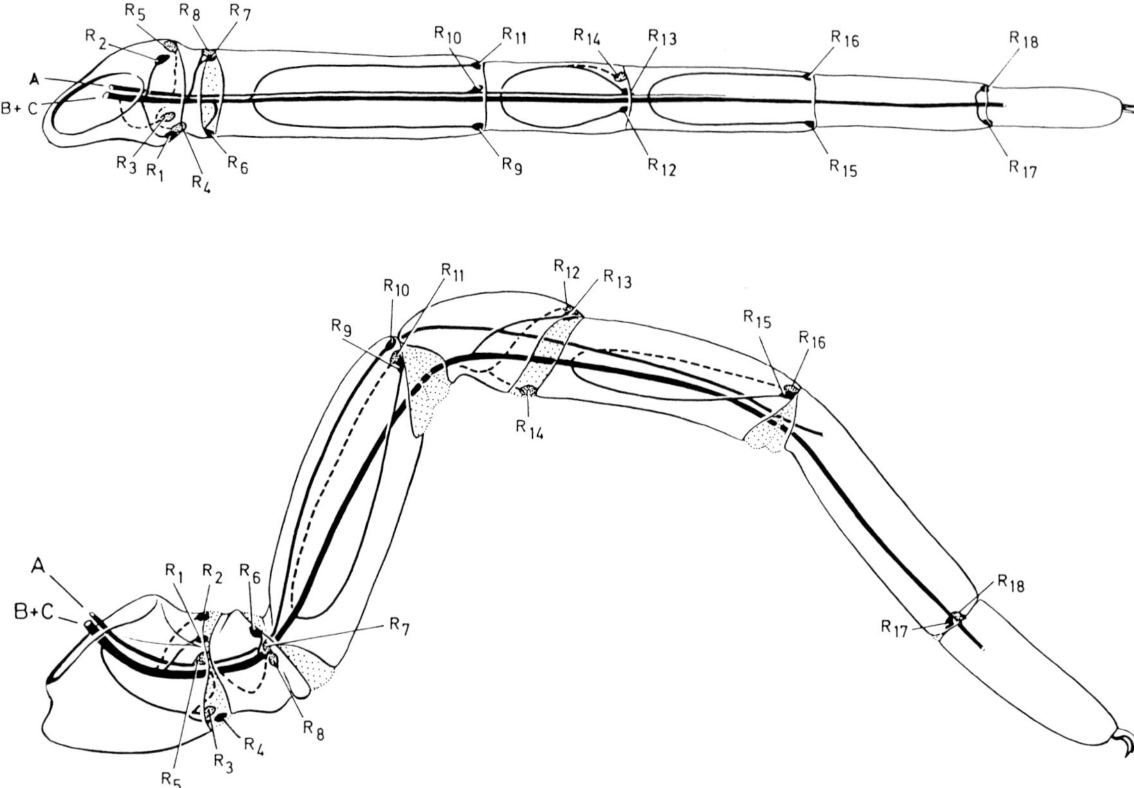

Fig. 15. Proprioreceptors in the leg of a bird spider, *Dugesiella hentzi*. Dorsal (*top*) and lateral (*bottom*) view. The different groups of sensilla at the different joints are numbered consecutively from proximal to distal (R_1 to R_{18}). *A*, *B*, and *C* are the three leg nerves. (Rathmayer and Koopmann 1979)

The question of the effective stimulus to the internal joint receptors can be answered simply and well only at the level of gross anatomy: they respond to the movement and position of the joint. In *Dugesiella* most of the cells were active during extension and only a few, during flexion of the joint (Rathmayer 1967). In *Ciniflo* Mill and Harris (1977) found that at a given joint they were divided into two groups, one responsive to flexion and the other excited by extension.

The events involved in transmission and transformation of the stimulus are largely unknown and have never been investigated as meticulously as might be wished. It seems plausible that the receptors are situated directly at the pivot points of the joints and hence at sites of maximal stress and strain, but this does not bring us much further toward an explanation. If we assume that the crucial factor is not strain ($\Delta l/l_0$) but rather the displacement of the dendritic fan, then it would make sense for them to be further away from the pivot points. Should we therefore conclude from their actual position that such displacement is unimportant? This does not seem likely, particularly in the case of an internal joint receptor with no tubular body.

Foelix and Choms (1979) studied an organ at the femur-patella joint (see R_{10} in Fig. 15) of *Zygiella x-notata* (Araneidae), and found that when the joint was extended, the joint membrane near the organ was displaced by about 40 μm toward the hypodermis. During this displacement, the sensory ganglion remained in the same place, and the length of the dendrite was also unaffected. However, the angle between microtubule bundles in the hypodermal cells and

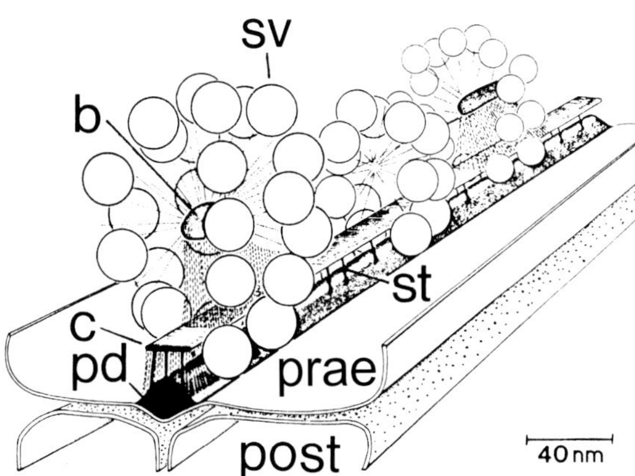

Fig. 16. Three-dimensional diagram of a presumably efferent synapse typical of the soma and dendrites of sensory joint receptor cells. Presynaptic side (*prae*) is highly structured; a triangular presynaptic density (*pd*) extends via fibers to an electron-dense bar (*b*). Filaments radiate from the dense bar and extend to the synaptic vesicles (*sv*). Above the triangular density there is a kind of covering membrane (*c*), which is connected to it by thin electron-dense struts (*st*). (Foelix and Choms 1979)

the dendrites decreased from 50° to 30°. The speculation: the dendrites, already flattened, are still further compressed by membrane displacement.

Whatever the effective stimulus to the dendrite may ultimately be, its effect on the conducted activity of the sensory cell has been observed in electrophysiological experiments (Rathmayer 1967; Mill and Harris 1977). Various cells are excited either during extension or flexion. Some give phasic-tonic responses and differ from one another in their levels of steady excitation; these have different working ranges and are suitable to function as position detectors. Other cells are pure movement detectors, which are excited either during flexion or during extension, to a greater degree, the higher the velocity of joint movement.

What do the internal joint receptors contribute to behavior? The response seems simple at first: they help control locomotion, of course! But on closer consideration this does not tell us very much. Is the information provided by the internal joint receptors necessary for the coordination of movement during "normal" walking, is it used for orientation, does it play a role only in the fine adustment of movements when obstacles are encountered or during complex courtship and threat behavior? All these questions remain open.

So far the only reliable conclusion to be drawn is that internal joint receptors at the tibia-metatarsus joint are involved in triggering a resistance reflex in the flexor muscles (*flexores metatarsi*) (Seyfarth and Pflüger 1984). The most interesting thing here is that the other proprioreceptors at the same joint, the slit and hair sensilla, are not needed to elicit this intrinsic reflex (see Chapter XXIV).

4 Muscle Receptor Organs

In view of all the preceding considerations about the abundance and diversity of proprioreceptors, it seems quite clear that *Cupiennius* and its fellow spiders in general must have available a highly detailed "picture" of their movements. With all these receptors firing in concert, it must be a real symphony of signals that surges into the central nervous system. And it may be that one or more additional types of proprioreceptive sensilla play in this concert.

We can now safely say that spiders and other arachnids have no scolopale organs. The existence of muscle receptor organs, however, remains a matter of discussion. Such organs have never been definitely demonstrated, but several lines of evidence place them in the realm of reasonable speculation. The legs of *Tegenaria atrica* (Parry 1960), *Dugesiella hentzi* (Ruhland and Rathmayer 1978) and *Cupiennius salei* (Seyfarth et al. 1985) are known to contain small muscles with fiber diameters of 18 to 30 μm, substantially less than those of the normal leg musculature. In *Cupiennius* the fiber bundle is ca. 1.2 mm long and 0.1 mm in diameter; it originates in the connective tissue at the distal patella, bridges the joint with the tibia and inserts proximally at the *flexor metatarsi longus*. Unfortunately, no sensory cells have yet been unequivocally associated with this muscle.

XI The Eyes

Spiders are not the first group of animals that would come to mind if we were asked which arthropods have especially highly developed vision. We would be much more likely to think of the large compound eyes of a dragonfly, the color vision of the honeybee or the ability of the desert ant to detect polarization. Nevertheless, spiders too are interesting with regard to their eyes.

Spider eyes, like those of other modern arachnids but not other arthropods, are ocelli rather than compound eyes. But this does not mean that their performance is poorer. That such an assumption would be false can even be demonstrated with *Cupiennius salei*, though vision certainly does not play such an eminent role in behavior here as it does in the day-active jumping spiders (Forster 1985, Land 1985), and even though the absolute sensitivity of its eyes does not equal that of the ogre-faced spider *Dinopis*, which hunts at night by spinning a small web that it throws at approaching prey, like a butterfly net. The jumping and ogre-faced spiders provide the most often cited examples of visual perfection: the spatial resolution of 2.4 min in the anterior median eye of *Portia* (Salticidae; Williams and McIntyre 1980), which is worse than in man by a factor of only 6 and better than in the "best" insect (a dragonfly) by a factor of 10 (Labhart and Nilsson 1995, Land 1997); the light-sensitivity of the large posterior median eyes of *Dinopis*, in which the responses of the receptor cells have already reached the half-maximal level when the light intensity is between those of starlight and moonlight (Laughlin et al. 1980), and the lenses of which have an aperture number $F = 0.58$ (Blest and Land 1977), a value much better than that of even very good camera lenses. The insider also recalls that the retinas of the anterior median eyes of the jumping spiders can be rotated by about 50° behind the fixed lens (Land 1969b). Eyes with such impressive abilities and such good imaging characteristics have so far been thought to belong exclusively to hunting spiders such as representatives of the Salticidae (jumping spiders), Thomisidae (crab spiders), Lycosidae (wolf spiders) und Sparassidae (huntsman spiders). Orb-web spiders (Araneidae), according to prevailing opinion, have a poorly developed visual sense with low spatial resolution and with lenses that cast a sharp image of a viewed object onto a plane *outside* that of the visual receptors (Land 1985; Yamashita 1985). In addition orb weavers lack the highly developed visual behavior typical of the day-active jumping spiders, which use their eyes to catch prey, to recognize conspecifics, and to distinguish between prey and a sexual partner.

We have long been underestimating what the visual sense of *Cupiennius* can do, because of the evident dominance of the mechanical senses, the fact that it is active at night, and the absence of behavior that is obviously under visual control. Now we know more about not only the optical properties of the eyes but also some of their electrophysiological characteristics and their fine structure. Furthermore, we now have an idea of the neuroanatomy of the visual centers in the spider brain, and at long last the first behavioral experiments are available that show *Cupiennius* actually making use of its visual system, which is clearly highly developed, albeit in a more modest way than in jumping spiders (see Chapter XXII). Because I personally am much more familiar with the mechanical senses, I have benefited greatly from the collaboration and advice of friends who are authorities on the arthropod visual system. The eyes of *Cupiennius*

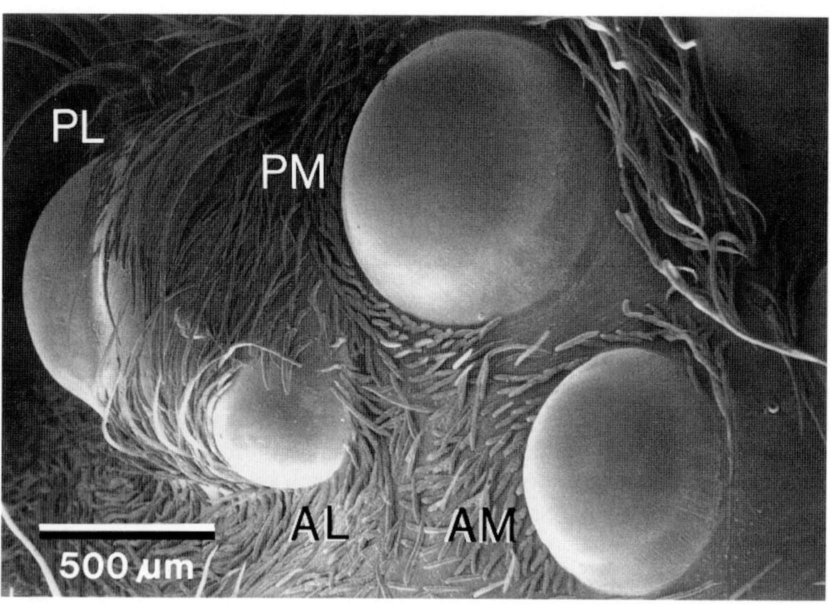

Fig. 1. Scanning electron micrograph of the four right eyes of *Cupiennius salei*. *AM* Anterior median eye (principal eye), *PM* posterior median eye; *Al* anterior lateral eye; *PL* posterior lateral eye

would have remained an object of third-rate interest for much longer without them: Mike Land, Brighton (visual optics), Nick Strausfeld, Tucson (neuroanatomy) and Eis Eguchi, Yokohama (electrophysiology).

Spider eyes are distinguished not only by their position but also, more thoughtfully, by their morphology and functional features. In this context, there are principal eyes and secondary eyes.

1 Position and Structure

Cupiennius, like most spiders, has eight eyes. They are arranged in pairs and named according to their positions: the anterior median (AM), anterior lateral (AL), posterior median (PM) and posterior lateral (PL) eyes. As mentioned in Chapter II, the position of the eyes is an important diagnostic character for the taxonomic identification of spiders. The eye position in *Cupiennius* (Fig. 1, see also Fig. 2 in Chapter II) is distinctly different from those in the Clubionidae and Lycosidae. Among the Pisauridae, the third family closely related to the Ctenidae, only the Thalassiinae have an eye arrangement similar to that of *Cupiennius* and other Ctenidae. And no other taxonomically related genus or family has eight circular eyes; at least the anterior lateral eyes are oval to kidney-shaped.

Fig. 2 A–F. The different eye types. **A** Organization of principal (*AM*) and secondary (*PM*, *AL*, *PL*) eyes. Secondary eyes have a tapetum, which reflects the incoming light. **B** Diagram showing the structure of the photoreceptor cells of principal (*left*) and secondary (*right*) eye. **C** Cross-section (light microscope) through the retina of a night-adapted *AM* eye (magnification 800×). The overwhelming majority of the receptive segments (*RS*) of the photoreceptor cells have four rhabdomeres (*RE*); no pigmented glial cells between the receptive segments. **D** Cross-section (light microscope) through the retina of a night-adapted *PM* eye (magnification 800×). The receptive segments of the photoreceptor cells form horizontal rows with two rhabdomeres each on the opposing sides. Extensions of pigmented glial cells (*PG*) separate neighboring horizontal rows from each other. **E** Rhabdom of a night-adapted *AM* eye (electron microscope, magnification 20,000×). Note the regular arrangement of the microvilli (*MV*). Only a few of the rhabdomeres show irregularities like those marked by *. **F** Rhabdom of a night-adapted *PL* eye (electron microscope, magnification 4250×). Each receptive segment (*RS*) has two rhabdomeres (*RE*). *PG*, *NPG* extensions of pigmented and non-pigmented glial cells. (After Grusch et al. 1997)

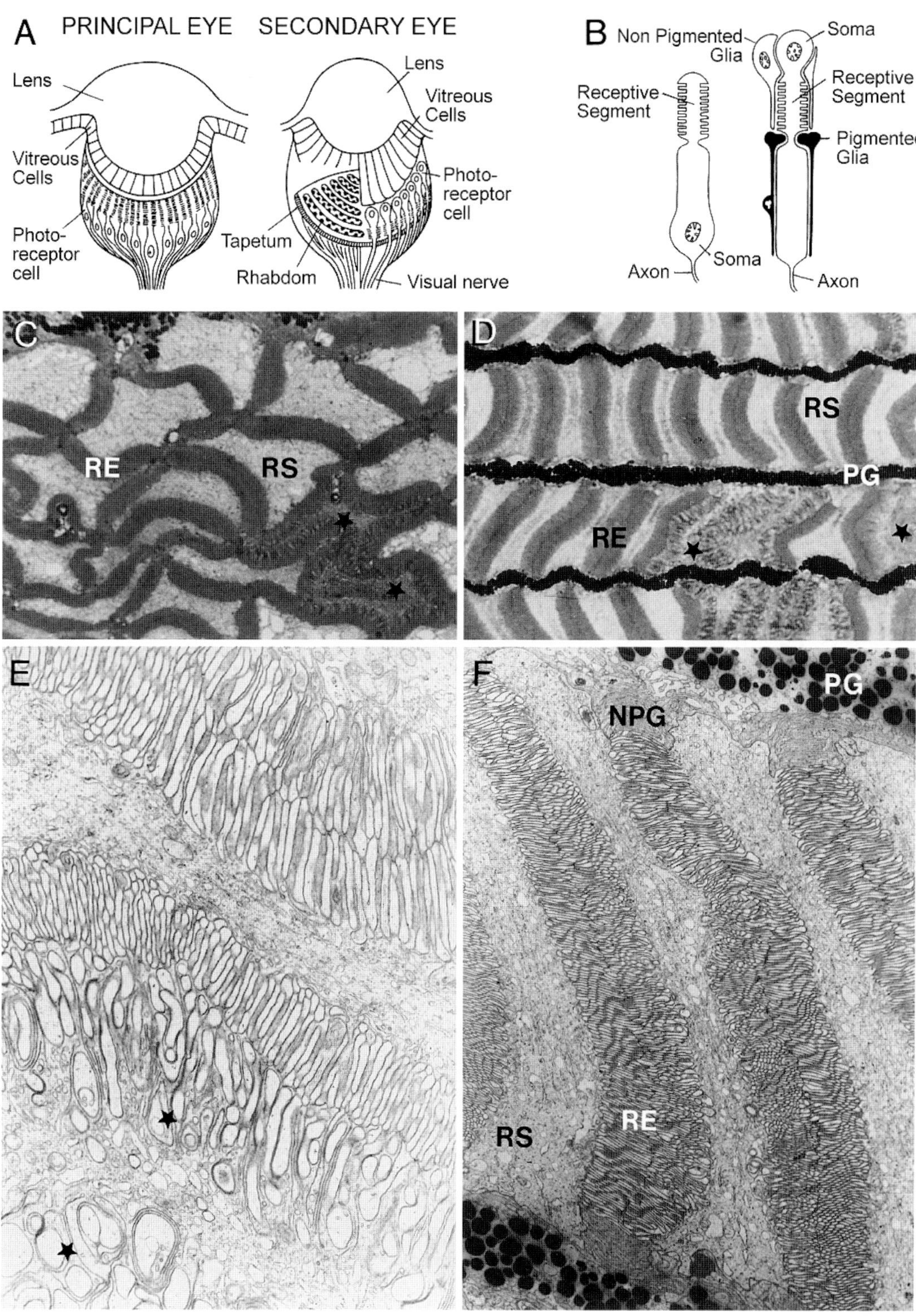
A
PRINCIPAL EYE
SECONDARY EYE
Lens
Vitreous Cells
Photo-receptor cell
Lens
Vitreous Cells
Photo-receptor cell
Tapetum
Rhabdom
Visual nerve
B
Non Pigmented Glia
Soma
Receptive Segment
Receptive Segment
Pigmented Glia
Soma
Axon
Axon
C
RE
RS
D
RS
PG
RE
E
F
PG
NPG
RS
RE

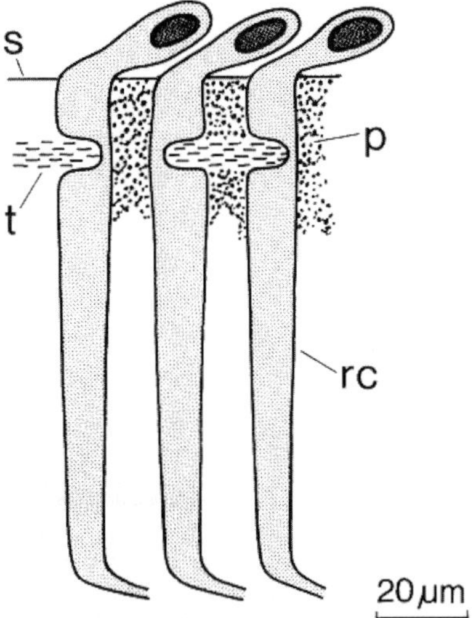

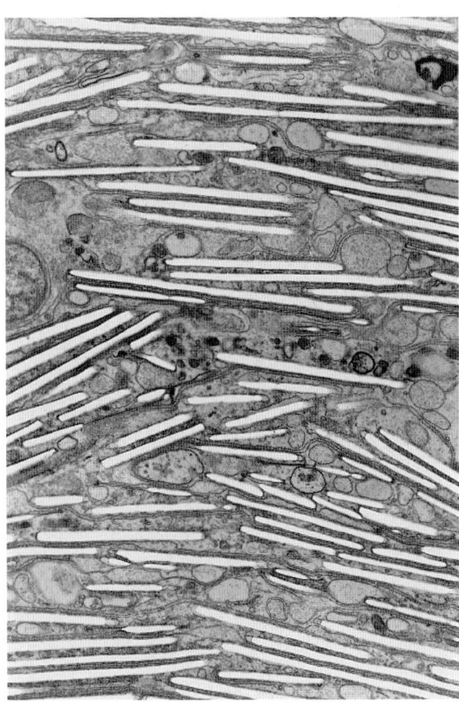

Fig. 3. The tapetum of the secondary eyes. *Above* Diagram showing the location of the tapetum (*t*) between the receptor cells (*rc*); *s* surface of retina; *p* pigment. *Below* Electron micrograph (magnification 15000×) of tapetal cell of PM eye. The light-reflecting crystals were washed out by the procedures necessary to obtain the section; the spaces they originally occupied are clearly visible. (After Land and Barth 1992 and Grusch et al. 1997)

The first to notice this was H. Grenacher, Professor of Zoology and Comparative Anatomy at the University of Rostock. In 1879 he published a famous work on "Investigations of the visual organ of the arthropods, in particular the spiders, insects and crustaceans", in which he wrote on page 40: "One of the most remarkable peculiarities of the eyes of the genuine spiders – that is, excluding *Phalangium* – resides only in an apparently widespread dimorphism. By this, to be sure, I understand not those long-familiar variations in external shape and size among the individual eyes of a given animal, but rather much more deep-seated differences, affecting particularly the retinal cells, which can go so far that one might often believe himself confronted with eyes of quite different genera or even families, though in fact they come from one and the same individual. This dimorphism, which must surely go hand in hand with a difference in the physiological performance of the individual pairs of eyes, seems not to have been observed by anyone before me". The first person to use (in German) the terms "principal" and "accessory" eyes was Philip Bertkau (1886) (see p. 39 in Chapter VII).

The principal eyes (AM) are always those in the anterior median position, even in the many spider species where they are not particularly well developed, either absolutely or in comparison to the other eyes (Plate 13). The few spider families with only 6 eyes lack the AM eyes altogether. All the other eyes (PM, AL, PL) are secondary eyes. All eight eyes contain a cellular vitreous body between the lens and the retina. How, then, do they differ from one another? As shown in figure 2, the differences are these (Land and Barth 1992; Grusch 1994):

1. First, the principal eyes have no tapetum, whereas the secondary eyes do. The tapetum, an interference reflector, reflects light from the back of the eye (Plate 14). This is very helpful to people searching for *Cupiennius* at night, because it makes the secondary eyes glow visibly in the beam of a flashlight many meters away, showing the way to the target (assuming that the searchers are in the right area to begin with, and know where to look!). The reflection is caused by crystals, presumably of guanine, that are aligned in the tapetum parallel to one another and perpendicular to the incident light (Fig. 3). The shape of the tapetum differs in the various spider families and is used as a taxonomic character. *Cupiennius*, like other hunting spiders (e.g. Lycosidae, Thomisidae, Sparassidae; Homann

1928), is characterized by the grid type: this consists of a series of parallel stripes (like a rack in an oven), which together appear as two ladderlike structures when one looks into the eye. In the PM and PL eyes these are approximately horizontal (that is, parallel to the longitudinal plane of the spider or to the substrate on which the spider is sitting). Each of the tapetum stripes bears two receptor cells, as shown in figure 2 (see also Plate 14).
2. Another difference: the principal eyes are everted eyes, the secondary eyes inverted. What does that mean? In the principal eyes the light-sensitive sections, the rhabdomeres, point toward the incident light and the cell nuclei are proximal to them (Fig. 2). The secondary eyes are the other way around, like the eyes of vertebrates: the rhabdomeres are turned away from the incident light, and the part of the cell containing the nuclei is closest to the lens. The receptive section of the cell is seated on the tapetum, so that it receives light both before and after reflection from the tapetum, which means that in theory its effective length is doubled (see Section 3 on "morphological" sensitivity).
3. Whereas the photoreceptor cells of the secondary eyes each comprise two rhabdomeres, which are located on opposite sides of the rectangular cross section of the cell with their microvilli oriented parallel to one another, the receptive segments of the photoreceptor cells in the principal eyes have a rhabdomere on each of the (usually four) sides of their cross section.
4. Finally, the retina of the principal eyes, and only these eyes, can be moved by a pair of eye muscles. They are described in Chapter XXIII.

On the whole, there is nothing surprising in the structure of the *Cupiennius* eyes. To a great extent it is the same as in the wolf spiders (Lycosidae) (Baccetti and Bedini 1964, Homann 1971). From now on, therefore, we shall concentrate on answering more specific questions, so as to get an idea about what the eyes can do and hence about how they might be used in behavior.

2 Visual Optics

To state that the best retina is of little use if the light-transmitting and image-producing optical apparatus is not of equally high quality may seem trivial. But just how are we to evaluate the quality of the optical apparatus in the eye of *Cupiennius*? The factors of particular interest are the F-number of the lens, the spatial resolution and the extent of the visual field. Before specific measurements related to these parameters are presented, a few simple but very effective methodological tricks will be described (Land 1985, Land and Barth 1992). Some of them are derived from the classic experiments of H. Homann (1928), whose more than 50 years of research profoundly enriched our understanding of spider eyes.

Focal Length. First, the measurement of lens focal length. A lens is dissected out together with a little of the surrounding cuticle and placed on the meniscus of a drop of physiological solution in such a way that the outer surface is exposed to the air, as it is in the intact animal, and the inner surface contacts the liquid (Fig. 4a). Now the image plane is determined with the microscope; that is, the eye is "looking" at an object. The focal length is found by measuring the sizes of object O and image J. If the object distance U is known, the distance of the image from the lens V can be calculated by the simple equation

$$V/U = J/O .$$

The posterior focal length f is then found from the lens equation

$$1/V + 1/U = 1/f .$$

Resolution. Because of their tapetum, the retinas of the secondary eyes can be observed directly in the living animal with an ophthalmoscope. In our work on the *Cupiennius* eyes Mike Land used an instrument specially adapted for examining small eyes (Fig. 4b). The ray path behind the objective lens is subdivided into an illumination beam and an observation beam. The

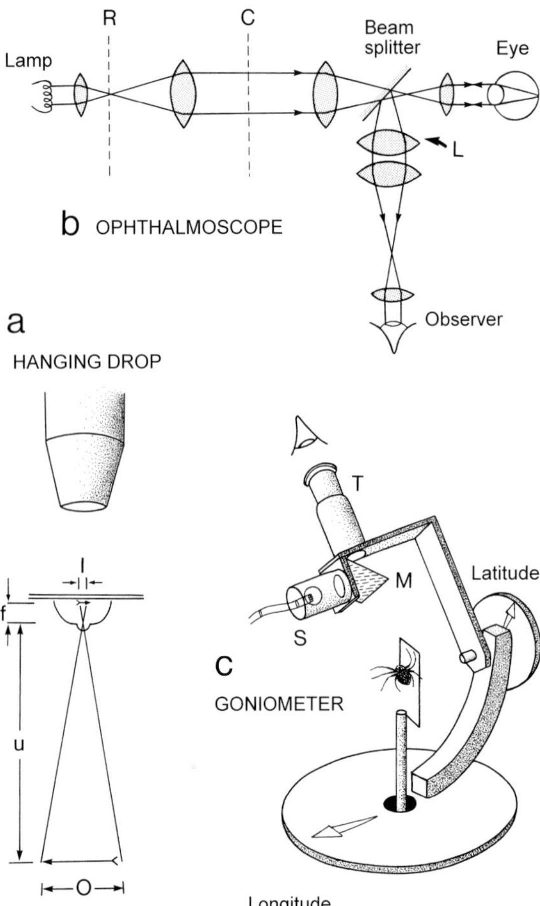

Fig. 4a–c. Methods used to determine the optical properties of spider eyes. **a** Hanging-drop method (Homann 1928) to determine focal length (*f*); *I* size of image; *O* size of object; *u* distance of object. **b** Ophthalmoscope for the observation of the living retina (Land 1969); *R* plane conjugate with infinity for objects to be imaged on retina; *C* plane conjugate with cornea for masking illuminating beam, *L* removable lens allowing observation of either retina or eye surface. **c** Cardan-arm goniometer used to measure the fields of view (Land 1985). *T* Telescope; *S* light source; *M* mirror (After Land and Barth 1992)

observation beam makes it possible to view the exterior of the eye with the microscope and also, by inserting an additional lens (L), to view the retina, which is conjugate with infinity. In the second configuration the field of view can be calibrated so as to obtain the angle between retinal structures and thus measure the resolution of the eye in vivo. With the help of the illumination beam an object brought into the plane R (see Fig. 4b) can be imaged on the retina by the lens of the eye. That is, objects of known angular extent can be projected onto the retina. The principal eyes are not suitable for this kind of ophthalmoscopic examination because they have no tapetum and hence do not reflect enough light.

Visual Fields. Finally, we consider the position and size of the visual fields of the individual eyes.

That this is something worth consideration was brought out long ago by W. Porterfield (1759), who gave the matter some thought while observing "house spiders" over 200 years ago and, according to one Dr. Power, wrote as follows: "The first eminent Thing we found in the House-spiders were theire Eyes, . . . This Eyes are placed all in the Fore-front of their Head (which is round and without any Neck) all diaphanous and transparent like a Locket of Diamonds, or a Set of round Crystal-beads etc. Neither wonder why Providence should be so anomalous in this animal, more than in any other we know of (Argus's Head being fixed to Arachne's Shoulder).

For Ist, Since they wanting a Neck cannot move their Head, it is requisite that Defect should be supplied by a Multiplicity of Eyes. 2dly, Since they were to live by catching so nimble a Prey as a Fly is, they ought to see her every way, and to take her perfactum (as they do) without any motion would have scarred away so timorous an Insect." (from Scheuring L 1914, p. 427).

That the divergence of the optical axes is greater in the walking and jumping spiders than in spiders that live in dark tunnels or crevices and have eyes close together (for example, *Atypus, Segestria, Mygale*) is another old observation, which is consistent with this line of reasoning (Dugès 1836).

The extent of a visual field is measured with a small telescope suspended on gimbals in a goniometer. In its center is the spider eye. A light source on the telescope illuminates the object in the direction of observation by way of a semitransparent mirror (Fig. 4c). As the telescope is rotated about the eye, the eye reflects a green light as long as the incident beam strikes the tapetum. The angular range over which the eye reflects light corresponds to the extent of the tapetum, which covers the same area as the retina. This method can also be used with the principal eyes, because here the color of the reflected light changes at the edge of the retina, from reddish brown to a darker brown.

Table 1. Dimensions (in μm) of the eyes of an adult female of *Cupiennius salei.* (Land and Barth 1992)

Eye	Diameter (*D*)	Radius of curvature		Focal length (*f*)	F-number (*f/D*)
		Anterior	Posterior		
AM	396	204	157	293	0.74
AL	256	147	–	148	0.58
PM	629	320	236	448	0.71
PL	625	320	–	432	0.69

AM, antero-median; AL, antero-lateral; PM, postero-median; PL, postero-lateral

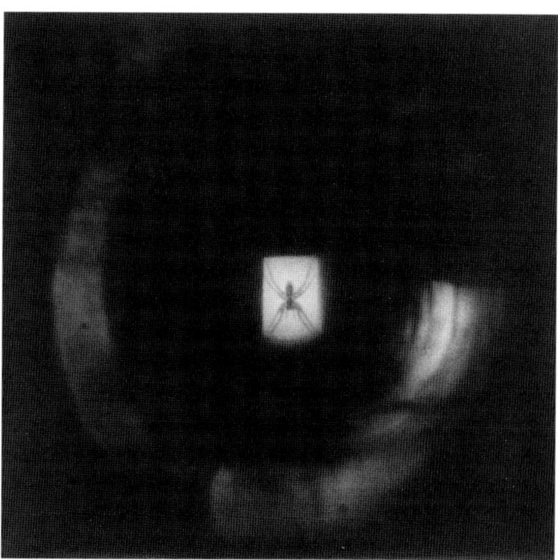

Fig. 5. Image of a lycosid spider photographed at the focus of the lens of a PM eye using the hanging-drop method (see Fig. 4a). (Land and Barth 1992)

Table 1 gives typical values for the diameter of the lenses, their outer and inner radii of curvature, the focal length and the F-number. The lens is biconvex, with both outer and inner surfaces approximately hemispherical, although the curvature of the inner surface is somewhat greater. The focal lengths f are proportional to the eye diameters D. Accordingly, they are smallest for the AL eyes (148 μm) and largest for the PM's (448 μM). The F-numbers (F = f/D) are all between 0.58 and 0.74. The lenses of spider eyes are thus far superior to the ordinary objective lenses of photographic equipment with respect to the brightness of the image. Furthermore, it is likely that the spider eyes are capable of mediating vision in very weak light, in twilight or even at night.

According to ophthalmoscopic observation and also histological examination, the retina has no fovea. It also does not consist of several layers with different types of receptors and different packing densities, as do the retinas of the principal (AM) eyes of jumping spiders (Land 1969a). The retina of *Cupiennius* is much more similar to those of the Lycosidae (Homann 1931, Baccetti and Bedini 1964, Land 1985). Because the receptor-cell mosaic is relatively coarse-grained, the resolution is not nearly as good as in humans. The finest resolvable striped pattern has a line separation of at least 2° (angular subtense of a line pair at the eye). For comparison, the inter-receptor angle in the human eye and the very high-performance eyes of *Portia*, a jumping spider, are 0.42 and 2.4 minutes of arc, respectively (Williams and McIntyre 1980). The optical properties of the *Cupiennius* eye in themselves would permit considerably better spatial resolution, with line separations down to 0.05°–0.11° (Land and Barth 1992). Accordingly, direct observation of the image cast by an isolated *Cupiennius* lens (Fig. 5) reveals a resolution substantially finer than the mosaic of photoreceptor cells in the retina.

More precise values for the "morphological" retinal resolution limits are summarized in Table 2. From the receptor-cell separation s and the focal length the inter-receptor-cell angle is found to be $(s/f)_{rad}$ or $57.3(s/f)_{deg}$ (1 radian = $180/\pi = 57.3°$). It can also be obtained by direct ophthalmoscopic observation (Land and Barth 1992). As Table 2 shows, the PM and PL eyes have the greatest resolution, with inter-receptor-cell angles of about 1° along the receptor rows (that is, in approximately the horizontal direction) and 2°–3° perpendicular to them. The AL eyes have distinctly poorer resolution, and the principal eyes (AM) have inter-receptor-cell angles of 3°. Another important factor affecting resolution, along with the fineness of the receptor-cell mosaic, is the refraction of light at the aperture of the eye. Refraction blurs the image of a point, producing an Airy disk, and thus places a

Table 2. Resolution of the retinas of an adult female spider (*Cupiennius salei*). Receptor spacing (*left*) was measured in histological sections, allowing 15% shrinkage. Measurements of angular separation of receptor cells are based on histological preparations as well as on ophthalmoscopic observation of the living eye. (Land and Barth 1992)

Eye	Receptor spacing measured histologically, allowing 15% shrinkage (μm)		Angular separation (degrees) from histology and ophthalmoscopy			
	Along rows (*a*)	Between rows (*b*)	57.3 *a/f*	57.3 *b/f*	Direct measurement	
AM	14.9	14.9	2.9	2.9	–	
AL	9.2	23.9	3.6	9.3	–	
PM	7.5	20.3	1.0	2.6	0.9	2.3
PL	–	–	–	–	1.0	3.0

AM, antero-median; AL, antero-lateral; PM, postero-median; PL, postero-lateral; *f*, focal length

limit on resolution that cannot be exceeded, however fine the receptor mosaic may be. The size of the Airy disk depends on the wavelength of the light (λ) and the lens diameter (D): the light intensity conforms approximately to a Gaussian distribution with a width at half height of λ/D radians. That is, the effect is diminished with decreasing wavelength and increasing lens diameter. This means that the stripes in a pattern can be closer together while still being imaged separately (Land 1985).

The calculation of resolution from the fineness of the receptor mosaic on one hand and the refraction effects on the other shows us that in the eyes of *Cupiennius* the receptor density is the limiting factor, and not refraction and thus the *optical* quality of the image. This also applies to the eyes with the smallest lens diameter (AL eyes) and can be readily seen from the quality of the spider image shown in figure 5. The only spider eyes so far known in which the receptor mosaic comes close to the limits set by refraction are the principal eyes of *Portia*, the champion regarding spatial resolution (Williams and McIntyre 1980).

According to Land (1981), the values so obtained can be used to calculate the minimal distance U of an object at which the spider eye still forms a sharp image:

$$U = f \cdot D/2s\ .$$

For the principal eyes (AM), U is about 4 mm. Therefore everything that is likely to interest a spider is seen in sharp focus. For the PM eyes U amounts to 7–19 mm, depending on whether s is measured along or between two neighboring rows of receptor cells. Mechanisms of accommodation – that is, of adjusting the optics according to the distance of an object (such as shifting the lens back and forth or changing its curvature) – are unknown, in *Cupiennius* as in other spiders. *Cupiennius* evidently does not need them.

Figure 6 shows that the fields of view of the PM and PL eyes together cover almost the entire upper hemisphere and extend about 40° below the horizon. Remarkably, there is a gap of 5° to 20° between the fields of these two eyes. The elongated shape of the visual field of the PM eyes is the same as that of their retinas. The AL eyes look downward, at an area just in front of the chelicerae. Their visual field is small and overlaps with the lower regions of the PM and PL fields. This arrangement is also typical of the lycosids. The visual field of the principal (AM) eyes corresponds approximately to that of the PM eyes. That is, the AM and PM eyes all look in the same direction. Why should two eyes view the same part of the surroundings? We suppose that they have different functions. This idea is supported by the differences in their structure, the fact that they send information to the brain by different routes (see Chapter XVI) and the remarkable circumstance that only the retinas of the principal eyes can be moved by muscles (see Chapter XXIII). The movements presumably prevent neural adaptation. In contrast to the secondary eyes, which are thought to be particularly suitable for keeping track of

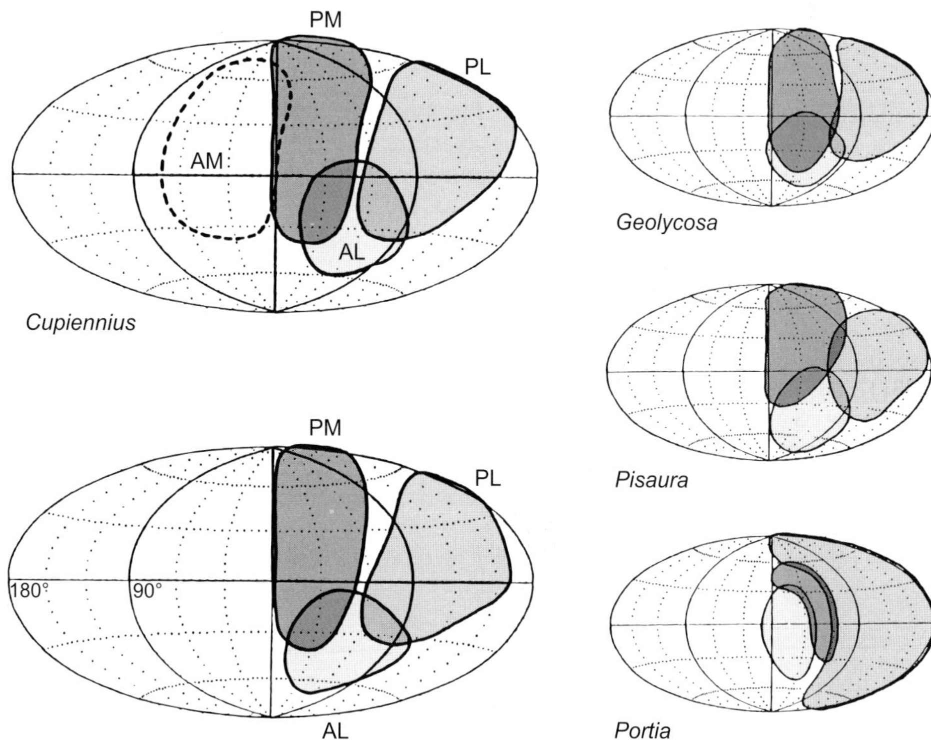

Fig. 6. Fields of view of the different eyes of *Cupiennius salei*, taking two adult females as representative examples, with three other species of hunting spiders for comparison. For the methods applied see Fig. 4c. The fields of view are plotted onto a globe with the spider at the center, and the projection depicts the whole of that globe, marked off at 90°, 30°, and 5° intervals. *Geolycosa* (Lycosidae) and *Pisaura* (Pisauridae) have visual fields similar to those of *Cupiennius*. In contrast, the fields of view of the jumping spider *Portia* (Salticidae) are quite different. (Land and Barth 1992)

moving objects, the principal eyes presumably analyze stationary objects. Similar relationships are thought to apply in the jumping spiders (Land 1972). The time course and amplitude of the movements of the principal eyes play a major role here. Florian Kaps recently investigated this question in *Cupiennius salei*, for his Master's thesis (1994) and subsequently his doctoral dissertation (1998). I shall describe his findings in Chapter XXIII, in connection with visual orientation and the discrimination of viewed patterns (Schmid 1998).

At this point a general consideration deserves mention. The compound eyes of insects are decidedly wide-angle eyes. Their retina is convex and their visual field not uncommonly encompasses the entire panorama. For small, mobile animals like insects, which are capable of binocular vision only in the immediate vicinity, at most a few centimeters away, this is a great advantage: for objects at larger distances the information contained in the movement of images over the visual field and produced by the animal's own movement can be fully utilized. Spider eyes have concave retinas with comparatively small visual fields. But spiders usually have eight eyes and thus considerably increase their overall visual field. For them, as for the insects, visual movement detection is extremely important (see Chapter XXIII).

3 "Morphological" Sensitivity

In addition to the lens diameter D and the focal length f (which together determine the aperture number F = f/D), the morphological features of the sensory cells themselves contribute to the sensitivity of an eye, in particular the diameter d_r and length l of their rhabdomeres. The amount of light absorbed by the photopigment (expressed as the absorption coefficient k) must also be taken into account. According to Land (1981) the largely morphologically determined sensitivity S of an eye (its ability to catch photons) can be estimated by the following equation:

$$S = \left(\frac{\pi}{4}\right)^2 \cdot \left(\frac{D}{f}\right)^2 \cdot d_r^2 \cdot (1 - e^{-kl}) .$$

For photoreceptors of the rhabdomeric type the value of k is generally taken to be 0.0067 (0.67% per μm) (Land 1981) and for eyes with a tapetum it initially seems reasonable to use double the actual rhabdomere length, because light that is not absorbed is reflected by the tapetum and sent through the rhabdomeres a second time. According to our electrophysiological experiments (see below), however, this doubling seems questionable. Therefore Table 3 also gives the S values for the case of a single rhabdomere length.

What is the point of this exercise? Sensitivity S is particularly interesting as a basis for comparing different eyes and different animals. An electron-microscopic analysis of the eyes of *Cupiennius salei* (Grusch et al. 1997) supplied us with very precise values for d_r and l and also allowed us to calculate the total surface area of the microvilli (Table 3) and the proportion of the total retinal area occupied by rhabdomeres (rhabdom occupation ratio). The crucial results of all these studies are as follows.

Table 3. Sensitivity S, number of microvilli per photoreceptor cell and area of rhabdomere membrane per photoreceptor cell of night-adapted eyes of *Cupiennius salei*. In the case of the secondary eyes the S-value given in brackets is based on a doubled rhabdomere length to take account of the presence of the tapetum. (Grusch et al. 1997)

Eye	S	Microvilli per photoreceptor	Area of rhabdomeral membrane per photoreceptor (μm^2)
AM eye	109	3.2×10^5	2.5×10^5
PM eye	78 (135)	2×10^5	2.1×10^5
AL eye	88 (152)	2.1×10^5	1.6×10^5
PL eye	86 (147)	2.1×10^5	2.3×10^5

1. That the secondary eyes are adapted to seeing at very low light intensities is indicated not only by the large F-number of the lenses but also the dimensions of the rhabdomeres, the slight optical isolation of the photoreceptors by screening pigment, the large rhabdom occupation ratio (40–65%) and the presence of a tapetum. For comparison: in night-active moths (Sphingidae) the rhabdomeres occupy 60% of the retinal area, whereas the proportion in day-active species is only 10–25% (Eguchi 1982). And Blest (1985b), to whom we are indebted for a number of particularly valuable and elegant papers on the functional morphology of spider eyes, was able to show that among the day-active jumping spiders (AM eyes) those in dark habitats have larger rhabdom diameters (and less screening pigment) at the expense of spatial resolution than the ones that live in sunny habitats.
2. The values for the "morphological" light sensitivity S of the *Cupiennius* eyes are 78–109 μm^2, greater by a factor of 10^3 than the value for the day-active jumping spider *Phidippus* (AM eyes; Land 1969a). The highest S value known so far, 387, was found for the large PM eyes of the night-active ogre-faced spider *Dinopis subrufus* (Blest and Land 1977; Blest 1978; Land 1985). The sensitivity of the AL eyes of *Olios* (Sparassidae) is the same as that of *Cupiennius*, and the S values for night-active insects, such as *Ephestia* with S = 82 (Land 1981), are also of the same order of magnitude. There is an obvious correlation with ecology, small values of S being found in day-active animals and larger values of S in night-active animals.
3. The rhabdom occupation ratio is not constant. As in other arthropods (Nässel and

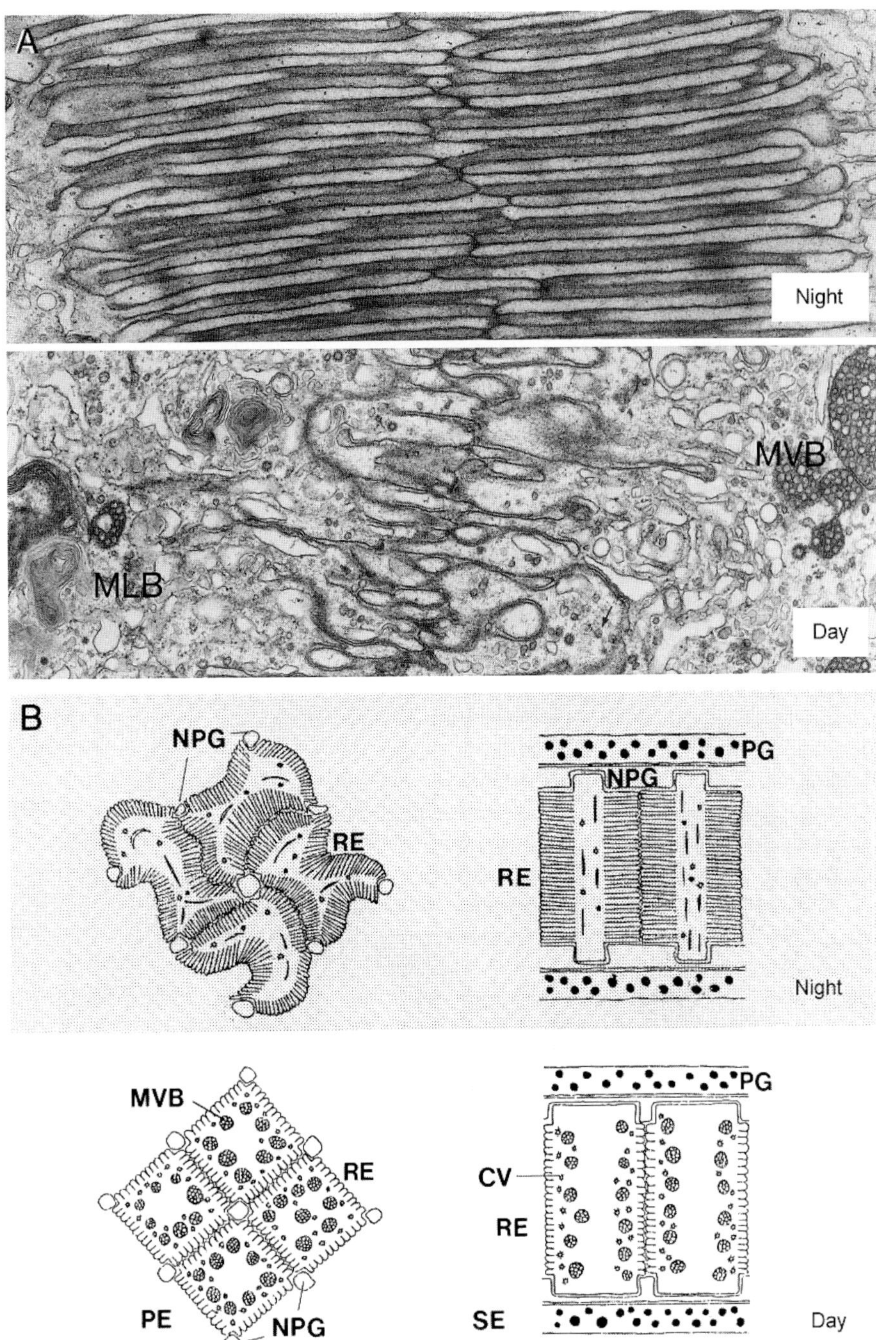

Fig. 7 A, B. Microvilli in the eyes of *Cupiennius salei* during daytime and at night. **A** Night- and day-adapted rhabdom of a PM eye, both at same magnification (×21,700); *MVB* multivesicular bodies; *MLB* multilamellar bodies; *arrow* points to coated vesicles, which are numerous in the cytoplasm of day-adapted receptor cells close to the rhabdoms. **B** Diagrammatic cross-sections of the receptive segments of night-adapted and day-adapted eyes. *Left* Principal eye; *right* secondary eye; *CV* coated vesicles; *NPG* nonpigmented glia; *PG* pigmented glia; *RE* rhabdomere. (Grusch et al. 1997)

Waterman 1979, Blest 1985a), it changes in a day/night rhythm. During the day, the rhabdomeres of the *Cupiennius* eyes are largely dismantled: only two hours after the light period has begun, 80% of the microvillar membrane area has disappeared (Fig. 7). This, too, is dazzlingly consistent with the picture of a night-active animal! This sort of membrane turnover has been particularly well investigated in another night-active spider, the ogre-faced spider *Dinopis* (Blest and Land 1977, Blest 1978). But it has also been observed in close relatives of *Cupiennius*, such as *Dolomedes* (Pisauridae) (Blest and Day 1978).

4. To judge by their morphology, the most sensitive of the four types of eye in *Cupiennius* are the principal eyes (AM), if the tapetum of the secondary eyes is neglected. The AM eyes have the most microvilli per photoreceptor cell: 320,000. The electrophysiological findings (see below) point in the same direction. What, then, is the significance of the tapetum? It gives the secondary eyes two chances to utilize the light, but in compensation the principal eyes have longer rhabdomeres and more microvilli. In the calculation of S, if the rhabdomere length is doubled for the secondary eyes to account for the tapetum, the S values obtained are well above those of the principal eyes. However, we must assume that the action of the tapetum is not equivalent to making the rhabdomeres twice as long. According to our recordings of the electroretinogram (Barth et al. 1993) all the eyes of *Cupiennius* are about equally light-sensitive. The intracellular recordings (Walla et al. 1996) show that individual receptor cells in the principal eyes are actually more sensitive than the most sensitive photoreceptors in the secondary eyes. Interestingly, the extremely light-sensitive posterior median eyes (that is, secondary eyes) of *Dinopis* have no tapetum.

4 Absolute and Spectral Sensitivity

Despite their explanatory value, morphological data cannot replace the physiological experiment. A direct answer to the question of the actual absolute sensitivity can be obtained only from electrophysiology. It is worth knowing not only the absolute but also the spectral sensitivity, if we want to understand the possible roles of the visual sense in the behavior of *Cupiennius*. Does *Cupiennius* see color? Or, to put it more cautiously, are the spectral sensitivities of the receptor cells such as to permit color vision?

Absolute Sensitivity

By recording electroretinograms (ERG), a rather archaic method but a very useful and simple one for the question of interest, we found that the eyes of *Cupiennius* are sensitive enough to enable the spider to see not only shortly after sundown but even under considerably dimmer conditions, such as by moonlight (Barth et al. 1993). For white light the threshold for the absolute corneal illuminance is distinctly below 0.01 lux. When a monochromatic stimulus light is used, a corneal illuminance of about 3×10^{12} photons $cm^{-2}s^{-1}$ is needed to elicit a half-maximal response and about 3×10^{9} photons $cm^{-2}s^{-1}$ for threshold excitation. Converted to retinal illuminance (Land 1981), this amounts to 5.9×10^{9} photons $cm^{-2}s^{-1}$. According to the ERG measurements, then, all the eyes are approximately equally sensitive at night (the activity period of *Cupiennius*). Despite the high light sensitivity of its eyes, *Cupiennius* comes in a clear second to the champion among the spiders: the ogre-faced spider *Dinopis*, the PM eyes of which are more sensitive by a factor of ca. 10^{7} (Laughlin et al. 1980).

The absolute sensitivity of individual photoreceptor cells is 10- to 100-fold higher (2.19×10^{7} to 56.3×10^{7} photons $cm^{-2}s^{-1}$) than that determined from the ERG. The greatest sensitivity was found in the principal (AM) eyes. We were surprised by this because of their lack of a tapetum; as discussed above, the tapetum is generally thought to enhance the utilization of light considerably. Knowing about the massive cyclic buildup and breakdown of the rhabdomeres (see above), however, we were not surprised to find that the sensitivity of the cells varies in the day/night rhythm. I shall return to this point later.

Spectral Sensitivity

The electroretinograms indicated that all the eyes are particularly sensitive between ca. 520 nm and 540 nm, the wavelength region that looks green to us. Furthermore, the ERGs of all eyes revealed a shoulder in the ultraviolet region (between 340 nm and 380 nm). This second peak is quite high, 65 to 80% of the maximum in the green region. A comparison between the wavelength dependence of the ERG and the theoretical absorption curve of the visual pigments, which has absorption maxima in the corresponding wavelength regions, reveals the possibility of trichromatic color vision based on green, blue and UV receptors. On the other hand, an absence of color vision would certainly not be unexpected in such a markedly night-active animal as *Cupiennius*. Recently intracellular recordings from individual visual cells (Walla et al. 1996) have demonstrated the presence of three photoreceptor types, with absorption maxima at 480 nm (blue), 520 nm (green) and 340 nm (UV) (Fig. 8). That is, the conclusions drawn from the ERGs have been confirmed. However, UV cells were found only in the secondary eyes, and even here only once in each eye. This is remarkable. One explanation could be that the UV cells are concentrated in a particular part of the retina that was inaccessible to the electrode – like the dorsal rim area in the compound eye of many insects. Of course, they could also be uniformly distributed but very rare.

Another question is, what are the *UV cells* good for? Although light coming directly from the moon does contain short-wavelength components, below 450 nm (Menzel 1979), it would be unusual to employ UV cells to detect this particular light. In other arthropods UV vision is used to see during the day, and its many receptor cells with maximal sensitivity in the blue or green region ought to be sufficient to enable *Cupiennius* to see by night. It may be that the UV cells have something to do with polarization vision. But here, too, the evidence is not very convincing. Görner (1958, 1962), in behavioral ex-

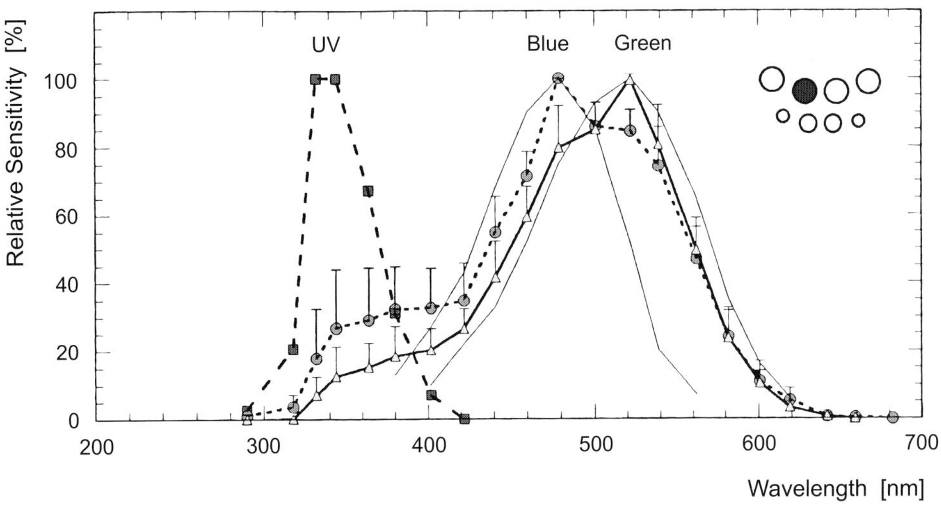

Fig. 8. Mean spectral sensitivity of single photoreceptor cells in the PM eye of *Cupiennius salei*. Data from 1 UV cell (λ_{max} at 340 nm), 16 blue cells (λ_{max} at 480 nm), and 15 green cells (λ_{max} at 520 nm). *Continuous lines* without measured values are theoretical absorption curves for pigments absorbing maximally at 480 and 580 nm (Dartnall nomogram). *Bars* give SD. (Walla et al. 1996)

periments on *Agelena labyrinthica*, was able to demonstrate an involvement of the principal eyes, and only these, in orientation by polarized light. In wolf spiders, which are closely related to *Cupiennius*, sensitivity of the principal eyes to polarized light has been documented electrophysiologically (ERG, Magni et al. 1962, 1965); furthermore, a possible morphological substrate has been identified in the form of rhabdomeres arranged perpendicular to one another in the periphery of the retina (Melamed and Trujillo-Cenóz 1966). Unfortunately, analysis of the fine structure of the *Cupiennius* eyes (Grusch et al. 1997) has not provided a comparable result. We are still groping in the dark for an explanation of the function of the UV cells. Can it be that *Cupiennius* has UV marks on its body that play a role in behavior, like the flowers that attract pollinating insects? Photographs taken in UV light show no UV-reflecting patterns (Barth et al. 1993), so this approach has also got us nowhere.

We have fewer problems with the *green receptors* found in all the eyes. The light reflected from plants and from the soil is dominated by wavelengths above 450 nm, which we perceive as green and yellow light (Menzel 1979).

Regarding the receptors particularly sensitive to *blue light*, which are present in the PM, PL and AM (principal) eyes of *Cupiennius salei* (Walla et al. 1996), those in the PM eyes have been found to become about ten times as sensitive in the dark, the activity period of *Cupiennius*. It is interesting that no such change occurs in the green receptors. The conclusion: the blue cells are particularly important for seeing at night or in dim light.

In comparison with the seemingly infinite number of studies on the eyes of crustaceans and in particular insects, little is known about arachnid eyes. There is still plenty of scope here for future research. However, data are available on the spectral sensitivity of representatives of four spider families, which differ widely from one another in both their life styles and the significance of vision in their behavior: web spiders (Argiopidae), ogre-faced spiders (Dinopidae), wolf spiders (Lycosidae) and jumping spiders (Salticidae) (for a review see Yamashita 1985).

Because of their physiological proximity to *Cupiennius*, the most interesting are the wolf spiders. De Voe (1972) found only one receptor type in the principal eyes (AM) of *Lycosa baltimoriana*, *L. lenta* and *L. miami*; it is most sensitive in the green region (510 nm) and has a secondary peak in the UV (360–370 nm). The same applies to the AL eye and presumably also to the other secondary eyes (ERG, De Voe et al. 1969), which is in clear contrast to our findings for *Cupiennius*. We must conclude that the lycosids have no color vision. The enormous PM eyes of *Dinopis*, used at night to catch prey with the "butterfly net" the spider spins, contain only green receptors (Laughlin et al. 1980). In fact, all the other secondary eyes of spiders that have been tested are the same, which is taken to mean that all of them are blind to colors. But as usual in biology, where diversity is a dominant principle, an exception is known. Several species of orb-web spiders of the genus *Argiope* resemble *Cupiennius* in possessing PL eyes with UV, green and blue receptors, much like the princi-

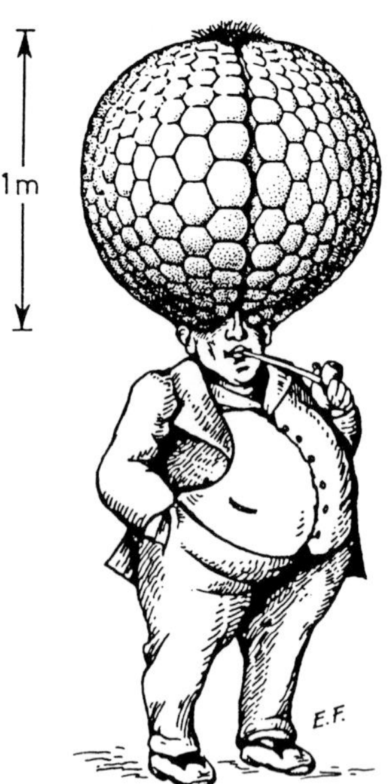

Fig. 9. A human being with a compound eye of the smallest possible size providing about the same spatial resolution as its own normal lens eye. The size of the individual facets is not to scale. Instead of the 100 facets for each eye there should be one million of them. (Kirschfeld 1976)

pal eyes (AM) of the same spiders and of jumping spiders (Salticidae), which have di-, tri- or tetrachromatic vision and hence presumably can also discriminate colors (Yamashita 1985).

The arachnids are the only group of arthropods in which the principal eyes are not compound but rather "camera" eyes. The presumed ancestors of the spiders, relatives of the Xiphosura (a group that includes the still extant horseshoe crab *Limulus*), and the extinct Eurypteridae had both simple and compound eyes. Whereas the principal eyes of the recent spiders are thought to have been derived from the single pair of simple eyes in their ancestors, it has been suggested that the secondary eyes were formed by subdivision of their compound eyes (Paulus 1979). How can it be that such a successful invention as the compound eye was abandoned by the spiders and other arachnids? We do not know, but there might have been good functional reasons. The simple or camera eye is optically the better solution, because its resolution is not limited by refractive effects at the lens as is the case for the small lenses of compound eyes. If the eye of a jumping spider were to be replaced by a compound eye with the same resolving power, the latter would have to be so large that it would surely not fit into the spider's prosoma (Kirschfeld 1976; Land 1985). As figure 9 shows, much the same thing would happen if we tried to replace our own eyes with compound ones.

Johannes Müller, the world-renowned Berlin physiologist, as recently as the beginning of the last century maintained that the lens eyes of spiders are comparable to the vertebrate eye and not to the compound eyes of arthropods. But soon the Dutch scientist Brants (1838, Annal. d. scienc. nat.: Observations sur les yeux simples des animaux articulés) contradicted this idea, and Franz Leydig of Würzburg wrote in 1855 ("Zum feineren Bau der Arthropoden", Archiv für Anatomie, Physiologie und wissenschaftliche Medicin, 376–480), rather disrespectfully, it seems to me: "I should like to indulge in the hope that the celebrated physiologist in Berlin will regard it as justified to assume that the rodlike structures in the facetted eye are analogous to the rods in the visual organ of the vertebrates". The logic continued with a circular argument: "The single important difference in structure between the facetted and so-called simple eyes is that in the former for every nerve rod there is an associated lens or at least subdivision of the cuticle, while here in the simple eye a single lens is assigned to all the nerve rods that are present. However, this may well also be the main difference between the facetted eye of the arthropods and the vertebrate eye, so that in trying to systematize the visual organs of the various animal species one ultimately goes around in a circle, . . ."

XII Chemoreception

Most spiders are really hairy creatures, and *Cupiennius* is certainly among their number. We have already met some of the many hair sensilla: the extremely movement-sensitive trichobothria (Chapter IX), the proprioreceptive hairs (Chapter X), and the vibration-sensitive hairs (Chapter VIII). The great majority of hairs undoubtedly have a tactile function, but chemoreceptors also take this form – which should come as no surprise, since chemosensitive hairs are well known in other arthropods, especially insects.

Hair Sensilla

Among all the other hairs of spiders, the chemoreceptive ones are notable for their steeper angle to the cuticular surface and in having a shaft that is slightly curved, in an S shape (Fig. 1). Electron micrographs reveal a hole at the tip of the hair (Fig. 2), which suggests that these are contact chemoreceptors (Foelix 1970; Foelix and Chu-Wang 1973b; Harris and Mill 1973, 1977; Drewes and Bernard 1976; Friedrich 1998). The interior of the hair is well equipped with sensory cells: 21 of them. Two can be identified as mechanoreceptive by the presence of a tubular body at the dendrite tip and the coupling to the base of the hair; these respond to deflection of the hair shaft (Harris and Mill 1973, 1977). The other dendrites extend to the "open" hair tip, as would be expected of a gustatory hair. Gustatory

Fig. 1. Pedipalpal tarsus of a young spider (*Cupiennius salei*, 7th molting stage) in dorsal view. Contact chemoreceptive hair sensilla are marked by *arrows*. (Photo O. Friedrich 1998)

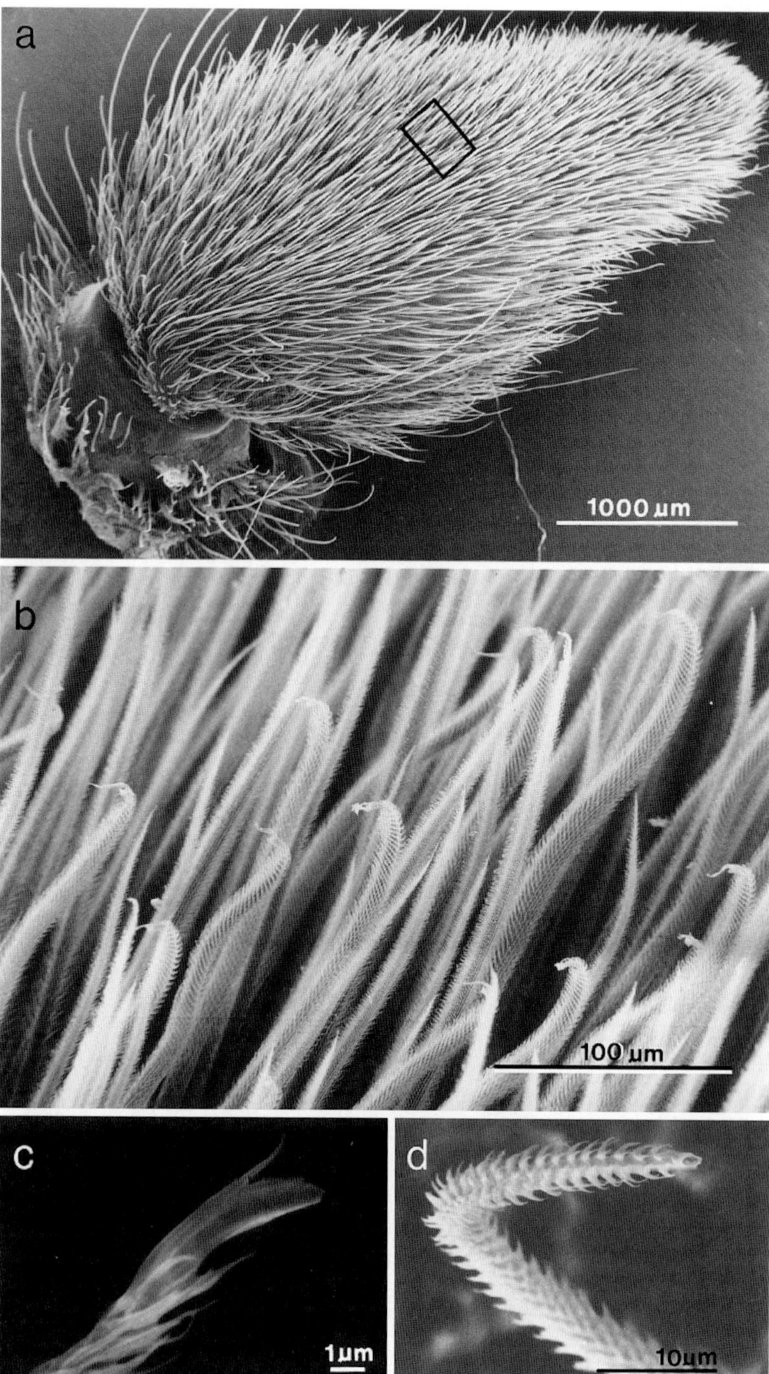

Fig. 2a–d. Contact chemoreceptive hair sensilla on the pedipalpal tarsus (dorsal side) of an adult male *Cupiennius salei.* **a** Overview of pedipalp with dense cover of sensory hairs. Among these there are many contact chemoreceptors. **b** Chemoreceptive hairs typically form a large angle with the cuticular surface and are often slightly bent like an s; they are further characterized by a typical apical cone with a terminal pore and an apical spur. Enlarged view of the area marked in **a**. **c, d** Tip of hair at higher magnification. (Photos O. Friedrich, F.G. Barth)

hairs are most abundant on the distal sections of walking legs and pedipalps.

In insects olfactory hairs are characterized by fine pores in the wall of the shaft; such "wall-pore" sensilla have not yet been found in spiders. It has been suggested that the tarsal organ has an olfactory function (Dumpert 1978). However, its responses to moisture and thermal stimuli have since proved to be at least as strong, and because these have been thoroughly investigated in *Cupiennius salei*, the tarsal organ will be considered in the next chapter (XIII).

It should be frankly admitted at the outset that analysis of the physiological bases of chemoreception in spiders is in a poor state, in particular when it comes to olfaction. Only a very few electrophysiological recordings have been made from taste hairs of *Ciniflo* (Dictynidae) and *Tegenaria* (Agelenidae), and the authors themselves emphasized their preliminary nature (Harris and Mill 1977). Responses were obtained to stimulation with certain salts (NaCl, KCl, KBr, choline chloride) and to hydrochloric acid (HCl), but neither here nor in any other case are reliable response spectra or characteristic curves available.

In the case of *Cupiennius*, too, there is a major gap in our knowledge when it comes to the physiological characteristics of chemoreceptive hairs. Figure 3 shows the distribution of chemoreceptive hairs on the tarsus of the right pedipalp of a small *Cupiennius salei* that had just undergone its first molt outside the egg sac. Already 11 such hairs are present. In the adult state this animal will have about 200 of them on the same, by then much larger segment, and they will be 300–500 μm long rather than about 60 μm.

In the scanning electron micrograph of figure 2 the approximately 0.4-μm-wide pore at the end of the hair is clearly visible. In other hairs of the same group this pore is subterminal, the end of the hair being occupied by a conspicuous cuticular spur that projects a few μm beyond the pore. All the evidence indicates that these sensilla are gustatory hairs, like the corresponding tip-pore sensilla of insects. In contrast to the general ignorance about the physiology of chemoreceptive sensilla in spiders, many reports

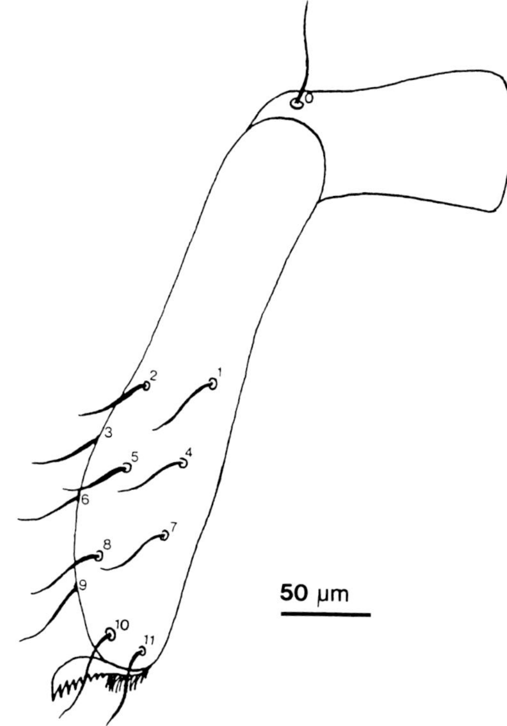

Fig. 3. Number and arrangement of chemoreceptive hairs on the pedipalp of a young *Cupiennius salei*, shortly after its first molt outside the egg sac. (Gingl 1998)

have been published about the significance of chemical signals in behavior, especially courtship (Tietjen and Rovner 1982, Pollard et al. 1987).

A Female Pheromone

In just this context, however, there is something new to say about *Cupiennius*. Its courtship has occupied us for a long time. It includes a "chemical phase" (see Chapter XX) during which the male like those of other species (Tietjen and Rovner 1982), becomes sexually excited as a result of making contact with a female's dragline. First, the pheromone released by the female along with the silk has now been identified. This required exhaustive behavioral experiments (Tichy et al. 2000) combined with appro-

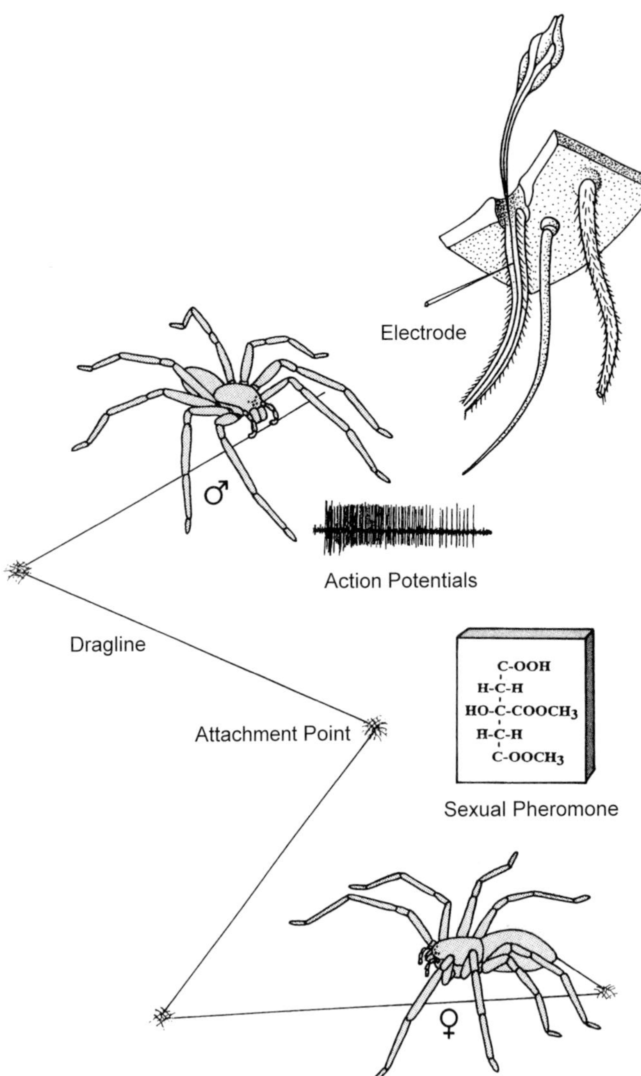

Fig. 4. The chemical phase of courtship in *Cupiennius*. The male comes across a female dragline that contains a female sexual pheromone, now known to be the S-dimethyl ester of citric acid, and examines it with his pedipalps. Action potentials of the contact chemoreceptive hairs are recorded by inserting an electrode into the shaft of the hair. (After Gingl 1998)

priate chemical analyses (Schulz et al. 2000). That such a chemical identification has been achieved for another spider is noteworthy; there has been only one previous case. A few years ago Schulz and Toft (1993) showed that (R)-3-hydroxybutyric acid and its dimer (R)-3-[(R)-3-hydroxybutyryloxy]-butyric acid are sex pheromones of the sheet-line weaving spider *Linyphia triangularis* (Linyphiidae).

A male *Linyphia triangularis*, without the long, cautious foreplay typical of many other web-building spiders, pushes in under the sheet-like web of his reproductive partner, which is slightly convex upward and resembles a canopy. Males are attracted by the pheromone released by the female, and evidently can enter without risk. Then something remarkable happens: the male destroys the female's web, while the female shows no aggressiveness at all, which can be interpreted as signalling her readiness to mate. Sometimes the web is reduced to a few threads within minutes. Then courtship and copulation follow (van Helsdingen 1965, Rovner 1968, Watson 1986).

By their spectacular act of destruction the males reduce the pheromone source and hence make it less likely that rivals will be attracted. Sensibly, the webs built by females that have already copulated do not contain pheromone, and have no attractant action. Furthermore, when a male is put into such a web, he leaves it entirely unharmed.

Now the second case is *Cupiennius salei.* As Stefan Schulz and Miriam Papke of the Institute for Organic Chemistry of the University of Hamburg discovered, in cooperation with Harald Tichy of our laboratory in Vienna, the pheromone on the silk spun by the females of "our" spider is the S-dimethyl ester of citric acid, or more precisely, the S isomer of 3,5-dimethoxycarbonyl-3-hydroxy-1-valeric acid. In honor of *Cupiennius* it was called *cupilure.* Although the sodium salt of this substance was ruled out as a pheromone component in a behavioral test on males, it does enhance the action of the dimethyl ester of citric acid. Chemical considerations suggest that under the humid conditions in the natural habitat of *C. salei* the dimethyl ester would be accompanied by its salt.

While preparing his Master's thesis Ewald Gingl, in collaboration with H. Tichy, took another important step. He was the first to reliably identify a pheromone receptor in spiders. It was already known that male *Cupiennius* examine the female draglines with the tarsi of their pedipalps (see Chapter XX), on which there are about 300 chemoreceptive hair sensilla (Fig. 4). Electrophysiological recordings of the responses of these "tip-pore hairs" (Foelix and Chu-Wang 1973) when they were brought into contact with female threads or synthetic pheromone would clearly be of great interest. Gingl managed to record action potentials by carefully inserting the fine tip of a tungsten electrode into the shaft of the hair with a nanostepper, which advances the electrode in 1 μm steps at a rate of 13 mm/s. Such high precision is essential, because the diameter of the shaft, through which the dendrites of 19 receptor cells pass, is only 2 μm. When the hair tip was touched with a piece of filter paper containing pheromone, or with silk spun by a female, several of the receptors became active (Fig. 5).

The sensilla on the pedipalps respond only to the pheromone (and related compounds), and to saccharose and NaCl. The hair sensilla on the tarsi of the walking legs respond to a broad range of olfactory stimuli: various aldehydes, alcohols, amines, carbonic acids, esters and ketones. That is, on the walking legs there are tip-pore sensilla that can be stimulated both by

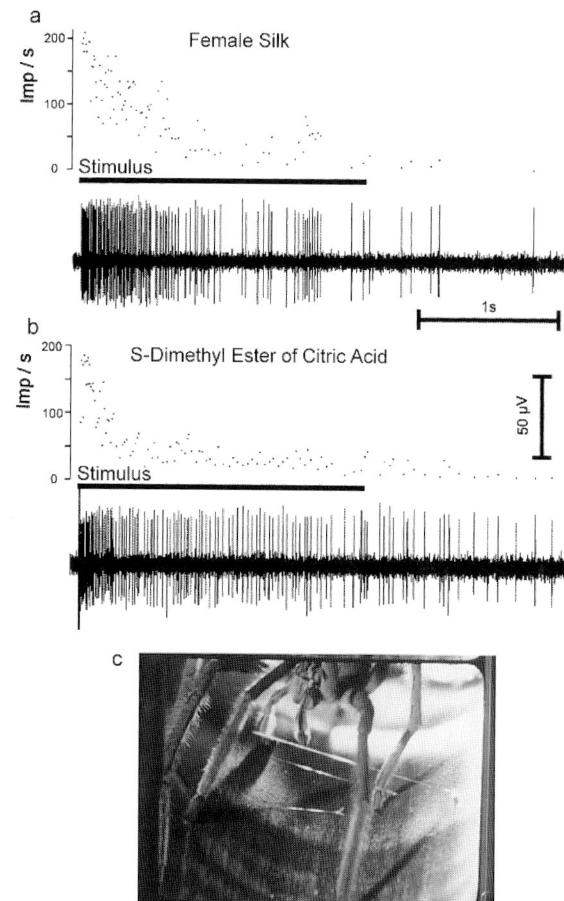

Fig. 5 a–c. The response of individual sensory cells of a tip-pore sensillum on the male pedipalp (*Cupiennius salei*) upon contact with **a** female silk and **b** the S-dimethyl ester of citric acid (the synthetic female pheromone). Stimulus duration (see *bar* above action potentials) 2 s. **c** Male examining a female dragline with his pedipalps. (**a** and **b** after Gingl 1998)

contact with pheromone and by airborne odors. Gingl also found a group of sensilla that respond only to airborne stimuli.

Perhaps this is the solution to the puzzle of why spiders have no wall-pore sensilla, the typical olfactory organs of insects. Of course, it could also be that the spider phenomenon (lack of obvious olfactory sensilla and dual function of the same hair as taste and odor receptor) is more widespread than we have thought. Städler and Hansen (1975) have described it for tip-pore sensilla of caterpillars of the tobacco moth. However, the olfactory stimuli he used, freshly cut leaves of the caterpillars' food plant, had to be extremely close to the sensillum tip: they were ineffective at a distance of only 600 μm. In the experiments on *Cupiennius* odor stimuli were presented by the classical method, with air pulses.

XIII Hygro- and Thermoreception: Blumenthal's Tarsal Organ

This is the tarsal organ mentioned as possibly olfactory in the preceding chapter. The "tarsal" part of its name is self-explanatory: there is one such organ on the tarsus of each walking leg and the two pedipalps, in the middle of the dorsal surface near the end of the segment. Heinz Blumenthal was the first to describe it in detail in 1935, while working in the spider department of the Berlin Zoological Museum. As his doctoral thesis, he identified it in specimens of 300 genera, belonging to 46 families. Friedrich Dahl (1883) had noticed the organ about 50 years earlier but erroneously interpreted it as the socket of a deteriorated "auditory hair" (trichobothrium) and never mentioned it again. A number of simple behavioral experiments, in particular, led Blumenthal to regard the tarsal organ as both a hygroreceptor and an olfactory receptor. After the tarsal organs had been plugged with Vaseline, the spiders not only stopped turning toward a drop of water but also no longer turned away from a drop of a strong-smelling substance such as terpineol, pyridine, "clove oil" or oil of wintergreen.

It took almost forty more years until research on the tarsal organs was again advanced by publication of a description of their morphology and fine structure in *Araneus diadematus* (Foelix and Chu-Wang 1973b). These authors showed that the capsule, an open pit in the cuticle, contains six or seven sensilla. In each of them a receptor-cell dendrite runs without branching to a pore at the apex of the sensillum (hence the name "tip-pore" sensillum), where it is separated from the outside air only by a thin layer of optically amorphous material. These findings have since been confirmed for *Cupiennius salei* (Anton and Tichy 1994). Here the capsule of the tarsal organ is an air-filled cavity in the cuticle about 18 μm long and 12 μm deep, which opens to the exterior by way of a hole measuring about 4×7 μm; standing on its floor are 7 nipple-shaped sensilla (Fig. 1). Six of these sensilla are innervated by 3 receptor cells, each with one dendrite. The remaining sensillum has only 2 receptor cells (Fig. 2). The diameter of the tip pore, where all dendrites terminate, is 0.7 μm.

What do these structural features imply regarding the function of the tarsal organ? This is hard to say, although one is immediately aware of parallels with contact chemoreceptors of insects. That the sensilla are enclosed in a capsule with a small opening suggests olfactory rather than gustatory reception, and an additional or alternative hygroreceptive function cannot be ruled out. Therefore an electrophysiological experiment is needed.

The first electrophysiological recordings from a tarsal organ were achieved by Klaus Dumpert in 1978, in our laboratory at the University of Frankfurt am Main. He used *Cupiennius salei* for these experiments, and his most exciting finding was that one of several cell types in the tarsal organ of males responds to the odor released by living females. Neither males nor females responded to male odor. Conclusion: the tarsal organ houses pheromone receptors. This interpretation is all the more attractive since behavioral experiments show that female pheromones are involved in the courtship of *Cupiennius* (see Chapter XX). However, Dumpert saw no electrophysiologically detectable response to the threads spun by the female, which do affect behavior (Barth and Schmitt 1991). Some of the cells responded to formic acid, valeric acid, caproic acid, t-2-hexenal, hexanone and/or tobacco smoke by increasing their discharge rate above

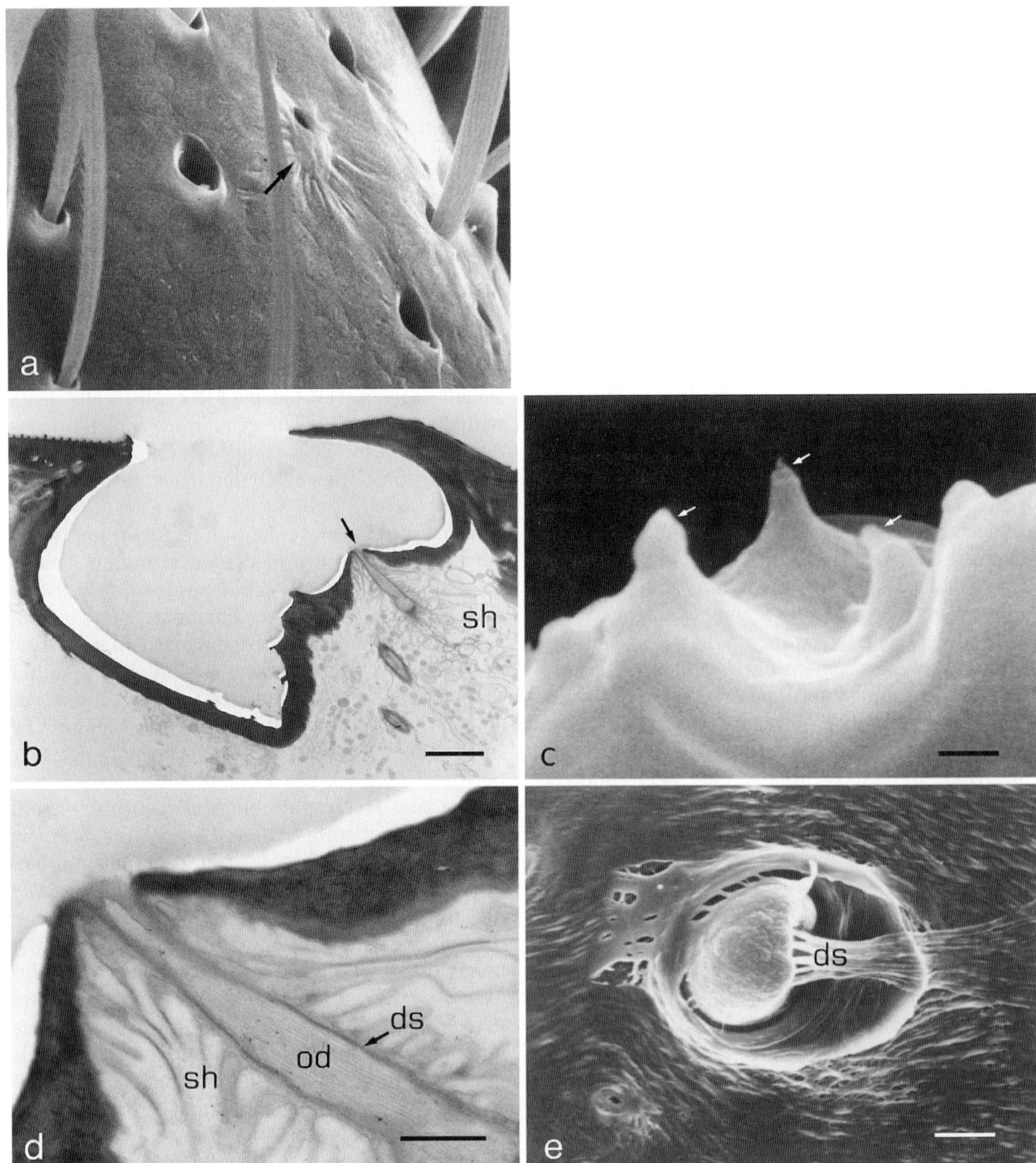

Fig. 1a–e. The Blumenthal tarsal organ and its tip-pore sensilla. **a** Scanning electron micrograph of the Blumenthal organ (*arrow*). **b** Longitudinal section through the cuticular capsule with its aperture pointing toward the tip of the leg. *Arrow* marks the pore at which the dendrites of sensory cells terminate. *sh* Sheath cells; *bar* 2.5 μm. **c** After the cuticular capsule has been opened, the sensilla (*arrows*) projecting from its bottom can be seen; *bar* 5 μm. **d** Longitudinal section through pore of tip-pore sensillum. *Od* Outer dendritic segment extending close to the surface of the pore, *ds* dendritic sheath; *bar* 1 μm. **e** Tarsal organ seen from inside (exuvia); *ds* dentritic sheaths connected to the cuticular capsule; *bar* 10 μm. (Anton and Tichy 1994)

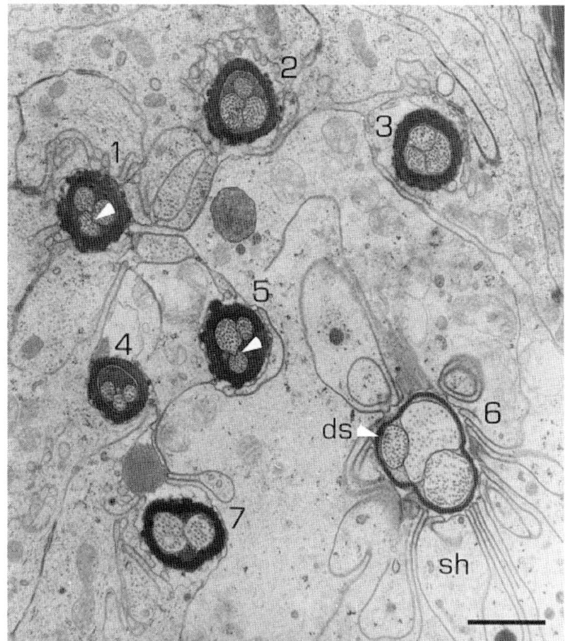

Fig. 2. The Blumenthal tarsal organ. Cross-section closely below cuticular capsule showing all seven groups of dendrites. Six groups (*1–6*) consist of three unbranched dendrites each, whereas group *7* is made up of only two dendrites. The dendritic sheaths (*ds*) are surrounded by the lamellae of the sheath cells (*sh*). *Arrows* point to interdigitation commonly seen among neighboring dendrites. *Bar* 1 μm. (Anton and Tichy 1994)

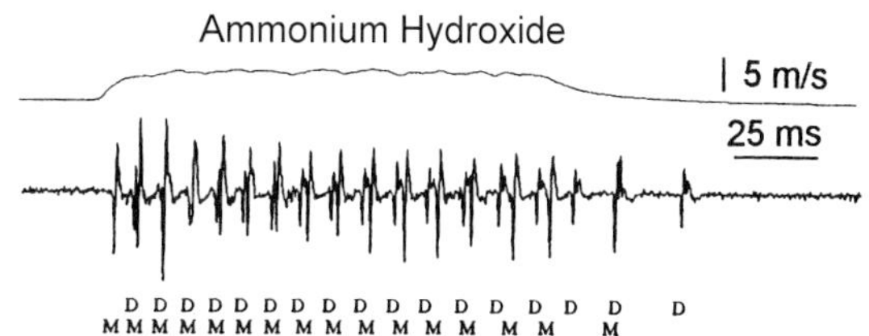

Fig. 3. Responses of a moist (*M*) and of a dry cell (*D*) to a puff of ammonium hydroxide (dilution to 3×10^{-3} in demineralized water). *Upper trace* shows velocity of stimulating air flow. (Ehn and Tichy 1994)

the spontaneous level, and to polyamine by either increasing or decreasing it. No response at all was obtained to stimulation with benzoquinone, geraniol, humid air or the odor of living or crushed prey (fly, cockroach). We had often noticed that *Cupiennius* does not locate its prey by means of olfaction (see, for example, Seyfarth and Barth 1972); as long as a prey animal stays very still, the spider will walk right past it.

At this stage we would deduce that the tarsal organ is an olfactory receptor, and that hygroreception is unlikely. But then some more electrophysiological studies were done on the tarsal organ of *Cupiennius salei*, and it was a new ball game: thermo- and hygroreception clearly established, chemoreception not ruled out. How did that happen?

Rudolf Ehn and Harald Tichy (1994, 1996a,b), 16 years after Klaus Dumpert, made a new analysis of the tarsal organ in our laboratory at the University of Vienna. They (see below) found no pheromone receptor cells, but with confidence-inspiring regularity recorded from cells responsive to moisture, dryness and heat. The responses were characterized quantitatively, leaving no doubt that the tarsal organ functions as a hygro- and thermoreceptor. But there was an interesting oddity: the hygroreceptors were unlike all the known hygroreceptors of insects (see references in Ehn and Tichy 1994) in that they also responded to certain odor substances, both natural mixtures and synthetic odorants. When stimulated with CO_2, however, they never responded. The dry cell and the moist cell behaved synergistically, responding with progressively decreasing strength to the following substances: ammonium hydroxide (Fig. 3), aliphatic amines with short chain length (C_2 to C_4), formic acid, acetic acid, longer-chain aliphatic amines (C_5 to C_9; decreasing efficacy with increasing chain length). There was no response to stimulation with aliphatic alcohols (C_4 to C_9), aldehydes (C_4 to C_8), carboxylic acids (C_3 to C_9), esters (C_4 to C_8), ketones (C_4 to C_9) and mercaptans (C_3 to C_8). Nor could any response be elicited when the stimulating air stream was passed over living males or females of *Cupiennius*, over their threads or feces, or over living domestic crickets (*Acheta*) and flies. But the moist and dry cells did respond to the odor emitted by rotting meat and by a living cockroach (and its feces).

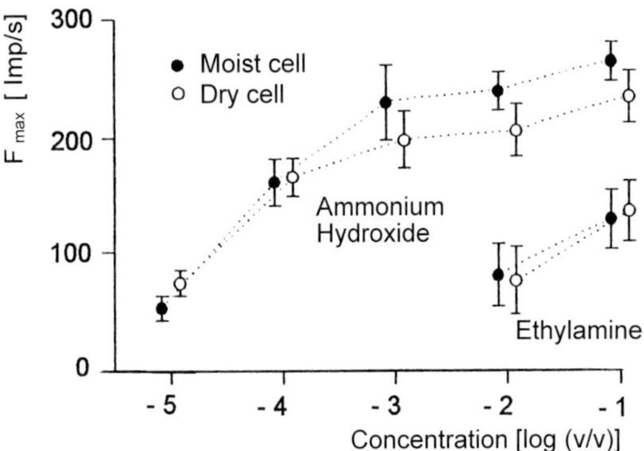

Fig. 4. Simultaneously recorded responses of a moist cell and of a dry cell to a range of concentrations of ammonium hydroxide and ethylamine, dissolved in aqua demin. Each value is the mean of six stimulus applications; *bars* give SD. For clarity, values for the same concentration are drawn *side by side*. (Ehn and Tichy 1994)

Having found that the moist cell and the dry cell respond in the same or at least a similar way to odors that we perceive as stinking and offensive, Ehn and Tichy (1994) tested the two most effective chemicals, ammonium hydroxide and ethyl amine, to see whether there is some "range fractionation" with regard to stimulus intensity. As shown in figure 4, this is not the case: threshold, range of effective intensities and shape of the response-vs.-intensity curve are all so similar that one is inclined to infer the same mechanism of action in both cells. The warm cells do not respond to chemical stimuli at all.

Hence the question about the chemoreceptive function of the tarsal organ has not yet been exhaustively answered, and we must return to it again. First, however, a more detailed analysis of the responses of the tarsal organ to moisture and temperature stimuli.

1 Hygroreception

It is difficult to design precisely quantified moisture stimuli, in particular because the temperature must be kept constant. Richard Loftus (1976), pioneer of neat, quantitative research on hygro- and thermoreception in arthropods, gives detailed advice on how to solve the problem in his publication about the dry receptor on the cockroach antenna. Harald Tichy (1987), a colleague of Richard Loftus for many years at the University of Regensburg, brought all this know-how along when he came to Vienna, developed it further and in collaboration with Rudolf Ehn constructed a new, considerably more powerful version of the elaborate stimulating machinery. The moisture stimuli are presented by way of two airstreams at the same temperature (±0.03 °C) that can be aimed at the tarsal organ in alternation and (if desired) in rapid succession. The two streams, flowing at 2.4 m/s, differ in their relative humidity, which can be adjusted with ±3% accuracy by means of a hygrometer.

Although they behave the same when presented with chemical stimuli (Fig. 3), the sensilla respond antagonistically to moisture stimuli: moist and dry cells can be distinguished. A moist cell increases its discharge rate when the relative humidity rises, and a dry cell does so when the humidity falls (Fig. 5). The maximal discharge rate in a response to a stepwise change in humidity is linearly related to the size

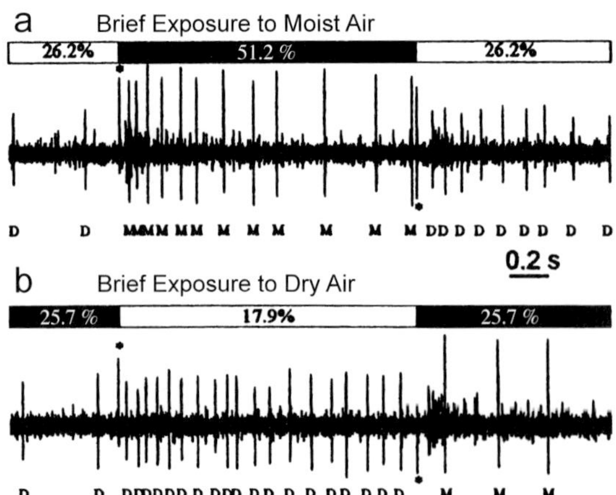

Fig. 5 a, b. Antagonistic responses of a moist (*M*) and a dry cell (*D*) to a 1.3-s exposure to moist (**a**, relative humidity 51.2%) and dry air (**b**, 17.9%) administered by quickly shifting from an adapting air current (directed onto the tarsal organ for 3.5 min) to a second air current of lower or higher constant humidity. *Asterisks* mark artifacts due to the fast change of the air jets' position. (Ehn and Tichy 1994)

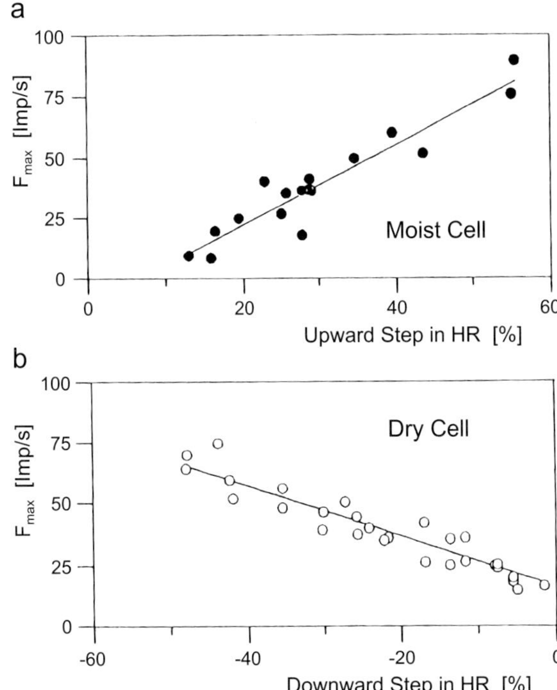

Fig. 6a, b. Simultaneously recorded responses of a moist cell (**a**) and of a dry cell (**b**) to step-like changes in humidity towards higher and lower humidity values, respectively. (Ehn and Tichy 1994)

of the step, and is independent of the relative humidity to which the cell had adapted before the step occurred (Fig. 6).

The discharge rate changes by about 0.8 Hz to 2 Hz per 1% change in relative humidity. These values in themselves indicate a high resolution, though resolution also depends on the variability of the response. According to the method described by Loftus and Corbière-Tichané (1981), it can be calculated that the moist cell reliably discriminates two humidity levels if they differ by at least 11% relative humidity. The dry cell is very similar, distinguishing stimuli that differ by 10%.

The Blumenthal's tarsal organ of *Cupiennius salei* has recently again raised the question about the primary processes in arthropod hygroreception (Tichy and Loftus 1996). The discussion still revolves around three main models, and it could ultimately turn out that more than one of them applies, in different moist cells. In the case of *Cupiennius* the "electrochemical hygrometer" is the most attractive model. It is based on the hypothesis that the humidity of the ambient air changes the electrolyte concentration of the liquid surrounding the dendrite terminals: the dryer the air, the higher the rate of evaporation and hence the greater the change in electrolyte concentration, which determines the electrical potential across the dendritic membrane. In contrast to the hygroreceptive sensilla of insects, in the tarsal organ of spiders the apical pores are "open" (Fig. 7). Apparently this enables the slight outward flow of lymph that the model requires, the existence of which also seems plausible in view of the deposits of material that normally accumulate over the pore. This lymph flow ought not to be arbitrarily massive; in fact, it must be only slight if the humidity is to affect the electrolyte concentration. A limit may be placed on lymph flow in the tarsal organ of *Cupiennius* by the small space available for movement of fluid between the dendrites and the dendritic sheath. In the spider as in insects, however, one major remaining question is why three dendrites surrounded by the same medium – the two antagonistic humidity receptors and the warm receptors – respond differently to the same stimulus. The reason can really only lie in differences between the dendritic membranes. Unfortunately, there are no experimental data that bear on this question.

For the sake of completeness, here in brief are the two other models for hygroreception. (i) The mechanical hygrometer model. The central postulate here is the involvement of hygroscopic materials that swell and shrink, such as the cuticle. For the insects this idea, that moisture sensitivity is mediated by mechanical changes, is currently the favorite (see references in Tichy and Loftus 1996). (ii) The more complicated psychrometer model. Here the hygrometers are not basically mechanoreceptors but rather thermoreceptors. A psychrometer is a wet- and dry-bulb hygrometer, which measures the relative temperature drop caused by evaporation at a thermometer that is kept wet; in the case of an arthropod, it would be lymph that evaporates (Lax and Synowietz 1967; see Tichy and Loftus

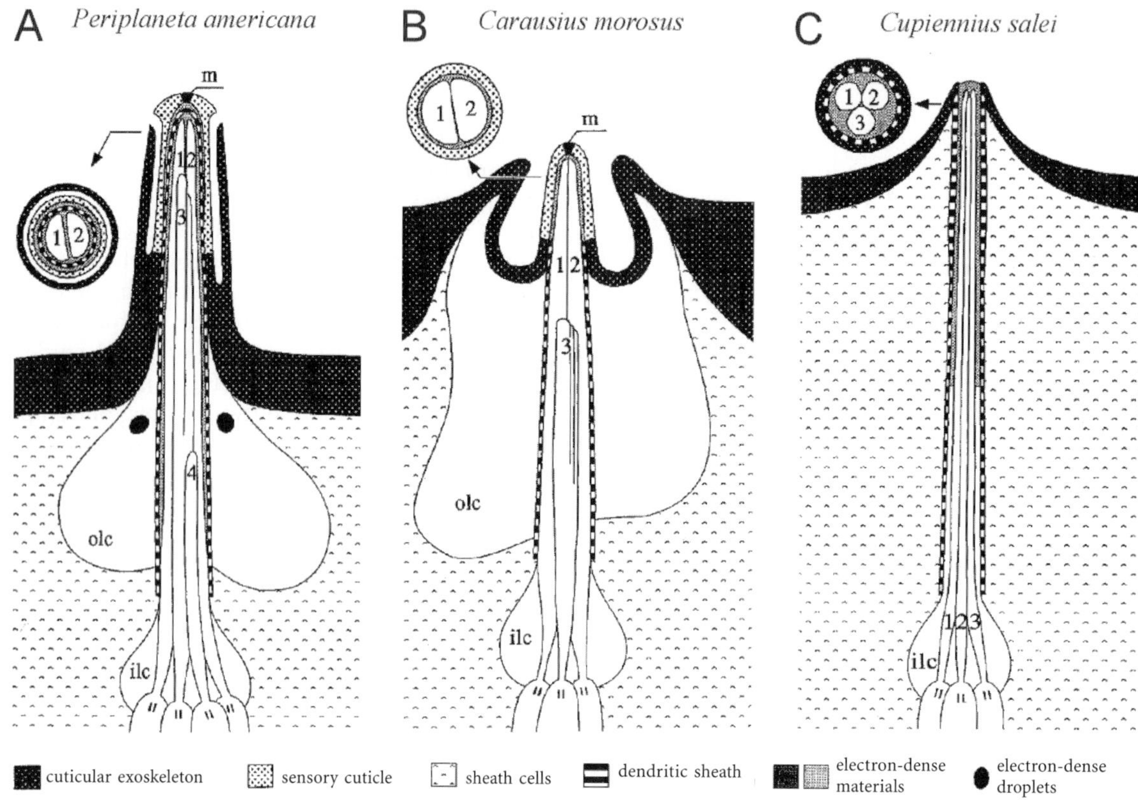

Fig. 7 A–C. The Bauplan of hygro-thermoreceptors in different arthropods. Cockroach (*Periplaneta americana*): sensillum capitulum; stick insect (*Carausius morosus*): sensillum coeloconicum; spider (*Cupiennius salei*): tip-pore sensillum, Blumenthal tarsal organ (see text). (Tichy and Loftus 1996)

1996). This model faces the particular difficulty that two independent temperature measurements must be made in the ambient air, one with a dry "thermometer" to compare with the reading of the thermometer that has a wet surface. The difference between the two readings is a measure of cooling by evaporation. Despite all the problems a psychrometric measurement would have to overcome (Tichy and Loftus 1996), it is currently favored both for the silkmoth *Bombyx mori* and for the body louse *Pediculus humani corporis* (Steinbrecht and Müller 1991, Steinbrecht 1994, see Tichy and Loftus 1996).

2 Thermoreception

The moist and the dry cell form a triad with the warm cell. To stimulate the warm cell, again two airstreams are used, which can be rapidly aimed onto the tarsal organ or turned away from it (Fig. 8). The warm cell raises its (peak) discharge rate in direct proportion to the magnitude of a temperature step, up to about 4 °C step size. In response to larger steps, the discharge rate becomes only slightly higher (Fig. 9). On average, the temperature of the airstream must be increased by 0.03 °C in order to increase the impulse frequency by 1 Hz. The resolution limit has been found to be 0.4 °C, on the basis of the peak frequency of the response.

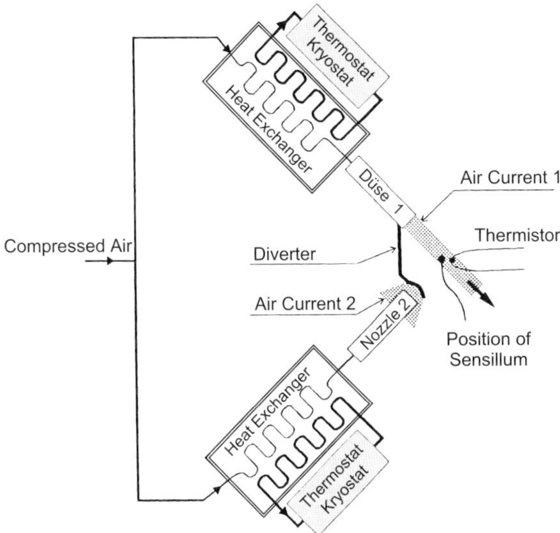

Fig. 8. Design of apparatus for the application of quick step-like and slowly oscillating temperature changes as well as stationary temperatures. Quick changes: *air current 1* is replaced by *air current 2* by diverting *nozzle 1* with magnetic valves. Slowly oscillating and stationary stimuli: only air current 1 is used. (Ehn 1995)

Our ideas about heat are formed to such a great extent by our own sensations that we often fail to realize that the quality *temperature* is distinct from the quantity *warmth*. The sensation of warmth provides information about the amount of heat being supplied or extracted per unit time and area. In contrast, the temperature is a parameter independent of amount, though conclusions about amounts of heat can be drawn indirectly from changes in temperature. A temperature gradient in turn is a prerequisite for heat flux, which occurs spontaneously only in the direction from a higher to a lower temperature.

In later experiments Ehn and Tichy expanded the stimulus spectrum and compared the responses of the warm cell to constant, slowly oscillating and abruptly changing temperatures. This program in turn enabled a comparison with similarly tested cold cells of insects, which led to some interesting interpretations of the functional significance of certain features in the fine structure of the tarsal organ (Ehn and Tichy 1996a). This is what they found.

Rapid Temperature Changes

The sensitivity to stepwise stimuli depends only slightly on the initial temperature. At 18.5 °C a temperature step of 0.023 °C suffices to raise the peak impulse frequency by 1 Hz, whereas at 34.6 °C an increase of 0.043 °C produces the same effect. The mean sensitivity is 35.2 Hz/°C (Fig. 10).

Slow Temperature Oscillations

The activity of the warm cell in the tarsal organ follows slow (±0.015 °C/s) and small (<1.5 °C) temperature changes with astonishing fidelity (Fig. 11), although with a phase lead of +35° with respect to the momentary temperature and a delay of -45° with respect to the velocity of the temperature change. The much greater sensitivity to the rate of change of temperature (°C/s) than to the momentary temperature (°C) is expressed quantitatively as follows: 105 (imp/s)/(°C/s) as compared to 1.94 (imp/s)/°C. Put another way: a 1-Hz increase in impulse frequency can mean both an increase in momentary temperature by 0.5 °C (constant rate of change of temperature) and an increase in the rate of change by +0.01 °C/s.

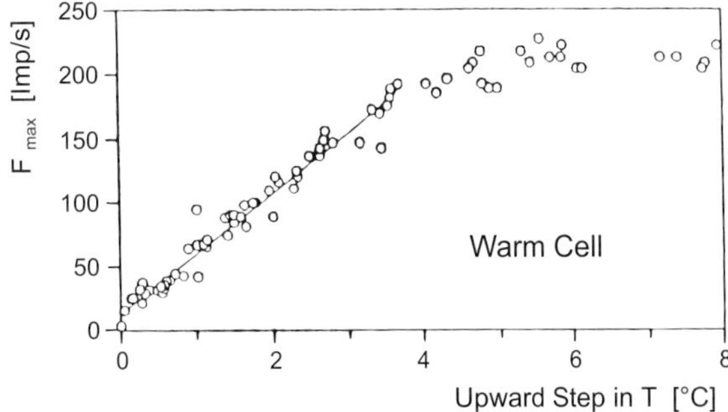

Fig. 9. The response of a warm cell to upward steps in temperature after 10 min of adaptation to a given temperature between 21 and 28 °C. A different constant higher temperature was presented every 3.5 min for 1 s. (Ehn and Tichy 1994)

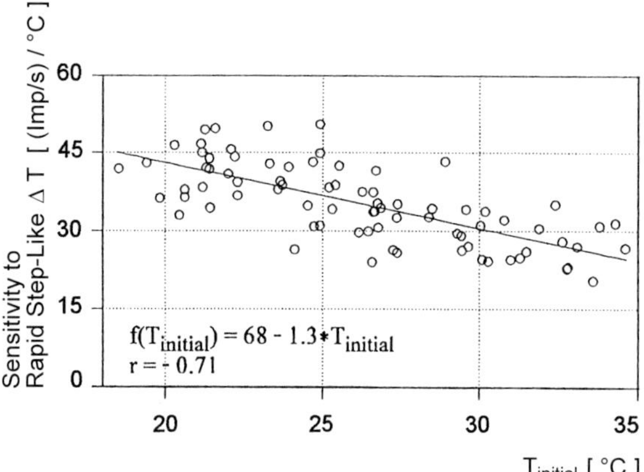

Fig. 10. Sensitivity of 21 warm cells to rapid step-like warming, as a function of initial temperature. Each value of sensitivity is given by the slope of the regression line approximating the relationship between peak frequency and step size. The mean sensitivity decreases linearly with increasing initial temperature. (Ehn 1995)

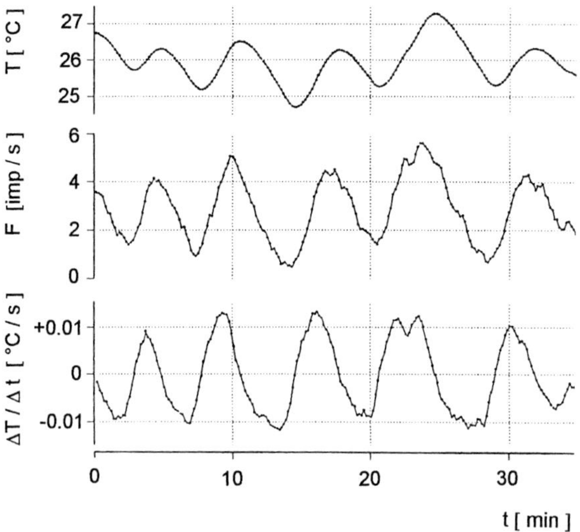

Fig. 11. The response of a warm cell to slow temperature changes. *Top* Instantaneous temperature (*T*); *middle* action potential frequency of warm cell (*F*); *bottom* rate of temperature change. Impulse frequency is ahead of instantaneous temperature but lags behind the rate of temperature change. (Ehn 1995)

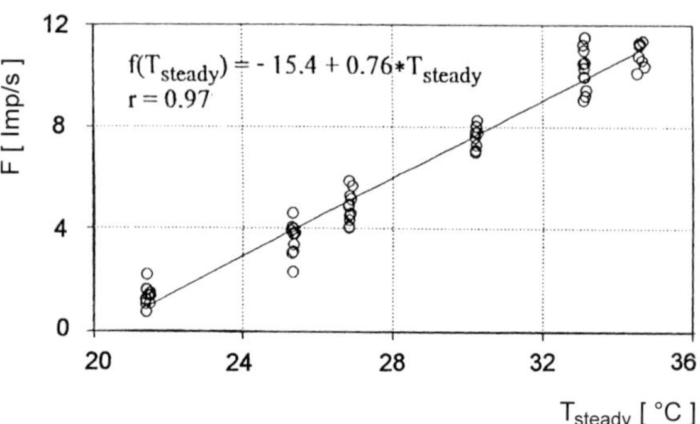

Fig. 12. Response of a warm cell to steady temperatures. (Ehn 1995)

Constant Temperatures

As would be expected, the discharge rate of the warm cell also depends on the level of a constant temperature. For instance, it can be 0.79 Hz at 21.3 °C but 10.9 Hz at 34.6 °C (Fig. 12). On average, the discharge rate rises by 1 Hz for a 1.5 °C difference in constant temperature.

Comparison with Insects and a Speculation about the Primary Process

Now it would be nice to know what these numbers tell us in the context of temperature detection in general (Ehn and Tichy 1996). How does the sensitivity of *Cupiennius* compare with that of other arthropods, and what are the implications?

Figure 13 compares what can be quantitatively compared. Only three insect thermoreceptors have been examined in experiments identical to and hence just as precise as those involving *Cupiennius*: cold receptors on the antennae of the cockroach *Periplaneta americana* (Loftus 1968, 1969), of the migratory locust *Locusta migratoria* (Ameismeier and Loftus 1988) and of the moth *Antheraea pernyi* (Haug 1986; Ameismeier and Loftus 1988). In addition, similar experiments have been performed with the cave beetle *Speophyes lucidulus* (Loftus and Corbière-Tichané 1981, 1987; Corbière-Tichané and Loftus 1983). Because the slopes of the curves for

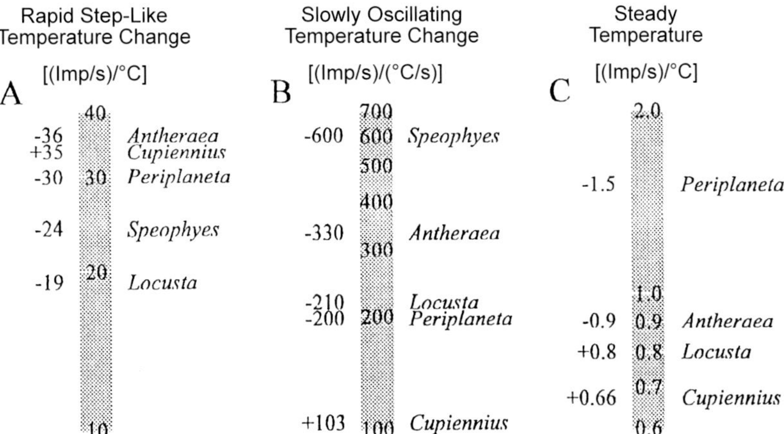

Fig. 13A–C. Sensitivity of thermoreceptors (see values given on *left side of columns*) of different arthropods to quick, step-like changes in temperature (**A**), to slowly oscillating temperature changes (**B**), and to stationary temperatures (**C**). *Periplaneta* (Loftus 1968, 1969), *Speophyes* (Loftus and Corbière-Tichané 1981, 1987; Corbière-Tichané and Loftus 1983), *Antheraea* (Haug 1986; Ameismeier and Loftus 1988), *Locusta* (Ameismeier and Loftus 1988), *Cupiennius* (Ehn 1995). (Ehn and Tichy 1996a)

warm and cold cells have opposite signs, absolute values are used for the comparison.

Cupiennius takes second place for sensitivity to abrupt step stimuli, but comes last in the other two categories. The cave beetle is the most sensitive of the five animals to slowly oscillating temperature changes but relatively insensitive to step stimuli, which it would also be unlikely to encounter in its very special biotope. Such differences in sensitivity to various kinds of thermal stimuli are less pronounced in the other insects. On the basis of their familiarity with the morphological details of the stimulus-receiving structures of the various temperature receptors (Corbière-Tichané 1971; Zimmermann 1991; Anton and Tichy 1994), Ehn and Tichy (1996a) formulated an interesting speculation: differences in the area of the dendritic membrane (these are considerable; Fig. 14) and in the position of the tip of the dendrite relative to the outside world are associated with functional differences. Regarding membrane area, the extremes of the spectrum are represented by the cave beetle *Speophyes*, where the ends of the dendrites branch to form a giant fan, and the spider *Cupiennius*, with a simple, small dendritic cylinder (diameter ca. 0.3 μm). Hypothetically, then, the length of the outline of a dendrite cross section could be taken as a relative measure of sensitivity; this length is ca. 500 μm for the cave beetle and only ca. 1 μm for *Cupiennius* (*Antheraea* ca. 20 μm, *Periplaneta* ca. 6 μm, *Locusta* ca. 4 μm). The rank order found in this way corresponds to the rank order of sensitivity to slowly changing and constant temperatures. That the cockroach *Periplaneta* is more sensitive than the moth *Antheraea* and to this extent is out of order may be explained by the fact that the data for the cockroach were obtained from the most sensitive of all its cold cells (Ameismeier and Loftus 1988).

Sensitivity to rapid temperature changes is correlated not with the surface area of the dendrite but with the degree of exposure of the dendritic membrane. In this respect *Antheraea* ranks ahead of *Speophyes*. The possible structural reason: in *Speophyes* the dendrite ends below the base of the sensillum, whereas in *Antheraea* it runs through a slender peg and ends ca. 60 μm above the antennal surface. Given equal heat conductance, therefore, after a temperature change the dendrite of *Antheraea* should reach the new temperature more rapidly than the dendrite of *Speophyes*. The warm cell of *Cupiennius* supports this line of reasoning: its dendrite ends in a pore at the tip of a ca. 5-μm-high cuticular peg; its sensitivity to quick temperature changes is correspondingly high. And the lower sensitivity of *Periplaneta* and *Locusta*? The dendrites end below the surface of the antenna!

To estimate the threshold for the rapid temperature changes, to which the warm cell of *Cupiennius* is so sensitive, Ehn and Tichy (1996b) determined the frequency-dependent noise. On

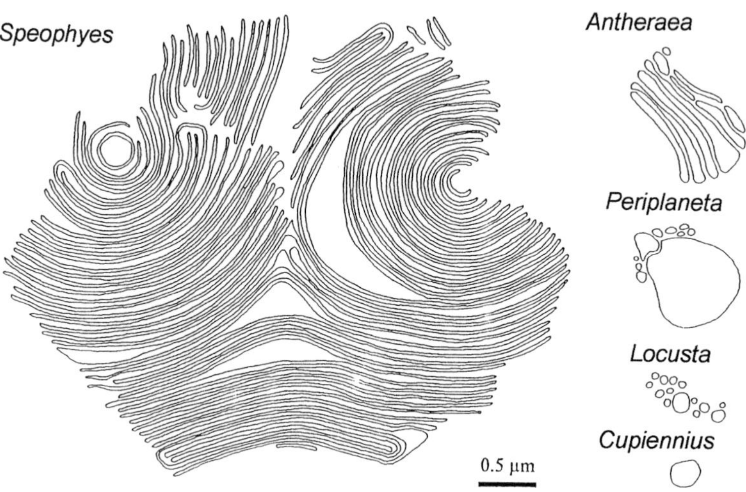

Fig. 14. Cross-sections of dendrites of thermoreceptive cells of different arthropods at same scale. Note the enormous differences in membrane area. *Speophyes* (Corbière-Tichané 1971), *Antheraea* (Zimmermann 1991), *Periplaneta* (Yokohari 1981), *Locusta* (Ameismeier and Loftus 1988), *Cupiennius* (Anton and Tichy 1994). (Ehn and Tichy 1996a)

the assumption that the signal-to-noise ratio is 1 at the threshold (which is not necessarily the case; it could be smaller) and that the upper limiting frequency is 10 Hz, then the just resolvable temperature change would be about 0.6 °C (initial temperature 25 °C).

The thermal sensitivity of the whole animal depends not only on the sensitivity of individual receptor cells but also on their total number and the way they are connected in the central nervous system. In all its tarsal organs *Cupiennius* has a total of about 70 warm cells. With adequate convergence, the behavioral threshold could be distinctly lower than the threshold for excitation of single sensory cells. In any case, it appears that *Cupiennius salei* can measure the temperatures and temperature fluctuations to be expected in its habitat. The annual mean temperature in its native Guatemala and Mexico is 20 °C (Barth et al. 1988). Because of convection, fluctuations as large as 1 °C/s are likely to be superimposed on the basic temperature. These rapid changes are certainly above threshold for *Cupiennius.* It would be interesting to measure with high temporal resolution the temperature changes that actually occur at the animal under natural conditions in the field. That it is important for *Cupiennius* to adjust its behavior to the ambient temperature can be inferred from the fact that temperature is closely related to humidity, which must be above a certain level for these spiders to survive (see Chapter III). Ecologically, then, the presence of thermo- and hygroreceptors and their close proximity to one another in the Blumenthal organ make good sense.

The Route to the Central Nervous System

As yet hardly any electrophysiological data are available for temperature-sensitive interneurons of arthropods (Nishikawa et al. 1991; *Periplaneta americana*). In an anatomical study, however, my colleagues Sylvia Anton and Harald Tichy (1994) traced the route of the primary afferents into the subesophageal mass. Here some of them terminate in a special area of neuropil which the authors named in honor of the man who first described the tarsal organ (Blumenthal 1935; Dahl had already mentioned it in 1883, but thought it was a rudimentary cup of a trichobothrium). Blumenthal's neuropil will be described in Chapter XV (Fig. 8). We do not yet know the extent of parallel central processing of the various stimulus modalities to which Blumenthal's tarsal organ responds. And we are also still faced with the mystery of the responses to chemical stimuli. How significant are they? Are they artefacts, especially given that the effective substances mostly have an obnoxious (to us) stench? It is easy to imagine that their vapors can reach the receptor cell, through the pore at the tip of the tarsal-organ sensilla. Behavioral experiments are urgently needed here.

It may help to consider an observation we commonly made in Central America while looking for *Cupiennius*: in civilized surroundings such as the vicinity of farmhouses, the spiders were often found in outdoor privies. Is that because of the strong odors there, or because of the prey insects such odors might attract? As detailed in Chapter XII, we now know that chemoreceptive hair sensilla are also sensitive to female courtship pheromone.

The hygroreceptor cells of insects do not respond to pungent chemicals (Tichy 1979), and the associated cuticular sensilla, unlike the sensilla of the tarsal organ, have no pore. However, insects do resemble *Cupiennius* in that the hygroreceptors are characteristically combined with another modality: a typical triad comprises a moist, a dry and a cold cell. Warm cells in insects have been found only in combination with cold cells, not with hygroreceptor cells (Altner and Prillinger 1980; Altner and Loftus 1985).

C The Central Nervous System and Its Peripheral Nerves

XIV The Central Nervous System

The experienced reader will expect any chapter about a central nervous system and a brain to end with the admission, "we still know far too little!" The spiders are no exception; there is more to ask here than to report. But science never really comes to a halt and, as in the evolution of life forms, the good is continually being replaced by the better. Instead of being discouraged by our ignorance, we can take comfort in the thought that our modest findings will eventually open the way to a deeper and broader understanding. Fortunately for the present purpose, of all the spiders it is *Cupiennius* in which the CNS has been most comprehensively studied. This applies mainly to its structure; we know only a little about the electrophysiological properties of central neurons even in *Cupiennius*, and nothing at all about other spiders.

Within the overall problem of what happens to the information the sense organs send to the CNS and how it is used to shape behavior, the sensory tracts are of particular interest. To make these comprehensible and to work out what is special about spiders, we have to make a relatively wide excursion.

1 The General Structure of the CNS

As in the other arthropods, the structure of the spider nervous system is closely related to the subdivisions of the body and the appendages that are present. Most spiders are relatively simple in this respect: the body comprises only a prosoma and an opisthosoma, there are no antennae and the only appendages on the opisthosoma are the spinnerets. Therefore we should not expect to see much of a primitive "rope-ladder" nerve cord, and this expectation is confirmed. The nervous tissue is enormously concentrated (Fig. 1). However, the CNS is like that of other arthropods in that one part is above the esophagus and another below it; these supra- and subesophageal masses are connected together by esophageal connectives.

The supraesophageal region is the actual brain, from an ontogenetic viewpoint, and it also incorporates the cheliceral ganglia, which migrate forward during ontogeny. That the cheliceral ganglia were originally in the region of the subesophageal ganglion is evident, for example, in the postoral position of the commissures between them. To reflect this conjunction of the cheliceral ganglia with the supraesophageal ganglion, the specialists speak of a "syncerebrum" in spiders. The subesophageal ganglionic mass is the entire remainder of the central nervous system and consists of two pedipalp ganglia, eight leg ganglia and the fused abdominal ganglia.

The eight optic nerves enter the brain from the front (Fig. 1); the thickest are those of the posterior median and lateral eyes, which have retinas comprising ca. 2300 and 2150 photoreceptors and are thus considerably larger than the principal eyes (650) and the anterior secondary eyes (200). When the dorsal surface of the brain is viewed, it is easy to discern in the anterior region the two optic lobes, in which the visual information from the eyes is processed. Below the optic nerves on either side a sturdy nerve runs to the musculature of the chelicerae. The pharynx and the poison gland are also innervated from the cheliceral ganglion.

Turning to the subesophageal ganglion, the most obvious external features are the imposing nerves to the pedipalps and to the walking legs.

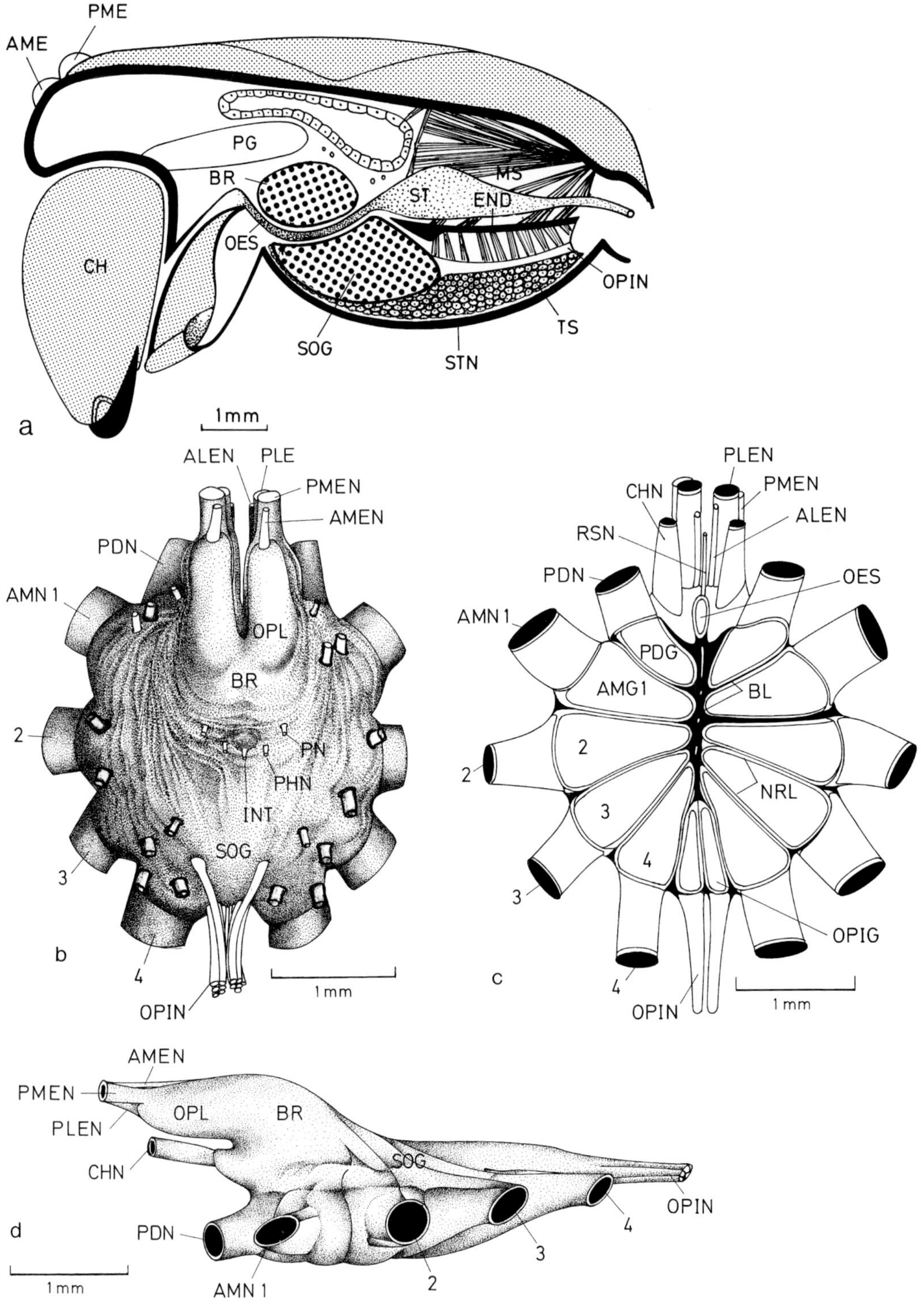

PME
AME
PG
BR
MS
ST
END
OES
CH
OPIN
SOG
TS
STN
a
1mm
ALEN
PLE
PMEN
AMEN
PDN
AMN 1
OPL
BR
2
PN
PHN
INT
SOG
3
4
b
1mm
OPIN
PLEN
CHN
PMEN
ALEN
RSN
PDN
OES
AMN 1
PDG
AMG1
BL
2
NRL
3
4
OPIG
c
1mm
OPIN
AMEN
PMEN
OPL
BR
PLEN
CHN
SOG
OPIN
d
PDN
4
3
1mm
AMN 1
2

The stomatogastric nervous system comprises an unpaired nerve that leaves the ganglion ventrally and runs forward, and the visceral nerves, which emerge dorsally and run backward.

Before we take a look into the nervous system and identify the various processing centers and tracts, we should consider a question that has long concerned anatomists and phylogeneticists: what can be said about homology between the spider CNS and that of other arthropods? The segmentation of the spider brain is still in debate (Weygoldt 1985), and it does not improve matters that its ontogen has been far less well studied in spiders than in other arthropods, especially the insects. It is necessary to assume that the Chelicerata are indeed related to the other arthropods, and that the Mandibulata and the Arachnata together constitute the Euarthropods, in order for there to be any point in comparing the segmental composition of the anterior part of the spider prosoma with the anterior end of other arthropods. This assumption does in fact seem to be justified (see references in Weygoldt 1985).

The chelicerae of the spiders are usually regarded as homologous with the second antennae of crustaceans (but see Damen et al. 1998) and the cheliceral ganglia, correspondingly, with the tritocerebrum of the Mandibulata. It follows that the region with the chelicerae is considered a true segment, the tritocephalon. Where the parts of the CNS anterior to the tritocephalon belong, however, is controversial. Presumably this region is also a true segment, the deutocephalon, and not a presegmental acron. Furthermore, there are indications of a prosocephalon situated anterior to the deutocephalon, and in the opinion of some authors the central body, described below, originates from the prosocephalon. The absence of antennae in spiders is consistent with the absence of a deutocerebrum, although another interpretation is that the deutocerebrum has completely fused with the protocerebrum. The brain of an adult spider thus presumably consists of a proto- and a tritocerebrum. All the large neuropil regions such as the optic ganglia and the so-called mushroom bodies, as well as the central body, are parts of the protocerebrum. According to counts of the cell somata, the protocerebrum and the cheliceral ganglia in *Cupiennius* together consist of about the same number of neurons as the subesophageal region, ca. 50,000 in each. In this respect our spider is well behind the insects: the estimate for the brain of the migratory locust is at least 360,000 cells (Burrows 1996), and for the honeybee brain (supraesophageal ganglion) 960,000 cells (Witthöft 1967).

In a recently published paper on head segmentation in arthropods, Damen et al. (1998) applied genetic techniques and, by way of the patterns of expression of Hox genes, obtained interesting new information about an old problem. Hox genes are involved in establishing the specificity of segments along the long axis of the arthropod body and have been especially thoroughly studied in *Drosophila*. According to the new findings for *Cupiennius salei*, as a representative of the Chelicerata, the expression pattern of the Hox genes contradicts the view that the Chelicerata lack the first antennal segment. In this revised interpretation, the cheliceral segment of spiders would be homologous with the first antennal segment of other arthropods and the pedipalp segment would be the homolog of their second antennal segment (intercalary segments of insects). This would also imply that the arthropods have only one kind of head segmentation and are a monophyletic group.

◀

Fig. 1 a–d. The central nervous system of *Cupiennius salei*. **a** Topography: lateral view of longitudinally sectioned prosoma. *AME* Anterior median eye; *BR* brain (supraesophageal nerve mass); *CH* chelicera; *END* endosternite; *MS* musculature; *OES* esophagus; *OPIN* opisthosomal nerves; *PG* poison gland; *SOG* subesophageal nerve mass; *ST* stomach; *STN* sternum; *TS* tissue below subesophageal nerve mass. **b, c,** and **d** Dorsal, ventral and lateral view of central nervous system. *ALEN* Anterior lateral eye nerve; *AMEN* anterior median eye nerve; *AMG1–4* ganglia of walking legs (neuromeres); *AMN1–4* walking-leg nerves; *BL* blood vessels; *CHN* cheliceral nerve; *INT* intestinal nerve; *NRL* neurilemma; *OPIG* opisthosomal ganglion; *OPL* optic lobes; *PDG* pedipalpal ganglia; *PDN* pedipalpal nerve; *PHN* pharyngeal nerve; *PLEN* posterior lateral eye nerve; *PMEN* posterior median eye nerve; *PN* principal nerve; *RSN* rostral nerve. (Babu and Barth 1984)

2 Neuropil Regions, Tracts, and Commissures

The inside of the CNS looks complicated, and the brain appears quite different from the subesophageal region. Following a general rule for arthropods, the cell bodies are at the periphery of all the ganglia, while the interior of each ganglion consists of nerve fibers. Whereas the cell bodies in the brain are in frontal dorsal and lateral positions, in the subesophageal region they are ventral and ventrolateral. The synaptic contact sites between the cells are restricted to the fibrous interior of the ganglion and to special neuropil regions; there are none in the outer layer of cell bodies. The fibrous areas in both brain and subesophageal region are highly organized, and many longitudinal and transversal tracts are visible.

If we look a little closer, we can see that *Cupiennius* conforms in a number of other details to what has long been known for spiders (Hanström 1921, 1923, 1928; see the summary by Babu 1985).

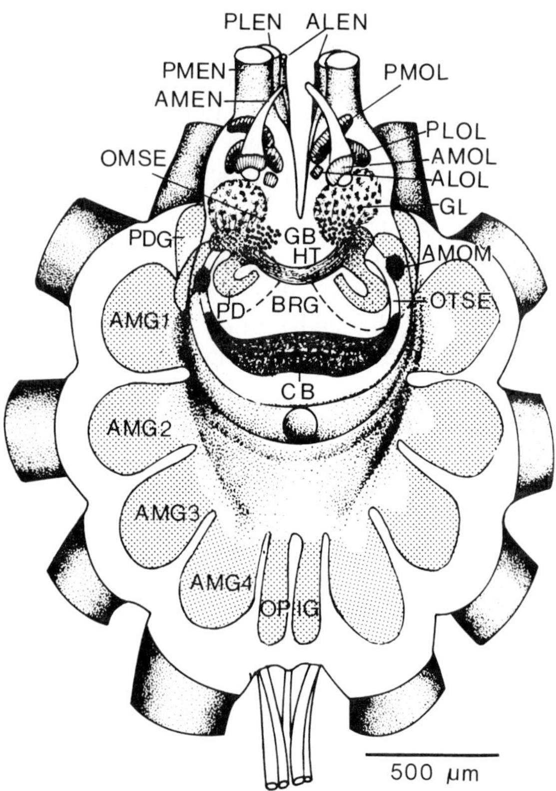

Fig. 2. Important centers in the brain of *Cupiennius salei*. *AMOL* Optical lamella of the anterior median eye; *AMOM* optical mass of anterior median eye; *BRG* bridge; *CB* "central body"; *GB* globuli cells; *GL* glomeruli; *HT* haft; *OMSE* optical mass of secondary eyes; *OTSE* optical tract of secondary eyes; *PD* peduncle; *PLOL* optical lamella of posterior lateral eye; *PMOL* optical lamella of posterior median eye. (Babu and Barth 1984)

The Brain (Protocerebrum)

In *Cupiennius*, as in the other arachnids, the morphologically most conspicuous neuropil areas are in the protocerebrum. Because these are to a very great extent part of the visual system, some of the fine points revealed by Golgi staining will be reserved for Chapter XVI. Here we are concerned with the preliminary, less refined picture we see by using Palmgren's (1948) silver-staining method (Babu and Barth 1984).

"Mushroom Bodies". Figure 2 shows the most important centers of the brain of *Cupiennius*. In the anterior, medio-lateral region, at the base of the optic lobes, the most noticeable features are two large, compact groups of so-called globuli cells (Fig. 3). Their delicate neurites (diameter ca. 0.4 µm) form the structures called mushroom bodies or corpora pedunculata, although (as discussed in relation to the visual system in Chapter XVI) they are presumably not homologous with the structures of the same name in insects and crustaceans. In addition to the globuli-cell neurites these "mushroom bodies" contain the glomeruli (Fig. 4), numerous knots of fine fibers together called the haft that run toward the midline and in combination with their contralateral counterparts form a commissure called the bridge, and the peduncles (see Fig. 2).

It is characteristic of the "mushroom bodies" that they are connected directly to the optic ganglia of the secondary eyes but not to those of the principal eyes, which makes them equivalent to the 3rd optic ganglion of the secondary eyes (see Chapter XVI). In addition, the "mushroom bodies" receive inputs from the subesophageal ganglia. In comparison to other arachnids, the "mushroom bodies" of *Cupiennius* seem simple. In the other forms, not only do the fibers from the visual centers converge on this structure but also fibers from the "central body" and from the longitudinal tracts, so that here the "mushroom body" can be regarded as a higher-order inte-

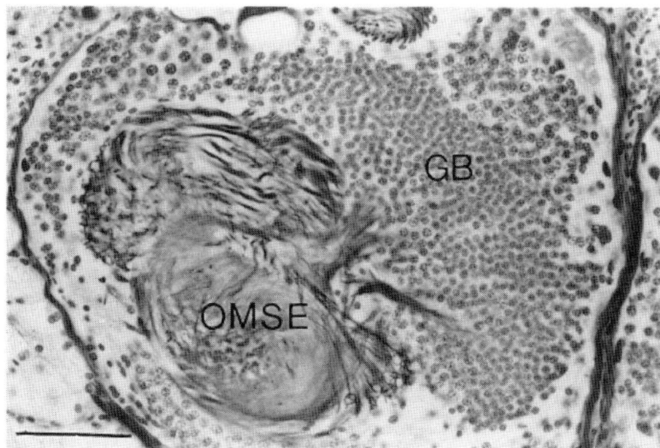

Fig. 3. Globuli cells seen in cross-section of protocerebrum; *GB* globuli cells; *OMSE* optical mass of secondary eyes. *Bar* 100 µm. (Babu and Barth 1984)

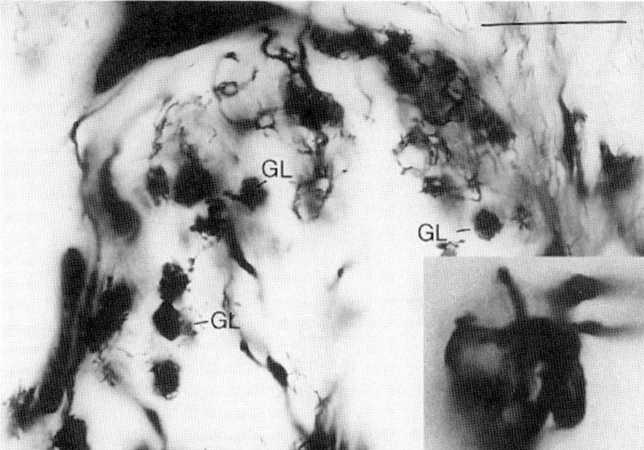

Fig. 4. Glomeruli associated with the posterior lateral eye after Golgi stain. They represent spherules consisting of densely interwoven fiber endings (see *inset*). *GL* Glomeruli. *Bar* 50 µm. (Babu and Barth 1984)

gration center. Apart from its role in the visual system, we have no idea what the functional significance of this arrangement may be.

"Central Body". In the posterior dorsal region of the brain is a second large neuropil area, which is called the central body even though it is presumably no more homologous with the structures of that name in other arthropods than is the "mushroom body" (see below). The cell bodies here are mostly at the dorsal and posterior periphery, and to a lesser extent are ventral. The main mass of the structure consists of fibers packed together to form a massive dorsal lobe and, behind it, a ventral lobe (Fig. 5). Like the "mushroom bodies", the "central body" receives input from the eyes, both principal and secondary. It also seems to be connected to the subesophageal ganglion, in particular by motor fibers; hence it is also interpreted as a motor center. The apparently most compelling evidence of a motor function (web building) was the claim that the "central body" is especially well developed in web-building spiders (Hanström 1928), but this has not stood the test of comparative volumetric analysis (Weltzien and Barth 1991). The proportion of the total brain volume occupied by the "central body" is very similar in *Nephila clavipes*, an orb-web spider, in *Cupiennius salei*, a hunting spider, in *Phidippus regius*, a jumping spider, and in *Ephebopus* sp., a bird spider: ca. 3–5% in all cases. Furthermore, the "central body" in *Nephila* does not, as has been claimed for *Argiope aurantia* (Babu 1975), differentiate in such a way that the typical bilobate structure appears only when the spider shows signs of web-building behavior; instead, as in *Cupiennius*, this structure is clearly discernible while the spiderling is still in the egg sac.

The spider brain, unike those of insects and crustaceans, receives direct inputs from only one sensory system, the eyes. Accordingly, the brain of *Cupiennius* is predominantly occupied with processing visual information. This is evident not only in the connections of the classical association centers "mushroom body" and "central body", but also in the presence of optic ganglia – two for each eye – through which the visual information passes on the way from the principal or secondary eyes to these association centers. More about this in Chapter XVI.

Figure 6 shows the relative contributions of various neuropil areas to the total brain volume, with a comparison of four spiders that belong to different families and have different modes of life. Whereas the proportion occupied by the central body varies only between ca. 3% and 5%, and the cheliceral ganglion also always accounts for ca. 2% of the volume of the brain, there are great differences between the contributions of the optic ganglia 1 and 2 (see Chapter XVI). As would be expected, this value is largest

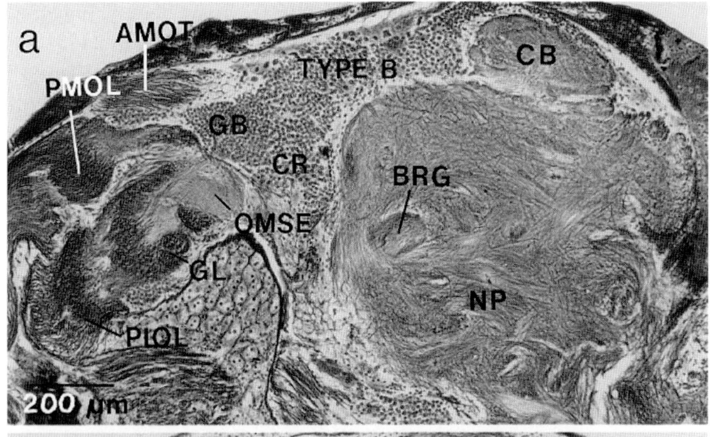

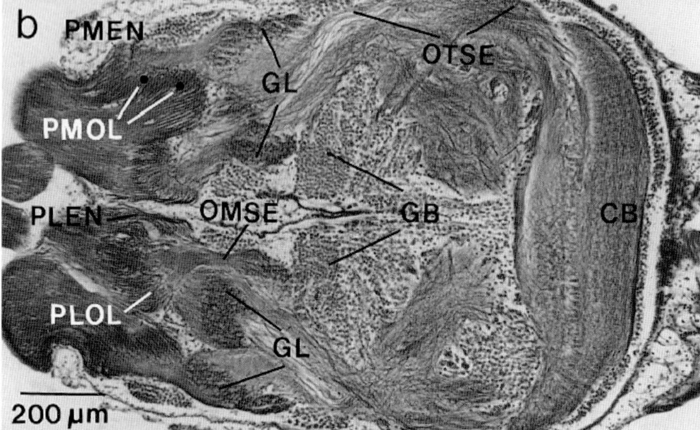

Fig. 5 a, b. The protocerebrum of *Cupiennius salei*. **a** Sagittal section. **b** Horizontal section. *AMOT* Optical tract of anterior median eye; *BRG* bridge; *CB* "central body"; *CR* cellular rind; *GB* globuli cells; *GL* glomeruli; *NP* neuropil; *OMSE* optical mass of secondary eyes; *OTSE* optical tract of secondary eyes; *PLEN* posterior lateral eye nerve; *PLOL* optical lamella of posterior lateral eyes; *PMEN* posterior median eye nerve; *PMOL* optical lamella of posterior median eye; *Type B* type of nerve cell. (Babu and Barth 1984)

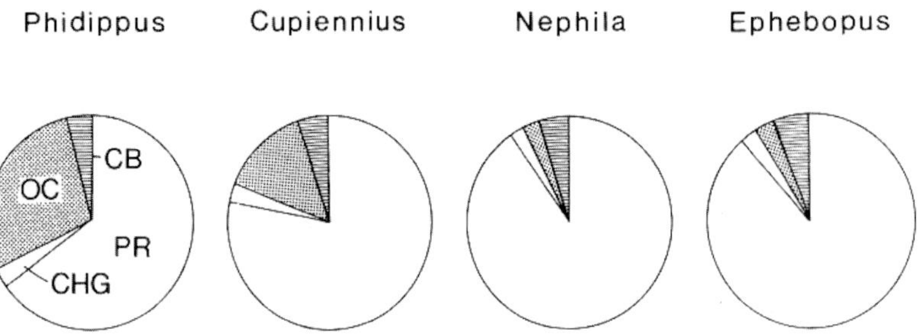

Fig. 6. Relative volumes (% of whole brain) of various parts of spider brain. *CB* "Central body"; *CHG* cheliceral ganglion; *OC* optical center; *PR* diffuse protocerebral lobes. (Weltzien and Barth 1991) (with permission of John Wiley and Sons Inc., New York)

(31%) for the jumping spider; its behavior is especially dependent on vision. The score for *Cupiennius* is still fairly high, 20%, but *Nephila* and *Ephebopus* come in at only ca. 2%. These numbers indicate not only that the "central body" is not particularly well developed in the web-building spiders, at least not in terms of size, but also that there is no inverse relationship between its size and that of the optical ganglia (Hanström 1926, 1928).

The various regions of the protocerebrum communicate both with one another and with the subesophageal ganglia. Figure 7 shows the details (Babu and Barth 1984). The "mushroom bodies"

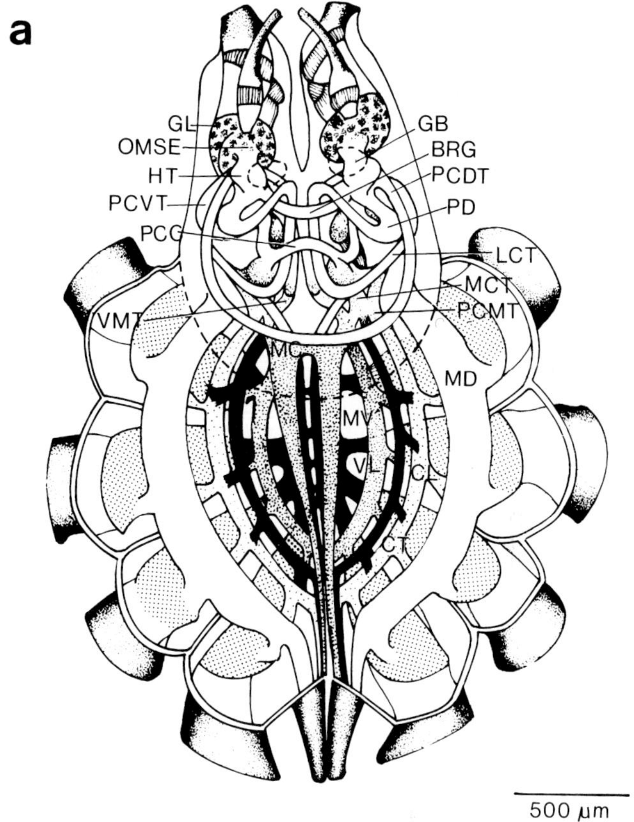

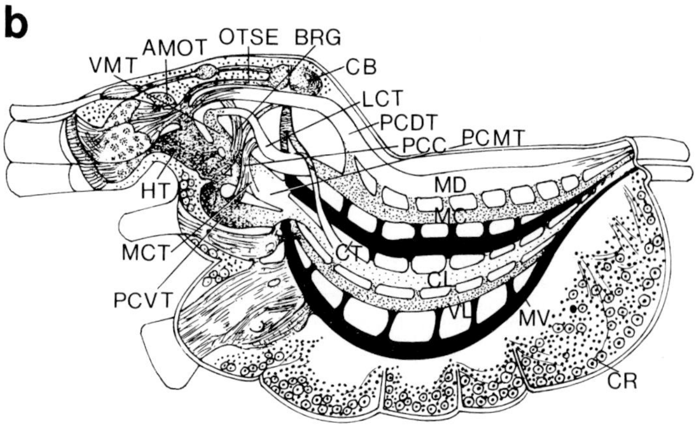

Fig. 7 a, b. Important centers and fiber tracts in the central nervous system of *Cupiennius salei*. For clarity optical centers and "central body" not shown in **a** (dorsal view). **b** Sagittal view. *AMOT* Tract of anterior median eye; *BRG* bridge; *CB* "central body"; *CL* centro-lateral tract; *CR* cellular rind; *CT* central tract; *GB* globuli cells; *GL* glomeruli; *HT* haft; *LCT* lateral cerebral tract; *MC* mid-central tract; *MCT* median cerebral tract; *MD* mid-dorsal tract; *MV* mid-ventral tract; *OMSE* optical mass of secondary eyes; *OTSE* optical tract of secondary eyes; *PCC* protocerebral commissure; *PCDT* protocerebral dorsal tract; *PCMT* protocerebral median tract; *PCVT* protocerebral ventral tract; *PD* peduncle; *VL* ventro-lateral tract; *VMT* ventral median tract. (Babu and Barth 1984)

seem to have the most massive connections to the tracts from the subesophageal region, which corroborates the hypothesis that they serve as association and integration centers.

The Subesophageal Mass

The subesophageal ganglionic mass in *Cupiennius* accounts for no less than about 85% of the volume of the central nervous system, and thus is considerably larger than the protocerebrum (15%). It is produced by fusion of all the ganglia associated with the pedipalps, legs and opisthosoma, the fibrous components of which are difficult to distinguish in the center of the mass. However, traces of the original ganglia (neuromeres) can still be well discerned in the arrangement of the cell bodies around the periphery and the organization of the tracts in the interior (Fig. 8). The distribution of the blood vessels and the neural lamellae, which partially separate the neuromeres from one another, also reliably indicate its composite nature. The somata of the neurons are arranged in several layers, forming a cellular rind on the ventral surface and also, with a thickness as great as 300 μm, lateral to the neuromeres. There are no somata around the roots of the nerves to the appendages. The somata typically form groups that give rise to bundles of fibers.

In our neuroanatomical analyses we were always particularly attentive to the arrangements of the sensory pathways, and by now we have quite a bit to say on that subject (Chapter XV). Before trying to clarify the routes that lead from the individual types of sensillum at the periphery into the interior of the subesophageal mass,

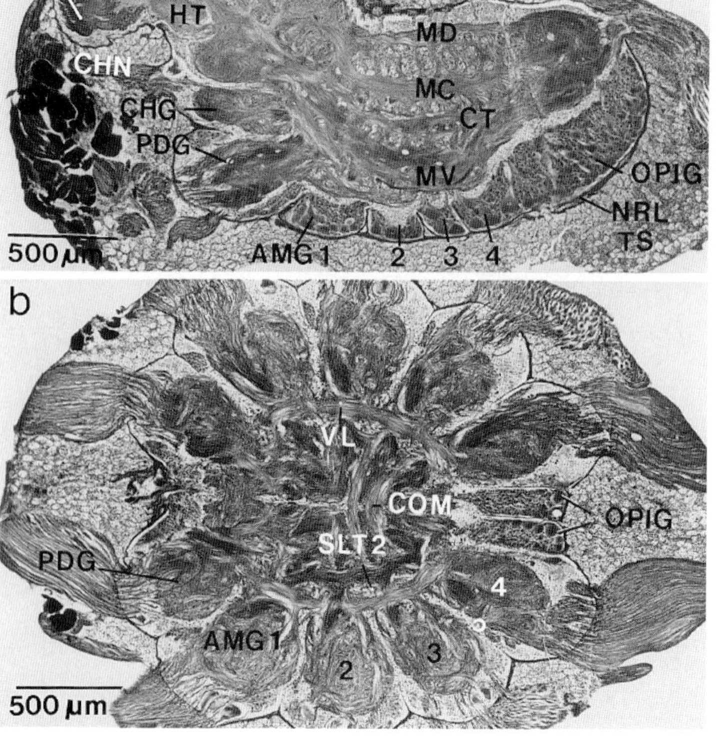

Fig. 8 a, b. Subesophageal ganglionic mass in the central nervous system of *Cupiennius salei*. **a** Sagittal section, **b** Horizontal section. *AMG1–4* Cells of the walking-leg ganglia; *AMON* optical mass of anterior median eye; *CB* "central body"; *CHG* cheliceral ganglion; *CHN* cheliceral nerve; *COM* commissural tracts; *CT* central tract; *HT* haft; *MC* mid-central tract; *MD* mid-dorsal tract; *MV* mid-ventral tract; *NRL* neurilemma; *OMSE* optical mass of secondary eyes; *OPIG* cells of the opisthosoma ganglion; *PDG* pedipalpal ganglion; *PLOL* optical lamella of posterior lateral eye; *SLT2* sensory longitudinal tract 2; *TS* tissue below subesophageal nervous mass; *VL* ventro-lateral tract. (Babu and Barth 1984)

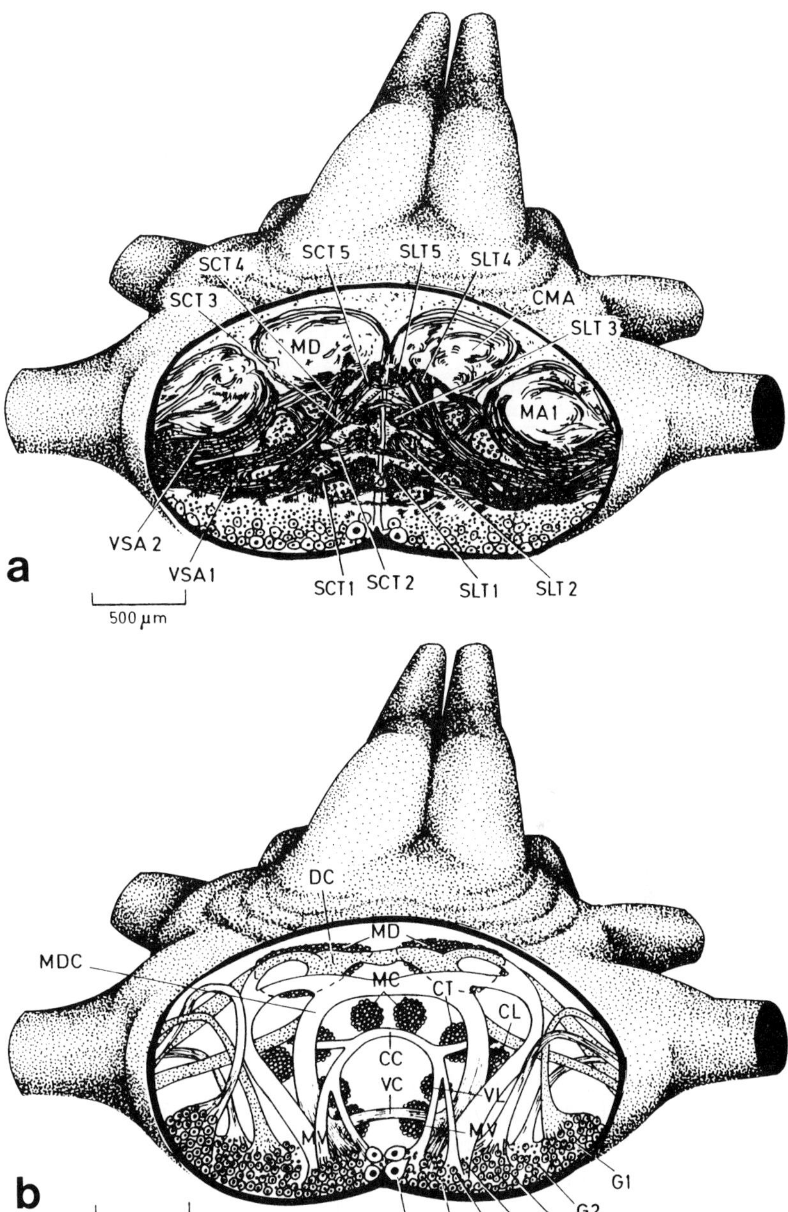

Fig. 9a–c. Organization of sensory and motor pathways; cross-section through subesophageal nervous mass at the level of the second pair of walking legs. **a** Sensory pathways. **b** Motor pathways in the mid-central region. **c** Dorso-ventral arrangement of sensory tracts. *CC* Central commissure; *CL* centro-lateral tract; *CMA* central motor association area; *COM* commissural tracts; *CR* cellular rind; *CT* central tract; *DC* dorsal commissure; *G1–3* three cell groups whose axons enter the leg nerves; *G4* cell groups that contribute axons to the commissural and longitudinal tracts; *MA* motor association region of leg ganglion; *MC* mid-central tract; *MD* mid-dorsal tract; *MCD* mid-dorsal commissure; *MV* mid-ventral tract; *SCT1–5* sensory commissural tracts 1–5; SLT 1–3 sensory longitudinal tracts; *VC* ventral commissure; *VL* ventro-lateral tract; *VSA1*–2 ventral sensory association areas (Babu and Barth 1984). (Figure **9c** see page 174)

I must briefly describe the conspicuous transversal and longitudinal tracts. This may be somewhat tedious, but it is a necessary exercise for everyone who wants to be at least roughly oriented within this labyrinth. Together with figures 7 and 8, figure 9 will be helpful here.

In the terminology of neuroanatomy, a tract is a structure that transmits information, whereas a neuropil is a region where information is processed. However, this distinction is not absolute; there are intermediate forms. We already know that the primary sensory afferents have endings in the sensory tracts, so it follows that these tracts contain synapses. The difference in the integrative properties of

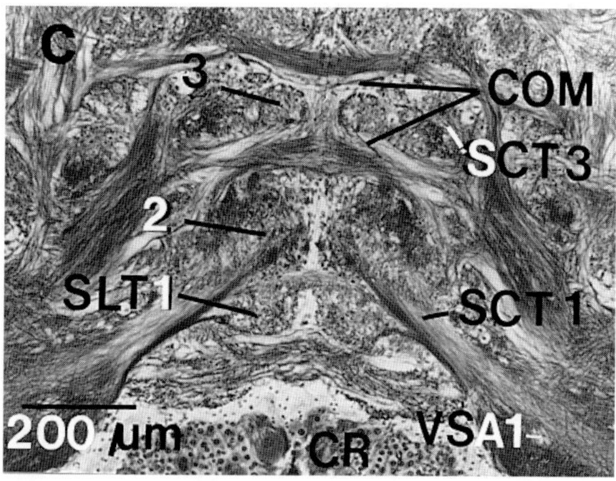

Fig. 9c

tract and neuropil is clearly manifest in the number of synapses, however: in an electron-microscopic study of *Cupiennius* (Zottl 1994) we found that their density (number per unit cross-sectional area) was significantly lower in the sensory longitudinal tract 4 (SLT 4, see Chapter XV), which averages 13/100 μm^2 (n = 15, SD ± 3.4), than in the third optic ganglion (ON3), a "classical" neuropil region in the brain (see Chapter XVI), where the value for the "head" was 89/100 μm^2 (n = 6, SD ± 18.8) and that for the region called "haft" was 133/100 μm^2 (n = 6; SD ± 22.7).

Similar findings by Pauls and Schürmann (1993) for the prothoracic sensory tracts and the ventral association area of a cricket (*Gryllus bimaculatus*) support the view that tracts do not differ fundamentally from neuropil regions with respect to their integrative functions. By the way, spiders differ from insects in that there are synapses between adjacent fibers even in the peripheral nerves (Foelix 1975, 1985b; see Chapter XV).

Longitudinal Tracts. In the longitudinal direction six pairs of main tracts run through the subesophageal nerve mass. At the anterior end they are connected to the brain; posteriorly they become thinner and near the site of origin of the abdominal nerve they merge with one another. On the way from front to back these tracts maintain their positions relative to one another, always passing through corresponding areas of the various neuromeres. Thus these longitudinal tracts not only connect various centers in the subesophageal ganglionic mass from anterior to posterior but also link all of these to higher centers in the brain. On the basis of fiber diameter and certain staining properties, we infer that these tracts are primarily motor pathways. In each neuromere a tract is joined by fibers originating in the posterior part of the neuromere, which can project either forward or backward. That is, the longitudinal tracts contain both ascending and descending pathways. The names of the individual tracts are shown in figures 7 to 9.

Near the middle of the subesophageal ganglion mass are another five pairs of longitudinal tracts, stacked in the dorsoventral direction. These have a predominantly sensory character and hence are called sensory longitudinal tracts SLT 1 to SLT 5. Fibers from all the neuromeres contribute to the SLT's, projecting forward and/or backward; we shall return to these in the next chapter.

Transversal Tracts. First, though, the transversal tracts must be described. They are present in all neuromeres and can be classified as either sensory or motor tracts. The primary afferents, according to light-microscopic observations, form two ventrally situated sensory association areas (VSA 1 and 2) after entering the neuromere. VSA 1 splits into five sensory commissural tracts, which run mediad, form dense, compact fiber masses, contribute to the sensory longitudinal tracts SLT 1–5 and send out fine bundles of fibers to make connection with the contralateral side. As shown in figure 9a, the five pairs of commissural tracts are arranged above one another. The ventral sensory association area VSA 2 is somewhat dorsal to VSA 1 and ends ipsilaterally in the motor regions of the neuromer.

The arrangement of the motor commissures and the associated cell somata can be seen in the illustration of the second-leg neuromere in figure 9b. It is a labyrinth of impressive complexity. The original publication (Babu and Barth 1984) can be consulted for further details, but at present it suffices that we have assembled a framework for reference when, in the next chapter, we trace the route of sensory information from the sensilla into the CNS. Then in the chapters on neuroethology of behavior, we shall draw upon electrophysiological recordings from interneurons to explain central nervous processes.

XV The Route for Afferent Inputs to the CNS

The ventral regions of the fiber mass in each neuromere are sensory, and the dorsal regions are motor. Degeneration studies have shown that in the leg nerves as well, at the level of the coxa the relatively thick motor fibers run dorsally while the ventral part of the nerve contains the thinner sensory fibers. The sensory fibers are subdivided into separate bundles, each of which is associated with a particular leg segment; that is, the leg nerve has a somatotopic organization like that found in insects (Zill et al. 1980; Brüssel and Gnatzy 1985) (Fig. 1).

1 The Peripheral Nerves

When a neuroanatomical or electrophysiological research project requires that the axons of particular sensilla be cobalt-stained or the receptor activity recorded, it is essential to know exactly the structure and course of the peripheral nerves. The same applies when a sensillar input is to be eliminated selectively by cutting nerve branches. For data on the course of the peripheral nerves in the walking legs of *Cupiennius* we have to thank Ernst-August Seyfarth and his coworkers (1985). As long ago as 1966 Rathmayer (see also Parry 1960) described the walking-leg nerves in a bird spider, as follows. Of the three main nerves A, B and C, the smallest is A; it is a mixed sensory-motor nerve, whereas nerve B is purely motor and C, by far the thickest nerve, is purely sensory. In *Zygiella x-notata* Foelix and coworkers (Foelix et al. 1980, Foelix 1992) counted the fibers in electron-microscopic cross sections of the nerves: nerve A, two motor and 51 sensory axons; nerve B, 126 axons; nerve C, around 5000 axons (first leg 7000, fourth leg 3900, pedipalp 2200). Foelix (1975, 1985b) also discovered an exciting difference between the peripheral sensory nerves of insects and those of spiders: in the spiders they contain many *synaptic contacts*. Unfortunately, very little is known about their function. These synapses are particularly common far in the periphery, near the sensory cells, and are relatively rare in the main nerves close to the body.

In a recent paper (Fabian-Fine et al. 1999a,b) from the laboratory of Ernst-August Seyfarth in Frankfurt am Main, immunocytochemical labeling was used to identify peripheral chemically transmitting synapses at the lyriform organ VS-3 (on the anterior surface of the walking-leg patella), touch-sensitive hair sensilla, trichobothria, and internal joint receptors of *Cupiennius salei*. These experiments employed monoclonal antibodies against synapsin, a protein that binds to the cytoplasmic surface of synaptic vesicles and can therefore be used to reveal presynaptic endings. Synapsin was found to have a punctate distribution and to be concentrated in the parts

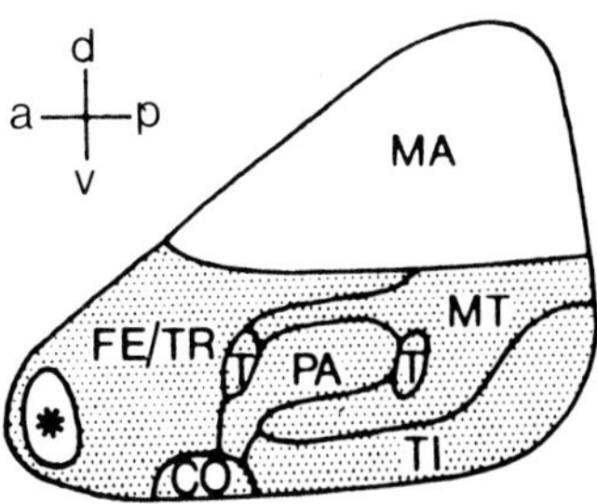

Fig. 1. Representation of afferent nerve fibers of the various leg segments in the cross-section of the leg nerve of *Cupiennius salei*. *CO* Coxa; *FE* femur; *MA* motor axons; *MT* metatarsus; *PA* patella; *T* tarsus; *TI* tibia; *TR* trochanter; *asterisk* thick axons in ventral region; *a* anterior; *p* posterior; *d* dorsal; *v* ventral. (Brüssel and Gnatzy 1985)

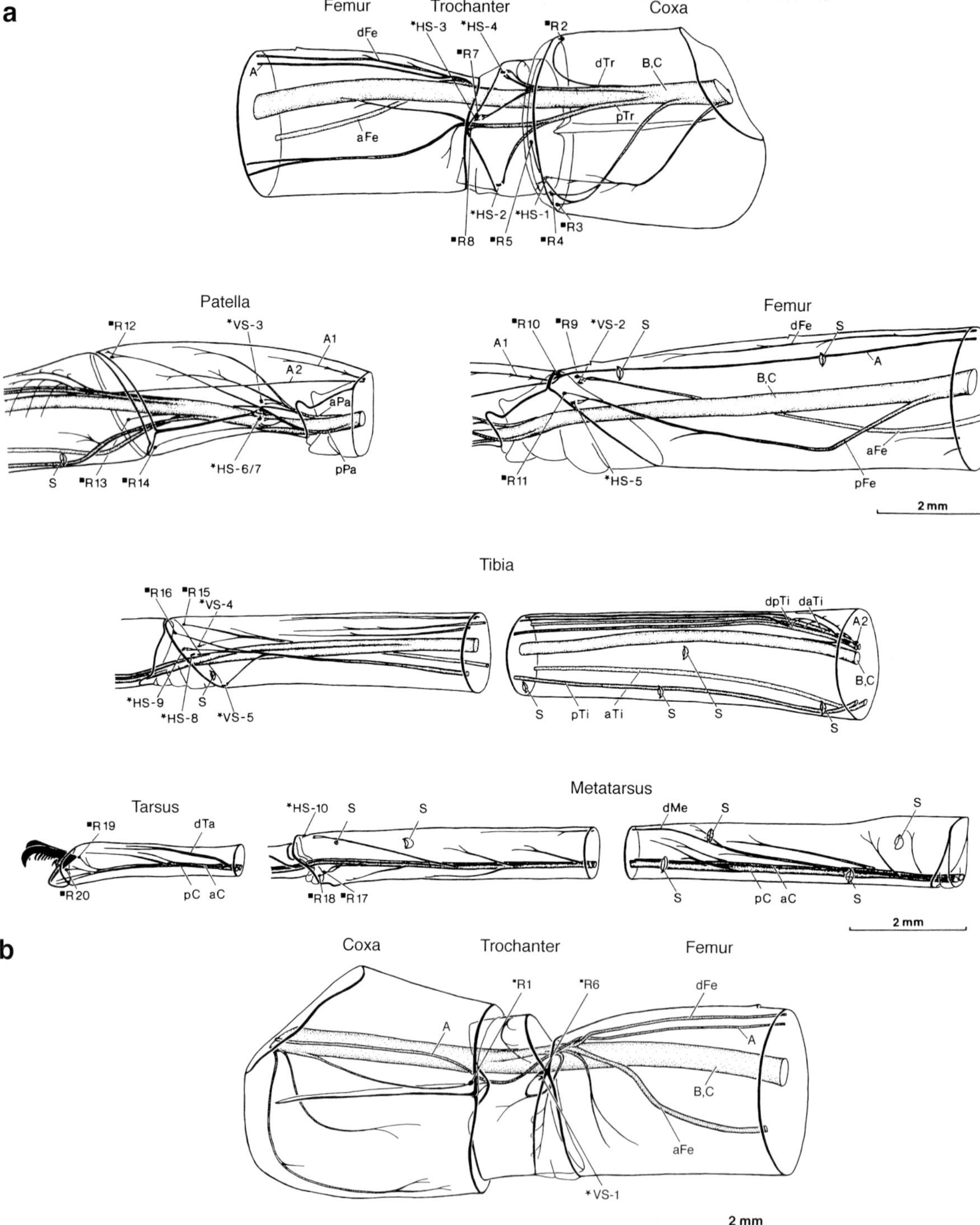
a
Femur
Trochanter
Coxa
dFe
*HS-3
*HS-4
R2
R7
dTr
B,C
A
pTr
aFe
*HS-2
*HS-1
R3
R8
R5
R4
Patella
R12
*VS-3
A1
A2
aPa
S
R13
R14
*HS-6/7
pPa
Femur
R10
R9
*VS-2
S
dFe
S
A1
A
B,C
aFe
R11
*HS-5
pFe
2 mm
Tibia
R16
R15
*VS-4
dpTi
daTi
A2
B,C
*HS-9
S
*HS-8
*VS-5
S
pTi
aTi
S
S
S
Metatarsus
Tarsus
R19
dTa
*HS-10
S
S
dMe
S
S
R20
pC
aC
R18
R17
S
pC
aC
S
2 mm
b
Coxa
Trochanter
Femur
R1
R6
dFe
A
A
B,C
aFe
*VS-1
2 mm

of the axons near the somata of the sensory cells (and also at glial cells). This finding was confirmed by electron-microscopic examination.

For the present we can only speculate that in spiders and other arachnids, neural integration is probably occurring far out in the periphery. Some of the synapses are related to thin afferent fibers that presumably control the sensory cells. Because many of the peripheral synapses are part of fibers showing GABA-ergic immunoreactivity, it can be concluded that their effect on the sensory cells is inhibitory (Fabian-Fine et al. 1999). There is also evidence for the presence of glutamate in one of the several types of synaptic vesicles. The structural richness of the synaptic microcircuits now known strongly suggests a complex efferent control of sensory activity (Fabian-Fine et al. 2000). It may be that this is a phylogenetically old trait which the spiders still share with primitive arthropods such as *Limulus*, known to have diffuse peripheral nerve networks (Foelix 1985b).

Another case of synapses in the sensory periphery of an arachnid should be mentioned here. The pectens of some species of scorpions carry up to 10^5 chemoreceptive sensilla with 10 to 18 sensory cells each. The somata of the sensory cells together with their proximal axons form a lamina where the axons have many synaptic contacts with each other (Foelix 1985b). Gaffin and Brownell (1997) presented electrophysiological data in support of the idea that chemosensory information is already integrated by inhibitory and excitatory interactions in the periphery, within the sensilla, before it is passed on to the central nervous system.

Fig. 2a, b. Main leg nerves, major sensory nerve branches and proprioreceptors in the leg segments of *Cupiennius salei*. **a** Posterior view of a left leg; *A* small leg nerve A; *B* and *C* main leg-nerve bundle; sensory nerves: *dTr, dFe, dMe, dTa* dorsal branches of trochanter, femur, metatarsus, and tarsus; *dpTi, daTi* posterior and anterior dorsal branches in the tibia; *pTr, pFe, pPa, pTi* posterior branches in respective segments; *aFe, aTi, aPa* anterior branches; *pC, aC* posterior and anterior part of the main leg nerve in distal leg segments. Proprioreceptors: (■) internal joint receptors (nomenclature after Rathmayer and Koopmann 1970); *R19* and *R20* newly described receptor groups in membranous "joint" between tarsus and pretarsus. (*) Lyriform slit sense organs (terminology of Barth and Libera 1970); *VS* anterior organs, *HS* posterior organs; *S* sockets of large spines drawn as landmarks. **b** Anterior view of left leg, showing the sensory nerves and the arrangement of the proprioreceptors proximally in the leg. (Seyfarth et al. 1985)

To return to the peripheral nerves of *Cupiennius*: by means of cobalt impregnation and examination of histological serial sections, Seyfarth and coworkers (Seyfarth et al. 1985) put together a very detailed picture (Fig. 2). It shows the arrangements of the main leg nerves (A, B, C) as well as of the larger sensory nerve branches associated with the lyriform organs and internal joint receptors, because in this project the authors were especially concerned with proprioreceptors near the joints. The axons of neighboring sensilla each form small bundles to which, as the nerve proceeds proximally through successive segments, more fibers are continually added. In each segment there is at least one dorsal sensory branch, which collects the axons of the dorsally situated sensilla, and at least one pair of lateral branches comprising the axons of the lateral and ventral sensilla. These ventro-lateral nerves are particularly thick in the tibia and the femur; they do not merge with the sensory leg nerve C until they have passed into the proximally adjacent joint. The main leg nerve itself is bifurcated from the tarsus into the distal third of the tibia, and its two branches are purely sensory until the nerve is within the metatarsus because there are no muscles in either the tarsus or the distal half of the metatarsus. Nerves B and C in *Cupiennius* pass through the leg as a single bundle.

Toward the proximal end of the leg the courses of the sensory nerves become more complicated. At the back of the leg the axons of the ventrolateral sensilla of the femur join those of the ventral sensilla of the trochanter before merging with the main leg nerves (B, C) in the coxa. At the front (Fig. 2) the relatively large dorsal nerve and the anteriorly situated femoral nerve become united with the fibers of the trochanter, which – just to complicate things further – form a loop at the femur-trochanter joint, before the junction with the main leg nerves (B, C).

And what about the small leg nerve A? Let us follow it outward from the base of the leg. In the coxa it runs anterior to the main nerves B and C, which at this stage form a single bundle; then near the trochanter-femur joint it bends dorsalward and finally, at the femur-patella joint, it bifurcates to form the nerves A_1 and A_2. Whereas A_1 innervates the large internal joint receptor R_{10} and a few patellar hairs, it may be that A_2 is composed of only two motor axons. These run through the patella, approach the main nerve again in the proximal third of the tibia, and finally innervate the claw elevator muscle.

2 Projections of the Various Types of Sensilla

Now we return to the sensory tracts in the CNS and to the ways the various types of sensilla are related to these. Here a number of clear rules can be established, which presumably have substantial functional significance.

To repeat: in the central region of the subesophageal ganglionic mass five main sensory longitudinal tracts (SLT 1–5) are stacked up. Most of the fibers in the sensory transversal tracts (STT 1–5) join the ipsilateral longitudinal tracts (the connections to the ventrolateral STT's are comparatively slight, which is why the originally chosen term "commissural" was rejected in favor of "transversal"; Babu and Barth 1989). The transversal tracts are also stacked one above another (Fig. 3); within them, the primary sensory afferents pass from the leg nerves to the longitudinal tracts. Cobalt filling experiments have revealed that the projections of tactile hairs of the coxa branch in another sensory longitudinal tract, the hair SLT or HSLT (which can also be regarded as a discrete part of SLT 1) and in a longitudinal sensory association tract LSAT. The term LSAT indicates that axons of the STT 3 and STT 4 end in this tract in each neuromere, as do axons of the median and central vertical tract (MVT, CVT), so that this area consists of remarkably densely packed arborizations of sensory fibers. Additional connections between sensory terminals and postsynaptic neurons in the LSAT have been described by Babu and Barth (1989). Both the fine branches of the afferents from various sensilla and the presence of "blebs" suggest that the LSAT is a neuropil and in fact serves as an important integration center in the subesophageal ganglionic mass.

The structure and course of the primary afferents of the various cuticular sensilla provide an interesting and comprehensive neuroanatomical picture of this sensory system. We know the details for a great many types of sensilla: lyriform organs, trichobothria, tactile hairs, coxal hair plates, chemoreceptive hairs and the tarsal organ, which is in particular thermo- and hygroreceptive. A long list, which as far as I know is not equalled in completeness for any insect or any other arthropod.

There are *two main types* of afferent projections: those with only a local projection area in the leg ganglion and those that arborize not only in the leg ganglia but also further inside the CNS. The first type includes the afferents of long smooth hairs on the coxa and from the coxal hair plates (Eckweiler et al. 1989; Seyfarth et al. 1990). To the second type belong lyriform organs and tactile hairs on the coxa, trichobothria, Blumenthal's tarsal organ and hair-shaped contact chemoreceptors (Eckweiler and Seyfarth 1988; Babu and Barth 1989; Anton and Barth 1993).

Figure 4 shows two examples to illustrate the difference between these two types of afferent projections.

The projections of (i) the *coxal hair-plate sensilla* run in small nerves parallel to the main nerve, entering the associated neuromere ventrolaterally. There they first proceed into its interior near the anterior connective-tissue septum, then ascend into medio-ventral neuropil regions and branch horizontally in the typical fork configuration. All the hairs have axons with a very similar course and the same mode of branching, and all are restricted to the ipsilateral leg neuromere. (ii) The branching pattern of an afferent from the *lyriform organ* HS-2 on the trochanter is considerably more complicated. This axon continues as far as the medio-central region of the subesophageal ganglion mass. On the way it sends out a number of collaterals, which either end in the surrounding neuropil or contribute to the formation of longitudinal tracts. Collaterals near the site of entry into the neuromere (B_1–B_3, see Fig. 4) end with "blebs" in the region of

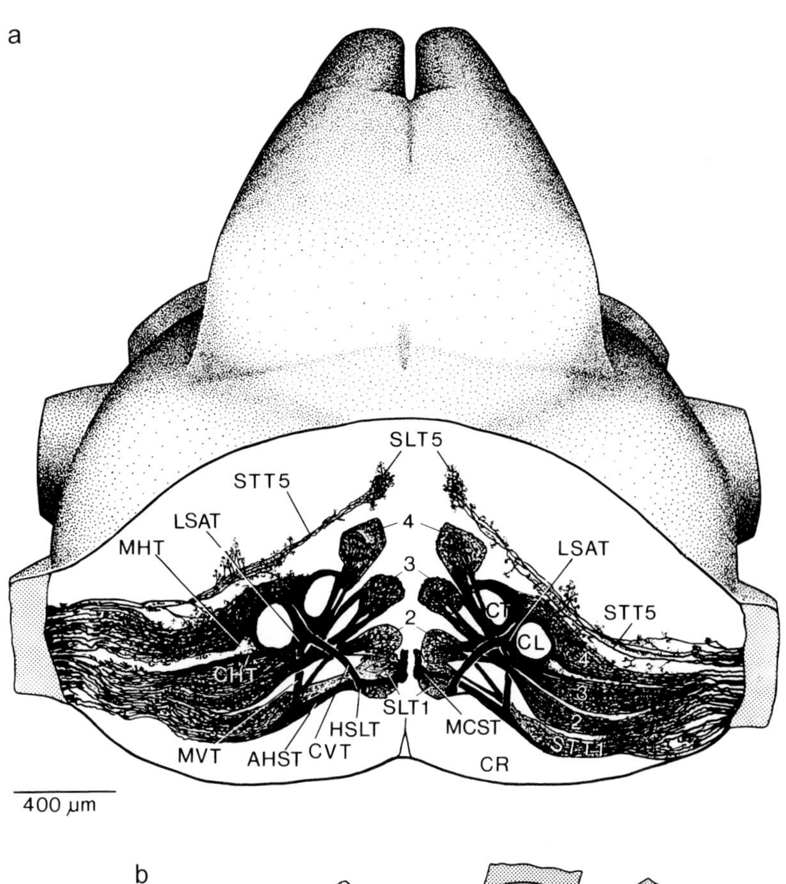

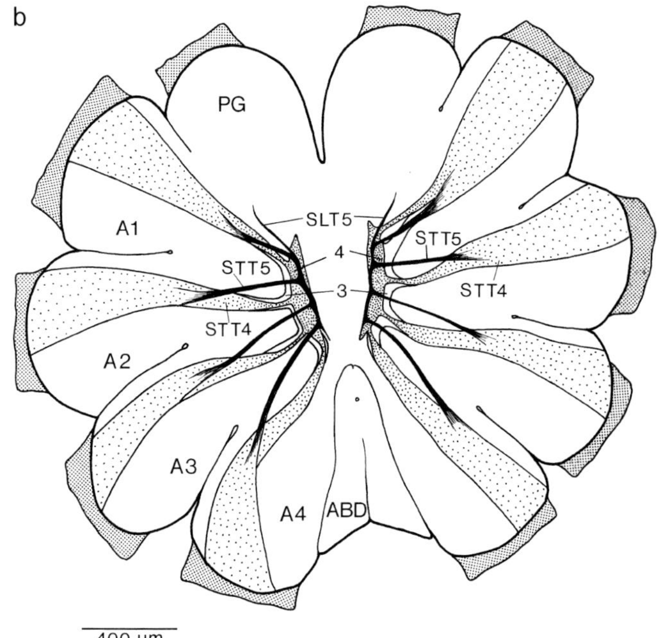

Fig. 3 a, b. Sensory pathways in the subesophageal nervous mass of *Cupiennius salei* in cross- (**a**) and horizontal section (**b**). **a** Cross-section at the level of the second pair of walking legs according to Golgi stains and cobalt fillings of identified sensilla on the leg. **b** Dorsal view of the major transverse (*STT4* dotted, *STT5* black) and longitudinal (*SLT3* white, *SLT4* dotted, *SLT5* black) sensory tracts with afferent fibers of lyriform slit sense organs. *A1-4* Leg neuromeres; *ABD* opisthosomal ganglia; *AHST* anterior sensory hair tract; *CHT* central hair tract; *CL* centro-lateral tract; *CR* cellular rind; *CT* central tract; *CVT* central vertical tract; *LSAT* longitudinal sensory association tract; *MCS* mid-ventral sensory tract; *MHT* median hair tract; *MVT* median ventral tract; *PG* pedipalpal ganglion; *SLT1-5* sensory longitudinal tract 1–5; *STT1-5* sensory transverse tracts 1–5. (Babu and Barth 1989)

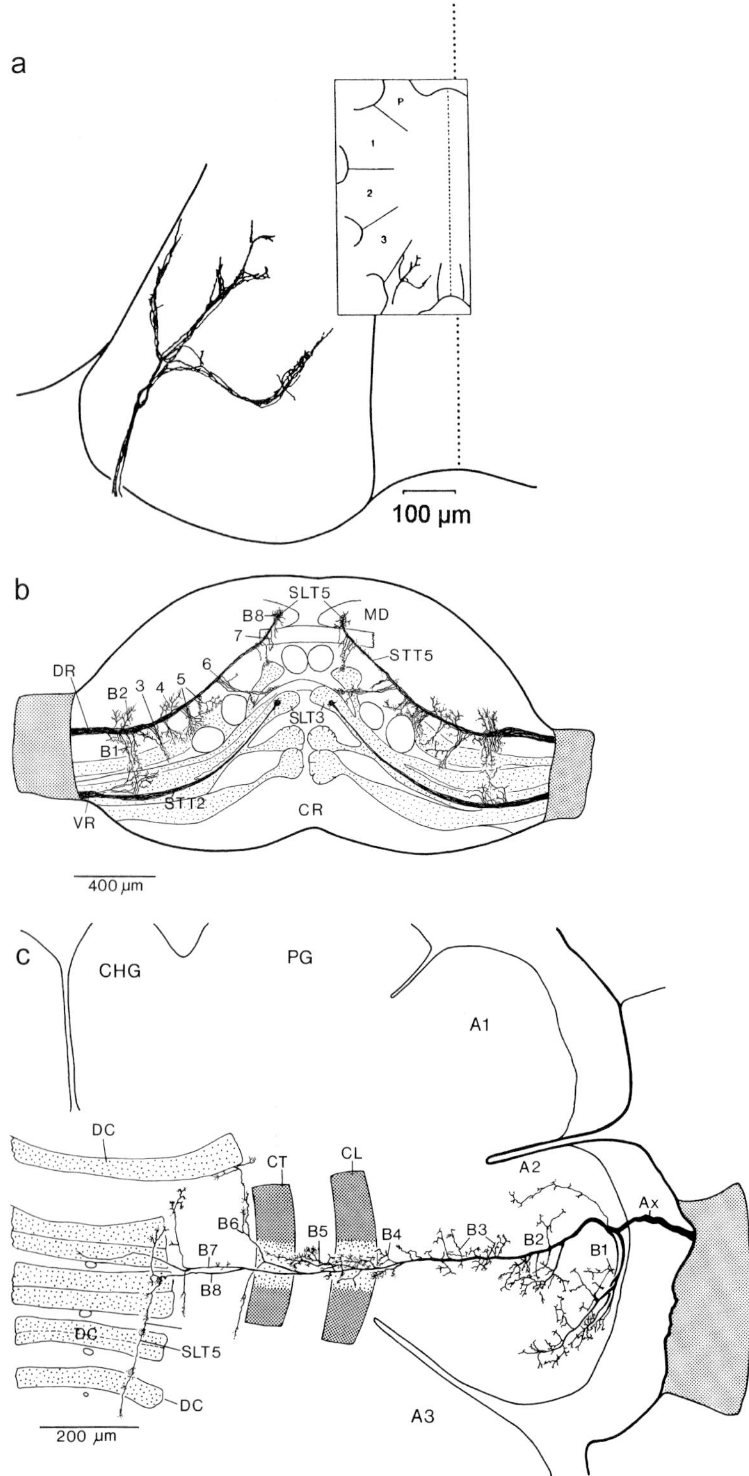

Fig. 4a–c. Projections of sensory afferents into the subesophageal nervous mass of *Cupiennius salei*. **a** Central projections of coxal hair-plate sensilla in dorsal view; dotted line indicates longitudinal midline of CNS. **b** Endings of sensory cell axons of lyriform organs HS-2 of second pair of walking legs; the axons from both organs form a dorsal (*DR*) and ventral (*VR*) root when entering the leg ganglia. Note the extensive branching (*B1–8*) of dorsal root fibers. *STT2/5* Sensory transverse tracts, *SLT2/5* sensory longitudinal tracts, *CR* cellular rind. **c** The branching pattern of an afferent fiber of the dorsal root of lyriform organ HS-2. Proximal branches *B1–3* terminate close to the dendritic areas of ipsilateral motor neurons. Branches *B4* and *B5* terminate at ipsilateral longitudinal tracts, branches *B6-8* at dorsal commissural tracts (*DC*). *A1,3* Walking-leg ganglia; *CHG* cheliceral ganglion; *PD* pedipalpal ganglion. (**a** Seyfarth et al. 1990; **b, c** Babu and Barth 1989)

the dendritic zones of ipsilateral motoneurons. The many branches in the middle (B_4 and B_5) end on ipsilateral longitudinal tracts (CT, central tract; CL, centro-lateral tract) and contribute to ipsilateral intersegmental pathways. The terminal branches (B_6–B_8), finally, project in the long direction of the animal and together with other afferents form the sensory longitudinal tract SLT 5 and, in some cases, contribute to SLT 4. From SLT 5 fine fibers also pass to the contralateral side. In addition to the dorsal roots just described, the afferents of the lyriform organs also form "ventral roots" (Fig. 4b). Their structure is simpler. In the case of the lyriform organ HS-2 they enter the neuromere by way of STT 2, arborize moderately on the way to the center and finally terminate in branches that run anteriorly and posteriorly as components of the SLT 3.

Now, by imagining that the afferents of the various sensilla on the various legs and the pedipalps are all stained at the same time, it is possible to understand the spatial structure and arrangement of the *sensory longitudinal tracts.* Figure 3b shows SLT 3–5 and the tracts leading to them, STT 3–5, in which the afferents of the lyriform organs HS-1, HS-2 and HS-4 run. The terminals of the afferents of identical lyriform organs of the eight legs form longitudinal tracts that can be U- or O-shaped, depending on the organ, or consist of two parallel fiber bundles. The corresponding sensory longitudinal tracts, which are joined by the afferents of tactile hairs, are shown in figure 5 (see especially HSLT, hair sensory longitudinal tract, and LSAT, longitudinal sensory association tract). These are positioned ventral to SLT 3–SLT 5.

A more detailed examination of the primary afferents of the trichobothria then brought us a considerable step forward (Anton and Barth 1993): the central projections of not only the trichobothria but all mechano- and chemosensitive sensilla so far investigated are organized *somatotopically.* Sensilla situated proximally on the appendages are represented in the dorsal sensory

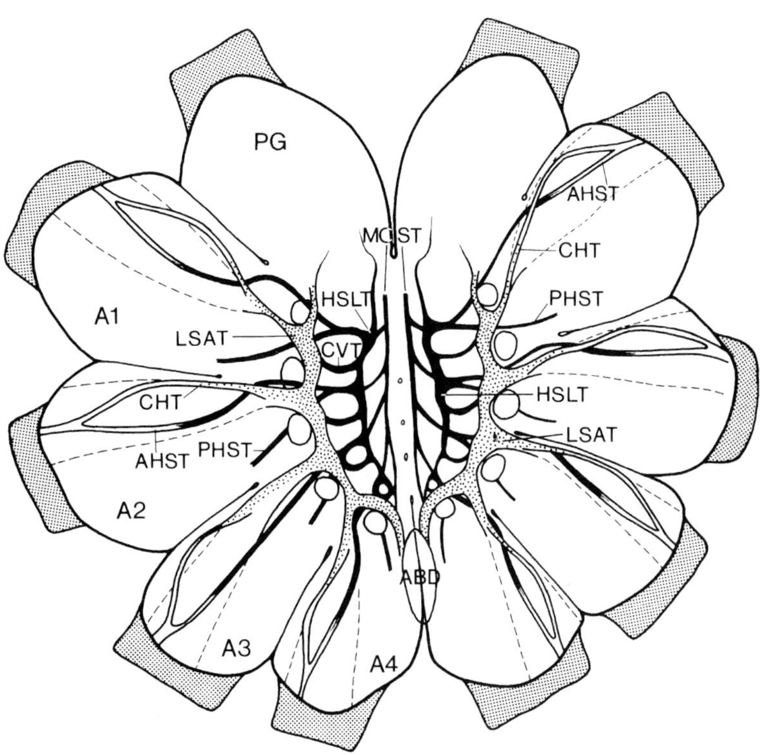

Fig. 5. The pathways of the sensory axons of tactile hairs on the coxa after entering the subesophageal central nervous system; *black* sensory tracts *AHST, CHT, HSLT*; *dotted* sensory tract *LSAT.* Tracts *PHST, CVT,* and *MCST* presumably are associated with hair sensilla of different leg segments. *A1-4* Walking-leg ganglia, *ABD* opisthosomal ganglia, *PG* pedipalpal ganglion. (Babu and Barth 1989)

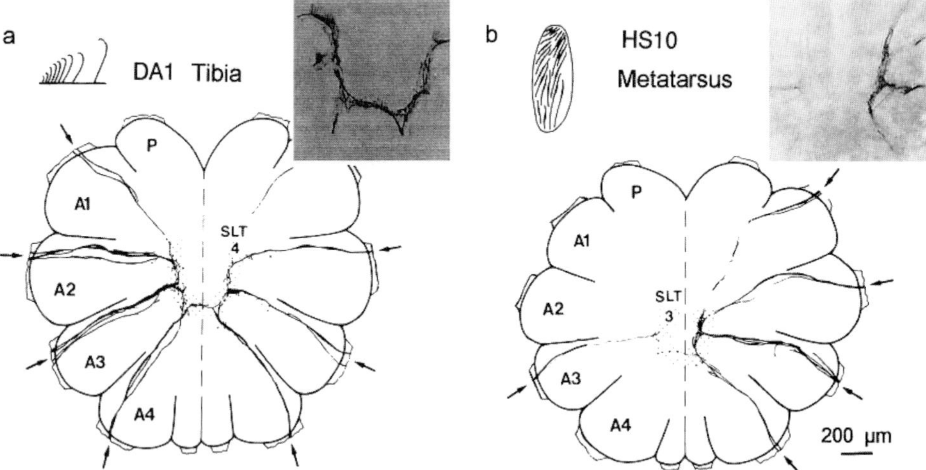

Fig. 6 a, b. Convergence of mechanoreceptor projections. Afferent fibers of trichobothria and lyriform slit sense organs of the same segments of all walking legs and pedipalps together form tract-like structures in the subesophageal central nervous system. **a** Projections of trichobothria of the tibial group *DA1* on all legs. **b** Projections of metatarsal lyriform slit sense organs of five different legs. (Anton and Barth 1993)

longitudinal tracts, whereas the terminal arborizations of the more distally located sensilla run in more ventral tracts. This also means that the afferent fibers from sensilla of the same modality in a particular appendage segment overlap in the same tract (Fig. 6). For the trichobothria in particular, a tonotopic organization in the CNS would also have been conceivable – that is, an organization according to best frequency, which varies with the length of the hair (Barth et al. 1993). But this is not the case.

The *convergence* of the afferents from sensilla of different modalities or submodalities in the same sensory longitudinal tracts is interpreted as an anatomical basis for functional interaction in behavior, which is thus already occurring at the level of the first synaptic junctions. That such interaction is important is not merely an obvious assumption but can be directly demonstrated, in particular for substrate vibrations and air-current stimuli (see Chapter XVIII). Similarly, the convergence of projections from the receptors on the same segments of different legs, like somatotopy, is an organizational pattern that could have a lot to do with a spider's orientation by mechanical or chemical stimuli; because of such organization, the CNS is potentially provided with information about both the position of the stimulated sensilla and the direction of the stimulus. More will be said on this subject in Chapters XVIII and XIX. In any case, the movements of *Cupiennius* during stimulation with directional substrate vibrations and air cur-

Fig. 7. Schematic drawing of all known sensory projections to the CNS in *Cupiennius salei*. Four types of primary afferent projections can be seen. (1) The afferent fibers from trichobothria, lyriform slit sense organs, tactile hairs and chemoreceptors branch not only in the leg neuromeres but also (and even more so) in the sensory longitudinal tracts *SLT3–5* (see *a, b, c,* *). (2) Projections of the tactile hairs on the coxa branch within the sensory longitudinal hair tract *HSLT* and within the sensory longitudinal association tract *LSAT* (see *a*). (3) The proprioreceptive long smooth hairs and the coxal hair-plate sensilla only project into the neuropil of the corresponding ipsilateral leg ganglion, not, however, into the sensory longitudinal tracts (see *d, e*). (4) The afferent fibers of the Blumenthal tarsal organ project not only into the sensory longitudinal tracts and into the leg and pedipalp neuromeres but also into the Blumenthal neuropil *BN* (see *f*). Note somatotopic organization of fibers in sensory longitudinal tracts. *AMN 1R–4R, 1L–4L* Right and left leg nerves; *BN* Blumenthal neuropil; *CHN* cheliceral nerve; *EN* eye nerves; *OPIN* opisthosomal nerve; *PNR* right pedipalpal nerve; *PR* right pedipalp. (Anton and Barth 1993)

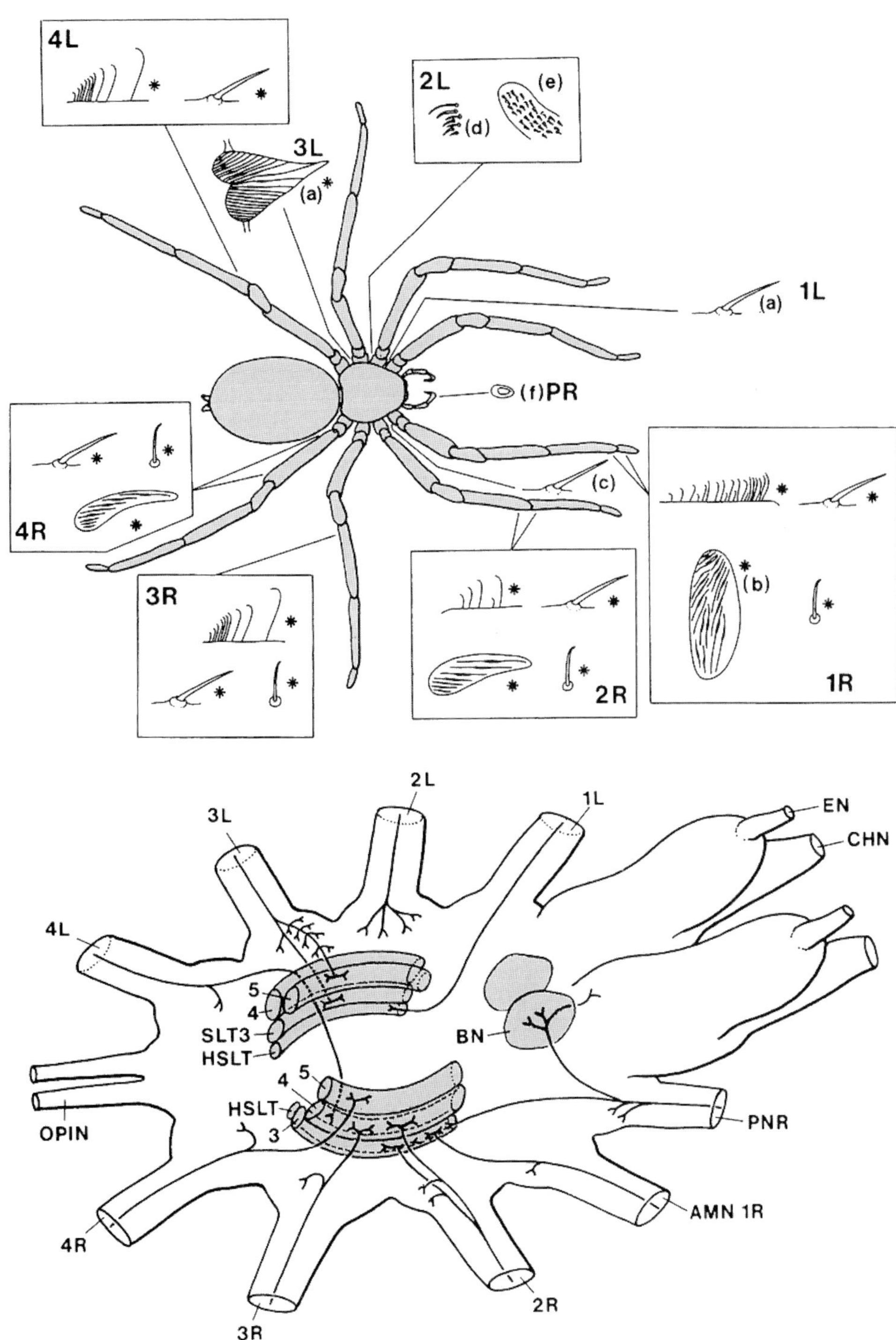
4L
3L
(a)*
2L
(e)
(d)
(a) 1L
(f)PR
4R
(c)
3R
(b)
2R
1R
2L
1L
3L
EN
CHN
4L
5
4
SLT3
HSLT
BN
4 5
HSLT
3
OPIN
PNR
AMN 1R
4R
2R
3R

rents are definitely oriented with respect to stimulus direction.

The lack of *tonotopic organization* in the case of the trichobothria might also be very relevant to behavior. It may be that the crucial function of the differential frequency tuning of individual trichobothria is not to allow precise identification of particular frequencies but rather to ensure high sensitivity over a broad range of frequencies.

Figure 7 summarizes the above findings. Among other things, it shows that whereas at the outset we distinguished two main types of *afferent branching patterns*, this should now be increased to four. (i) Trichobothria, lyriform organs, tactile hairs and contact chemoreceptors arborize both in the leg ganglion and also, more extensively, in the sensory longitudinal tracts SLT 3–5. (ii) The projections of the tactile coxal hairs end in the hair sensory longitudinal tract HSLT and in the longitudinal sensory association tract LSAT. (iii) The long smooth hairs and the hairs of the coxal hair plates project only into the neuropil of the ipsilateral leg ganglion, and not into the sensory longitudinal tracts. (iv) Finally, the thermo- and hygroreceptive tarsal organ: its afferents end both in longitudinal tracts and in the neuromere, as well as in a newly discovered Blumenthal's neuropil (Anton et al. 1992; Anton and Tichy 1994).

Anterograde cobalt filling of *tarsal organs* unfortunately has not enabled us to associate the three target regions of the primary afferents with individual sensory cells, especially since their projections overlap to a great extent. Therefore it remains conceivable but still unproven that the multiple central representation of the tarsal organ indicates a parallel processing of the sensory information received by way of different modalities. Occasionally a tarsal organ appears to have only one projection area, presumably because not all of the sensory cells have been stained. This at least hints that different cells project to different areas, or contribute differentially to the various projections.

The newly discovered paired *Blumenthal's neuropil* is situated below the esophagus, between the pedipalp ganglia and the cheliceral ganglia. It consists of 7 columnar subunits (Fig. 8). The parts of the individual columns that face toward the middle are dominated by projections of the tarsal organs on the pedipalps; the basal parts receive inputs predominantly from those on the walking legs. However, the separation of the projections from different legs into different columns seems to be more pronounced than the sorting of pedipalp and walking-leg afferents into different zones of a given column. In any case, the neuroanatomical relationships suggest that information about the spatial distribution of the moisture and thermal stimuli over the various appendages is retained in Blumenthal's neuropil.

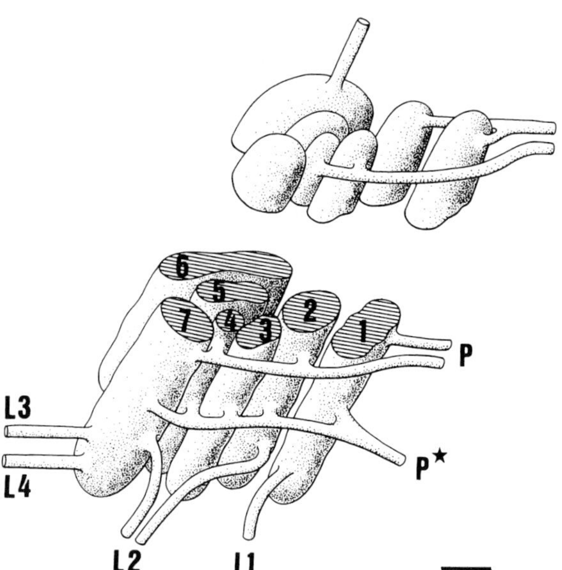

Fig. 8. The Blumenthal neuropil forming seven distinct columns in three-dimensional reconstruction. *L1–4, P* input from tarsal organs of left walking legs and pedipalp. Note that there are also fiber bundles connecting the apex of the columns to other neuropils. For clarity, connections between the columns are omitted in the diagram. *Bar* 20 μm. (Anton and Tichy 1994)

What do we learn from a *comparison with insects*?

- *Convergence.* A good example of this is given by the grasshopper *Tettigonia cantans*, in which information about airborne sound and vibration is brought together in the CNS. By referring to the vibratory inputs as well as the song, the female can orient more accurately to the courting male (Kalmring 1983).

- *Somatotopy.* This has been described for mechanoreceptive afferents in several insects and crustaceans. It has been thought to be associated with the ability to orient toward a stimulus source (Pflüger et al. 1981; Bacon and Murphey 1984; Kondoh and Hisada 1987; Murphey et al. 1989).
- *Chemoreceptors.* In contrast to the situation in dipterans (*Phormia*, *Drosophila*), the endings of the mechano- and chemosensitive cells of the contact chemoreceptors in *Cupiennius* do not pass into different areas of the CNS (Murphey et al. 1989). Perhaps a dependence on both chemo- and mechanoreception during dragline following by spiders has something to do with this (Tietjen and Rovner 1982).

The males of some wolf-spider species follow a female's dragline once they have been put into the right mood by a pheromone that adheres to the thread (see Chapter XX on courtship). Tietjen (1977) and Tietjen and Rovner (1980) studied the details of this behavior in *Lycosa rabida* and *Lycosa punctata*. After the first contact with the silk the males stop wandering around, examine the silk with their pedipalps and position themselves in such a way that the thread passes below the body, running between the two pedipalps. Then the males walk along the female's thread, keeping the pedipalps in contact with it. In addition to chemical stimuli, mechanical stimuli play a role that evidently varies in importance in the different species. Perhaps the term "contact chemoreception" has a special literal meaning in the context of dragline following, inasmuch as the contact can also be regarded as a mechanical stimulus. In this case the mechanical sensitivity that chemoreceptive hairs also possess (a mechanosensitive receptor cell inserts at the base of the hair) would have a plausible biological significance. This idea should be pursued, in the behavior and also as a potentially interesting case of multimodal interaction of sensory information.

- *Tonotopy.* In insects, primary auditory neurons seem always to be tonotopically organized if the animal has a need to identify certain frequencies accurately, in communication and for catching prey (Römer 1987; Oldfield 1988; Brodführer and Hoy 1990). In contrast, there is no tonotopy in the filiform hair system of *Acheta domesticus*; this system allows the cricket to detect air movements (produced by the reproductive partner, a predator or parasites) over a broad range of frequencies (Gnatzy and Heußlein 1986; Stephard et al. 1988). The resemblance to the trichobothria of *Cupiennius* is obvious.
- *Thermo- and hygroreceptors.* The projection patterns of thermo- and hygroreceptors in insects have not yet been studied. Nishikawa et al. (1991) described temperature-sensitive interneurons in the deutocerebrum of the cockroach (*Periplaneta americana L.*). Whereas the local interneurons have multiglomerular arborizations in the antennal lobe (and only there), the output neurons they examined have only uniglomerular arborizations and their axons run into the protocerebrum.

XVI Two Visual Systems in One Brain

Although so far we know very little about visually controlled behavior of *Cupiennius* (see Chapter XXIII), the capabilities of the eyes (see Chapter XI) provide the foundation for a well-developed sense of vision and we might well expect the animals to exploit this potential. Our expectation is strengthened when we look at the size and structure of the centers in the brain that receive input from the eyes and process this visual information. The situation is quite different from that in insects: in spiders the actual brain is almost completely devoted to vision, receiving only the optic nerves and containing only the optic ganglia and some association centers. A few years ago we examined the anatomy of these ganglia and discovered a number of interesting things (Babu and Barth 1984, 1989; Weltzien and Barth 1991; Strausfeld and Barth 1993; Strausfeld et al. 1993).

The distinction made at the periphery between principal and secondary eyes (Plates 13 and 14) persists in the brain. The two types of eyes each have their own visual pathway, with two separate sets of neuropil regions. To use more fashionable terminology: this is evidently an instance of parallel processing of the visual information provided by the principal and secondary eyes. With respect to the functional distinction between the two processing pathways, much remains obscure. However, we can make the preliminary assumption that the secondary eyes are specialized for viewing the horizontal movement of objects, whereas the principal eyes are especially suitable for the detection of shape and texture (see Chapter XXIII). This interpretation is partly based on findings in jumping spiders (Land 1971, 1972, 1985) and in *Cupiennius* is supported not only by the structure of the eyes but also by that of the visual interneurons in the CNS.

1 The Secondary Eyes

As figure 1 shows schematically, for each secondary eye there are two optic ganglia (ON1 and ON2); even the very early investigator Hanström (1925, 1926, 1928) suggested that these are comparable to the lamina (ON1) and the medulla (ON2) of the visual system of insects. The third optic neuropil (ON3) is common to all the secondary eyes. Hanström called this neuropil the mushroom body, suggesting a homology with the mushroom body of insects, but this interpretation is not supported by our results (see pp. 193, 201). The mushroom bodies of the insects receive their major input from olfactory centers (Homberg et al. 1989), process multimodal information and play a special role in learning and the formation of memories (Hammer and Menzel 1995). In contrast, the ON3 of *Cupiennius* is more like retinotopic neuropil regions of insects such as the lobula plate, which have large output neurons that mediate the generation of optomotor responses.

Let us follow the visual pathway, starting from the eyes (Fig. 2).

The following description is based on a modification of the silver-staining method according to Bodian (1937) and Palmgren (1948), Golgi impregnations and a few injections of Lucifer Yellow into single cells. For details of the methods see the original literature (Babu and Barth 1984, 1989; Strausfeld and Barth 1993).

The axons of the visual cells and processes from glial cells together form the optic nerve, which in the adult animal is about 4 to 5 mm long. Each retina projects exclusively into its own ON1. The size of this first optic ganglion is correlated with that of the associated eye; hence the ON1 of the PM eyes is the largest. Furthermore,

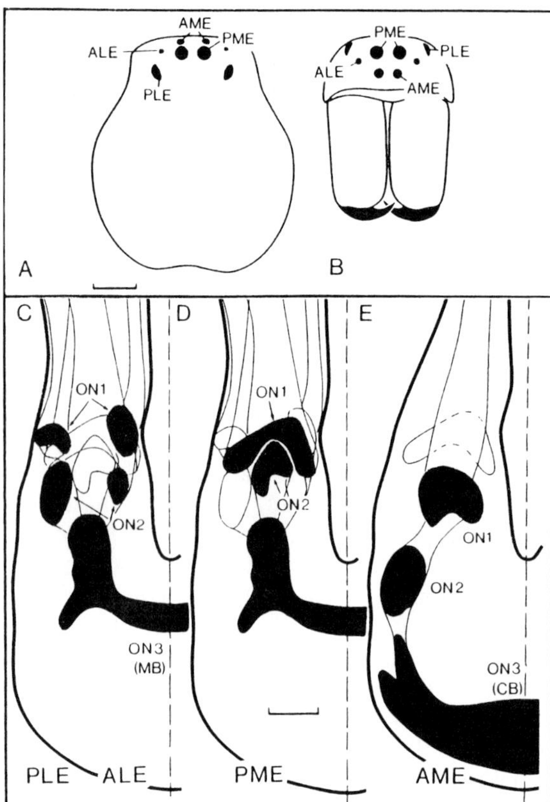

Fig. 1 A–E. The visual systems of the principal and the secondary eyes of *Cupiennius salei*. **A** and **B** Dorsal and frontal view of prosoma. *AME, PME* Anterior and posterior median eyes; *ALE, PLE* anterior and posterior lateral eyes. **C** and **D** Dorsal views of the visual system of the secondary eyes; *ON1* lamina, *ON2* medulla, *ON3 (MB)* "mushroom body". **E** Visual system of principal eyes; *ON3 (CB)* "central body". *Bar* in **A** 500 μm, in **B–E** 250 μm. (Strausfeld and Barth 1993) (with permission of John Wiley & Sons Inc., New York)

each of the three ON1's has its own characteristic shape (Fig. 1). The ON1 of the PM eyes, for instance, has a striking boomerang-like cross section. The first optic neuropils of all the secondary eyes project into separate second-order optic neuropils (ON2). In the terminology used for these pathways in insects, we would say that for each secondary-eye lamina there is a higher-order secondary-eye medulla, which in turn has a specific structure depending on the particular eye concerned.

The Cellular Organization of ON1

The cellular organization of ON1 can be seen in figures 2 and 3. First, four types of neurons can be distinguished.

The Terminals of the Photoreceptor Cells. The receptor endings in ON1 are brush-shaped and are arrayed side by side like small columns. The radially arranged axon collaterals exhibit varicosities and mostly are associated with a terminal of the thick type (terminal diameter 5–8 μm; 30–36 collaterals, about 1–2 μm thick), although a few arise from thinner terminals (diameter <5 μm). The "brush" formed by a terminal, which is circular except in the peripheral zone of the ON1, in each case overlaps with the endings of about 6 adjacent receptor terminals. We have never found any receptor axons that run straight into ON2 rather than synapsing with second-order axons in ON1, such as exist in some insects (Meinertzhagen 1975; see Strausfeld and Barth 1993).

Small Output Neurons. Between the receptor-cell endings are the endings of lamina cells (L cells), which connect the ON1's of the secondary eyes to the associated ON2's. The axons of these cells originate as a T-shaped bifurcation of the neurite (in insects the corresponding cells are unipolar). One of the branches runs into ON1 and the other, into ON2. Within ON1 the dendritic endings have either of two characteristic shapes, which distinguish narrow-field L cells (relatively small brush- or basket-shaped arborizations) from broad-field L cells (arborization area about as large as that of a receptor-cell terminal). These receive information from the visual cells and pass it on to the ON2.

Tangential Neurons. At the periphery of each secondary-eye lamina (ON1) there are numerous tangentially oriented processes covering large fields, in which they receive input from as many as 50 receptor-cell terminals. The endings of these Tan L neurons also connect ON1 to ON2. Spine-like specializations indicate the dendritic nature of the Tan L neurons and suggest comparison with centrifugal tangential neurons in insects and crustaceans, which connect the me-

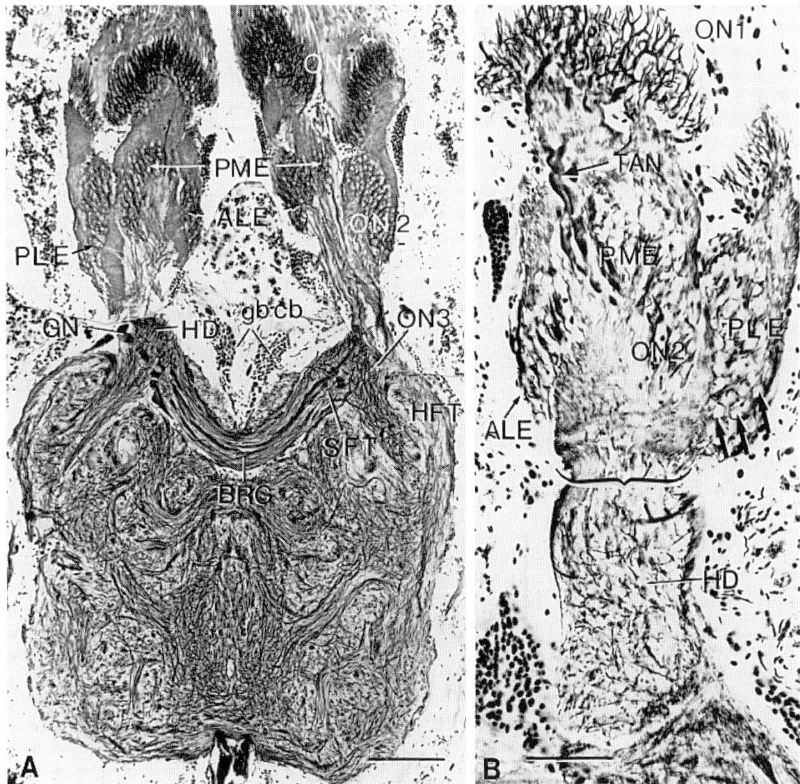

Fig. 2A, B. A Bodian-stained sections showing the relationship between visual neuropils *ON1* and *ON2* of posterior median eyes (*PME*) and anterior and posterior lateral eyes (*ALE, PLE*) and visual neuropil *ON3* ("mushroom body"). **B** Bodian stain showing *ON1* of the posterior median eyes (*PME*) supplied by large tangential neurons (*TAN*), the axons of which pass across *ON2*. *Arrows* in *PLE ON2* indicate spherical modules (spherules); *bracket* common pathway from ON2 neuropils to the head (*HD*) of *ON3*. *BRG, HFT, SFT* bridge, haft, and shaft neuropils of ON3; *gb cb* somata of globuli cells; *GN* giant ON3 neurons. (Strausfeld and Barth 1993) (with permission of John Wiley & Sons Inc., New York)

dulla to the lamina (Strausfeld and Nässel 1980; Pfeiffer-Linn and Glantz 1991).

Large-Field Neurons. Near the ventral surface of the ON1 of the PL eyes there arise several very thick axons, 9–20 μm in diameter. Each of these sends out several centrifugal branches, which form a kind of tangential plexus below the inner surface in all three laminae (see Fig. 2b and Fig. 4). From this plexus processes run outward, through the synaptic neuropil. The literature (Trujillo-Cenóz and Melamed 1967; Yamashita and Tateda 1981; Blest 1985a; Yamashita 1985) already contains several indications of an efferent innervation of the retina and probably also the lamina. In scorpions such neurons are involved in the endogenous control of the circadian changes in retinal sensitivity (Fleissner and Fleissner 1985).

Two other types of tangential neurons with thick axons and very extensive arborizations in the ON1 have been described by Strausfeld and Barth (1993). One of these is presumably centrifugal; the broadly distributed processes of its axon penetrate the ON1 of all three secondary eyes. Its cell body is ventrolateral to the "central body". The other type is centripetal, with several large axons that branch distally, project under all three laminae and send processes into them.

The Cellular Organization of ON2

What happens next? The three secondary-eye medullae (ON2) are positioned above the heads of the "mushroom bodies" (ON3) (Figs. 1, 2, 3). They comprise three layers and, as revealed by Bodian staining, are subdivided into round modules. Through each of these spherules runs a pair of axons of L cells (see Fig. 3), which connect ON1 and ON2 by way of an optic chiasma.

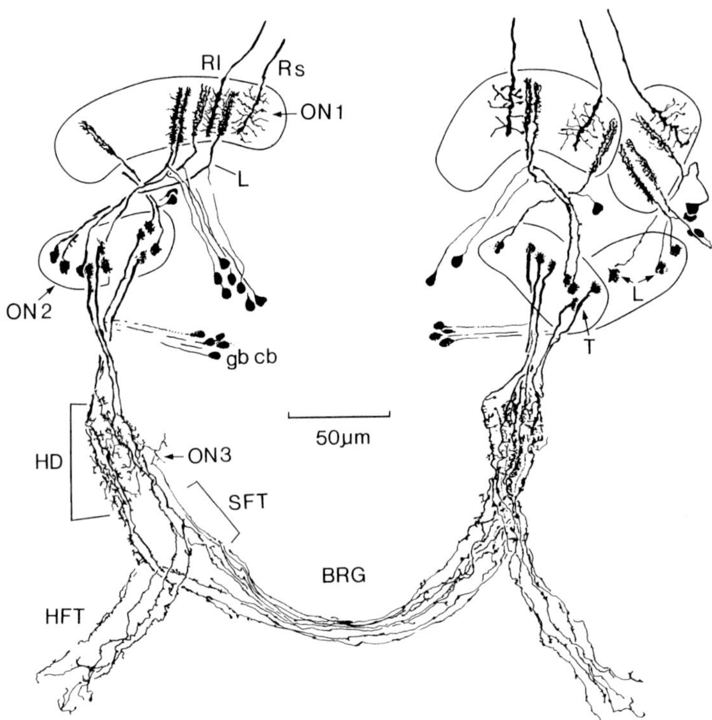

Fig. 3. Neural organization between *ON1* and *ON3*. *Left* Neurons linking posterior median eye lamina (*ON1*), medulla (*ON2*) and the *ON3* ("mushroom body"). *Right* Neurons linking posterior median and posterior lateral *ON1, ON2*, and the *ON3*. Large (*Rl*) and small (*Rs*) photoreceptor terminals are contacted by dendrites of brush- and basket-shaped L-cells (*L*) the axons of which cross over the outer chiasma into *ON2*. *L*-cells terminate among dendrites of bulbous and wide-field *T*-cells originating from groups of medial cell bodies (*gb cb* globuli cells). *T*-cell axons have short collaterals in the head of the "mushroom body" (*HD*) and branch in the shaft (*SFT*) to enter the haft (*HFT*) and bridge (*BRG*). (Strausfeld and Barth 1993) (with permission of John Wiley & Sons Inc., New York)

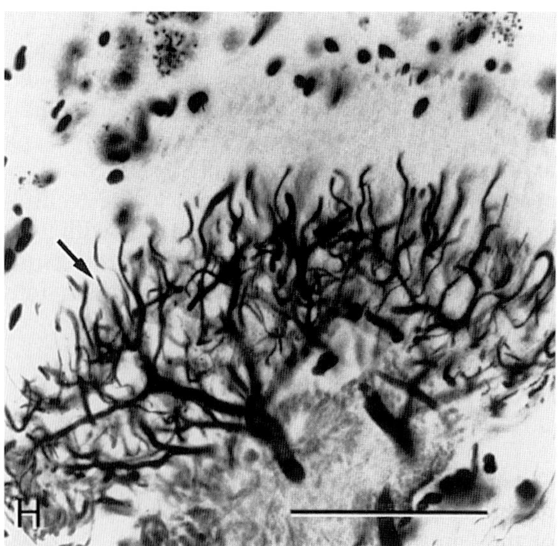

Fig. 4. Tangential cells in the lamina (*ON1*) (Bodian stain). *Bar* 50 μm. (Strausfeld and Barth 1993) (with permission of John Wiley & Sons Inc., New York)

We assume that the linear arrangement of the L-neuron axons persists after they are rotated by ca. 180°.

The spherical endings of the L neurons in ON2 mingle in the spherules with the knobby endings of T cells, which connect ON2 to ON3. Often the L-cell terminals so clearly enclose the terminals of these bulbous T cells that the inference of a functional relationship seems inescapable.

Presumably still other cells are also functionally involved in the spherules (Fig. 5). This is suggested, for example, by the shape and size of rosette-like endings of so-called bilateral T cells. These cells form one to three layers of bilateral arborizations, each ending in a rosette that is closely associated with an L-neuron terminal or the ending of a bulbous T cell.

Another form of T cell (again characterized by the approximately right-angled bifurcation of the neurite), which tends to be rare, is notable because of the broad distribution of its dendrites, which penetrate all three layers of the

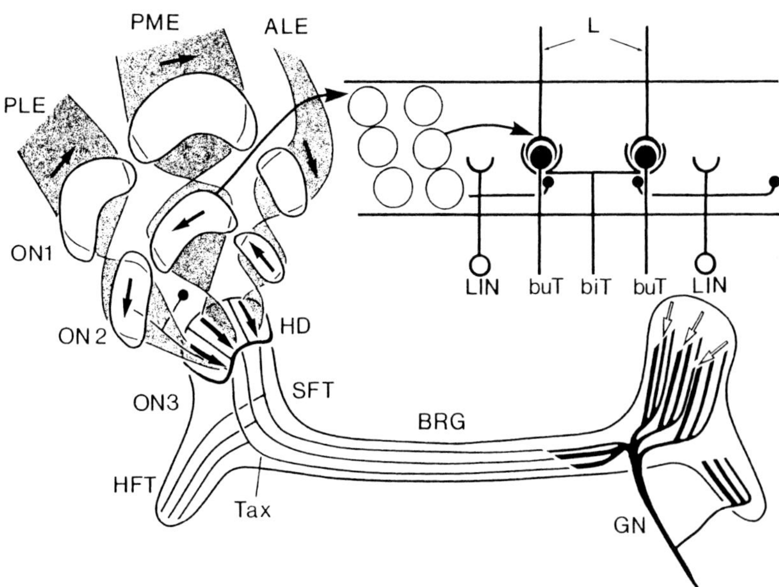

Fig. 5. Summary diagram of secondary-eye visual pathways and *ON2* and *ON3* organization. The laminae (*ON1*) of the secondary eyes (*PLE, PME, ALE*) send retinotopic projections to the medullae (*ON2*), forming chiasmata. The L-cell axons of ON1 rotate through ca. 180°, thereby inverting their arrangement (see *thick arrows*). T-cells in ON2 pass the inverted retinotopy on to ON3 while they further rotate by 45 to 90° between ON2 and ON3. The representations of the ON2s of the three secondary eyes come to lie side by side in the head neuropil (*HD*) of ON3, so that the visual environment is represented in three parallel stripes. T-cell axons (*Tax*) pass the visual information on to the haft (*HFT*), shaft (*SFT*) and the bridge (*BRG*) of ON3. The T-cells also contact the dendritic trees of the three giant output neurons (*GN*). The *inset* is a simplified diagram of ON2 of a PM eye. Rows of spherules are interpreted as retinotopic units: L-cell endings (*L*) meet the bulbous dendritic ends of T-cells (*buT*); both are surrounded by the endings of bilateral T-cells (*biT*). Local interneurons (*LIN*) connect the spherules with each other laterally. Additional cell types not shown. (Strausfeld and Barth 1993) (with permission of John Wiley & Sons Inc., New York)

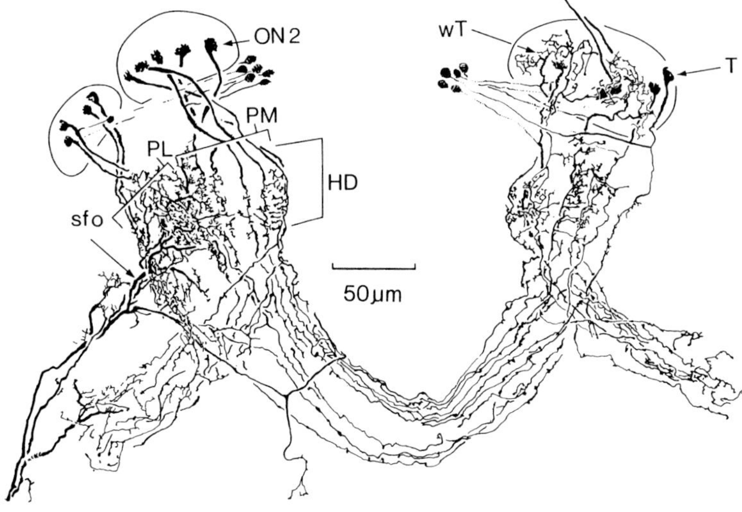

Fig. 6. Projections to the ON3 from posterior lateral (*PLE*) and posterior median eye (*PME*) ON2s. Together with small T-cells and bilateral T-cells, wT-cells contribute short stratified or diffuse arrangements of collaterals to the head neuropil. T-cell axons project to discrete zones in the head neuropil (*HD*) of ON3 (*PL* posterio-lateral and *PM* posterio-medial zones). *wT* Wide-field T-cells; *sfo* small-field output neuron arising from left head neuropil. (Strausfeld and Barth 1993) (with permission of John Wiley & Sons Inc., New York)

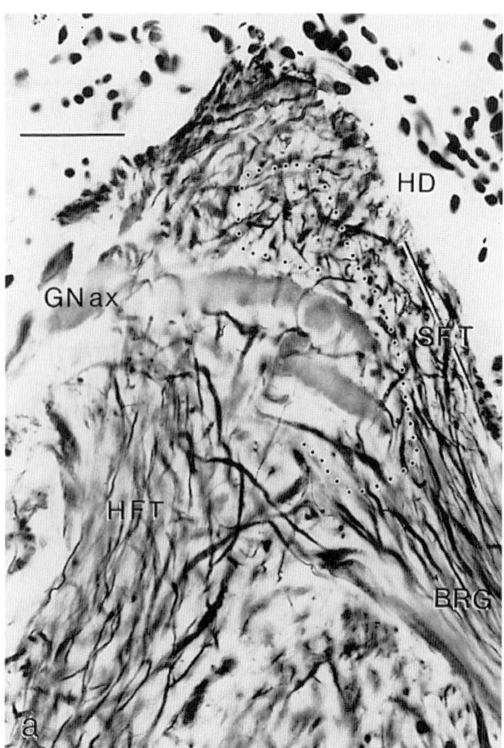

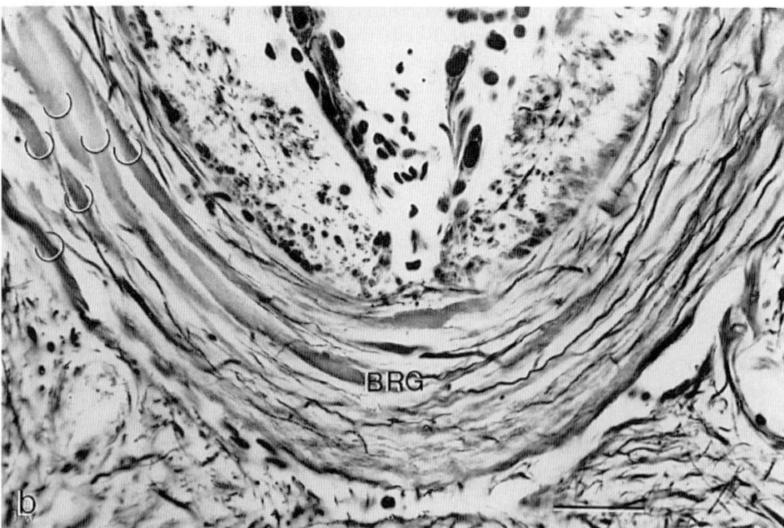

Fig. 7 a, b. ON3 organization. **a** Bodian-stained section across the junction of the shaft (*SFT*), haft (*HFT*), and origin of the bridge (*BRG*), showing outgoing axons from giant neurons (*GNax*). **b** Bodian-stained ON3 bridge (*BRG*), showing the arrangement of many small fibers running parallel to six giant neuron profiles (*ringed*). *Bars* 25 µm. (Strausfeld and Barth 1993) (with permission of John Wiley & Sons Inc., New York)

medulla and in each of them reach several spherules. We have called these cells wide-field T cells (WT); they are shown in figure 6.

In addition to all these T neurons (narrow-field or bulbous T cells, bilateral T cells, wide-field T cells), which are projection neurons that send long axons to the "mushroom body", the ON2 of the secondary eyes also contains local interneurons with short axons, which can be designated amacrine cells, and tangential cells, at least some of which provide centrifugal connections between ON2 and ON1.

The Cellular Organization of ON3

As a physiologist, I seriously doubt whether anyone will ever completely disclose the functional secret of all these neurons and their connections. Nevertheless, we must move on to the third processing station for visual inputs. Although we prefer the neutral name ON3, with reservations we also use the term "mushroom body" applied by Hanström (1921, 1928, 1935) to this bilaterally symmetric neuropil. In its "head" region, one on each side of the brain, it receives inputs from the T cells of the ON2 of each of the three secondary eyes (see Figs. 1 and 5). As output neurons of the ON3, giant fibers project into the midbrain.

Inputs. First let us consider the inputs by way of the T-cell endings in the head neuropil of the "mushroom body". Typically, the axons of T cells of all three secondary eyes enter the outer layer of the head neuropil, where they send out many short collaterals. They continue on into the "shaft", where they branch and then send individual axonal branches into the "haft" and the "bridge". The delicate collaterals of the T cells in the head neuropil bear both spines and blebs, which suggests that here they are both pre- and postsynaptic. In contrast, the structure of the T-cell axons in the rest of the ON3 implies solely a presynaptic function.

Outputs. Unlike the head neuropil, the shaft and bridge neuropil of ON3 is characterized by a parallel arrangement of the axons. Among the T-cell axons are very conspicuous giant fibers (Fig. 7), which serve as output neurons. Strictly speaking, unilateral giant output neurons should be distinguished from bilateral ones, as follows.

The *unilateral* giant axons are up to 20 μm in diameter, the largest fibers in the brain of *Cupiennius*. Their projections are ipsilateral, and their collaterals end in descending tracts that run to the segmental ganglia of the subesophageal region. The dendrites of these neurons are substantially restricted to one side of ON3 – hence the name. The axon divides into two main branches, which run into the haft or bridge and the shaft, respectively. Secondary dendrites originating in the shaft form a strikingly dense plexus of short processes, which extend into the head neuropil of ON3 (Fig. 8).

The *bilateral* giant output neurons of the ON3, as their name implies, form dendrites in both the right and the left shaft and in the bridge. Figure 9 shows the reconstruction of one such Golgi-impregnated neuron. The at least 6 giant fibers passing through the bridge presumably correspond to the three bilateral giant neurons found on each side of ON3 (Fig. 7b).

In addition to the giant neurons, the ON3 gives rise to *small* output neurons, the dendrites of which are limited to the head neuropil on the ipsilateral side. Figure 6 shows some details. These neurons are hard to stain, so that their axons can be followed only for short distances.

There remains a question that is very important, from a functional point of view. How are the three secondary eyes represented in the ON3, given that axons from all three ON2's come together here? Arrays of T-cell axons form a partial chiasma between ON2 and ON3, twisting through about 90° (Fig. 5) and evidently retaining their linear arrangement in the process. A number of indications (see Strausfeld and Barth 1993) suggest that the three ON2 neuropils are represented separately but next to one another in the ON3, as shown in figure 5.

The Homology Problem

Our work on the neuroanatomy of the visual systems of *Cupiennius* raised a question that applies to spiders in general: how well justified is it to homologize the structure called "mushroom body" (ON3) here with those of the same name in other arthropods, as was proposed by Hanström? There are certainly some superficial similarities. For instance, the three lobes on each side of the spiders' "mushroom body" resemble the peduncle, the α-lobe and the β-lobe of the paired insect mushroom body. Furthermore, the densely packed small somata near the head of the "mushroom body" in spiders might

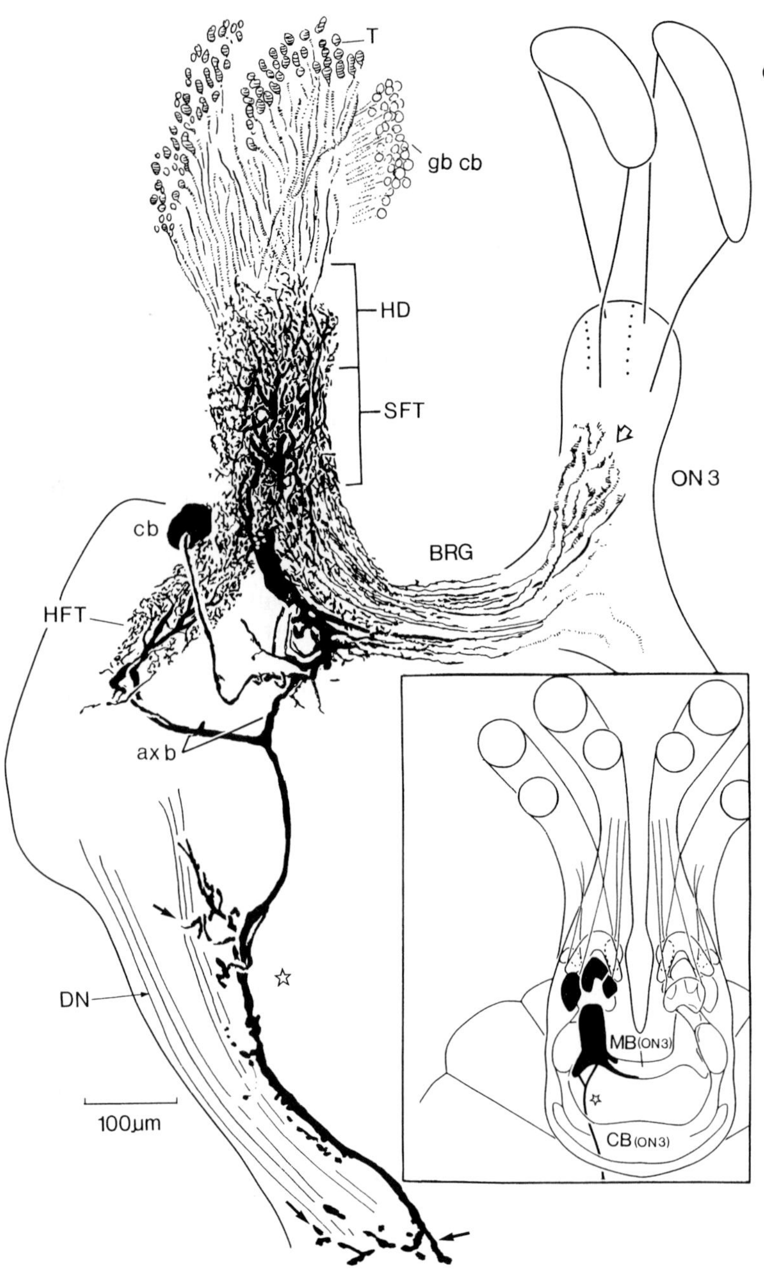

Fig. 8. Lucifer Yellow-filled unilateral giant neuron in ON3. Two dendritic trees from axon branches (*ax b*) invade the head (*HD*), shaft (*SFT*), and haft (*HFT*) of one side, with dendrites extending as far as the contralateral shaft. The cell body (*cb*) of the giant neuron lies dorsally. Its axon (*star*) extends postero-laterally to branch among descending neuron tracts (*DN*). *gb cb* Cell bodies of globuli cells. *Inset* Diagram illustrating the role of the "mushroom body" (*MB*) as the third optic neuropil *ON3* of the secondary eyes. *CB* "Central body" (ON3 of principal eyes). (Strausfeld and Barth 1993) (with permission of John Wiley & Sons Inc., New York)

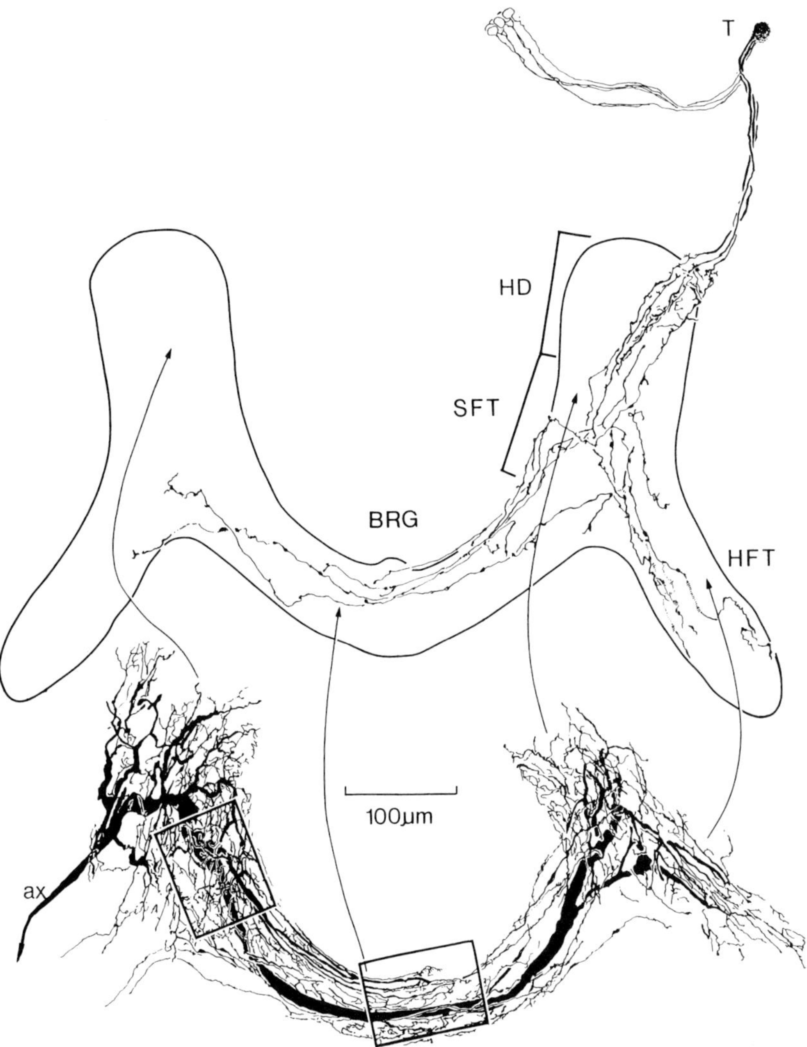

Fig. 9. Bulbous T-cell (*T*) input and giant neuron output (*ax*) of secondary eye ON3. *Above* ON3 showing its divisions into head (*HD*), shaft (*SFT*), haft (*HFT*), and bridge (*BRG*). A row of bulbous narrow-field T-cells in the posterior lateral ON2 undergoes a rotation through 90° to enter the *HD*. Axons project roughly in parallel through the shaft and bridge, but undergo some mixing in the haft. *Below* Golgi-impregnated bilateral giant output neuron. *Arrows* indicate relation of giant neuron to ON3 outline. (Strausfeld and Barth 1993) (with permission of John Wiley & Sons Inc., New York)

initially be regarded as homologous with the small basophilic cell bodies in and near the calyces of the insect mushroom body. But on closer examination, as was hinted in the preceding text, serious doubts as to their homology arise. The reasons can be summarized as follows (Strausfeld and Barth 1993).

1. Non-equivalence of the connections. It is an especially important criterion for homology of structures in the CNS that the *connections* be the same, by which is also meant that the neurons concerned should be the same ontogenetically. Far too little is known about the ontogeny of the CNS of spiders, in particular regarding cell lines; therefore any comparison must be based on adult animals.

Both the mushroom body of insects and the ON3 of spiders are connected to sensory interneurons, but in quite different ways. In insects projections run from the antennal lobes into the calyces of the mushroom bodies, from which Kenyon cells send parallel axons into the peduncle and the lobes. The orderly arrangement of the dendritic trees in sensory neuropil, which might be evidence of odortopy (glomeruli in the olfactory lobes) or of retinotopy (columns in the lobula) is thus lost (Gronenberg 1986, Kanzaki et al. 1989).

According to what is known so far, all afferents to the spider "mushroom body" come from the secondary-eye neuropils ON2. There is no such thing as a calyx in *Cupiennius*, and the globuli cells identified by Hanström are the somata of T cells that mediate between ON2 and ON3.

The *position* of the "mushroom body" with respect to the segmentation of head and brain also presents problems. These are largely associated with the fact that very little can be said with certainty regarding segmental homologies between the brains of spiders (Chelicerata) and insects (Weygoldt 1985). In insects the mushroom body is located in the protocerebrum and the calyx receives visual inputs from interneurons higher than 4th order. In contrast, the "mushroom body" of the spiders is relatively far back in the brain and receives input from 2nd-order visual interneurons.

2. There seem also to be important *functional differences* between the mushroom bodies of insects and spiders. The mushroom bodies of insects are olfactory centers, which play a crucial role in learning and memory (Erber et al. 1987, Hammer and Menzel 1995). They also receive sensory inputs of various other modalities, including vision and mechanoreception (Schildberger 1981; Homberg 1984; Kaulen et al. 1984; Gronenberg 1986; Mizunami et al. 1992), and subserve multimodal integration. The organization of axons that run in parallel in the "mushroom body" of *Cupiennius* seems similar to that in the insect mushroom body (lobes and peduncle) only on very cursory examination. A decisive difference is that in the insects the parallel afferent fibers are those of Kenyon cells (Kenyon 1896), whereas in the spider these fibers, as far as is currently known, are only the terminals of fibers from the medullae (ON2) of the secondary eyes.
3. When the *visual pathway* of insects such as calliphorid dipterans is compared with that of *Cupiennius*, similarities can be discerned between the lobula plate and the "mushroom body" of the spiders. In the insects the raster points of the retina are represented individually by serially arranged retinotopic neurons in a path that runs through columnar modules of the medulla into the lobula and lobula plate (Braitenberg 1970; Meinertzhagen 1975). In *Cupiennius*, too, the neuroanatomy suggests the presence of such modules in the ON2, although here they are spherical rather than columnar. However, it has not yet been demonstrated directly that the spherules of the spider medulla represent "visual sampling points". A further analogy also deserves mention. The most conspicuous output neurons of the lobula plate are giant cells that respond to vertical or horizontal movements in the visual field (Hausen 1984). The ON3's of the secondary eyes of *Cupiennius* also contain giant output neurons, on the dendrites of which the terminals of relay neurons of the ON2's end. Given that the photoreceptor cells in the secondary eyes of *Cupiennius* form horizontal rows (Land and Barth 1992) and that horizontal rows of T cells in the ON2 project onto the giant output neurons of ON3, the latter are comparable to the HS cells (responsive to horizontal movements) in the lobula plate. According to Land and Barth (1992) and Grusch (1995), the spatial resolution of the secondary eyes of *Cupiennius* is greatest in the horizontal direction. Because spiders do not fly, and live in a "flat world", the analysis of horizontal movements may be especially important for them. Confirmation or rejection of these hypotheses must await future experiments, in particular an electrophysiological characterization of the responses of the various types of neurons.

For a comprehensive recent review on current knowledge and interpretations of the arthropod mushroom body (spiders – Araneae – largely excepted) see Strausfeld et al. (1998).

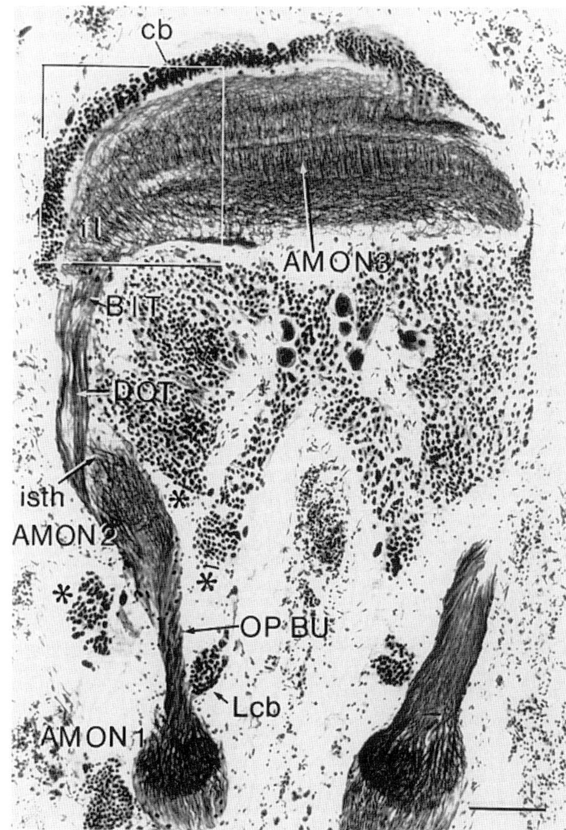

Fig. 10. The principal eye neuropils (Bodian stain). *AMON1–3* Lamina, medulla, and "central body"; *DOT* dorsal optic tract; *isth* isthmus; *cb* cell bodies of columnar neurons of ON3; *BIT* branch leading to the inferior tract; *OP BU* optic bundle carrying the chiasma between *AMON1* and *AMON2*, *Lcb* cell bodies of lamina output neurons; *asterisks* cell bodies associated with *AMON2*. *Bar* 100 µm. (Strausfeld et al. 1993) (with permission of John Wiley & Sons Inc., New York)

2 The Principal Eyes

Three visual neuropil regions are also associated with the principal eyes of *Cupiennius*. The ON3 of these eyes is not the "mushroom body" but rather the structure called "central body" by Hanström (1935). To avoid confusion with the neuropils of the secondary eyes, we call the lamina of the principal (anterior-median) eyes AM-ON1, their medulla AM-ON2 and the third station AM-ON3 or CB (central body). Figures 1 and 10 give an overview of this arrangement.

The two AM-ON1 lie close to the dorsal surface of the brain. From them thick fiber tracts run to the two olive-shaped AM-ON2's. A kind of isthmus connects AM-ON2 to the dorsal optic tract (DOT), which enters AM-ON3 laterally. This "central body" in spiders (Weltzien and Barth 1991) is crescent- to horseshoe-shaped. It is situated dorsally in the middle of the posterior brain section and consists of two lobes, the long axis of which is perpendicular to that of the brain and which in turn are each composed of several discrete layers (Fig. 11). Along its concave inner surface run fiber bundles that we call inferior tracts (IT). The IT's project to the opposite side and are joined by fibers from the DOT, small-field afferents that pass into the AM-ON3. Let us follow the route of the visual information from the beginning on, as we did for the secondary eyes (Fig. 12).

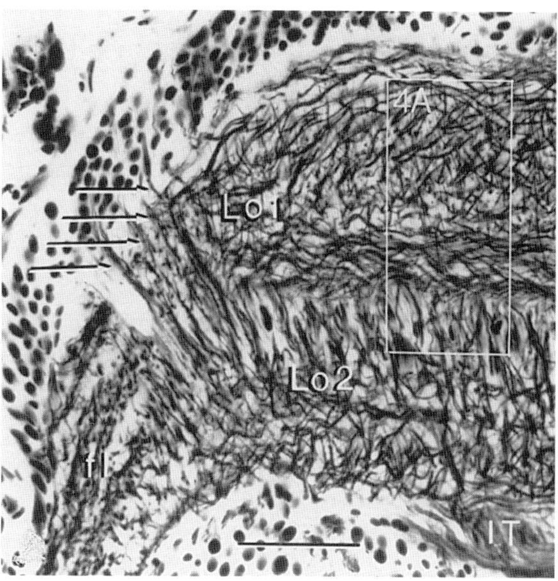

Fig. 11. Principal eye ON3 ("central body"). The two lobes *Lo1* and *Lo2* are separated by a serpentine tangential layer and are penetrated by columns, four of which (*arrows*) are parallel to the plane of section. *fl* Flange of ON3; *IT* inferior tract. *Bar* 50 µm. (Strausfeld et al. 1993) (with permission of John Wiley & Sons Inc., New York)

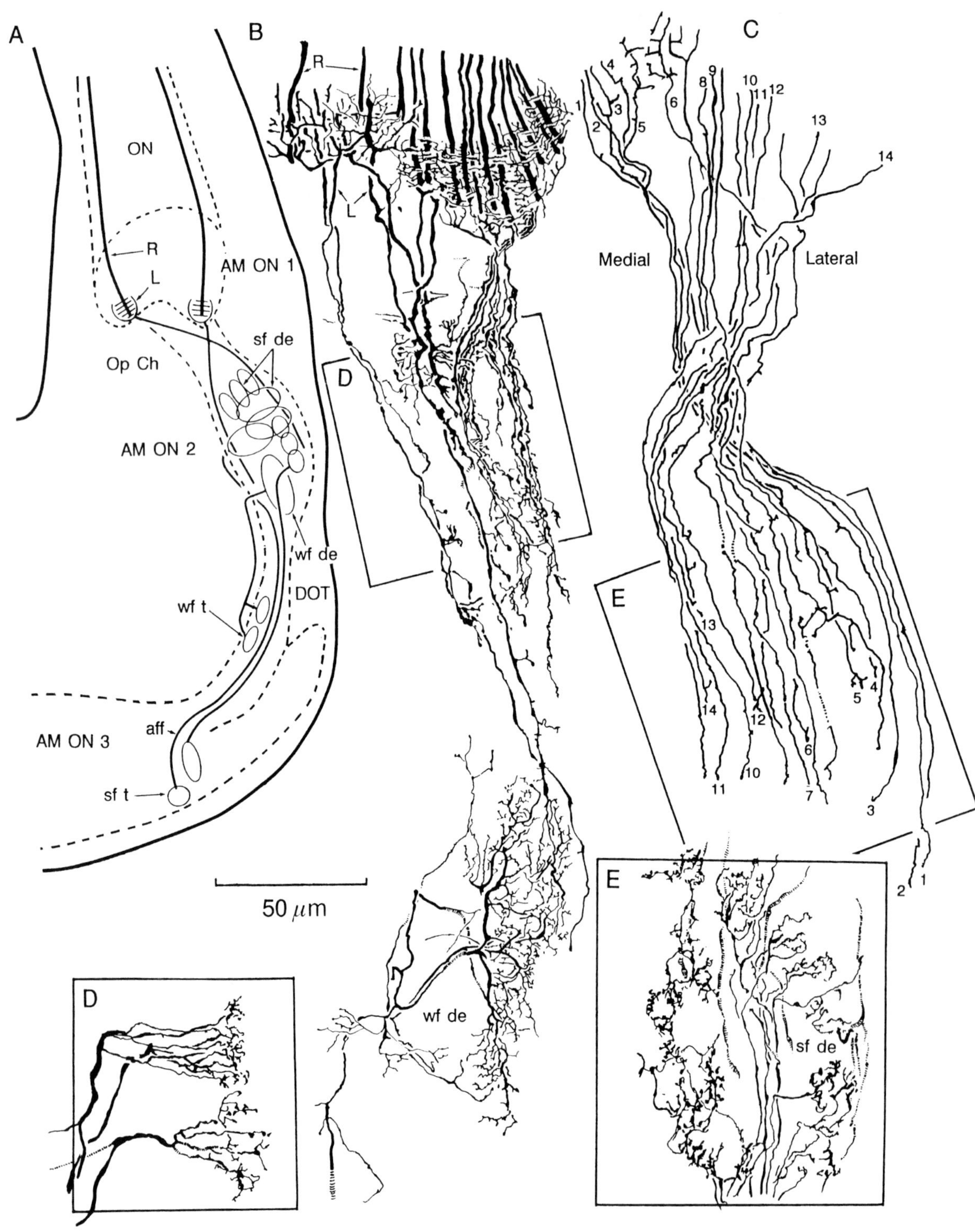
A
ON
R
L
AM ON 1
Op Ch
sf de
AM ON 2
wf de
DOT
wf t
AM ON 3
aff
sf t
50 μm
B
R
L
D
wf de
C
Medial
Lateral
E
sf de
D

Cellular Organization of AM-ON1

Like the first optic neuropil of the secondary eyes, the AM-ON1 contains a uniform population of photoreceptor terminals with radially arranged collaterals; here, again, these form a brushlike end structure. The narrow-field L cells (output neurons) so common in the ON1 are absent here. Instead there are only broad-field L cells, which – as in the ON1's of the secondary eyes – are T-shaped, with somata very probably assembled in small groups near the chiasma between AM-ON1 and AM-ON2. The amount of space occupied by the dendrites of the L cells in the lamina suggests that each L cell receives information from at least 9 photoreceptor-cell endings. As far as we could determine, the wide-field tangential cells of the ON1, like the narrow-field L cells, are not present in the AM-ON1.

In- and Outputs of the AM-ON2

The structure of the principal-eye medulla is far more complicated than that of the secondary-eye medulla and is less well understood. AM-ON2 merges with neuropils and fiber tracts (DOT) that run to the ipsilateral lateral border (flange) of the "central body". There is no subdivision of the AM-ON2 into spherical modules, the structure typical of the ON2.

The horizontal rows formed by the axons of the L cells are twisted through 180° on the way from lamina to medulla, but not with easily discernible precision (Fig. 12C). The terminals of the L cells lie at various levels in the medulla, which we interpret as indicating different functional categories. There are no structural signs of such differentiation in the ON2.

Output neurons leave the AM-ON2 neuropil at approximately a right angle to the terminals of the L neurons (Fig. 12A, D). They are of at least two types: (i) small-field elements, the axons of which enter the DOT and pass into the central body either laterally or by way of the inferior tracts, and (ii) wide-field neurons with endings in the DOT.

This is certainly not a complete picture of the principal-eye medulla; additional reconstruction of individual cells is needed.

◀

Fig. 12A–E. Visual pathways of principal eyes. **A** Summary of neural organization. Receptor axons (*R*) project through the optic nerve (*ON*) into *AM ON1*, where they terminate among wide-field lamina projection neurons (*L*). Rows of L neuron axons cross over in the optic chiasma (*Op Ch*) and terminate at various depths in *AM ON2*. They are visited by the dendrites of small-field projection neurons (*sf de*, represented as oval outlines). Projection neurons send axons through the dorsal optic tract (*DOT*) into the *AM ON3* where their afferent axons (*aff*) end as small-field terminals (*sf t*). Wide-field projection neurons (*wf de*) arise deep in *AM ON2* and give rise to endings (*wf t*) in the *DOT*. **B** Branched receptor axons (*R*) ending among L-neurons (*L*), the axons of which terminate at various levels in the secondary eye medulla (*ON2*). *wf de* Dendrites of wide-field projection neurons to the *DOT*. **D** Three small-field projection neurons leading from *AM ON2* to *ON3*. **C** Tracings of axons from the principal eye lamina to the medulla. The distal order within a row of L-neurons in the lamina (numbered *1–14*) is imprecisely mapped onto the principal eye medulla. **E** Dendrites of small-field output neurons (*sf de*). (Strausfeld et al. 1993) (with permission of John Wiley & Sons Inc., New York)

The AM-ON3, the So-Called Central Body

The 3rd optic neuropil of the principal eyes is characterized by the palisade-like arrangement of many thousand axons, which project perpendicularly through consecutive layers of tangential fibers (Fig. 13). As mentioned above, the "central body" of *Cupiennius* receives afferents chiefly by way of inferior tracts on its concave side, and also in its lateral regions from tracts originating in AM-ON2 and running through the DOT. The endings of such afferents occupy different layers of the "central body".

Groups of 10 to 30 efferent neurons run straight through the central body from back to front; being approximately equidistant from one another (8 to 10 μm apart), the groups look like columns (Fig. 14) forming a striking palisade that extends through the whole AM-ON3. The somata of these cells have a diameter of 10 to 15 μm and are arranged in a dense, conspicuous layer above and behind the central body (Figs.

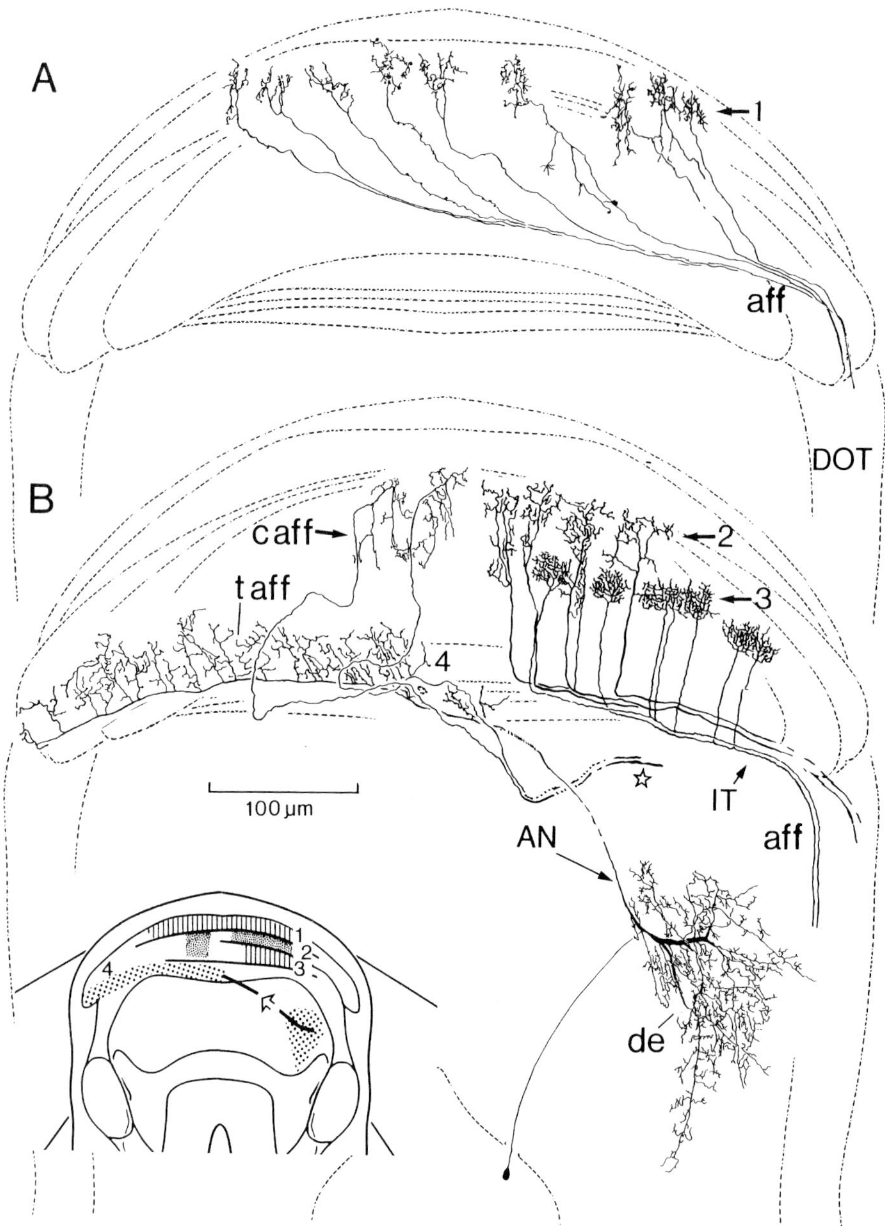

Fig. 13 A, B. The "central body" ON3. **A** Climbing afferents (*aff*) entering a distal layer (*1*) of ON3 from the dorsal optic tract. **B** Tufted afferents invading levels *2, 3* from the inferior tract (*IT*) and the *DOT*. Diffuse climbing afferents (*c aff*) also arise from *IT* (at *star*). *AN* afferent neuron (*de* dendrites; *t aff* terminals) in proximal stratum (*4*) of ON3. *Inset Numbers* in *A* and *B* refer to *numbered hatched and stippled areas in the inset*; *upwards* is direction toward posterior end of spider. (Strausfeld et al. 1993) (with permission of John Wiley & Sons Inc., New York)

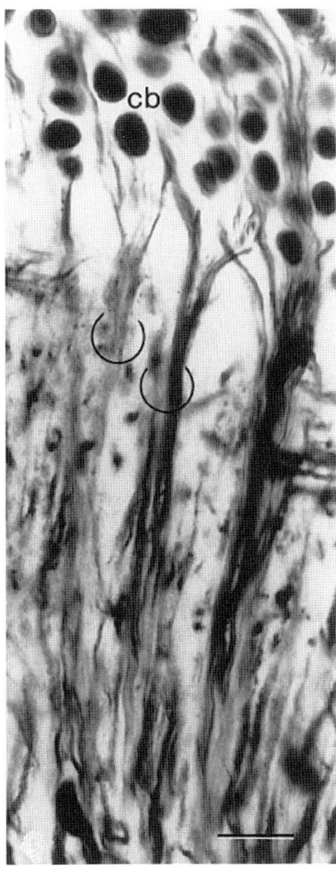

Fig. 14. Columns (ringed) in AM-ON3. *cb* cell bodies of neurons projecting anteriorly. *Bar* 10 μm. (Strausfeld et al. 1993) (with permission of John Wiley & Sons Inc., New York)

10 and 11). After passing through several layers of tangential fibers, the axons of the column cells turn into the ventral midbrain (between the anterior concave side of the "central body" and the haft and shaft of the "mushroom body"). In the central body most of the column axons send out remarkably few branches, which may indicate that the input synapses are on the axons themselves. In addition to the column cells, there are a number of tangential output neurons of the AM-ON3. Their dendrites bear abundant varicosities and spines, implying intense synaptic interaction. Within their projection areas in the midbrain, too, these neurons have richly arborizing terminals (Fig. 15). Like the giant neurons of the "mushroom body", output neurons of the AM-ON3 end in tracts of descending fibers, which project from the brain into the segmental aggregations of motoneurons that supply the musculature of the appendages.

The shapes of intrinsic neurons of the "central body" vary widely. They can be classified roughly as intrinsic column cells and intrinsic tangential cells. They ensure an extensive interlinking between columnar groups, and they themselves receive inputs from AM-ON2 (Fig. 16). However, the "central body" also receives inputs from outside the AM-ON2.

As in case of the "mushroom body" of the spiders – and even more so! – we are still far from even a moderate understanding of the "central body". The early interpretation put forth by Rádl (1912), namely that it is a visual center, has been fully confirmed by our studies of *Cupiennius*. On the other hand, Hanström's (1926) view that it is an association center particularly related to web construction in the spiders with that behavior cannot be upheld. A morphometric study in which web-building spiders were compared with other spiders provided no support for such an idea (Weltzien and Barth 1991). In *Cupiennius* the "central body" seems to be essentially a visual center, the architecture of which is reminiscent of complex retinotopic neuropils of insects and crustaceans, such as their medulla and lobula (Strausfeld and Nässel 1980). However, it should be kept in mind that the "central body" is not connected only to the principal eyes. In both *Cupiennius* and *Nephila* connections also exist to the secondary eyes (shown by Bodian staining), so that the central body appears to link the visual pathways of the two types of eye (Babu and Barth 1984, Weltzien 1988). In addition, the central body receives inputs from neurons in the midbrain, which might possibly provide information from other sensory pathways for multimodal integration.

Again the Question of Homology

In conclusion, we must again ask whether the structures in the various arthropods that since Hanström (1921, 1926, 1928, 1935) have been known as the central body are in fact homologous. As in the case of the "mushroom body" of spiders, there are important counter-arguments

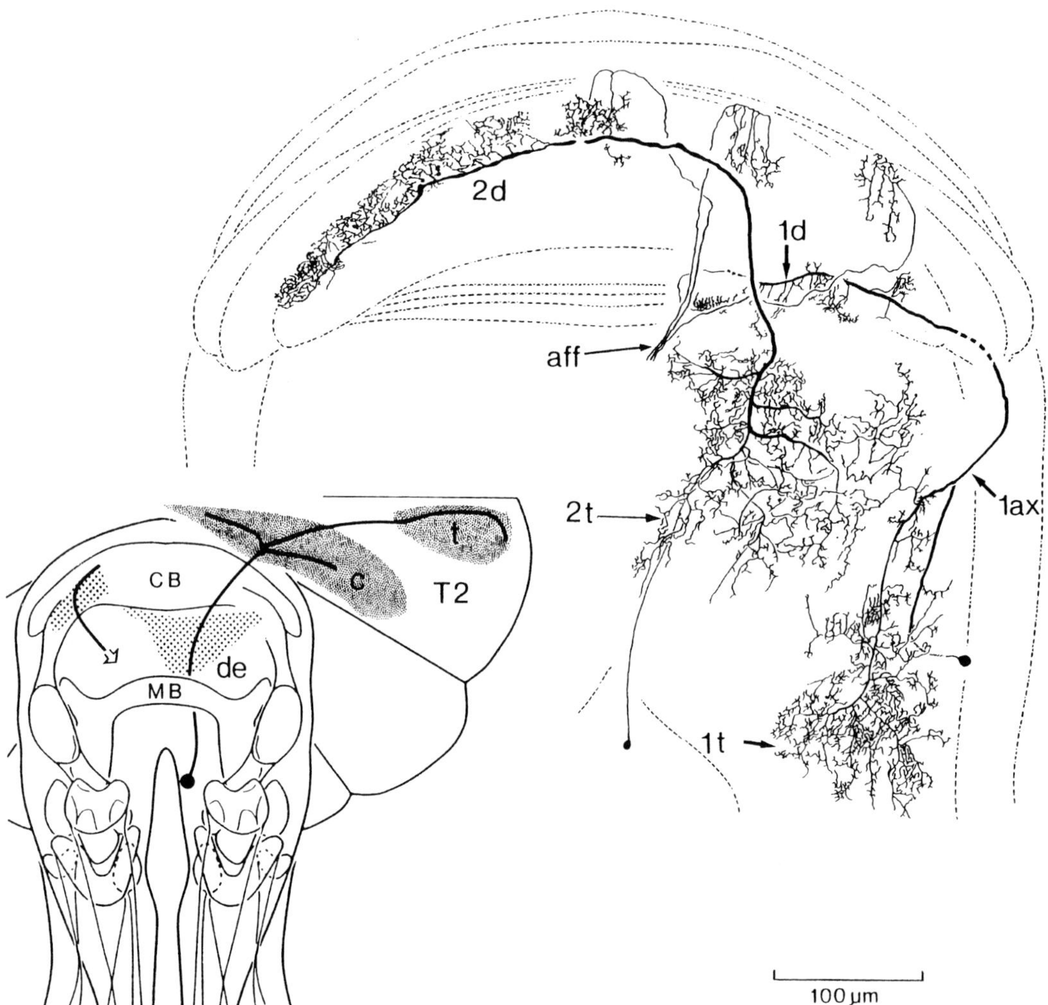

Fig. 15. Efferent neurons of the AM ON3 ("central body"). Two tangential efferents (cells *1*, *2*; *t* terminal, *ax* axon) shown with three small-field afferents (*aff*) originating from the inferior tract. *Inset* Relation of an afferent neuron and of a descending neuron to other parts of the brain and second prosomal neuromere (*T2*). *CB* "Central body"; *MB* "mushroom body"; *shaded* collateral (*c*) and terminal (*t*) fields of afferent neuron; *stippled* (in *CB*): dendritic field (*de*) of descending neuron. (Strausfeld et al. 1993) (with permission of John Wiley & Sons Inc., New York)

regarding the "central body". These have been explained in detail by Strausfeld et al. (1993). Here it suffices to say that the main evidence lies in the descending connections that link the retinotopic neuropil to various brain regions. The insect central body is a conglomerate of several neuropils, which have many connections to the midbrain but are not directly connected to any of the optic neuropils, whereas in spiders they are (AM-ON2). If we consider the spider central body as the primary target of the efferents from AM-ON2, then it occupies in the visual system a position corresponding to that of the lobula in insects. Furthermore, the central body of insects also lacks the many thousands of somata that are so typical of the spider "central body".

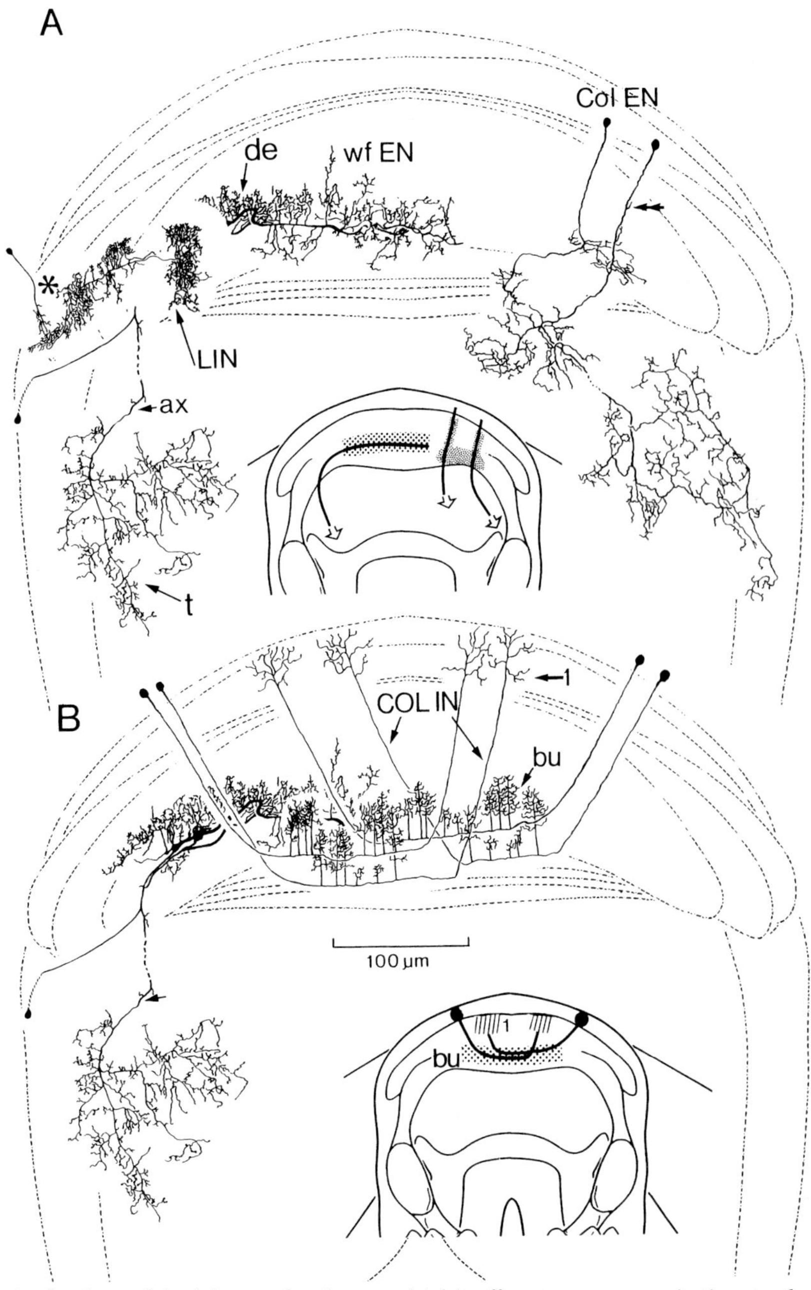

Fig. 16 A, B. Neuronal arrangement in the "central body" (AM ON3). **A** *Col EN* columnar efferent neurons. *Double arrows* indicate occasional dendritic branch; *wf EN* wide-field efferent neuron, the axon (*ax*) of which connects dense dendritic arbors (*de*) with a diffuse varicose terminal (*t*) in the central brain neuropil. *LIN* Local intrinsic neuron which gives rise to two symmetric fields, one dendritic in nature (*asterisk*). *Inset* Layer relationships between tangential (*left*) and columnar (*right*) efferent neurons projecting to the mid-brain. **B** *COL IN* columnar local interneurons with tufted dendrites distally (*1*) and bushy arborizations (*bu*) at deeper levels. *Inset* Diagram of trajectories of columnar intrinsic neurons from layer 1 proximally. Posterior is *upward* in the drawings. (Strausfeld et al. 1993) (with permission of John Wiley & Sons Inc., New York)

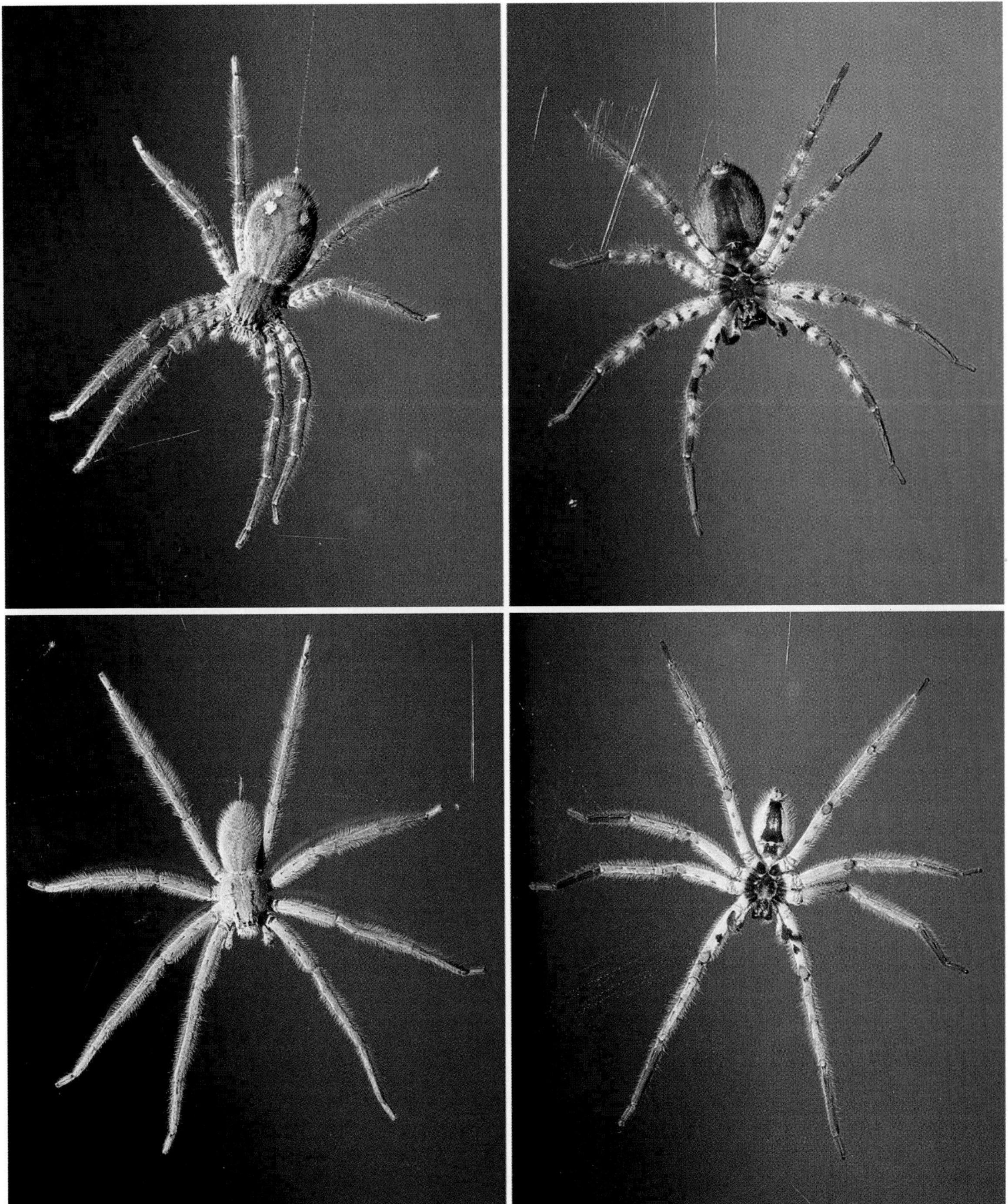

Plate 1. *Cupiennius salei* (Keyserling 1877). Dorsal and ventral view of adult female (above) and adult male (below). Body length (excluding legs) is 3.3 cm (female) and 3.2 cm (male), respectively

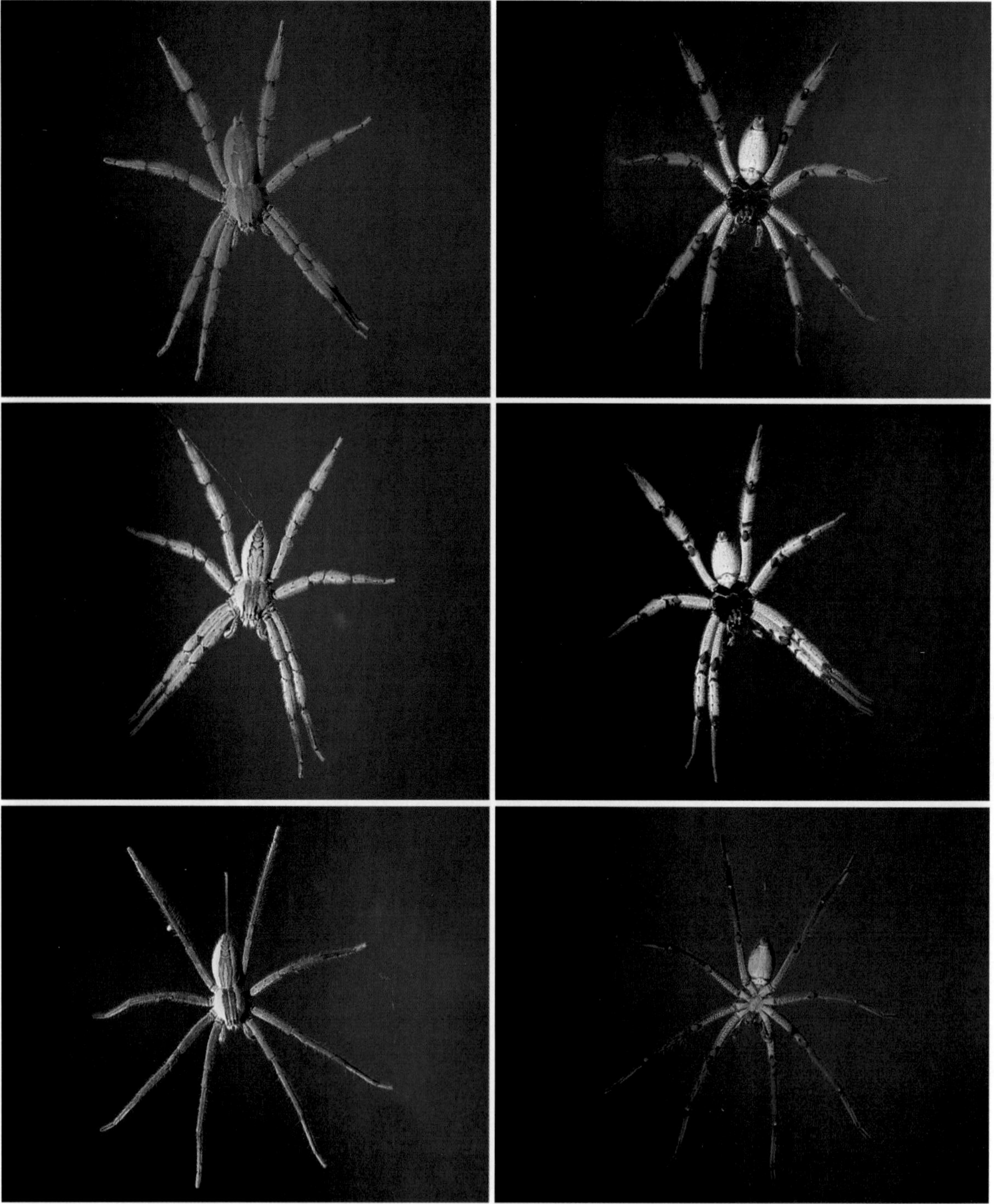

Plate 2. *Cupiennius getazi* (Simon 1891). Dorsal and ventral view of the orange (*top*) and gray (*middle*) varieties of adult females and of an adult male (*bottom*). Body length is 2.8 cm (orange female), 2.9 cm (gray female), and 2.3 cm (male)

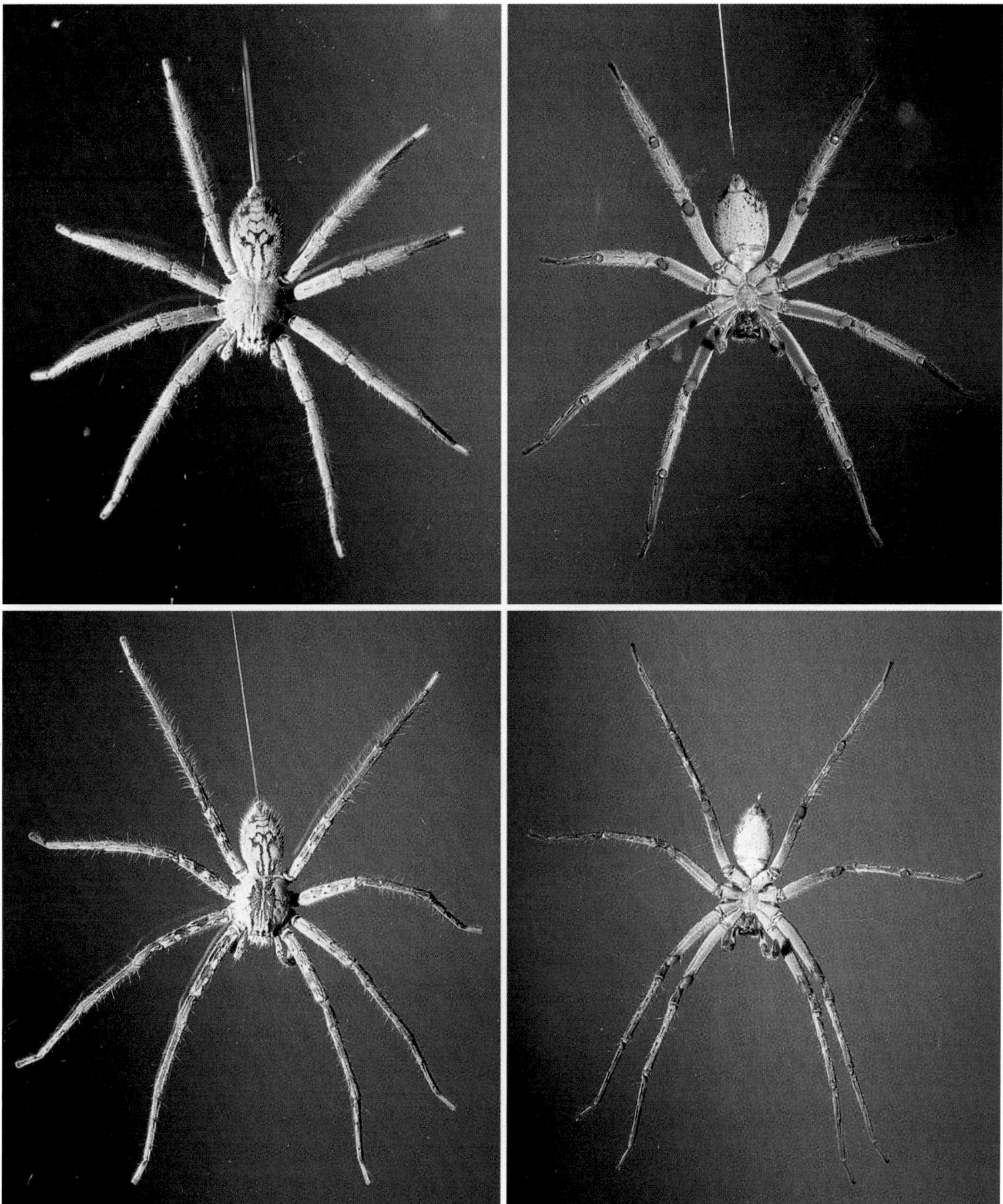

Plate 3. *Cupiennius coccineus* (F. Pickard, Cambridge 1901). Dorsal and ventral view of adult female (*above*) and adult male (*below*). Body length is 2.8 cm (female) and 2.9 cm (male), respectively

Plate 4. *Above* An adult male of *Cupiennius salei* with its body slightly raised and its legs spread out more or less evenly around its body. The spider typically assumes this posture when waiting for prey or when aroused by prey or courtship vibrations received through the plant. To the zoological observer this posture tells that the spider is ready for action. *Below* An adult female of *Cupiennius salei* injecting venom into a grasshopper successfully overwhelmed just seconds before the photograph was taken. Both pictures taken in Mexico

Plate 5. *Above Cupiennius coccineus* female during the last of its 11 molts outside the egg sac. When molting, spiders of the genus *Cupiennius* (not unlike wolf spiders, Lycosidae) hang from a molting thread attached to their host plant at night with their dorsal side pointing downward. This female has almost completely shed its old cuticular exoskeleton. It will soon hang from the exuvia with all its legs stretched out and then go through a phase of "gymnastics", flexing and extending its legs before the new cuticle is stiff enough to support normal locomotion. The exuvia nicely shows the red femora typical of female *C. coccineus*. Photograph taken in Costa Rica. *Below* A female *Cupiennius salei* seen from behind in its retreat in a banana plant in Mexico. Even late on a sunny morning the spider was covered by water droplets, indicating the high humidity prevailing in its retreat

Plate 6. A female *Cupiennius cubae* carrying its egg sac with many hundreds of eggs. Photograph taken in a gallery forest above Rio Ariguanabo on Cuba

Plate 7. A female *Cupiennius coccineus* during its nightly meal, sitting on a bromeliad. Like the other species of the genus *Cupiennius*, this female is not specialized for a particular kind of prey. Instead it rather unselectively takes any insect it can get. Sometimes the large species of *Cupiennius* can even be seen with a frog or a lizard in their fangs

Plate 8. *Above* A female of the newly described species *Cupiennius remedius* from the highlands of Alta Verapaz in Guatemala. Note the conspicuously patchy coloration typical of this species. *Below* A female of *Cupiennius panamensis* from Costa Rica, the other newly described species of the genus. Whereas the large species of *Cupiennius* typically use largely closed retreats provided by monocots such as bromeliads, banana plants, and heliconias, the small species are often seen to simply use the underside of any leaf for protection

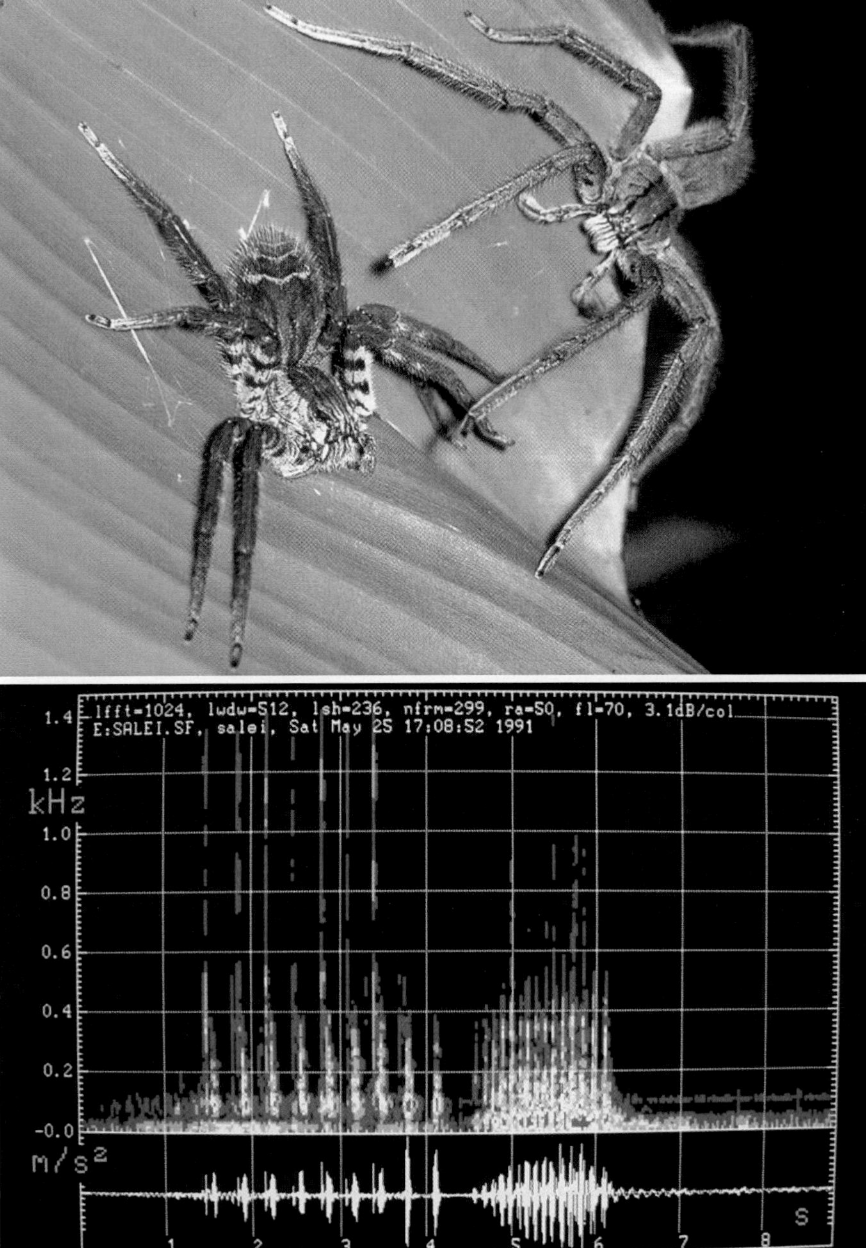

Plate 9. In *Cupiennius*, courtship behavior is a complex ritual. Both the male and the female produce vibratory signals to communicate with the other sex. The dwelling plant is used as a communication channel over distances of up to several meters. *Above* Male and female *Cupiennius salei.* In this species the female is stationary during courtship. The photograph shows the beginning of the tactile phase of courtship after the successful approach of the male guided by the female's vibrations (Photo JS Rovner). *Below* Oscillogram (lower trace) of a series of male vibratory courtship signals followed by a female vibratory response; signal amplitude given as acceleration of the plant leaf. Above the oscillogram the corresponding "sonagram" (vibrogram) is presented to show the frequency contents of the signals. *Small red areas* indicate energy maxima at very low frequencies (male, around 100 Hz; female, around 40 Hz). (Data from D. Baurecht)

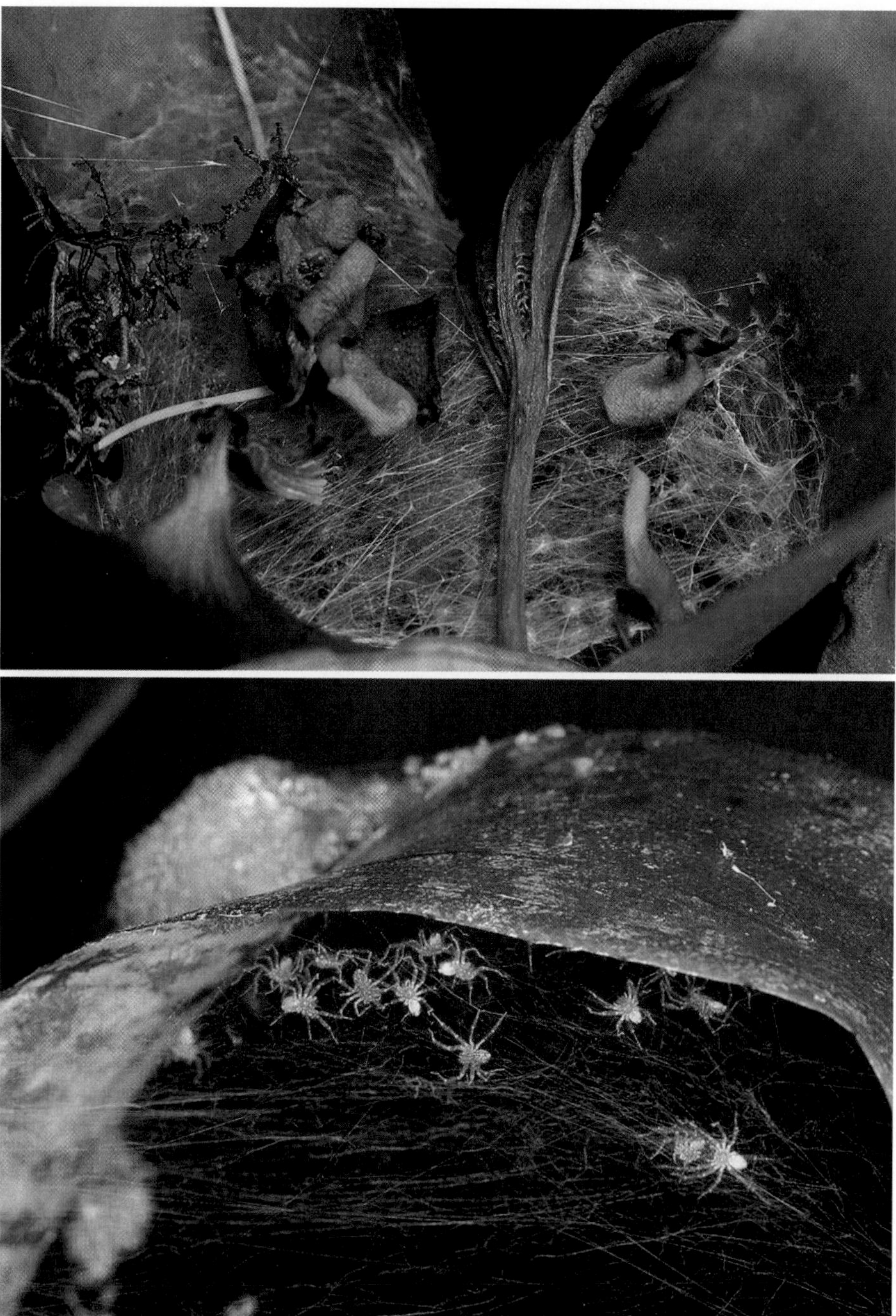

Plate 10. *Above* A few days before a female *Cupiennius*, in particular when belonging to one of the large species, actually builds its egg sac after copulation, it withdraws into a retreat which it typically closes with a densely woven sheet web for protection. The female remains in its closed space until the young emerge from the egg sac. The photograph was taken in Costa Rica. It shows a sheet web which covers the funnel in the center of a bromeliad with a female carrying its egg sac. *Below* Hundreds of spiderlings leave their egg sac after having spent about 25 days in it. They then spend about 10 days in a confusion of threads mainly woven by the female before her young begin their individualistic lives and disperse by drop and swing behavior

Plate 11. *Above* A particularly large protective sheet web woven by a female *Cupiennius coccineus* a few days before building its egg sac. Sometimes such sheet webs were seen to attract several male spiders simultaneously who then engage in male competition behavior. *Below* *Phoneutria boliviensis* photographed in Costa Rica in an area where it lives sympatrically with *Cupiennius coccineus*. The genus *Phoneutria* represents one of the few spider groups whose venom is dangerous even for man. Its taxonomic placement within the same family (Ctenidae) together with *Cupiennius* is now considered doubtful on the basis of DNA analysis

Plate 12. A female *Cupiennius coccineus* on the pseudostem of a banana plant waiting for prey in the dark. Both vibrations reaching the spider through the plant and air movements are detected by sensitive mechanoreceptors. The signals received from potential prey animals need to be distinguished from background noise. The metallic device above the spider is part of an anemometer designed for field use in the behaviorally relevant situation

Plate 13. Portrait of a female *Cupiennius salei.* Note the position of the eyes and the circular outline of their lenses, both characteristic features of the genus

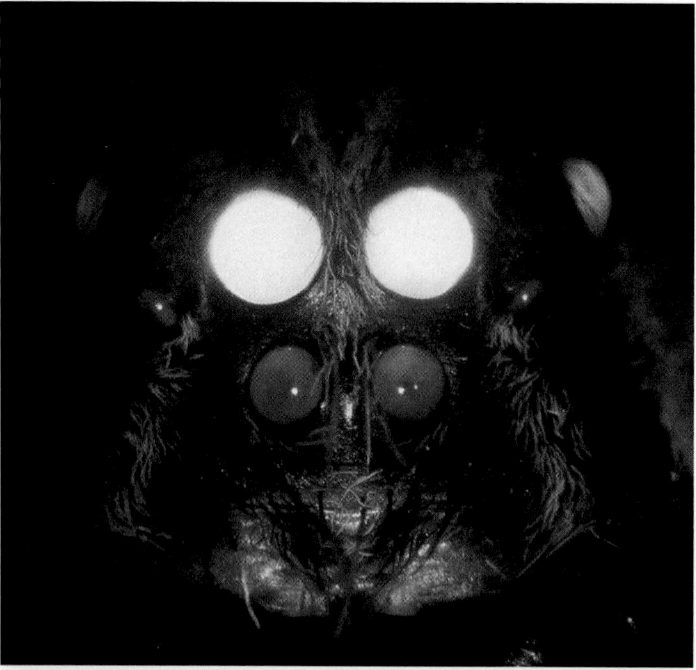

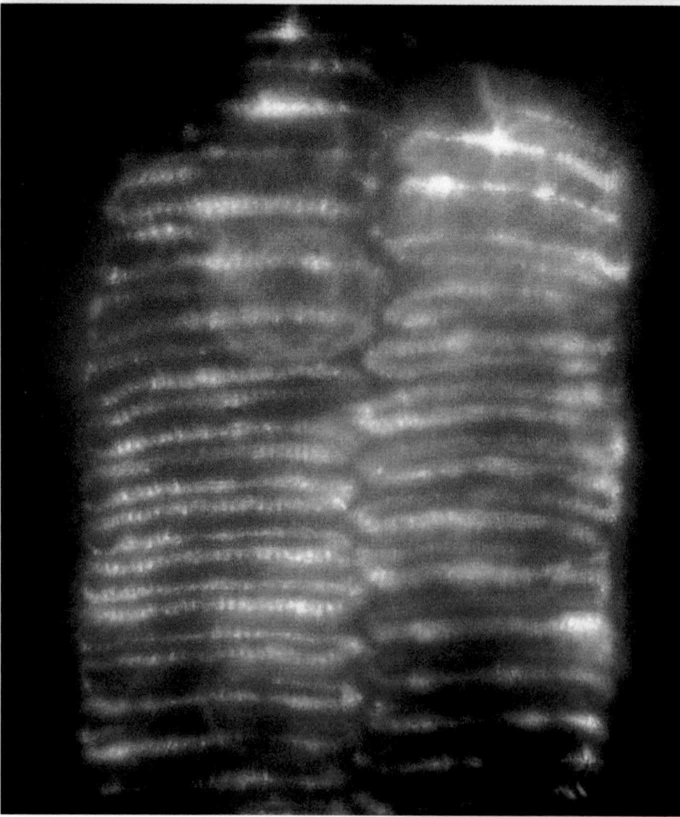

Plate 14. *Above* When viewed under the ophthalmoscope the tapeta of the spider secondary eyes reflect light like the eyes of a cat in the shine of a car's headlights. The photograph shows the reflection from the posterior median eyes of *Cupiennius salei*. When one is searching in the field at night with a flashlight this reflection often is a valuable guide to a spider. *Below* The retina of a posterior median eye seen through the eye's own lens. The incident light is reflected by the tapetum between the photoreceptor cells. Note the arrangement of the photoreceptor cells in horizontal rows. (Photo Ch. Becherer)

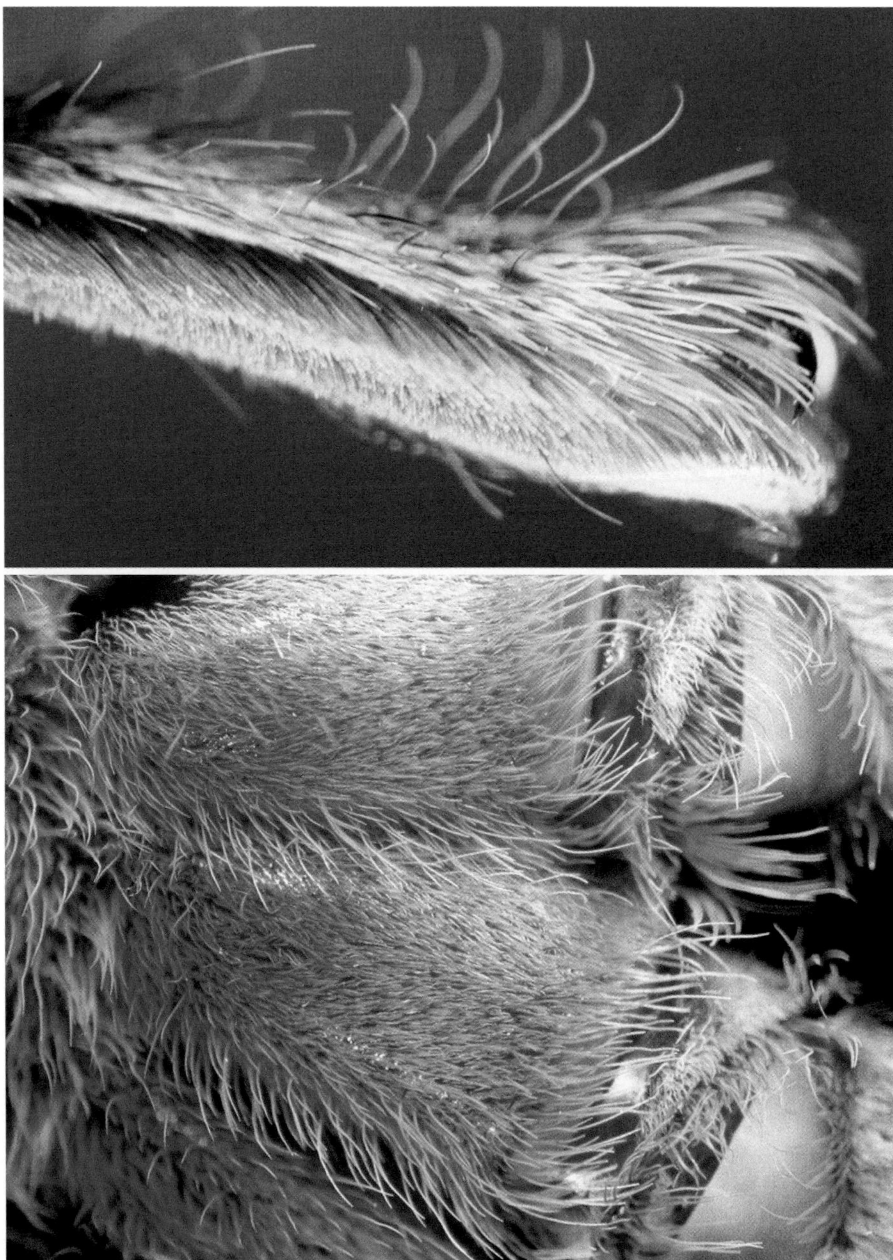

Plate 15. The body surface of *Cupiennius salei* and of the other species of the genus is covered by hundreds of thousands of cuticular hairs. Typically, these hairs are innervated and serve different mechanoreceptive functions, ranging from tactile and proprioreceptive sensitivity to air movement detection. *Above* The last segment (tarsus) of a walking leg with grayish scopular hairs ventrally, many short and fewer very long tactile hairs and trichobothria dorsally. Trichobothria are easily recognized here by the bend of the distal part of their hairshaft. They are effectively stimulated by the lightest movement of air. *Below* Ventral view of the proximal segments (*from left to right* coxa, trochanter, proximal end of femur) of two neighboring walking legs. Note the rich supply of the joint regions with relatively long hairs, most likely serving a proprioreceptive function

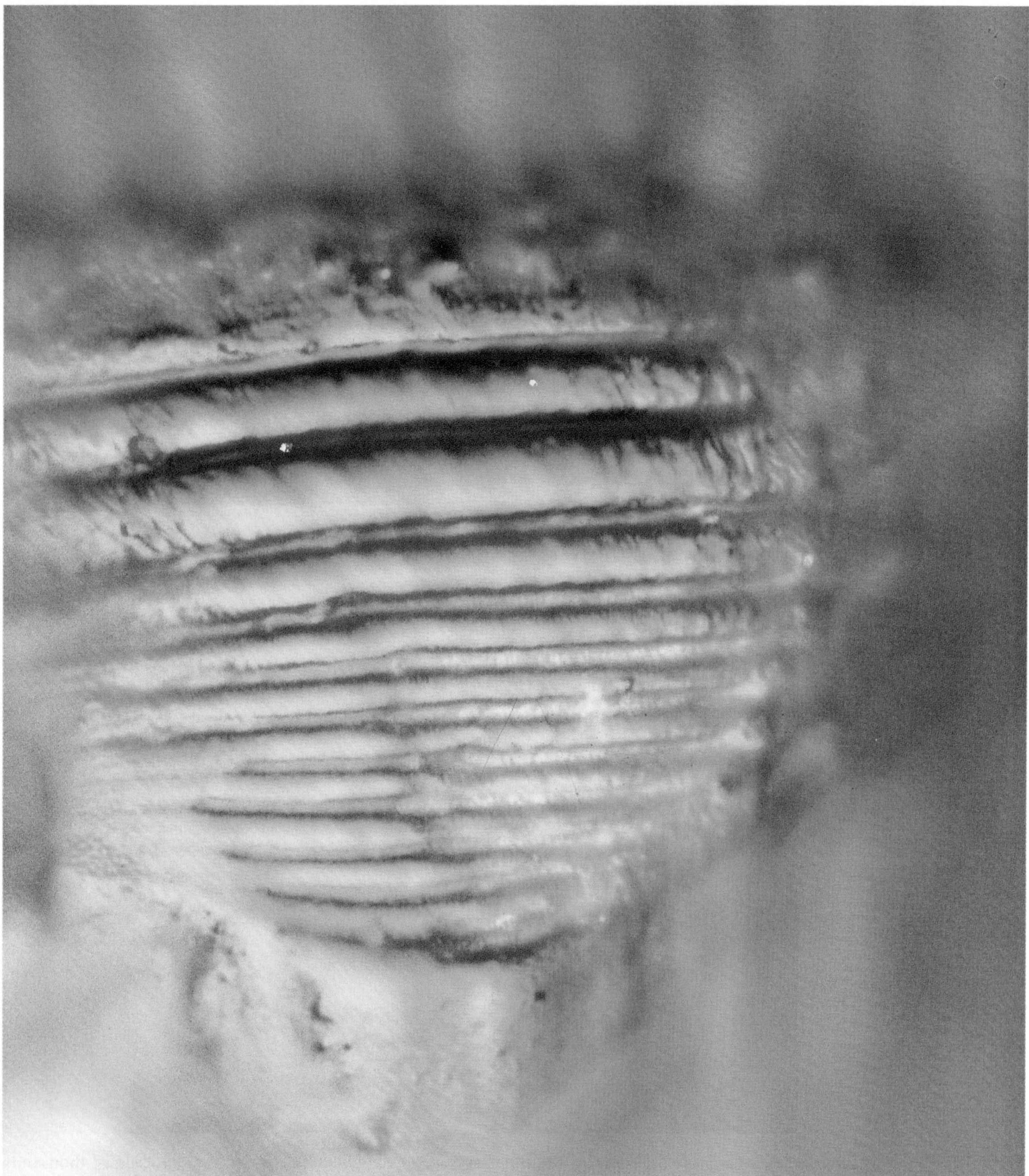

Plate 16. The spider exoskeleton is richly supplied with sensory devices. From a physical point of view the slit sensilla are of the opposite kind to the trichobothria. Instead of being movement detectors they are biological strain detectors. They measure minute strains in the exoskeleton due to muscular activity, hemolymph pressure, and gravity. The most sensitive sensor for substrate vibrations also is a slit sense organ, though a specialized one. The photograph shows a lyriform organ on the walking-leg tibia of *Cupiennius salei* composed of 11 innervated slits arranged in parallel. It measures roughly 50 μm in the vertical direction. (Nomarski interference contrast in incident light)

XVII Neurotransmitters and Neuromodulators

After a long history of nervous-system investigations preoccupied with electrochemical events and the analysis of electrical potential differences, it has now become clearer than ever that nervous systems are also, or perhaps even mainly, chemical machinery. Today the analysis of "neuroactive" substances – that is, the neurotransmitters, neuromodulators and neurohormones – is well advanced. Neurotransmitters are short-lived molecules, which are released at synapses and change the ion permeability of the postsynaptic membrane at restricted sites. Neuromodulators are released near synapses and have pre- and/or postsynaptic actions. And neurohormones, finally, are long-lived in comparison to neurotransmitters, are released into the bloodstream (hemolymph) like other hormones, and accordingly act on any of various target tissues. All these substances are eminently significant in releasing behavior and determining what form it will take.

We can take it for granted that all this also applies to spiders. Furthermore, spiders could well be of interest not only in comparative studies of the nervous system but also with respect to mechanisms that underlie the motivational changes they so evidently display in their behavior. These are manifest, for example, in the circadian activity rhythm (see Chapter IV) and in the transition from prolonged motor inaction to a lightning-quick capture of prey (see Chapter XIX). Presumably neuroactive substances are involved in modulating the activity of central neurons, in their synchronization or desynchronization, and in altering the threshold for excitation of sensory and motor cells as well as, perhaps, peripheral receptor cells.

Such considerations, vague as they were at the time, motivated us (in particular Axel Schmid) to dive into this deep well of new complexities. By the time we first came up for air, it was clear that the CNS of *Cupiennius salei* houses a rich collection of neuroactive substances. When these had been identified and localized with immunocytochemical methods, some aspects of the anatomy of the CNS had a completely new look. But the crucial next step, to analyze the relationships of these substances to specific forms of behavior, has not yet been taken, and it will be difficult. The histamine-immunoreactive system of six giant fibers described below may well be the best point of departure for such a study.

In arthropods the biogenic amines are a particularly important group of neuroactive substances (Klemm 1985). Special attention has been paid to the biogenic monoamines serotonin, octopamine and histamine, and to the polyamine neuropeptides GABA (γ-aminobutyric acid), FMRFamide (F, phenylalanine; M, methionine; R, arginine), proctolin und myotropin I. The following description is based chiefly on the excellent Master's thesis by Matthis Duncker (1992) and the resulting publication (Schmid and Duncker 1993).

1 Serotonin (5-HT, 5-hydroxytryptamine)

This is a neurotransmitter and neuromodulator widely distributed in insects and crustaceans and thoroughly investigated there (Klemm 1985; Nässel 1988; Bicker and Menzel 1989; Harris-Warrick et al. 1989). Its actions are exerted at the most diverse levels (from the afferents of the receptor cells through central nervous integration to motor coordination), including associative functions of the CNS such as learning and memory (Erber 1989; Kloppenburg and Erber 1989; Kravitz 1991; Hammer and Menzel 1995).

In *Cupiennius salei* serotonin-immunoreactive (SIR) neurons are found in all neuromeres. Each neuromere in the subesophageal region contains three to six such neurons. Their neurites run from the ventrally situated somata dorsolaterally and enter longitudinal tracts, in particular the centrolateral tract (CL) and central tract (CT) (see Chapter XIV). From these longitudinal fibers the SIR neurons send processes into the segmental, ventrally located sensory neuropils, where they form dense arborizations. Considerably less serotonin immunoreactivity is found in the dorsal motor neuropil, and there is none at all in the nerves that emerge from the subesophageal region.

Each of the cheliceral ganglia also contains three SIR neurons (somata). In the protocerebrum, finally, there are several groups of SIR somata in front of the "central body". The primary neurites of these cells all run ventrally into the central protocerebral neuropil, where they send out fine branches. In the visual system only the ON1 and AM-ON1 (that is, the first optic ganglia) exhibit serotonin immunoreactivity. None can be demonstrated in the "mushroom body".

The picture of the SIR cells obtained in *Cupiennius salei* is consistent with what is known for articulates in general. The serially homologous arrangement and the small number of such neurons are typical (in insects: 0.01–0.03%; Nässel 1988) and confirm the findings of Seyfarth et al. (1990) in *Cupiennius salei.* The most conspicuous way in which the serotoninergic neurons of insects and crustaceans differ from those of spiders (insofar as *Cupiennius* typifies the spiders) is that there are none in almost all the peripheral nerves of spiders. It follows that serotonin has no direct modulatory action at the periphery in *Cupiennius salei.* The only exceptions are the cheliceral nerves and their target sites. In the CNS, however, the distribution of SIR neurons is the same as in other arthropods. In view of the target sites of these neurons, it seems likely that their main function is to modulate information transfer from the primary afferents of the mechano- and chemoreceptive sensilla (Anton 1991).

2 Octopamine

In insects octopamine acts as both a neurotransmitter and a hormone. It has been found in many arthropods and plays a role in many behavioral contexts (Orchard 1982; Hoyle 1985). For example, it is present in DUM (*D*orsal *U*npaired *M*edian) neurons and is involved in the modulation of the myogenic rhythm of the skeletal musculature (Evans and O'Shea 1978). In scorpions octopamine participates in the control of the dramatic circadian change in sensitivity of the photoreceptor cells (Fleissner and Fleissner 1985). In crustaceans an action of octopamine on mechano- and chemoreceptive afferents has been demonstrated (Arneson and Olivo 1988; Pasztor and MacMillan 1990). Often octopamine and serotonin act antagonistically (Kravitz 1991).

What do we find in *Cupiennius*? The answer is provided by Duncker (1992) and Seyfarth et al. (1993). Octopamine-immunoreactive fibers project into all neuromeres of the CNS and are especially numerous in the central neuropil of the protocerebrum. They ascend to this region in the protocerebral median tract (PCMT) from the subesophageal ganglion mass. Somata with octopamine immunoreactivity (OIR), however, are absent in the protocerebrum; in each of the other neuromeres there are 2 to 12 such somata in a serial arrangement, giving a total number of as many as 90 (Fig. 1). In these neuromeres they are situated in the dorsal region of the ventral cell-body layer (cf. Figs. 8 and 9 in Chapter XIV), at the median neuromere septum. Unfortunately, the primary neurites of these cells are not labelled, which means that the cells cannot be completely reconstructed. Both the sensory longitudinal tracts (SLT 1 and 2) and the ventrolateral tract (VL) contain octopamine-immunoreactive fibers. The VL fibers project into the sensory neuropil of the pedipalps and walking-leg neuromeres and then ascend through the cir-

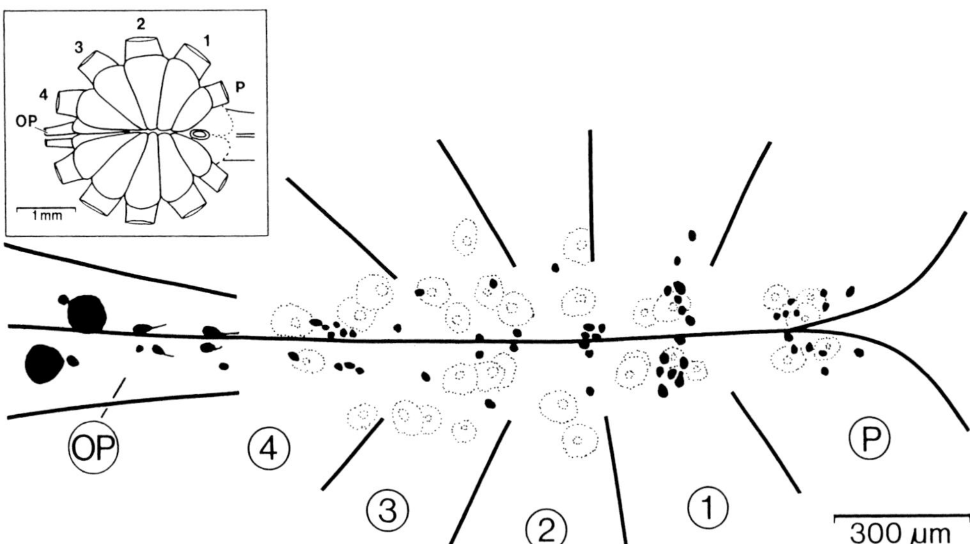

Fig. 1. Octopamine-immunoreactive somata in the subesophageal nervous mass of *Cupiennius salei*. Horizontal view of the ventral cellular rind. *Black dots* represent heavily stained somata, whereas *dotted* somata stained only weakly. *Continuous black lines* indicate borderlines between neuromeres. *Inset* Ventral aspect of central nervous system with neuromeres of pedipalps (*P*), walking legs (*1–4*) and opisthosoma (*OP*). (Seyfarth et al. 1993)

cumesophageal connectives into the central protocerebral neuropil. OIR fibers of the central tract (CT) and those of the middle dorsal tract (MD) send ipsilateral processes into the motor neuropil as well as contralaterally , to the protocerebral dorsal tract (PCDT). In the peripheral nerves of the subesophageal ganglion mass no OIR fibers were found.

The largest aggregation of OIR somata (15 cells) is in the cheliceral neuromeres, which also contain widely arborizing branches of several OIR fibers. In contrast, the optic neuropils and the eye nerves exhibit no octopamine immunoreactivity.

The "chemical picture" of the CNS of *Cupiennius* has already become quite complicated, and it is time to consider the possible relevance of the findings (Table 1; Figs. 2 and 3). One important result seems to be the absence of equivalents or even homologs of the DUM neurons of insects, which in those animals are likely to be involved in a variety of functions. Neuromodulation of the activity of the skeletal musculature was mentioned above. In addition, DUM neurons have also been found to serve as photomotor neurons to the firefly lantern (Carlson and Jalenak 1986) and as neurons mediating the release of hormone from the corpora cardiaca (Orchand and Loughton 1981) or of octopamine from neurohemal organs to control the activity of the antennal heart (Pass et al. 1988). The DUM neurons of insects are a relatively small group of nerve cells with bilaterally symmetrical projections. Ontogenetically they all arise from the so-called median neuroblasts (Goodman and Spitzer 1979). No such neuroblasts are present during the development of the spider CNS, which proceeds differently (Weygoldt 1985; Chabaud 1990).

Another difference between *Cupiennius* and insects is the absence of efferent OIR fibers in the peripheral nerves. Such fibers are characteristic of the DUM neurons and have also been found in *Limulus* (Lee and Wyse 1991). Therefore it also seems unlikely that the skeletal musculature of *Cupiennius* is regulated by way of octopamine.

In the protocerebrum of chelicerates, so far as is known, there are no OIR neurons. In contrast, a conspicuous group of such neurons is present

Table 1. Intensity of staining in the different areas of the central nervous system and associated nerves of *Cupiennius salei* after application of seven different antisera against various biogenic amines. Size of dot is a measure of intensity of staining. AMG leg neuromeres, AMN leg nerves, OPIG and OPIN opisthosoma ganglion and nerve. The four dot sizes represent >1000, 100–1000, 10–100, and 1–10 stained cell somata (within the ganglia) or axons (in the nerves). *Open circles* denote uncertain staining; *bars* indicate lack of staining. (Duncker 1992)

Neuromere Nerve / Amines	Proto-cere-brum	Eye nerves	Chel. gln.	Chel. nerve	Ped. gln.	Ped. nerve	AMG 1	AMN 1	AMG 2	AMN 2	AMG 3	AMN 3	AMG 4	AMN 4	OPIG	OPIN
GABA	⬤	–	●	•	●	•	●	•	●	•	●	•	●	•	●	•
Serotonin	•	–	·	○	·	–	·	–	·	–	·	–	·	–	·	–
Octopamine	–	–	•	○	·	–	·	–	·	–	·	–	·	–	·	–
Histamine	·	⬤	–	–	–	–	–	–	–	–	–	–	–	–	–	–
Proctolin	●	–	•	○	•	·	•	·	•	·	•	·	•	·	•	·
FMRFamide	●	–	•	·	•	–	•	–	•	–	•	–	•	–	•	–
Myotopin I	●	–	•	–	•	–	•	–	•	–	•	–	•	–	•	–

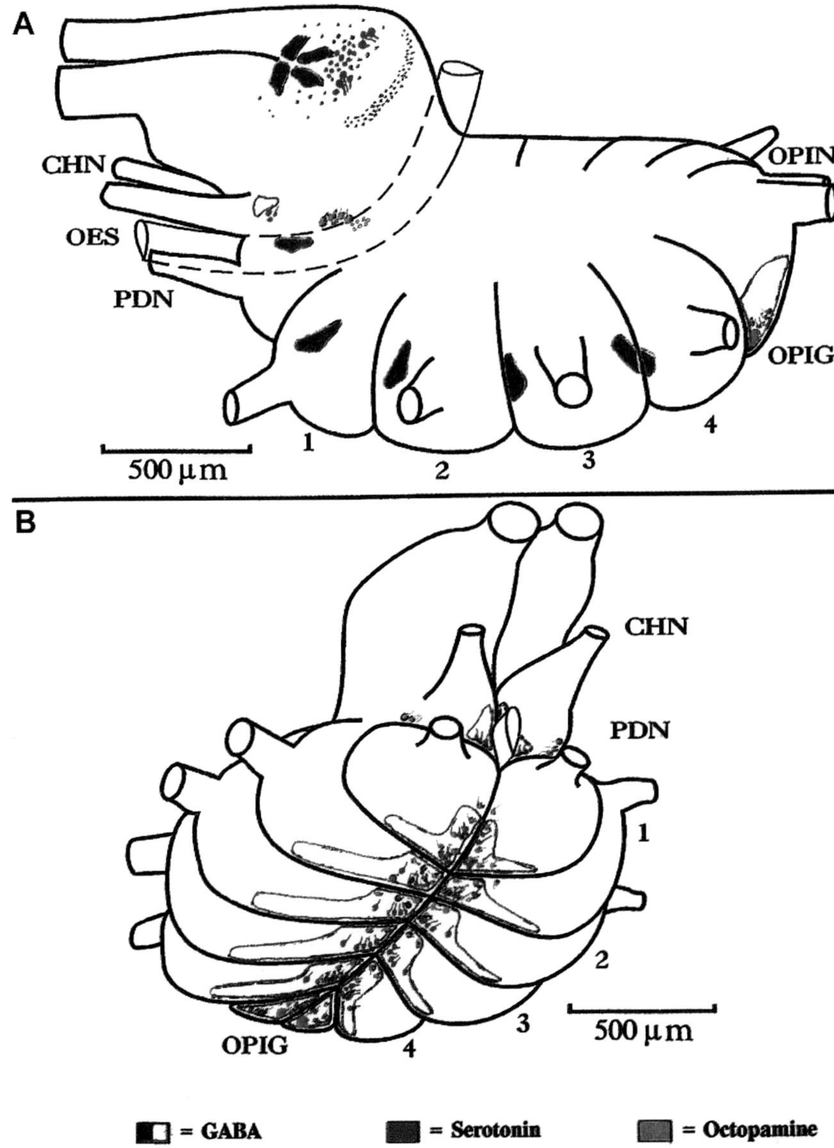

Fig. 2 A, B. Summary diagram of distribution of serotonin-, octopamine- and GABA-immunoreactive somata in the central nervous system of *Cupiennius salei*. **A** Dorso-lateral view. **B** Antero-ventrolateral view. *CHN* Cheliceral nerve; *OES* esophagus; *OPIG* opisthosomal ganglia; *OPN* opisthosomal nerves; *PDN* pedipalpal nerve, *1–4* walking-leg neuromeres. (Duncker 1992)

in the lateral tritocerebrum. In both scorpions (*Androctonus australis*) (Fleissner and Fleissner 1985) and the horseshoe crab (*Limulus polyphemus*) (Calman and Battelle 1991) this group comprises neurosecretory efferents that innervate all the eyes by way of axon collaterals and play a decisive role in altering their sensitivity. *Cupiennius* has a comparable group of OIR neurons. However, it is not yet clear whether these are efferents to the eyes and also share responsibility for their circa-

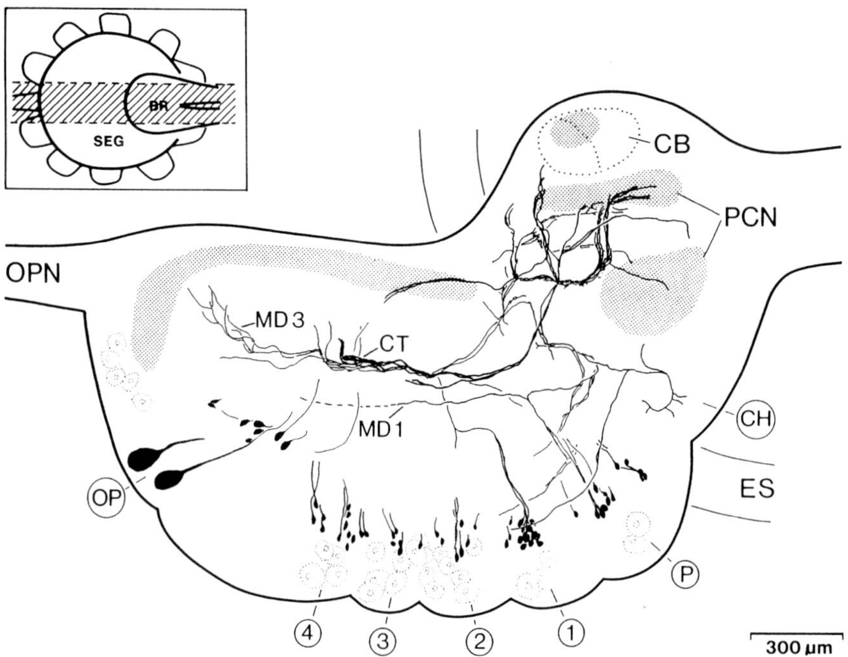

Fig. 3. Octopamine-immunoreactive somata in the ventral region of the subesophageal nervous mass of *Cupiennius salei* and their major fiber projections to octopamine-immunoreactive neuropil regions (*shaded areas*). Reconstruction of a region along both sides of the longitudinal midline of the CNS and 1.6 mm wide (see *inset*). Because primary neurites often are not stained completely, terminals cannot be attributed to specific cell somata. *CB* "Central body"; *CH* cheliceral ganglion; *CT, MD1, MD3* longitudinal tracts; *ES* esophagus; *OPN* opisthosomal nerves; *PCN* protocerebral neuropils. (Seyfarth et al. 1993)

dian modulation (see Chapter XI). In the cases of *Argiope* (Yamashita 1985) and *Lycosa* (Carricaburu et al. 1990) – that is, an orb-web and a wolf spider – the existence of an efferent innervation has already been demonstrated electrophysiologically; for *Cupiennius* we have only a few neuroanatomical indications (cf. Chapter XI).

Seyfarth et al. (1993) have pointed out that OIR varicosities are concentrated in the neuropil regions and near the hemolymph spaces of the CNS (Fig. 4). From this they infer that octopamine not only functions as a neuromodulator at central synapses, but is also released into the hemolymph and acts as a hormone in peripheral regions. The hemolymph does indeed contain octopamine (12–40 nM/l; Seyfarth et al. 1993). There is thus a good deal of evidence that octopamine in spiders, as in other arthropods, has a dual function as neurotransmitter/modulator and as a hormone.

Fig. 4. Hemolymph vessel system of *Cupiennius salei*, sagittal (*top*) and dorsal (*middle*) view. *H* Heart with three pairs of lateral arteries (*SA*) and one unpaired Arteria posterior (*APO*). The prosoma is supplied by the Aorta anterior (*AO*), which divides into two Trunci peristomacales (*TP*), which in turn branch into an Arteria cephalica (*ACE*) (projecting dorso-cranially) and an Arteria crassa (*ACR*) (projecting ventrally to the subesophageal nervous mass). The *ACR* forms the Sinus thoracalis (*ST*), where the leg arteries (*BA*) originate. *DQ* Dorsal transverse artery; *LV* lung vein; *O* ostium; *PA* pedipalpal artery; *PR* right pedipalp; *RCH* Ramus cheliceralis, *ROC* Ramus oculorum. *Bottom* Vessels in the second leg neuromere. Vessels invade the neuromeres from the septa and in particular supply the cell cortex close to the septum. In the inner regions of the neuromeres vessels are only rarely found. *AIM* Arteria interganglionaris medialis; *1–3* leg neuromeres. Section on right side represents level *c* as shown by inset on left side. (Kosok 1993)

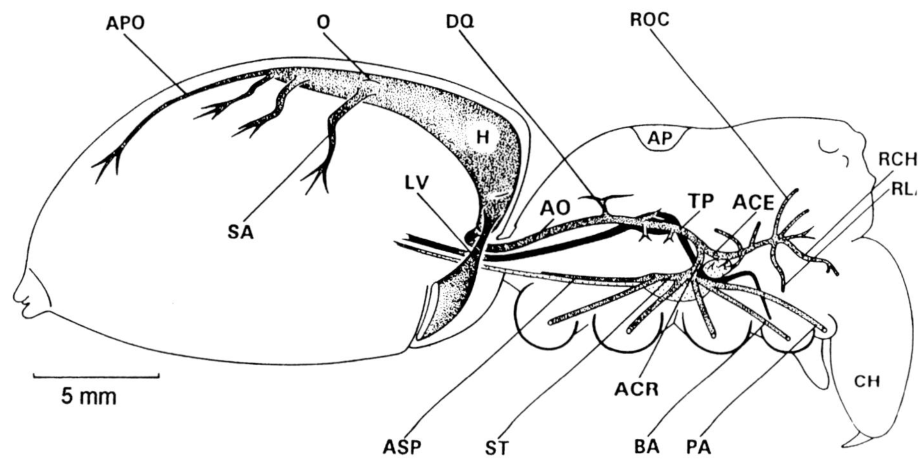
APO
O
DQ
ROC
AP
H
RCH
LV
AO
TP
ACE
SA
5 mm
ACR
CH
ASP
ST
BA
PA

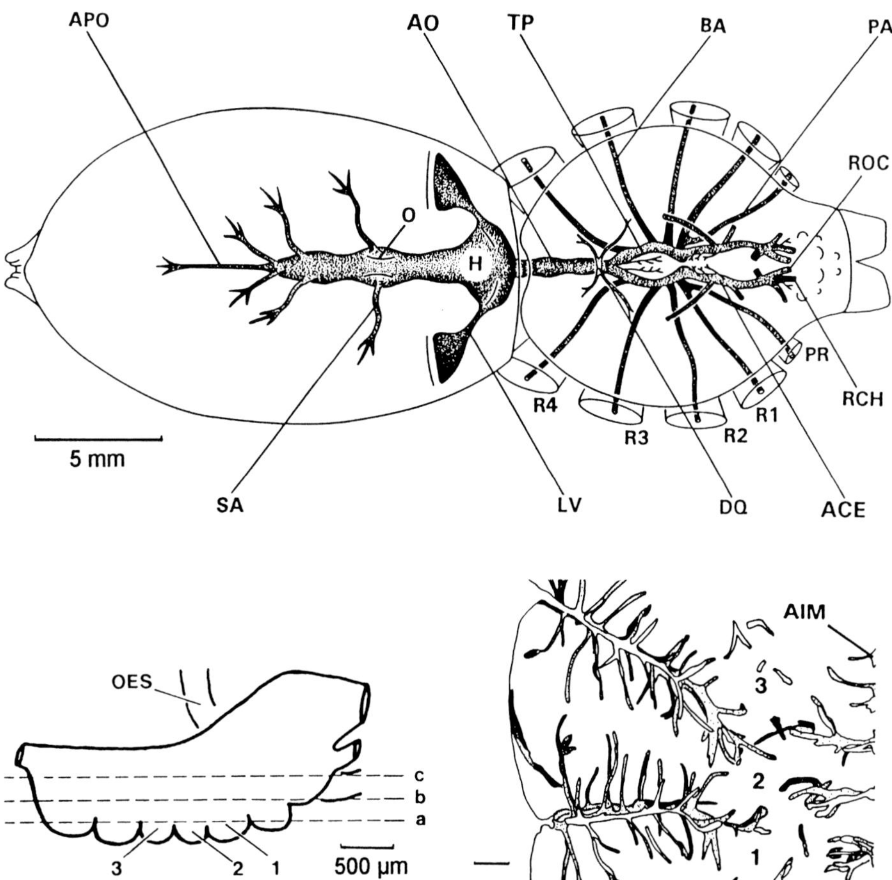
APO
AO
TP
BA
PA
ROC
O
H
PR
RCH
R4
R1
R3
R2
5 mm
SA
LV
DQ
ACE
OES
c
b
a
3
2
1
500 µm
AIM
3
2
1

As Gregor Kosok (1993) showed, while working on his Master's thesis in the laboratory of E.-A. Seyfarth in Frankfurt, the central nervous system of *Cupiennius salei* is well supplied with hemolymph. By injecting dyes into the circulatory system and filling parts of it with plastic, he was able to reveal in great detail the many vessels passing into the CNS and their abundant branches.

In regions with many neuronal somata, such as the parts of the leg neuromeres near the septum of the subesophageal ganglion complex, and also in the neuropil regions ("central body"; optic neuropils of the secondary eyes, especially ON3), there are capillary networks. In contrast, the neuronal tracts are less well supplied with vessels. A similar situation prevails in our own brain: the soma-filled gray matter is massively perfused with blood, whereas there is relatively little perfusion of the axon-packed white matter.

In the particularly well supplied regions of the central nervous system of *Cupiennius*, the diffusion distances are less than 60 μm (in some cases less than 25 μm), and – not unexpectedly – in the poorly vascularized regions they are as large as 120 μm. The capillary diameter is 7 to 14 μm, which is close to that in the vertebrates.

Figure 4 is taken from Kosok's thesis. It gives an overview of the hemolymph vascular system of *Cupiennius salei*, which is largely consistent with the basic organization known from other spiders. In the detailed inset the rich vascular supply of the neuromeres in the subesophageal ganglion complex can be seen. Because *Cupiennius* has only one pair of book lungs and few or no tracheae, a richly developed vascular system would *a priori* be expected.

3 γ-Aminobutyric Acid (GABA)

GABA, an amino acid known as an inhibitory neurotransmitter, is well represented not only at the neuromuscular synapse but also in the CNS of arthropods. GABA immunoreactivity has been demonstrated, for example, for the optic lobes of various insects (Meyer et al. 1986; Schäfer and Bicker 1986; Füller et al. 1989). Many individual, identified neurons in the CNS of crustaceans and insects are also GABA-ergic. Among these are the common inhibitors (CI neurons), which innervate whole groups of muscles (Honegger et al. 1990; Walrond et al. 1990). The papers of Meyer and his collaborators have provided good documentation of the presence and approximate distribution of GABA in several spider families (Meyer et al. 1980, 1984; Meyer and Poehling 1987). Its inhibitory action in arachnids has been demonstrated by several authors, with reference to the activity of the spider heart (Sherman 1985), to the walking-leg musculature of a bird spider (Brenner 1972), and to the spontaneous activity in the CNS of a scorpion (Goyffon et al. 1980).

The findings regarding GABA immunoreactivity in the CNS of *Cupiennius salei* fit well into this general picture. Of all the substances we looked for in our experiments, no other was so intensely labelled or had such a wide distribution as GABA (Table 1). About 10,000 somata, no less than ca. 10% of all neurons, exhibited GABA immunoreactivity! It appeared in all the neuromeres (in somata and/or neurites) and, apart from the eye nerve, was also present in all peripheral nerves. The latter fibers are presumably motoneurons, as they are situated dorsally and have large diameters.

In the neuromeres of the walking legs are two groups of GABA-immunoreactive somata (Figs. 2 and 5). (i) The first is located in the ventral cell-body layer (cf. Figs. 8 and 9 in Chapter XIV), in a broad region extending laterally from the midline. Here about 50% of all the somata (ca. 450) are GABA-immunoreactive. According to our previous neuroanatomical findings (Babu and Barth 1984), these cells are inter- and motoneurons. (ii) The second group is smaller than the first and is in a lateral and dorsal position, at the anterior edge of the neuromere (Fig. 5). In this area >90% of the somata (ca. 100) are immunoreactive.

Immunoreactive fibers can be found in high density throughout the subesophageal ganglion mass. Unfortunately, the primary neurites of the somata in the above two groups are often not labelled, or are too weakly labelled for their courses to be followed well enough. Therefore, the relationship between somata and the many fibers cannot yet be reliably reconstructed.

How does it look in the actual brain, the supraesophageal ganglion? Sumptuous! Each ganglion here contains several hundred GABA-immunoreactive somata, and the cheliceral ganglion has over a thousand (one group medioventral to the site of entry of the cheliceral nerves, a second in the lateral part of the ganglion). Of the many labelled fibers some are comparatively very thick, up to 15 μm in diameter. These run into the cheliceral nerves. The protocerebrum

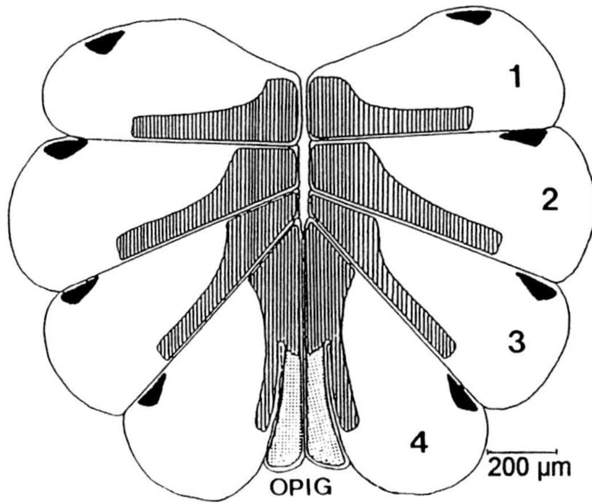

Fig. 5. GABA-immunoreactivity in the cell somata layer of leg neuromeres *1–4* of *Cupiennius salei*. Area I vertically striped, area II black. Immunoreactivity in opisthosomal ganglia indicated by *dotted shading* (*OPIG*). (Duncker 1992)

contains many scattered GABA-immunoreactive somata in the dorsal cell-body layer plus four groups (two per hemiganglion) each comprising several hundred labelled somata; these are situated medially, directly behind the non-immunoreactive globuli cells, and project into the central protocerebral neuropil. There is also a fifth, unpaired group on the dorsal surface of the "central body" (several hundred small somata, 7–8 µm in diameter). These neurons project into the upper half of the "central body" and its anterior layers; here the fibers become lost in the intense general labelling ascribable to GABA immunoreactivity of the visual projection neurons. The cells of this fifth group are presumably intrinsic neurons (Fig. 2).

In the visual system of the secondary eyes the heads of the "mushroom body" (ON3) are particularly strongly GABA-immunoreactive; these regions are connected to the bridge by thick (7–10 µm) fibers that are also labelled. The optic lamellae (ON1) contain projections of fibers that branch in the first optic chiasma and pass tangentially along the proximal side of the ON1, into which they send many fine fibers at right angles. Within the ON1 they run between the palisades of the photoreceptor terminals. Because the somata of the first- and second-order visual interneurons (see Chapter XVI) are not GABA-immunoreactive, these fibers very probably belong to other somata.

In summary. The immunocytochemical studies indicate that GABA plays a prominent role in the CNS of *Cupiennius*, as it does in other spiders and other arthropods. Inhibitory control of both the skeletal muscles and the cardiac ganglion by means of GABA seems entirely possible. GABA is also a widespread transmitter within the CNS of *Cupiennius*, where the optic neuropils are particularly notable for massive GABA-ergic innervation.

4 Histamine

Histamine, an indoleamine, like the amino acid GABA has been demonstrated in many studies of arthropod nervous systems. It acts as transmitter, modulator and hormone (Prell and Green 1986). The transmitter function is particularly well known for the visual system of arthropods. Here histamine is evidently the only transmitter of photoreceptor cells, with the single exception of the GABA-ergic R7 cells in the rhabdoms of some Diptera (Callaway and Stuart 1989; Batelle et al. 1991; Nässel 1991; Pollack and Hofbauer 1991). Outside the visual system, histamine is found in local and plurisegmental interneurons, where its neuromodulatory and neurohormonal role is under discussion (Nässel et al. 1990).

In *Cupiennius* Schmid and Duncker (1993) found histamine immunoreactivity in the photoreceptor cells and interneurons of the protocerebrum, but not in the subesophageal ganglion mass.

Visual System

Once the retinas of all eight eyes of *Cupiennius* had been found to be strongly histamine-immunoreactive (as well as the axons of the photoreceptor cells and their endings in the optic lamellae ON1 and AM-ON1) (Fig. 6), analogous re-

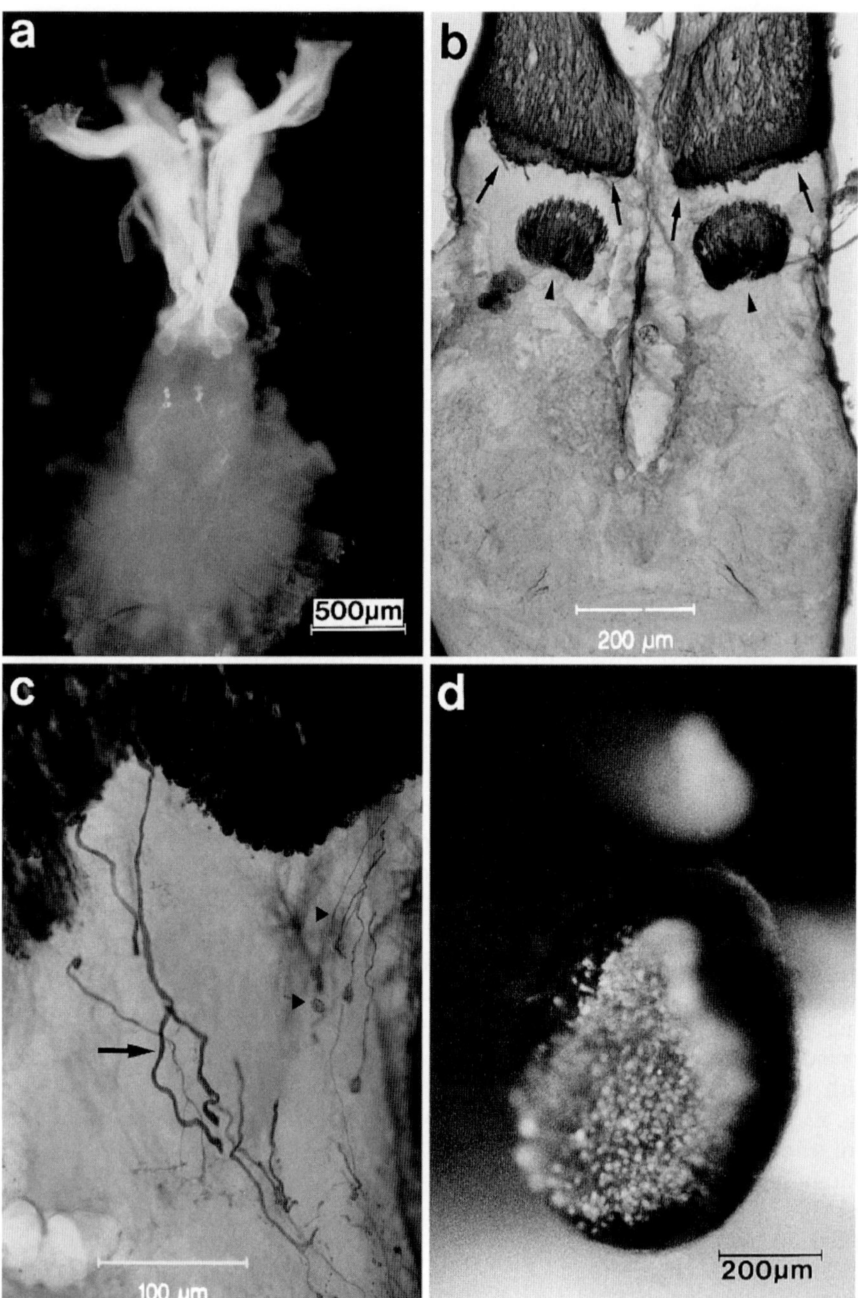

Fig. 6a–d. Histamine-immunoreactivity. **a** Dorsal view of whole-mount preparation showing intensive reaction in all eye nerves and optic neuropils (juvenile *Cupiennius salei*). **b** Horizontal section through dorsal protocerebrum. The optic lamellae (ON1) of the principal eyes (AM) (see *arrowheads*) as well as those of the secondary eyes (PM) (see *arrows*) are intensively stained due to the many photoreceptor cell terminals they contain. **c** Optic neuropils ON1 of PL- and PM-eyes ventral to the section shown in **b**. Note thick (*arrow*) and thin (*arrowheads*) histamine-immunoreactive fibers traveling to higher-order visual neuropils. **d** Retina of PM eye seen from the side of the lens; note numerous immunoreactive photoreceptor cells. (Duncker 1992; Schmid and Duncker 1993)

sults are now available for four classes of arthropods (Merostomata, Crustacea, Insecta and Arachnida). It thus does not seem premature to consider histamine as the neurotransmitter of photoreceptor cells of all arthropods. Our findings in *Cupiennius* (and also in *Nephila clavipes*) also provide an argument in favor of the monophyletic origin of the arthropods and their eye types (Paulus 1979), particularly in view of the fact that other invertebrate visual cells have not yet been found to contain histamine (Mollusca: Elste et al. 1990; Platyhelminthes: Wikren et al. 1990).

Of the three types of histamine-immunoreactive interneurons, two are in the visual system of the secondary eyes and one in that of the principal eyes. Because the input areas of these projection neurons are located between the intensely labelled visual-cell terminals in the first optic ganglia, we cannot yet reliably identify them as broad- or narrow-field neurons (see Chapter XVI); however, comparison of the fiber courses established immunocytochemically with those in Golgi preparations (see Chapter XVI) suggests that they are both, at least in the case of the secondary eyes. The three types of histamine-immunoreactive interneurons are distinguished as follows.

Type 1: nL neurons – that is, narrow-field projection neurons between ON1 and ON2 (see Chapter XVI) – which are especially numerous in the region of the PM eyes, less so in that of the PL eyes and not present in that of the AL eyes.

Type 2: These interneurons pass without branching through ON2 and the optic chiasma, ultimately projecting into the head of the "mushroom body" (ON3) or in some cases continuing on towards the "central body".

Type 3: This type is found in the neuropil of the principal eyes (AM) and runs dorsally, without branching, in the optic tract from AM-ON1 to AM-ON2, sending no projections into the "central body".

The "Giant-Fiber System" – DPHN and VPHN

This is a spectacular system. It comprises only six neurons, which branch widely to innervate both motor and sensory neuropil regions in all neuromeres. The term "giant fiber" for these multisegmental descending fibers refers to their diameter, which can reach 25 μm. Although these have the largest diameter of all fibers in the CNS of *Cupiennius*, they do not strictly conform to the definition of a "giant" fiber. They do not constitute a size class of their own. Therefore they are designated by the unattractive, but more correct abbreviations DPHN and VPHN: dorsal/ventral plurisegmental histaminergic neuron.

Figure 7 shows the details (Schmid and Duncker 1993). The six somata form two groups of three arranged with bilateral symmetry in the protocerebrum, in the dorsal cell-body layer directly in front of and above the "central body". On the basis of the courses of the fibers, the system is divided into a dorsal (2×2 neurons) and a ventral (2×1 neuron) system. The primary neurites of all the cells run just in front of the "central body", proceeding ventrolaterally in the protocerebral-dorsal tract (PCDT); at the lower side edge of the "central body" they send branches into the central protocerebral neuropil (to the back surface of the tritocerebrum and to its ipsi- and contralateral neuropil). The "mushroom body" is not included in the histaminergic system, and the "central body" receives only a few fine fibers that enter its dorsal posterior surface. Not until the thick fibers have descended through the circumesophageal connectives into the subesophageal mass does the division into a dorsal and a ventral system become evident.

The two pairs of fibers in the dorsal system run in the motor longitudinal tract MD2, from which they send out projections into the ipsilateral dorsal motor neuropil of all pedipalp and walking-leg neuromeres. Another projection from each fiber ends in the ventral sensory neuropil of each of these neuromeres. The DPHN fibers terminate in the fused opisthosomal neuromeres, where they branch abundantly.

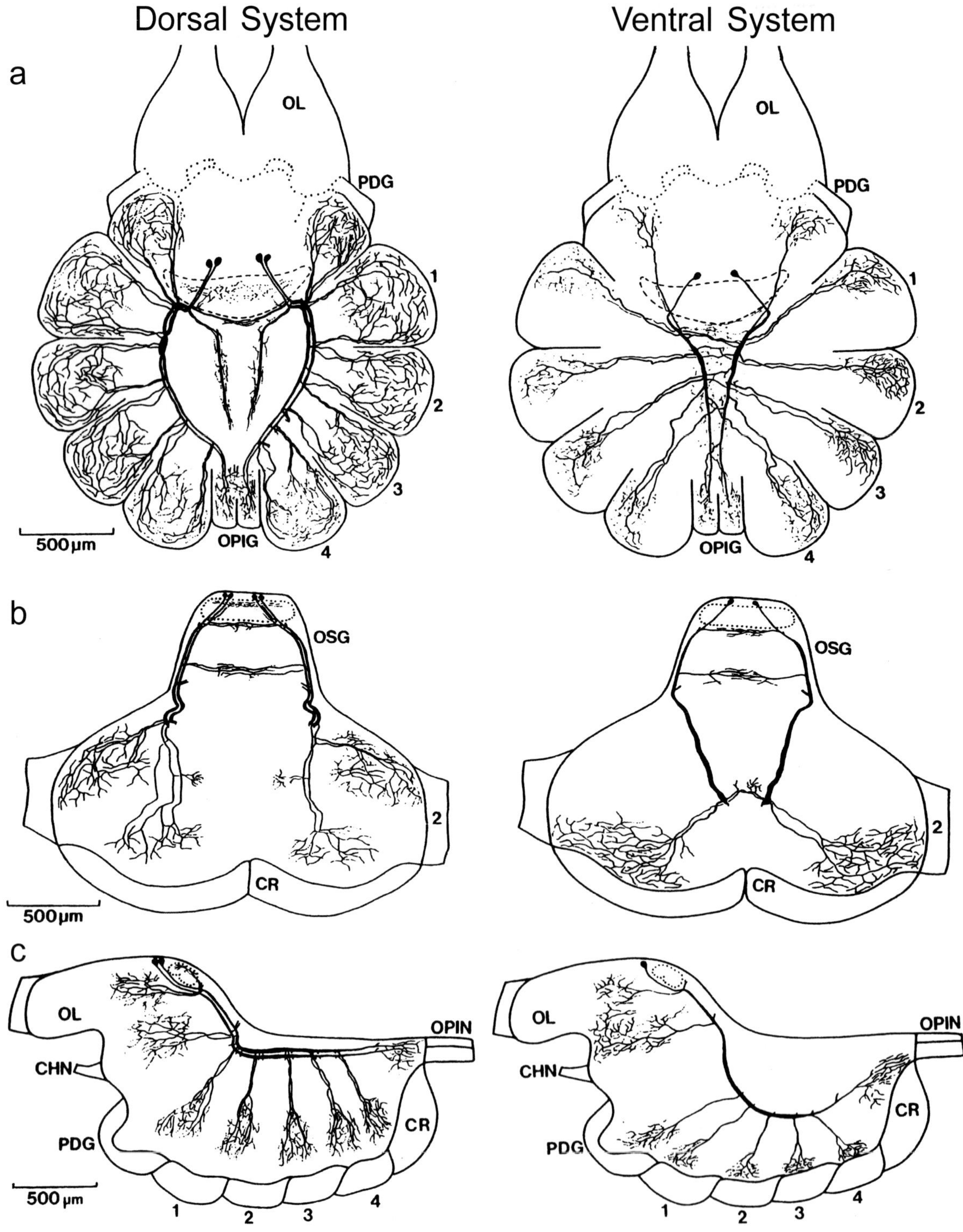

Dorsal System
Ventral System
a
OL
PDG
1
2
3
4
OPIG
500 µm
b
OSG
2
CR
500 µm
c
OL
OPIN
CHN
PDG
CR
500 µm
1
2
3
4

The two VPHN fibers pass through the circumesophageal connectives into the sensory longitudinal tracts SLT 3 and in each neuromere of the subesophageal ganglion send out ipsi- and contralateral projections. Each of the ipsilateral projections ends in the sensory neuropil, where the terminal arborizations of mechanoreceptive afferents are situated (Babu and Barth 1989; Anton and Barth 1993). The contralateral projections form a network of fine branches in the central neuropil, its ipsi- as well as contralateral regions (Fig. 7).

The extensive projections of the histaminergic system into sensory and motor neuropils immediately suggest a neuromodulatory function. The course of the ventral system in particular indicates that it may mediate the modulation of mechano- and chemoreceptive inputs. In contrast, the course of the dorsal system raises the possibility of a special involvement in modulating the activity of motoneurons and in large-scale systemic changes. Because of its size and arrangement the histaminergic system seems particularly suitable for a physiological analysis of such relationships. By the way, this system has been documented not only for *Cupiennius salei* and *Nephila clavipes*, but also for bird spiders (Theraphosidae) of the species *Psalmopoeus cambridgei*, for the funnel-web spiders (Agelenidae) *Tegenaria ferruginea* and *T. atrica*, and for the tarantula (*Lycosa tarentula*), which belongs to the wolf spiders (Lycosidae) (Becherer 1993).

Fig. 7 a–c. Histamine-immunoreactive giant fiber system of *Cupiennius salei* seen in sections along the three principal planes. **a** Horizontal view; for clarity the projections of the giant fibers into the proto- and tritocerebrum (cheliceral ganglia) not shown. **b** Frontal view; projections into proto- and tritocerbrum only slightly indicated. **c** Sagittal view; only the fibers on one side of the CNS shown. *CHN* Cheliceral nerve; *CR* cellular rind; *OL* optic lobe; *OPIG* opisthosomal ganglia; *OPIN* opisthosomal nerves; *OSG* supraesophageal nervous mass (brain); *PDG* pedipalpal ganglion; *1–4* walking-leg neuromeres. (Duncker 1992; Schmid and Duncker 1993)

5 Polyamine Neuropeptides

Along with the biogenic monoamines treated above, the polyamine neuropeptides are important neuroactive substances. In his Master's thesis, therefore, Matthis Duncker also described the occurrence of FMRFamide, proctolin and myotropin in the CNS of *Cupiennius salei.*

Before discussing these we should consider a question to which surely not all readers will have an answer: what kind of substances are these?

FMRFamide is a tetrapeptide named according to the single-letter code for the amino-acid sequence of its C-terminal group: Phe-Met-Arg-Phe-NH_2. It was discovered in molluscs, where it acts to stimulate the heartbeat. Since then many similar peptides have also been discovered in the CNS of arthropods. They change the conductivity of muscle-cell membranes and influence the release of transmitter especially in cholinergic neurons (Kobayashi and Muneoka 1989; Penzlin 1989; Fossier et al. 1990; Holman et al. 1990).

Proctolin is a pentapeptide that likewise has a myotropic action; it is produced in various ganglia of insects and released by special proctolinergic neurons (Starrat 1979). Its myotropic characteristic is expressed in its activation of intestinal musculature, the heartbeat, the muscles of the oviduct and other muscles. Proctolin has also been found in decapod crustaceans, where it is released into the hemolymph to accelerate the heartbeat (Orchand et al. 1989). In crustaceans proctolin also has a neuromodulatory action on sensory elements (proprioreceptive mechanoreceptors) (Pasztor and Macmillan 1990; Manira et al. 1991).

Myotropin I is an octopeptide which, like proctolin, was first described as a neurotransmitter in the visceral musculature of orthopterans. Again like proctolin, however, myotropin I has turned out not to be an actual transmitter but rather a co-transmitter that acts as a neuromodulator, controlling the effectiveness of the transmitter. In accordance with its myotropic function, myotropin, like proctolin, is found in excitatory motoneurons (O'Shea 1985; Bishop et al. 1991).

The results for *Cupiennius salei*, essentially reduced to key words, are as follows (Table 1 and Fig. 8).

FMRFamide

SEG: in all neuromeres between 15 and 65 immunoreactive somata, scattered in the ventral cell-body layer; serially homologous branching

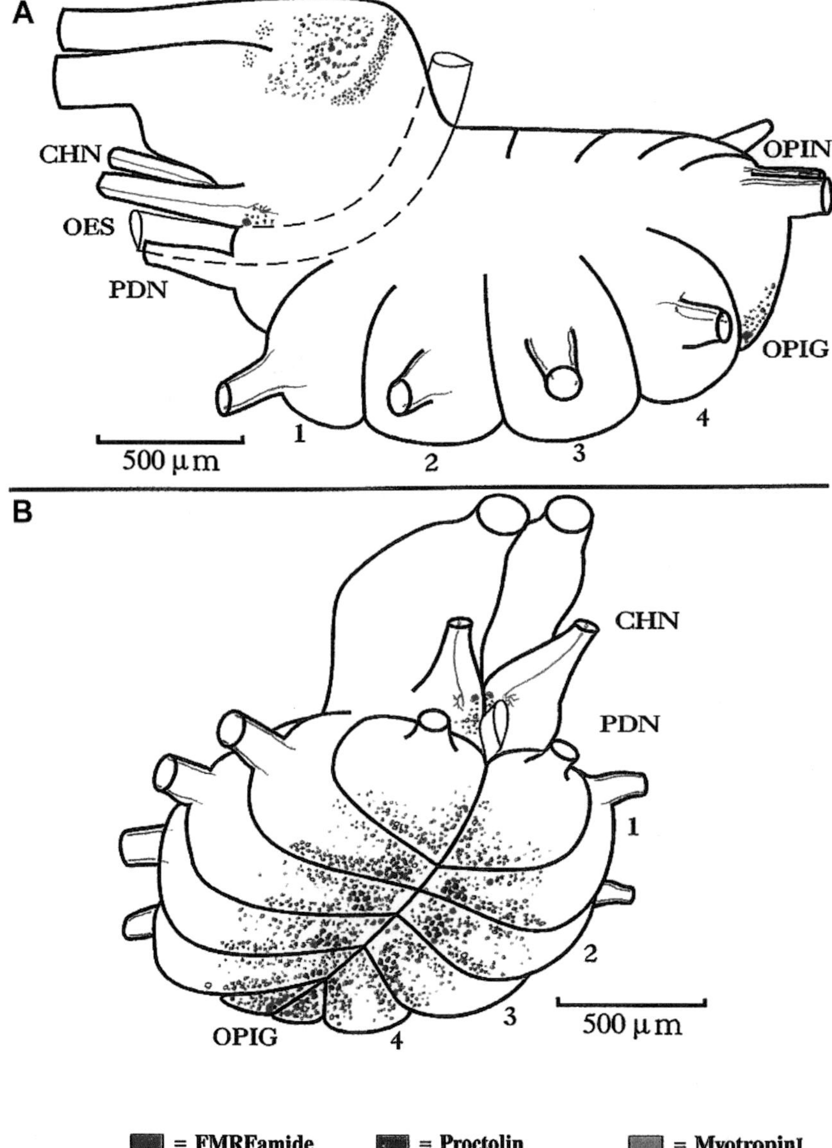

Fig. 8 A, B. Distribution of FMRF amide-, proctolin- and myotropin I-immunoreactive somata in the central nervous system of *Cupiennius salei.* Dorso-lateral (**A**) and antero-ventrolateral view (**B**). *CHN* Cheliceral nerve; *OES* esophagus; *OPIG* opisthosomal neuromeres; *OPIN* opisthosomal nerves; *PDN* pedipalpal nerves; *1–4* walking-leg neuromeres. (Duncker 1992)

in the dorsal motor neuropils of the walking-leg neuromeres, distinctly weaker labelling in the sensory neuropil; of the central tracts, only the presumably motor tracts MC (mid-central), MD (mid-dorsal) and the dorsal commissures (MDC, DC) distinctly labelled; no FMRF-IR fibers in the other tracts or in the peripheral nerves (exception: cheliceral nerves).

Brain: Ventral half of the "central body" very strongly labelled but its dorsal anterior half only weakly; several hundred immunoreactive intrinsic neurons at the back of the "central body", in the dorsal cell cortex in front of it also well over 100 somata; strong labelling also in the central protocerebral neuropil, in the optic tracts and the optic lamellae (ON1, AM-ON1) of all eyes.

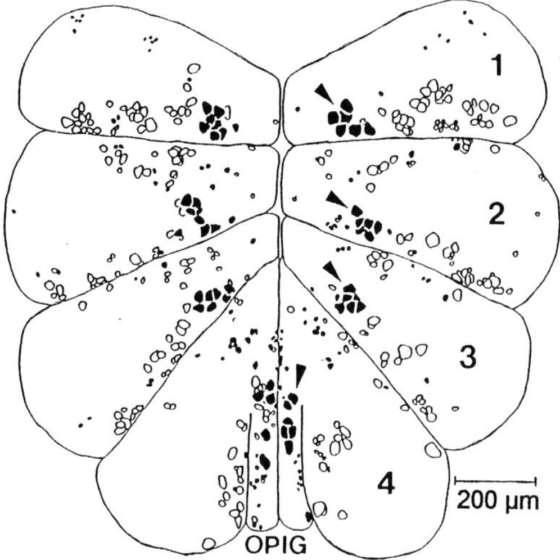

Fig. 9. Proctolin-immunoreactivity in the subesophageal ganglia of *Cupiennius salei.* Immunoreactive cell bodies in the leg neuromeres *1* to *4*. *Black dots* (*arrowheads*) Intensively stained cells (5 to 9) close to the posterior septum of neuromere; *open circles* weakly stained cells also close to the posterior septum of the neuromere. (Duncker 1992)

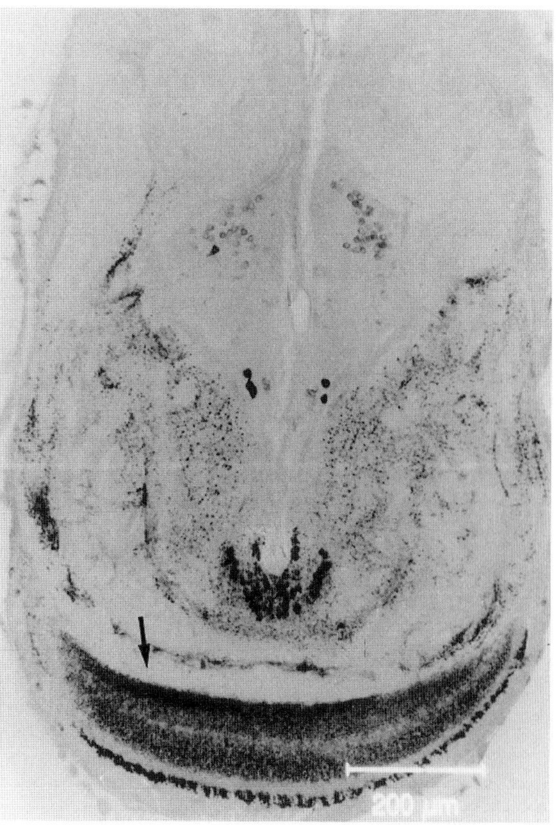

Fig. 10. Proctolin-immunoreactivity in the brain of *Cupiennius salei.* Dorsal region of protocerebrum with "central body" (*arrow*). (Courtesy A. Schmid)

Proctolin

SEG: In the ventral cell-body layer of all walking-leg neuromeres is a group of 6 to 9 large (ca. 40 µm diameter) somata in serially homologous arrangement; in addition, at least 40 scattered somata (Fig. 9); immunoreactivity in walking-leg neuromeres distinctly stronger in the dorsal motor region than in the ventral sensory region; two labelled fibers per walking-leg nerve; particularly conspicuous labelling in the dorsal neuropil of the opisthosomal neuromeres; in central SEG neuropil only a few, multisegmental proctolin-immunoreactive fibers in the large motor longitudinal tracts (CL, centro-lateral; MD, mid-dorsal), but strong labelling in their immediate vicinity; of the sensory longitudinal tracts, only SLT 2 exhibits clear proctolin immunoreactivity.

Brain: Again (cf. FMRFamide) the cheliceral neuromere is especially strongly immunoreactive, whereas the cheliceral nerve exhibits no proctolin immunoreactivity; "central body" and "mushroom body" are very strongly labelled (especially the area lateral to the peduncle) (Fig. 10); strong reactivity in the central protocerebral neuropil area, in the middle in front of the "central body"; also in the optic tract of the secondary eyes but not in the eye nerves or in the 1st optic ganglia.

Myotropin I

SEG: This too is found in all neuromeres, but in none of the nerves leaving the subesophageal ganglion mass; somata in the neuromeres (>100 each) scattered over the ventral cell-body layer (Fig. 8); in the neuropil labelling both in the dorsal motor and the ventral sensory part, distinct reactivity (similar to FMRFamide-IR); of the dorsal longitudinal tracts, MC (mid-central) is strongly immunoreactive, of the sensory longitudinal tracts, SLT 5.

Brain: In the "central body" and its dorsal cell-body cortex, strong immunoreactivity (especially ventral and posterior half); in the optic tracts of all eyes numerous myotropin-immunoreactive fibers, also in AM-ON1 and AM-ON3 distinct labelling; in the visual system of the secondary eyes only ON1 immunoreactive; no labelling in eye nerves or cheliceral nerves.

What does all this mean in functional terms? We must resort to analogies and to prognoses based on neuroanatomical information. Only the neuromodulatory action on the spider heart (myotropin I, *Tegenaria atrica*, *Coelotes atropos*: Richter and Stürzelbecher 1971; proctolin, *Argiope*, *Araneus*: Groome et al. 1991) is certain. The intensive FMRFamide immunoreactivity in the "central body" of *Cupiennius* corresponds to that in *Limulus* (Groome et al. 1990) and should probably be interpreted as signifying a considerable transmitter or modulator function. The proctolin immunoreactivity of large cell bodies in the walking-leg neuromeres and of axons in the leg nerves indicates an innervation of the skeletal musculature, the activity of which is known to be modulated by proctolin in *Limulus* (Rane et al. 1984). There is no reason to postulate a direct influence of myotropin I on the heart muscle or the heart ganglion, because the opisthosomal nerves exhibit no myotropin immunoreactivity.

D Senses and Behavior

XVIII Signposts to the Prey: Substrate Vibrations

"The spider gazed with malevolent, sparkling eyes at little Maya, looking patient in an evil sort of way and horribly cold-blooded. Maya let out a loud scream. It seemed that she had never screamed in such utter fear before. Death itself could appear no worse than this brown, hairy monster with its wickedly armored mouth and legs sticking out like the framework of a basket, in the center of which the plump body lurked."

The language used here to describe a spider reinforces prejudices about spiders, and does nothing to dispel them – even though Waldemar Bonsel's (1912) little bee Maya is ultimately rescued by a beetle. In reality, a spider catching its prey is no more immoral than any other creature that needs to eat in order to live, and during the process highly interesting things happen.

Spiders do not have such a bad reputation in all cultures. In particular, in the creation myths of the native North Americans spiders play an entirely positive role (Schwarz 1921).

For example, the Sia believed that at the beginning there was only a single animal, and that was a spider! This spider was in charge of two packages, and when it began to sing, from each package there emerged a woman. These were the two original mothers, who gave rise to all humans: from one of them the Indians descended, and from the other, everyone else. The production of clouds, lightning and thunder and the creation of the rainbow are also ascribed to the spider.

Among the Pima, another tribe indigenous to the southwestern USA, the weaving skill of a spider has a global significance. After an earthquake the earth, which the earth doctor originally made out of dust, no longer fits the heavens as it should, and it is the spider's job to spin them together and thus enable their further development.

The Mohawks tell a particularly nice story. Here a spider is one of four animals who joined forces to capture the sun, after it had disappeared and darkness had descended on the earth. They managed to find the sun resting in a tree. How could they get it down? The spider climbed into the tree, attached a rope to its top and lowered itself by the rope. Then the beaver gnawed at the trunk until the tree could be felled, whereupon the spider pulled on its rope. The tree fell. However, the hare seized the sun and leaped away with it before the other animals could decide what to do. But the sun should belong to everyone, so finally the animals threw it up into the sky, on a journey that continues to this day.

This is going to be a long chapter. It is concerned with a fundamental behavior: getting food. Anyone who has watched a spider catching prey will immediately understand that the role of the sense organs is an important consideration here, and why that is so. *Cupiennius* is basically a "sit and wait" hunter. It can stay motionless on its plant for hours, and then suddenly move like a flash to capture an insect that happened to come too close (Plates 4 and 7).

Here is a typical scenario in the field. Darkness has fallen, we can no longer see without a flashlight, the air is still warm and damp but it is not raining. *Cupiennius* left its retreat some time ago, half an hour after sunset, and now it is sitting fully exposed on its plant. By this time the wind has died down; not only is the air movement much slower than during daylight, it is steadier, with only slight fluctuations. Everywhere insects are emerging from their hiding places and beginning to crawl about on the plants. Many of them are earwigs and another species well known to zoologists from the laboratory: the cockroach *Periplaneta americana*. *Cupiennius* never moves from its spot, but by raising its body slightly it shows us that it has noticed the cockroach, at a distance of half a meter or more. It doesn't leap forward until the victim is definitely within reach, only a few centimeters away from the spider. Then things go very fast. Mechthild Melchers (1967) photographed prey capture by *Cupiennius salei* at speeds as high as 1000 frames per second and analyzed the "wild tangle of spider and prey" to

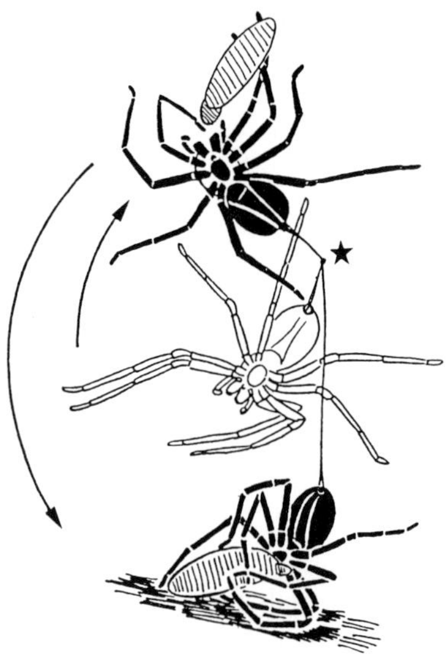

Fig. 1. Prey-capture behavior of *Cupiennius salei*. From its resting position (*center*) the spider first jumps upward to the cockroach and then falls down to the ground together with its prey. The safety line (★ attachment point) helps to avoid a hard impact of the opisthosoma. The time between the starting position (*center*) and the upper position is 96 ms, that between the upper position and the lower one 120 ms. (Melchers 1967)

identify the movements involved in this behavior. For technical reasons, these films had to be made in small cages. When confined in such a cage, *Cupiennius* always takes up a waiting position on one of the vertical glass walls, with its prosoma downward. The whole capturing procedure, from the first reaction until the prey is bitten, lasts between 200 and 700 ms (Fig. 1). Usually *Cupiennius* touches the prey insect only with the tips of its legs and uses only the first pair of legs to grip it. Immediately after these have seized the prey, by moving at speeds up to 3–4 m/s, the bite is administered. While catching prey *Cupiennius* is attached to its waiting site by the dragline, which supports the spider even if it falls off the leaf when jumping at its target (a not uncommon occurrence). *Cupiennius* does not begin to eat until the prey has stopped moving.

This complex behavior of *Cupiennius* is evidently based on a considerable amount of information. But what do we mean when saying this? We can break the problem down into a number of questions that are important in neuroethology in general.

- What signals are emitted by the cockroach, or any prey animal? And how are they transmitted from sender to receiver?
- To what extent do prey signals differ from other stimuli that are unimportant in this context, in particular signals of abiotic origin? In other words, can we find a physical basis for the discriminability of various signal types?
- Over what range can the sensory system used for prey capture detect signals?
- And finally: how does the spider know the direction from which the stimulus is coming? How does it orient itself toward the prey?

In the case of *Cupiennius* the problem is complicated by the fact that the signals of interest are not only the vibrations of the plant caused by prey activity but also, on some occasions, air movements. *Cupiennius* has actually been observed to jump toward a flying fly or, in experiments, toward a nozzle emitting an airstream that simulates such a stimulus. Futhermore, the messages sent out by the sense organs in response to these two types of stimuli interact in complicated ways in the central nervous system. This interaction is treated in the next chapter; here we are concerned only with substrate vibrations.

As might be expected from the fact that these spiders hunt at night, the visual sense plays no or only a very minor and subtle role in prey capture. When we blind *Cupiennius* by covering its eyes with paint, it can still catch its prey with no discernible difficulty, and with the usual precision. Observations in the field suggest that the only effect visual stimuli might have is to interfere with prey catching. But we know too little to say anything more about this (see Chapter XXIII).

Before we turn to the details of a study on the sensory, ecological and neuroethological aspects of predatory behavior, one simple question: what, actually, does *Cupiennius* eat? We first tried to answer this many years ago, while observing *Cupiennius salei* in Guatemala (Barth and Seyfarth 1979). Subsequent observations during many trips to other *Cupiennius* biotopes in various Central American countries confirmed the main finding: *Cupiennius salei* is not all that fussy about its food; it is a generalist, catching not only cockroaches and earwigs but also flies, crickets, grasshoppers and moths. Nentwig (1986) later found the same thing in laboratory experiments. In Guatemala cockroaches in particular were often captured. On several occasions we actually saw a *Cupiennius salei* with frogs between its chelicerae, and I recall an impressive cover picture on a copy of the Journal of Arachnology dated 1989 (Vol. 17), a photograph taken by Jerry Rovner in La Selva, Costa Rica; it shows a *C. coccineus* with a lizard (*Norops limifrons*) "in its mouth".

1 Vibratory Signals and Their Propagation

It has been amply demonstrated that the habitat of an animal has a strong influence on the evolution of its senses and their integration into behavior patterns, so when considering the vibratory sense we need to know what kinds of vibration the spiders encounter in their ordinary life. In exploring this question, as expected, we soon were immersed in problems that are likely to be as interesting to a physicist as to a biologist.

Sitting on its preferred plants in the Central American forest, without a web to help it catch prey, *Cupiennius* waits to be alerted by vibratory signals. In fact, it also actively emits such signals (see Chapter XX). In both cases, the plant is the channel through which the signal is transmitted. The spider does not only need to detect these signals with sufficient sensitivity, it must also be capable of distinguishing various kinds of vibration – for example, prey signals from courtship signals – and must not confuse significant vibration with the inevitable vibratory background noise. We shall see that the various types of vibration differ in both their spectral and their temporal properties, and the plants specifically chosen by *Cupiennius* are good transmission channels; that is, they conduct vibration with relatively little attenuation (Barth, Bleckmann et al. 1988).

Once we have explored these questions for plants as transmission channels, we shall move on to a comparison with other substrates relevant to spiders: the web and the surface of water.

Plants

We studied the ways in which vibrations spread through plant tissue in the field as well as in the laboratory, and field work in Mexico was not a simple matter. One of the most serious obstacles was getting a large crate of electronic apparatus through a Customs station in Mexico City on the way to Fortin de las Flores in the state of Veracruz. This town lies at the foot of the highest volcano in Mexico, the 5,653-m Pico de Orizaba, which the natives also call Citlaltépetl or mountain of the star.

Fortin de las Flores turned out to be a real stroke of luck for us. As so often, chance played its part. Since we had first encountered *Cupiennius salei* in Guatemala, that country had become too unsafe politically, and field trips into remote regions seemed inadvisable at the beginning of the 1980's. We discussed our problem with arachnologist friends and later heard from Jerry Rovner (University of Athens, Ohio), that George Uetz, of the Department of Biological Sciences at the University of Cincinnati, Ohio, knew of a place in Mexico where it was supposed to be possible to watch *Cupiennius salei* from a swimming pool. Knowing George's exuberant attitude to life in general, we were inclined to believe this rumor even before he confirmed it. Eight months later, in August 1983, Ernst-August Seyfarth and I arrived from Mexico City on our first visit to Fortin de las Flores, and moved into the Posada Loma. It was a pitch-black night and the rain was pouring down. Nevertheless, it took only a few minutes until, in an idyllic garden (right next to the swimming pool), our flashlight beams picked out the first specimens of *Cupiennius salei* in bromeliads.

Next morning we were handed a very helpful note from George, who had been here studying *Metepeira spinipes*, a social spider species that builds enormous community webs, and had left shortly before our arrival:

"Dear Fred/Ernst:

I am truly sorry that my timing on this trip is off, and that our paths will not cross here. ... I've enclosed a map of some of the local roads where we've been working. In general, the people are very friendly and will allow you to come onto their plantation to look for spiders. We first saw Cupiennius at the plantation on Rio Blanco. The people call them tarantulas or arañas grande de platano, and will be glad to show them to you in the leaf axils of the bananas. ... So – good luck. Enjoy your visit, and good spider hunting.

Best regards. George Uetz"

The spiders turned out to be especially plentiful in the banana plants on a coffee plantation near the road to Huatusco. Its owner, Delfino Hernandez Guitierrez, showed great interest in our peculiar (and nocturnal) activities. In later years we met him again on many occasions. It was his habit to turn up silently out of nowhere as soon as we parked our rented car by the side of the road, and every time he was delighted by our reunion – as were we.

On September 19, 1983 I wrote to George Uetz:

"Dear George,

We had a very nice, enjoyable, and rewarding time in Mexico and just arrived here in Frankfurt with lots of live Cupiennius. The Posada Loma was a real good suggestion. As you predicted everyone was very helpful and understanding, including the campesinos who very friendly tolerated our work in their plantations.

We had a look at the Metepeira colonies. The one across from the office is doing very fine and of the seeded ones along the fence the two towards the barking dog have developed beautifully. ..."

Biologists seem to have their own special way of experiencing foreign countries. Especially under the conditions of working in the field (without swimming pool!) it is quite apparent that to be a biologist is not only a profession but a way of life.

The laboratory experiments were done partly on a visit to Axel Michelsen in Odense, Denmark. At that time his laboratory was *the* pioneering site of a functional laser vibrometer.

- The main plant species we studied: *Musa sapientium* (banana plant), *Aechmea bractea* (bromeliad) und *Agave americana.*
- The electronic apparatus in the field: (i) a non-contact displacement detector that operates on the basis of eddy currents (Micro-ε, Multi-Vit KD 2300), has a sensitivity of 1 mV/µm and is particularly suitable for recording low-frequency vibrations; (ii) piezoelectric accelerometer (Bruel & Kjaer Type 4375, charge amplifier Type 2625), which is particularly good for recording high frequencies and, with a mass of 2.4 g, is considerably lighter than the leaves on which it is placed and weighs about the same as a fully grown female spider; for physical reasons it is important to add as little mass as possible; (iii) also an oscilloscope, a suitable recorder ... and all this battery-operated, including the detector electronics. Finally, devices are needed for spectral signal analysis.
- Measurements in the laboratory on the transmission properties of the plants: (i) an electrodynamic vibrator (Ling 106) with electronically stabilized frequency response (Bohnenberger et al. 1983) to induce controlled oscillations in the plant; (ii) non-contact laser vibrometer (DISA 55L66) together with spectral analyzer; (iii) piezoelectric accelerometers for simultaneous measurements at several points.

The procedures are described in detail by Barth, Bleckmann et al. (1988). Here we are interested only in the results.

Signals and Vibratory Background

According to our studies in Central America on the plants actually occupied by *Cupiennius salei*, the various types of signals differ distinctly in

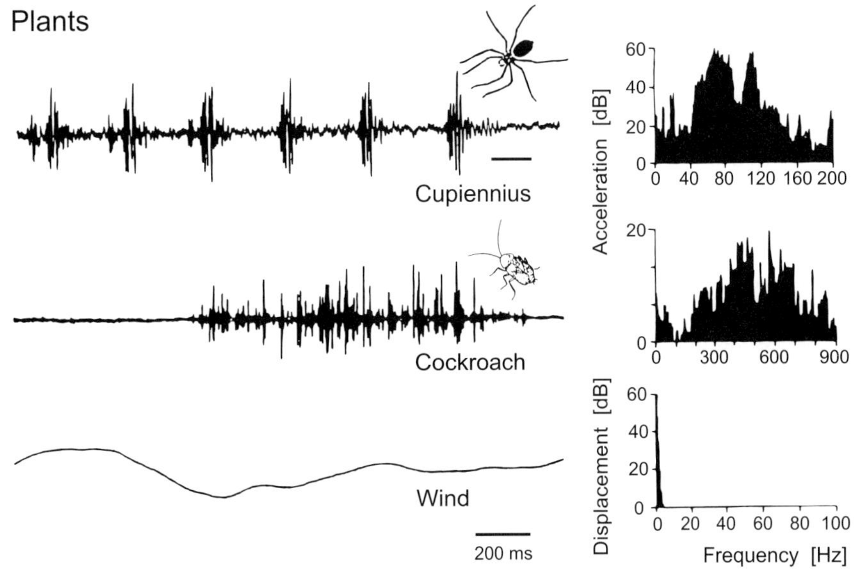

Fig. 2. The vibrational environment of *Cupiennius salei* in its natural habitat. *Left* Oscillograms; *right* frequency spectra. *Top* Male courtship signal on an agave; *middle* cockroach crawling on the pseudostem of a banana plant; *bottom* vibrations on the pseudostem of a banana plant due to wind. (Barth 1986a)

their frequency content. A very narrow bandwidth and very low frequencies are typical of the vibratory background, whereas prey signals are relatively broad-band and contain higher frequency components. Courtship signals are approximately between these two types with regard to frequency composition, and the male signal is distinguished by its marked temporal structure (Fig. 2).

In detail, it looks like this. The frequency spectrum of the *background noise*, which is caused by the wind and interferes with the prey and courtship signals, in banana plants and bromeliads is characterized by a peak at about 1.4 Hz (there are small variations, but the peak is usually well below 10 Hz). Often the spectrum contains no components above 10 Hz; if present, these are smaller by 40 to 60 dB (that is, by a factor of 100 to 1000) than the main peak. Even when the leaves bump into one another in a high wind, the spectrum of vibrations so induced has components only up to about 50 Hz at 20 dB below the main peak. However, such a strong wind is quite untypical of the situation in which *Cupiennius* is active on its plant, and we have never seen the spiders exposed to one in the field. Instead, they withdraw into their retreats, or else they do not emerge in the first place (see Chapter XIX).

Prey signals such as are produced by a cockroach – a common victim of *Cupiennius* – crawling on the pseudostem of a banana plant (a preferred place for a spider to sit in wait) are associated with a comparatively very broad-band frequency spectrum, the dominant peaks of which are between 400 Hz and 900 Hz. The –20 dB frequency bands extend from a few Hz to ca. 900 Hz. Anticipating a later section, it should be mentioned here that a number of potential prey animals escape the spider's attention (and hence its bite) by having a "vibrocryptic" gait, which induces only slight, extremely low-frequency oscillations.

The vibratory *courtship signals* of the male, which will be considered more extensively in Chapter XX, are characterized by a very regular temporal structure. Their basic element, the syllable, is produced in sequences, being repeated up to 50 times at intervals of about 350 ms. The most prominent frequency components of the syllable in *Cupiennius salei* are at ca. 75 Hz and 115 Hz, roughly intermediate between "wind" and "prey". The female courtship signal usually lacks this clear syllabic structure; it is temporally more irregular than the signal of the partner and, accordingly, its frequency spectum is broader with main components mostly between

20 and 50 Hz, lower than those of the male. In the chapter on courtship the neuroethological significance of all these quantities will be thoroughly illuminated.

These values are not exhaustive, but after much experience with vibrations in the field we know that they are typical and describe situations frequently encountered by *Cupiennius*. We also analyzed the vibrations caused by raindrops – and the vibratory "environmental pollution" imposed on a banana plant by a truck passing on a road about 50 m away. However, there is no need to go into that here because *Cupiennius* does not catch prey in the rain, and truck signals are neither very biological nor likely to be confused with prey signals by a spider. More details can be found in Barth, Bleckmann et al. (1988).

The Propagation of the Signals on the Plant

Before we can find a relationship between the vibrations of various biotic and abiotic origins and the spider's sensory capabilities and behavior, we need to know something about the physics of propagation of vibration in a plant. How severely are the signals attenuated on the way through the plant? Are there frequency-dependent differences in this attenuation? Of the various kinds of vibratory waves, which is present in plants?

The best way to answer these questions is to introduce specified band-limited noise into the plant at one place and compare it with the oscillations that arrive at positions progressively further away from that place. Damping values are then obtained for the individual frequencies. We did this with a banana plant, for frequencies between 12.5 Hz and 5000 Hz. The conclusion: the plant where the spider sits is excellently suited for the transmission of vibrations and thus also for vibratory communication.

Figure 3 shows the results of an experiment with the laser vibrometer. One finding is that the very low frequencies typical of oscillations caused by wind are conducted along the leaf into which they were introduced with an average attenuation of 0.32 dB/cm (longest distance measured 40 cm). To put it more simply: the signal has fallen to half its original amplitude after it has travelled about 18 cm. At 75 Hz, a major component in the male courtship signal, the average damping values are again between 0.3 and 0.4 dB/cm – remarkably, even if the signal has travelled over 80 cm from one banana leaf to another or onto the pseudostem. The main component of the female courtship signal (ca. 40 Hz) is even less attenuated. However, damping does not depend in a simple linear way on the distance travelled, presumably because of irregularities in the leaf mechanics, reflections and interference. It is striking that the transmission curve is relatively smooth up to about 100 Hz and between ca. 150 Hz and 1000 Hz becomes considerably more irregular. This is the frequency range that includes in particular the frequencies of the prey signals (see above). At frequencies over about 1000 Hz the transmission curve again becomes relatively smooth, and even at 5000 Hz the damping is only 0.35 dB/cm, no greater than at the other end of the frequency band, when averaged over the distance travelled from one leaf to another or to the pseudostem.

One thing that might seem unlikely at first is established by these measurements: the plants occupied by *Cupiennius* conduct vibrations very well over apppreciable distances and hence are a substantial factor in determining the range at which signals detectable by the vibration sense can originate, in the contexts of prey capture and communication with a sexual partner (see below). The measurements also imply that the frequency spectrum of a signal changes during propagation over the plant. This has biologically significant consequences, to which we shall return when considering the responses of the metatarsal organ to the courtship signals (see Chapter XX).

Bending Waves

One important question had never previously been answered: what kind of oscillations are we dealing with here? This question may seem cryptic to the biologist until the solid-state physicist lists all the various possibilities: longitudinal waves, quasilongitudinal waves, torsion waves, Rayleigh waves and bending waves. We know that as our waves travel over the plant,

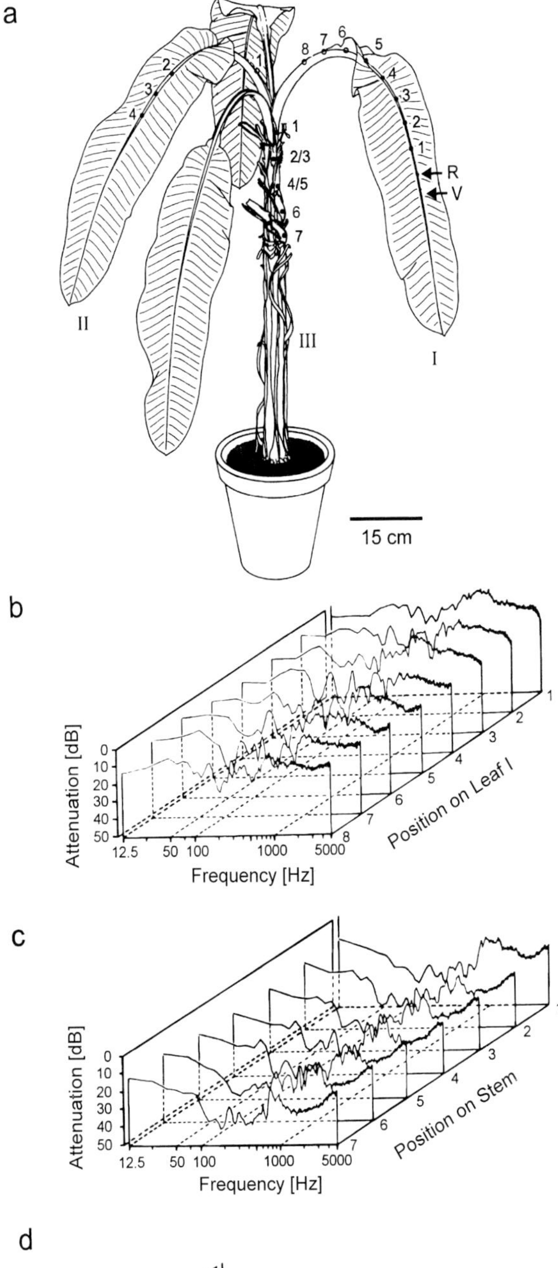

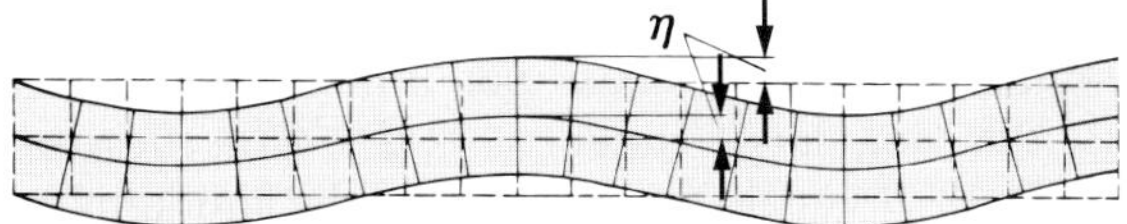

Fig. 4. Bending wave. Diagram showing deformation of a plate (cross-section) by a sinusoidal pure bending wave (wavelength large relative to cross-section); η transverse displacement. (Barth 1986)

they change their amplitude, their frequency composition and also their temporal pattern, because the propagation velocity also increases with increasing frequency (i.e., the propagation is dispersive). Fortunately, matters simplify because we can readily answer the above question: bending waves are the crucial type here. To establish this we examined banana plants and agaves, and benefited greatly from a pioneering earlier paper by Michelsen et al. (1982).

When a structure such as the leaf of a bromeliad is subjected to bending waves, it moves preferentially in a plane perpendicular to the surface and to the direction of propagation of the oscillations (Fig. 4). Because of the occurrence of stresses and strains in the longitudinal direction, however, bending waves cannot be categorized as transverse waves. Another good clue to their identity is their low (both absolutely and in comparison to the other types of oscillation), frequency-dependent rate of propagation. This can be calculated from a simple equation and then compared with the values actually measured (Cremer et al. 1973):

$$C_B = \sqrt[4]{B/m'}\sqrt{\omega}\ .$$

Fig. 3 a–d. Vibration transmission through a banana plant. **a** *V* site where vibrations (noise, flat amplitude spectrum over investigated frequency range of 12.5 to 5000 Hz) were introduced into the plant; *R* reference point used for comparison with the vibrations picked up with a noncontact laser vibrometer at various distances from *V* along the ipsilateral (*I*) and a contralateral (*II*) leaf and along the pseudostem (*III*). *Arabic numerals* mark measuring sites. **b, c,** and **d** Attenuation as a function of frequency for the different measuring sites on the two leaves (**b** leaf *I*; **d** leaf *II*) and the pseudostem (**c**). 0 dB attenuation corresponds to value at *R*. (Barth, Bleckmann et al. 1988)

where C_B is the group velocity, B is the flexural strength, m′ is the mass per unit length and ω is the angular frequency ($2\pi f$). The flexural strength B equals the product of the modulus of elasticity E and the axial geometrical moment of inertia J ($B=E \cdot J$). And for those who want to know exactly: the group velocity C_B refers to the envelope of the carrier frequency of a wave group and is distinguished from the phase velocity.

Our measurements on banana plants (*Musa sapientium*) gave values of propagation velocity below 50 m/s at 100 Hz and 500 Hz, for both the actual leaves and the pseudostem. The values found for an agave (*Agave americana*) were in good agreement with those calculated from the measured mechanical characteristics, namely between 5 m/s in the apical part of the leaf and 55 m/s in the basal part (Wirth 1984; Barth 1985 a).

The available data, including those published by Michelsen et al. (1982) and by Kämper and Kühne (1983) for quite different plants, thus indicate that bending waves play the chief role in transmission of the plant vibrations important for arthropods. But we shouldn't prematurely and generally rule out the possibility that other types of waves also contribute (Barth 1985).

Range of Vibratory Signal and Communication System

Among the selection pressures that affected the evolution of the sense of vibration, the physical properties of the information-transmitting channels must have been crucial factors. For one thing, they influence the distance over which a signal can be sent and thus place limits on the whole communication system. It is important to understand these relationships in order to interpret behavior.

The range of effective vibratory signals can be found from the properties of the signal originally emitted, the transmission properties of the medium and the sensitivity of the receiver's sensors. For the moment let us turn from prey capture to the courtship signals in order to give an example of range estimation. Courtship signals are much more stereotyped than those emitted by prey, and much more is known about them. We shall return to prey capture at the end of this excursion.

The vibratory communication of *Cupiennius* during courtship while sitting on a plant can be specified very precisely with respect to its active space. The acceleration measured for the vibrations emitted by a female of *Cupiennius salei* at a distance of about 10 cm is maximally about 1.6 m/s^2 (Schüch and Barth 1990). The slits of the metatarsal lyriform organ of the receiver have thresholds, for stimulation in the dominant frequency range of the female signal, as low as ca. 7 mm/s^2 (Barth and Geethabali 1982). Taking the mean value for attenuation of the signal on the way through the plant from female to male to be 0.3 dB/cm (Barth, Bleckmann et al. 1988), we find that the female's signal can travel about 150 cm on the plant before becoming subthreshold for the receiving organ and hence undetectable. This is what we would have expected from our observations of the behavior. The calculated maximum of 150 cm rises to almost 200 cm when we take into account that the receptor thresholds are as much as 10 dB lower when the stimulus is band-limited noise (bandwidth ca. 1/3 octave; $Q=0.35$) instead of a sine wave (Barth 1985 a).

Here is some good news for the biologists: according to more recent field observations, the spiders do distinctly better than our calculations implied. During our travels in the native lands of *Cupiennius* we tried on many occasions to establish the distance record for courtship communication. It turned out that the limiting factor here was not the spiders but whether we could find ladders long enough to put ourselves in position for viewing and recording the responses of a female seated far up on a 3- or 4-m-high banana plant, while the male was vibrating near its base. The longest distance we have found so far is 3.8 m (measured for *C. coccineus* in Costa Rica).

The discrepancy between this and the calculated values is presumably not associated with the fact that two different *Cupiennius* species are involved. The problem is that we do not know the threshold sensitivities of all the slits in the metatarsal organs, and some of the ones not yet measured are especially sensitive. Furthermore, behavioral thresholds often do not match the receptor thresholds. It is entirely possible for them to be lower, because of convergent circuitry, central nervous noise suppression or the like.

Because the male signal contains somewhat higher frequencies and has a somewhat lower amplitude, its range is shorter than that of the female signal. We shall return to this point in the chapter on courtship (Chapter XX). The prey signals, which are of most interest here, contain still higher frequencies and so have an even shorter range. This may be one of the reasons that *Cupiennius* attacks its prey only when it comes very close. The frequency spectra of vibrations generated by typical prey animals of *Cupiennius salei* (for instance, a cockroach walking on the same leaf) are, as already mentioned, broad-band with highest peak frequencies usually between ca. 400 and 900 Hz. A mean damping value measured on a banana plant is about 0.4 dB/cm (Barth, Bleckmann et al. 1988). That is, the difference from the attenuation of courtship signals on average is not dramatic. The threshold for excitation of the metatarsal organs is lower at the higher frequencies than at those in the courtship signals, as far as the amplitude of substrate deflection is concerned. For the reasons discussed in Chapter V, Section 1 on the vibration sense, however, in the frequency range typical of prey signals the metatarsal organ should be regarded as an acceleration detector. Then the situation looks rather different; that is, from the available data it is impossible to be certain that there is a threshold difference. From all this it follows that the range of the vibration sense is also large for prey signals, and undoubtedly in very many cases it is considerably greater than the distance a spider can jump and be sure of catching its prey. In courtship the receiver gives a clear response to indicate reception of the sender's message, but it is harder to tell that a signal has been received when observing predatory behavior. It is probably a good thing for the spider that it does not pursue its prey over long distances or run toward an approaching victim, as its own movements also generate vibrations. Most prey insects themselves possess a sensitive means of sensing vibration, and use it as an early warning system.

Does *Cupiennius* seek out plants on which the vibrations relevant to its behavior propagate particularly well? At present we know only that "*Cupiennius* plants" are good vibration channels, because of the low average attenuation, ca. 0.35 dB/cm, and the very homogeneous (with respect to amplitude and velocity) transmission, particularly in the range of frequencies of the courtship signals. However, it is questionable whether these plants are chosen primarily because of these transmission properties (see Chapter III on the significance of the hiding place). Quite different plants transmit substrate vibrations just about as well, although with more irregularities, as has been shown by the studies of Michelsen et al. (1982) and Kämper and Kühne (1983) on the vibratory songs of insects.

In Summary. The vibration sense of spiders is on the whole a short-range system, despite the several-meter record range for communication in *Cupiennius*. As such, there may be a number of advantages to using it rather than other senses. Vibratory signals are available even during the night, unlike visual signals (except for fireflies). The short range and the brief duration of courtship signals may be advantageous because they make the signals relatively private and less likely to attract the attention of enemies. Furthermore, the propagation of vibrations along plants or threads of silk is much less diffuse than the spreading of sound or chemical stimuli through the air. There is no danger that the signal will drift away on the wind, and it is unlikely to be blocked by obstacles – which makes it easier, for example, to orient to a vibration source at branch points.

Water Surface

Most spiders identify and locate their prey by means of vibration stimuli received through their web (see below) or another solid material such as the ground or, as in *Cupiennius*, a plant structure. But there is an interesting exception. Some species of fishing spiders (Pisauridae) and wolf spiders (Lycosidae) feed on insects that have fallen onto the water surface and are trying to free themselves by struggling, generating concentric surface waves in the process. These surface waves reach amplitudes of around 80 μm (peak-to-peak) and their upper frequency limit is 60 to 150 Hz (Lang 1980a). That backswimmers, water striders, whirligig beetles, surface-feeding fish and amphibians use these waves as pointers to prey has long been known (see the review by Bleckmann 1994). It had always been assumed that spiders do the same, but was never conclusively documented until Horst Bleckmann came to our laboratory and brought along his special expertise regarding the physics of the water surface and the measurement of surface waves, acquired during his work on surface fishes. Jerry Rovner and Jim Carico had previously, in Ohio and Virginia, given me a thorough introduction to the biology of the water spider *Dolomedes*. Now, for the first time, we were able to give serious experimental consideration to questions such as the role of visual and air-current stimuli in prey capture, the actual precision of prey localization, the sensitivity to water-surface waves, and the type of wave signal that most commonly triggers prey-capture behavior.

Although in this section we are concerned only with the signals themselves, it is worth stating in advance that *Dolomedes triton* to a certain extent also uses the visual sense, the air-current sense and the substrate-vibration sense in hunting. But when it is more than 10 cm away from the prey, water-surface waves play far the most important role (Bleckmann and Barth 1984). There is a biological justification, then, for taking the time to analyze these waves.

In a water-surface wave the water particles move along elliptical or circular paths (depending on the depth of the water) and, as in the bending waves of plants, in a plane perpendicular to the unmoved surface. On the water surface, too, the waves propagate with dispersion; the propagation velocity changes with frequency and hence with wavelength. Natural signals, with several frequency components, change in a complex manner on the way from transmitter to receiver, because high and very low frequency components increasingly take the lead over frequencies between about 5 Hz and 20 Hz. Components near 6 Hz move the most slowly. Their group velocity is only 17.5 cm/s. A 100-Hz component, for example, travels about three times as fast, at 55 cm/s.

The attenuation of water waves in general is considerably stronger than that of bending waves on plants and of oscillations in a spider web. Like the propagation velocity, it is frequency-dependent; high frequencies are more strongly damped than low frequencies. As an example for the biologically relevant range: 140 Hz, 8.57 dB/cm; 10 Hz, 1.67 dB/cm (both values refer to a distance of about 3 cm between measurement site and vibration source). The changes in a water wave with increasing distance from the source are thus due not only to the differences in propagation velocity but also to the markedly different attenuation of the various frequencies. These factors affect both the temporal structure of the wave and, even more, its spectral composition. After it has traveled only a few cm, all the frequency components above 200 Hz have disappeared, and after 10 cm the signal contains essentially no components above 100 Hz (Sommerfeld 1970; Bleckmann 1994).

The wave stimuli a spider is likely to detect while waiting on a water surface for prey to appear have been investigated in detail (Lang 1980a) and summarized by Bleckmann (1994). Measurements have also been made in a typical habitat of *Dolomedes triton*, a small lake near Athens, Ohio (Bleckmann and Rovner 1984). Although Dow Lake is an idyllic spot, my memories of it are not all pleasant. While wading for hours near the shore, together with Jerry Rovner, I had my first and more than sufficient encounter with chiggers. Fortunately this did not become evident until the excursion to become acquainted with *Dolomedes* on its own ground had been successfully completed.

As is the case on a plant, wind-induced vibrations on the water surface are characterized by low frequencies and a narrow frequency spectrum, with a maximum at a few hertz. No frequencies above 10 Hz are present, at least for a moderate wind (up to 3.2 m/s). In contrast, the water waves generated by prey animals contain frequencies up to about 150 Hz (Fig. 5): for instance, when a fly is scrabbling on the surface or a fish touches the surface from below. *Dolomedes* also encounters surface waves of abiotic origin, produced by the impact of falling leaves, seeds and raindrops. These are also characterized mainly by low frequencies and a narrow spectrum, although here frequencies up to 50 Hz may be represented. They typically resemble the waves generated by vertebrates contacting the surface in that they are brief, lasting less than 1.5 s. The surface waves produced by struggling insects are of considerably longer duration, from 3 s to more than 60 s (Bleckmann 1985a, 1994).

In the courting season, females of *Dolomedes triton* use sexual pheromones to attract males (Roland and Rovner 1983). A courting male taps the water with his first pair of legs so as to generate brief wave signals with an energy maximum between 8 and 15 Hz (Bleckmann and Bender 1987), which clearly distinguishes them from the waves generated by the usual prey animals. Presumably this is one reason that the female does not misidentify the male as something to eat.

The Spider Web

Finally, you'll be thinking, we have overcome our preoccupation with *Cupiennius* and arrived at the web. This can be regarded as an extension of the sensory space of the spider, a small woven world full of vibrations. Spi-

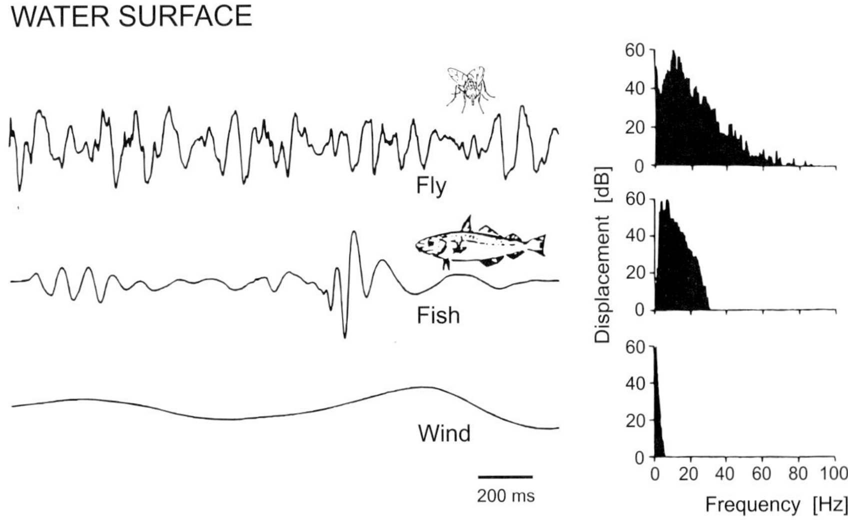

Fig. 5. Vibrations on the water surface. *Left* Oscillograms of displacement; *right* amplitude spectra of surface waves of biotic (fly trapped on surface and struggling; fish touching surface) and abiotic origin (wind). (Barth 1998)

ders both receive and send out vibrations in their web. They use it not only to obtain information about prey animals and potential predators, but also as a channel for communicating with their conspecifics, especially during courtship (see Chapter XX). From an evolutionary viewpoint spider webs are the result of a compromise, dictated on one hand by the demands of "architecture" (in particular, the web must be sufficiently stable to absorb the impacts of, for example, prey animals) and on the other hand by the need for efficient transmission of vibration. Because *Cupiennius* is a hunting spider and catches prey without a web, the first of these aspects will be neglected here; the interested reader can consult a recent review (Barth 1998). For present purposes, we are concerned only with the transmission of vibration.

Spider webs are lightweight constructions *par excellence*: those of the European orb-web spiders weigh only 1 to 2 mg (Masters and Markl 1981). Addition of the slightest mass changes their oscillatory properties, so that the only sensors suitable for most experimental studies must be of the non-contact type. Modern technology has come to the rescue with apparatus we could only dream about ten years ago, in the form of the laser vibrometer and associated processing electronics. Despite this major advance, such analysis is a demanding business, and far too few studies of the oscillatory behavior of spider webs have been undertaken. The orb web is the only type yet examined with faultless methodology; the species concerned are *Nuctenea sclopetaria* (now *Larinioides sclopetarius*) (Masters and Markl 1981; Masters 1984a, b; Masters et al. 1985) and *Nephila clavipes* (Landolfa and Barth 1996).

Orb webs are constructed of threads with differing mechanical properties. That is, they are mechanically heterogeneous structures (Wirth and Barth 1982). The radial threads are much more important for the transmission of vibrations than the spiral and adhesive threads. It makes sense, then, that while a spider is lying in wait it keeps the tips of its legs in mechanical contact with the radial threads. The most important reason for the lower degree of damping of vibrations in the radial threads than in the others is that the radials have a larger modulus of elasticity (radials: 3 to 20×10^9 N/m^2; spiral threads: 0.05×10^9 N/m^2; Denny 1976, Masters 1984a). The density of the thread material also plays a role, as does the number of intersections between radial and other threads. On the other hand, the tension in a thread does not affect signal attenuation (Frohlich and Buskirk 1982). For most biologists this will be counter-intuitive and hence remarkable. The orb web of *Nephila clavipes* is sometimes gigantic, having a capture region up to about 1.5 m in diameter. It differs from other orb webs, such as that of *Larinioides*, in that the spiral thread put in place at the beginning of web construction, to serve as a "scaffold" while the remainder of the web is built, is retained in the finished web, whereas other orb-weavers get rid of it once the job has been completed. An important consequence of the presence in the *Nephila* web of both a sticky capture spiral and a non-sticky spiral is a greatly increased mechanical redundancy (Karner and Barth, unpubl.), which makes the structure less vulnerable to mechanical damage. Accordingly, such a web often has a "lifetime" of weeks.

In a spider web, as on a plant, several wave types can be expected: lateral, transverse, longitudinal and torsional waves. Longitudinal waves, along the long axis of the thread, are presumably especially important for the spider because they are transmitted with the least attenuation.

In summary. We can draw the following simple conclusions from the available literature (*Larinioides*: Masters 1984a; *Nephila*: Landolfa and Barth 1996; see also *Zygiella x-notata:* Liesenfeld 1956, 1961; Graeser 1973, Klärner and Barth 1982).

(i) Longitudinal waves are transmitted by the radials with an attenuation of only about 0.2 dB/cm (*Larinioides*) or 0.3 dB/cm (*Nephila*). These values are close to that found for bending waves in a plant. The corresponding attenuation values for transverse and lateral waves are at least 5 to 10 times as large.

(ii) The oscillations caused by insects after landing in the web contain the most energy at frequencies below 100 Hz. Spectral peaks between 5 Hz and 50 Hz are typical of insects that struggle with their legs but not by beating the wings. Additional peaks, between ca. 100 Hz and 300 Hz, are introduced by fluttering flies and bees. In contrast, the predominant frequencies of wind-induced vibrations (*Larinioides*) are considerably lower (below about 10 Hz) (see Masters 1984b).

(iii) According to a number of observations and analyses, spiders can distinguish prey animals of different sizes and masses from one another (*Nephila clavipes*: Robinson and Mirick 1971; Robinson and Robinson 1976). Furthermore, an ability to discriminate various frequencies of vibration has been demonstrated for *Cupiennius* (Hergenröder and Barth 1983) and *Dolomedes* (Bleckmann 1985a). It follows that both amplitude and frequency can be used to identify a stimulus. For further clarification of this question we need quantitative experiments in which the behavioral response to artificial stimuli is examined, so that the physical characteristics of the vibrations and the conditions under which they are introduced to the web can be meticulously controlled. This is a broad and interesting field for future research! Such work should also include the many spider webs that are not circular and take into account the many tricks spiders have evolved for detecting vibrations. That is beyond the scope of this book, but Fig. 6 should give at least an impression of the many aspects that would reward closer investigation.

(iv) As a potentially very useful cue for orientation to a vibrating stimulus source the spider could use the gradients of signal amplitude between the tarsi of its various legs. In *Nephila* these amount to 20–30 dB, sufficient to indicate stimulus direction. In contrast, differences in the times of arrival of a stimulus at the different legs are less than 1 ms (*Nephila*). It is doubtful whether such short intervals can be resolved. In the case of spiders smaller than *Nephila*, with tarsi closer together, arrival-time differences will be even less and the problem correspondingly greater.

(v) The high "mechanical redundancy" of the web of *Nephila* is necessarily accompanied by a relatively inefficient transmission of vibration. If the vibration signals had a higher priority, in comparison with the other demands placed on the web mechanics, *Nephila* could have arranged the capture spiral more loosely in the web and removed the "scaffold" spiral – which many orb-web spiders in fact do.

2 Discrimination Among Signals

Because we are not interested just in theory but want to have a practical understanding of the behavior of *Cupiennius*, we must now ask whether the types of vibration described for *Cupiennius* plants really are distinguished by the spiders, and whether what we define as the characteristics of the prey signals are actually used to identify prey.

According to a considerable number of behavioral observations, the answer to both questions is "yes".

- In experiments prey-catching behavior can be elicited significantly more often with broadband signals than with sine waves. Even a narrow frequency band covering only 1/3 octave (Q=0.35) is more effective than sinusoidal vibrations at the mean frequency of the noise (Hergenröder and Barth 1983a). The same applies to *Dolomedes triton* on the water surface (Bleckmann 1994). And the neurophysiological findings are remarkably consistent with these. Both the sensory cells of the metatarsal organ and vibration-sensitive interneurons are significantly more sensitive to these broad-band stimuli than to purely sinusoidal vibration. The difference can be as great as 20 dB (Speck-Hergenröder and Barth 1987).
- Behavioral experiments in which either the frequency of vibration is varied while the deflection amplitude is kept constant, or the reverse, show that *Cupiennius salei* is capable of discriminating differences in frequency as

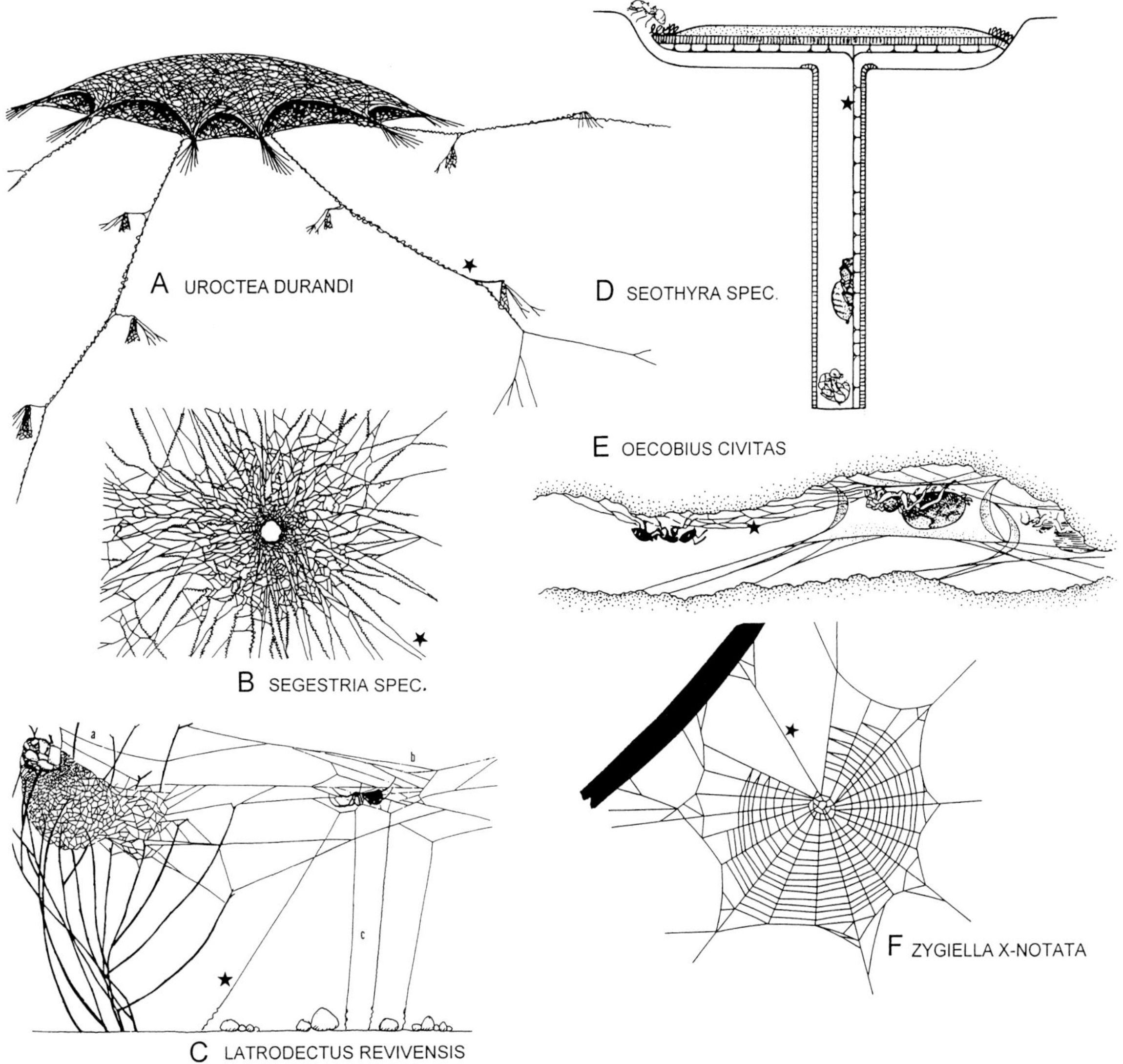

Fig. 6 A–F. Vibrations in spider webs differing in architecture. *Asterisk* Site of introduction of vibrations relevant for eliciting prey capture (see text). (Barth 1998)

well as in amplitude (Hergenröder and Barth 1983a).

- Behavioral observations in the field have provided specific evidence that the vibrations *Cupiennius salei* recognizes as prey signals are "noisy" – that is, non-sinusoidal, covering a relatively large frequency range and having an irregular temporal structure. We were very surprised when, while working near Fortin de las Flores in Mexico, we first noticed that fat potential prey sometimes passed close to a hungry spider and almost touched it, but went on its way unharmed (Barth, Bleckmann et al. 1988). The solution to the puzzle: insects thus spared produce vibrations of low amplitude and, in particular, without high-

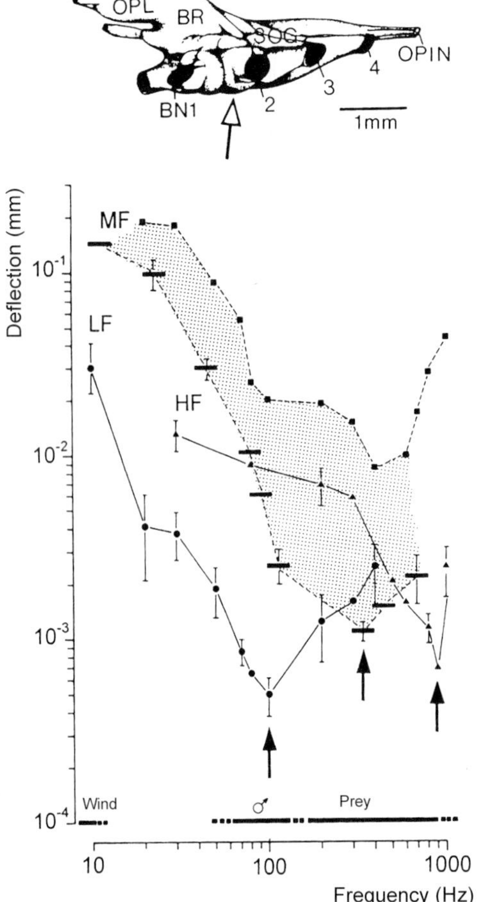

Fig. 7. Electrophysiologically determined threshold curves of vibration-sensitive interneurons in the second-leg neuromere (*BN2*). Stimulation by sinusoidal displacement of tarsus. *Arrows* point to best frequencies of low- (*LF*), middle- (*MF*), and high-frequency (*HF*) interneurons. *Shaded area* refers to MF-neuron which was stimulated by both purely sinusoidal stimuli (*squares*) and small frequency bands (*horizontal bars*; 1/3 octave); the latter stimulus is associated with a remarkable drop of the threshold curve. *Wind*, ♂, *Prey* indicate the frequency ranges typical of these vibrations; *BR* brain; *OPIN* opisthosomal nerves; *OPL* optic lobe; *SOG* subesophageal ganglionic mass. (Barth 1998)

frequency components. Their locomotion is very slow; they walk "sinusoidally", avoiding all rapid transients. Something quite similar can be observed in the case of the "klepto-parasitic" insects and spiders that sneak into the webs of other spiders, for instance *Nephila*, and steal prey previously caught and immobilized by the owner of the web (Vollrath 1979 a, b; Barth 1982). On the other hand, it is likely that the low frequencies and the absence of transients are the very properties that enable *Dolomedes* to detect "noisy" prey signals with deflection amplitudes of only a few µm on the water surface, even when they are superimposed on wind-induced waves several cm high.

- Finally, here is an argument from neurophysiology. The shape of the threshold curves of the metatarsal-organ slits (see Chapter VIII) suggests that the organ responds to the amplitude (as opposed to velocity or acceleration) of tarsal deflection at low frequencies and has very low sensitivity up to about 10–30 Hz (Barth and Geethabali 1982). It follows that low-frequency background vibrations are unlikely to be detected by the sensory system. The vibration-sensitive interneurons that have been investigated behave quite differently from the receptor cells: their threshold curves reveal band-pass rather than high-pass characteristics. And at least some of these interneurons have best frequencies in the range represented by the prey signals (or the courtship signals), while being unresponsive to low frequencies (Fig. 7) (Barth 1986, 1997; Speck-Hergenröder and Barth 1987). Everything we know so far implies that we are dealing here with a neuronal two-stage filter: its first components, the primary sensory cells, block the low frequencies while the second components, the vibration-sensitive interneurons, "pick out" the biologically important signals from the mixture of frequencies in a stimulus.

The main difference between courtship and prey signals lies in their temporal structure. We have synthesized male courtship signals, which are essential for vibratory species isolation, and systematically varied their individual parameters in order to work out the significance of each of them in identifying the conspecific reproductive partner. It turned out that some are much more important than others, as will be detailed in Chapter XX on courtship behavior.

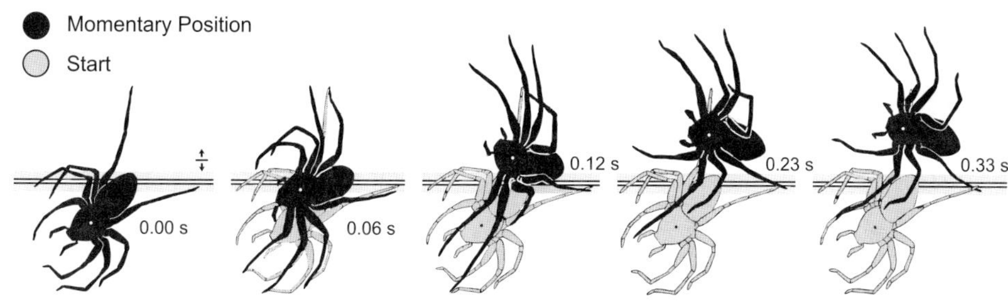

Fig. 8. Approach reaction of *Cupiennius salei* following vibratory stimulation of the fourth right leg. The vibrating platform is *above the shaded bar* in the drawing. Starting position indicated by *light spider*. (Hergenröder and Barth 1983a)

3 Orientation to a Stimulus Source

At first glance it seems trivial: once the presence of an intruder has been established by sensitive vibration detectors, and the intruder has been correctly identified as prey, the next step is to spring into action and seize the victim. This is logical enough, but how does the spider know where to jump – that is, where the vibrations are coming from? The same question applies to the courtship signals: where is the partner? When a hungry spider receives a vibratory signal representing prey, it turns through the correct angle so that it is facing the stimulus source. Evidently spiders can determine this angle regardless of whether they are looking for prey on a plant, in a web or even on the surface of water (Klärner and Barth 1982; Hergenröder and Barth 1983b; Bleckmann and Barth 1984).

The *Cupiennius* Plants

In our experiments to clarify the mechanisms *Cupiennius* may use for *angular orientation* we used a special kind of arena. It consisted of two plates, each measuring 5×10 cm, one of which was mechanically fixed while the other was movable and coupled to a vibrator so that it could be triggered electronically to vibrate, with controlled oscillations, whenever required. As long as the spider was standing so that one or more of its legs was in contact with the movable plate, when the plate vibrated in a "prey-like" manner, the spider turned toward it (Fig. 8). The angle through which it turned was related in a specific way to the particular combination of legs that were stimulated. That is, *Cupiennius* can infer the stimulus angle solely from the combination of simultaneously and equally strongly stimulated legs. With an arena in which each of the two halves was coupled to a vibrator of its own, we were able to study the influence on prey localization of the amplitude (plate-displacement) and time gradients between the stimuli arriving at the various legs. *Cupiennius* can also make use of these (Hergenröder and Barth 1983b).

Pattern of Legs Stimulated. First let us consider how the information about stimulus angle might be extracted from a combination of stimulated legs. By systematically varying this combination we were able to set up a diagram for the interactions of the inputs of the eight legs in the central nervous system that describes all our results correctly, predicting the spider's turning angle for all possible combinations. The picture of the model (Fig. 9) may look confusing at first, but it contains only a few important elements. (i) The behavioral experiment shows that the angular error of the turn (Fig. 10) is smaller for stimulation of the forelegs than for stimulation of the hindlegs. In the model, therefore, the stimulus angle (which equals the angle α between a leg and the long axis of the body in the resting position) is weighted by a leg-specific factor F (for the first leg it is 1.2; 2nd leg, 0.6; 3rd and 4th legs, 0.4). (ii) Legs on the same side of the body are connected by additive ipsilateral inhibition (I). This is directed from front to back and re-

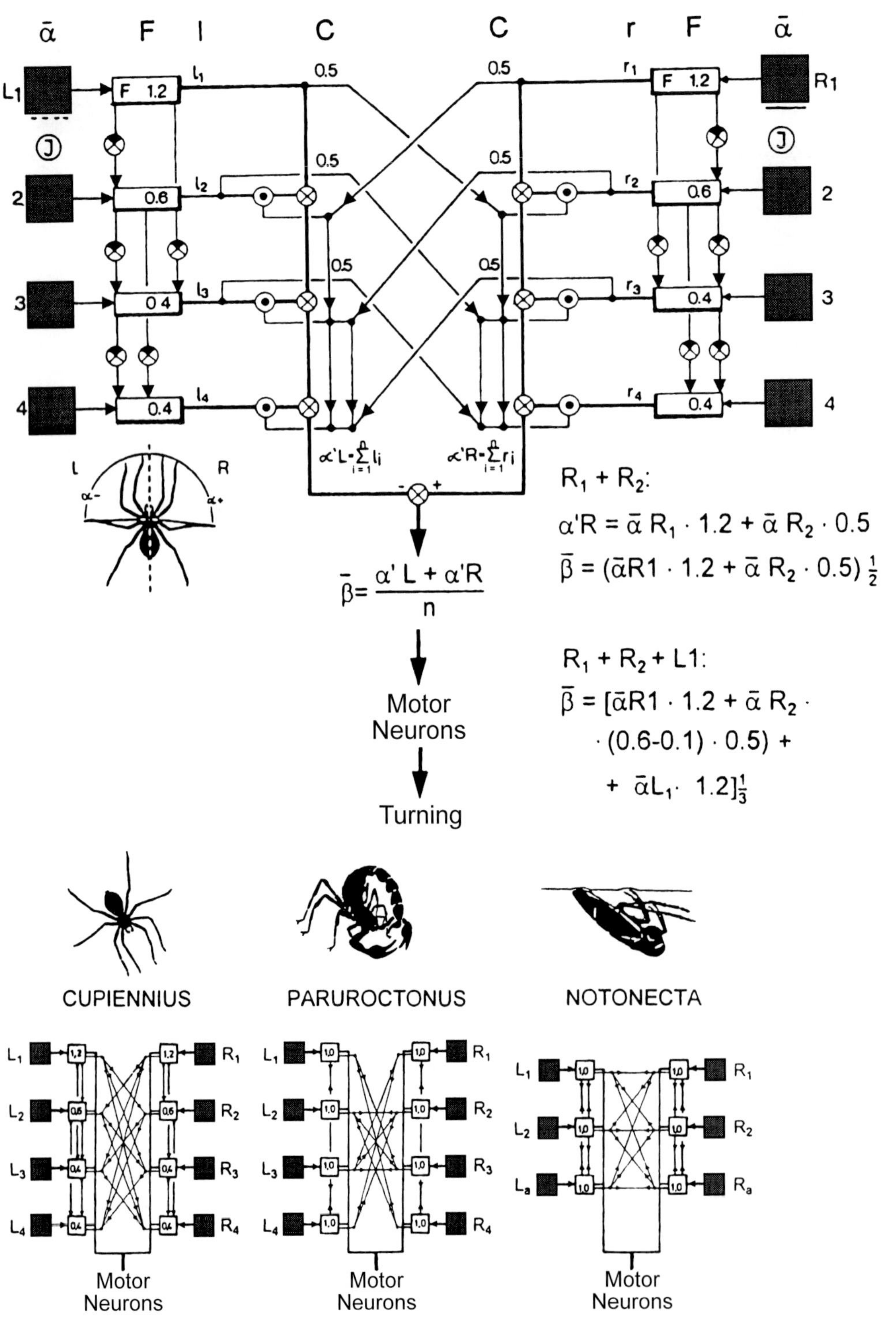

ᾱ F l C C r F ᾱ
L1 R1
2 2
3 3
4 4
F 1.2
0.6
0.4
0.5
$\alpha' L = \sum_{i=1}^{n} l_i$
$\alpha' R = \sum_{i=1}^{n} r_i$
$\bar{\beta} = \frac{\alpha' L + \alpha' R}{n}$
Motor Neurons
Turning
$R_1 + R_2$:
$\alpha' R = \bar{\alpha} R_1 \cdot 1.2 + \bar{\alpha} R_2 \cdot 0.5$
$\bar{\beta} = (\bar{\alpha} R1 \cdot 1.2 + \bar{\alpha} R_2 \cdot 0.5) \frac{1}{2}$
$R_1 + R_2 + L1$:
$\bar{\beta} = [\bar{\alpha} R1 \cdot 1.2 + \bar{\alpha} R_2 \cdot (0.6-0.1) \cdot 0.5) + \bar{\alpha} L_1 \cdot 1.2] \frac{1}{3}$
CUPIENNIUS
PARUROCTONUS
NOTONECTA
Motor Neurons
Motor Neurons
Motor Neurons

duces the weighting factor of each leg by 0.1. Stimulation of the first leg reduces the weighting factor of all the other ipsilateral legs. (iii) Legs on opposite sides of the body are connected by multiplicative contralateral inhibition (C). Thus the anterior legs act to inhibit the sensory inputs of those further back, on both the ipsilateral and the contralateral side. Quantitatively, the contralateral inhibition can be described as multiplication of the weighting factor F by C = 0.5. We were glad to learn that there are indeed anatomical ipsi- and contralateral connections between sensory neurons in the central nervous system, although no far-reaching conclusions can be drawn from that fact (Babu et al. 1985; Babu and Barth 1989; Anton and Barth 1993).

What happens when all eight legs are stimulated simultaneously and in the same way? The model predicts correctly: the spider will move forward. How does a spider behave when a leg that has lost its neighbor (an entirely "natural" situation) is stimulated? Again observation agrees with prediction. However, the model needs to be corrected for the case in which only one of the two first legs is stimulated. Then the model says the spider should turn too far, which it does not do. The correction: in this case F for the first leg is not 1.2 but rather 0.6. Biologically this seems reasonable; because the prey of *Cupiennius* can run away rapidly, the jump to catch it must be very quick. *Cupiennius* turns just far enough to seize the prey with its forelegs (Melchers 1967). This interpretation is supported by a comparison with web-building spiders and with *Dolomedes triton*, which hunts on the water surface; in these forms, the stimulus angle and turning angle are nearly the same (*Zygiella x-notata, Nephila clavipes:* Klärner and Barth 1982; *Dolomedes triton*: Bleckmann 1988). This again seems sensible to the biologist: here the prey is stuck either to the web or to the water surface, and in either case it cannot easily escape.

◄

Fig. 9. Diagram, inferred from behavioral experiments, of the central nervous interaction among the sensory inputs from vibration receptors of the eight legs by which the turning angle in response to substrate vibration is determined. See text for details. *Top* Relationships found for *Cupiennius* (Hergenröder and Barth 1983a); *bottom* comparison with a scorpion (after Brownell and Farley 1979) and a backswimmer, which uses its first and second pair of legs and vibration-sensitive hairs on its abdomen to localize a source of vibrations. (After Murphey 1973)

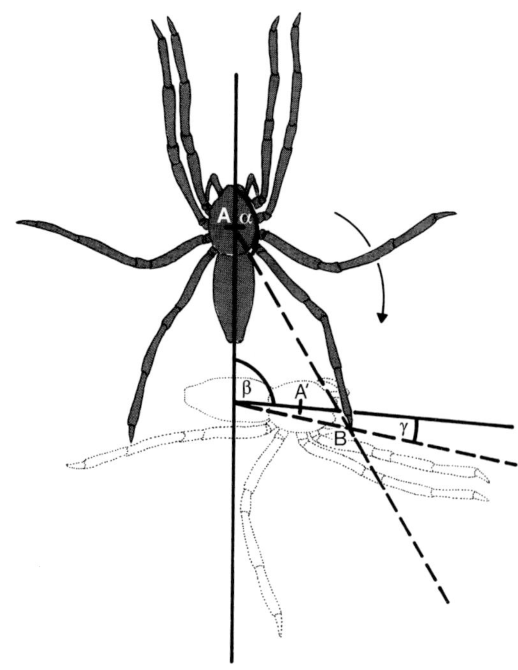

Fig. 10. Turning movement of *Cupiennius salei* elicited by stimulation of tarsus of right leg 4 (*B*) and the parameters measured to quantify it. *A, A'* Center of prosoma (dorsal view), α stimulus angle; *B* site of stimulation; β turning angle; γ error angle. (Hergenröder and Barth 1983b)

The model of the central nervous connectivity of the vibratory inputs from the eight legs of *Cupiennius* seems particularly attractive when compared with similar, though only qualitatively determined postulates for a scorpion (*Paruroctonus*, Brownell and Farley 1979) and the backswimmer (*Notonecta*, Murphey 1973). In all three cases both ipsi- and contralateral inhibition are required. However, the differential weighting of the sensory inputs from the various legs and the unidirectional (front to back) inhibition seem to be a specialty of *Cupiennius*. More details can be found in Hergenröder and Barth (1983b).

The central message of the platform experiments is that angular orientation can be achieved entirely on the basis of a specific connectivity of the sensory inputs from the eight legs, with no reference to differences in amplitude or time of arrival of the stimuli at the various legs. However, these differences are most probably present under natural conditions. Does *Cupiennius* make use of them?

Differences in Amplitude and Time. The arena with two movable plates gives the answer. In these experiments different legs are stimulated either at the same time with vibrations of different amplitude (Δd) or one after another with equal-intensity vibrations (producing time-of-arrival differences Δt). Values of 4 ms for Δt and 10 dB (smaller differences not tested!) for Δd have a distinct effect: *Cupiennius* turns as though the only leg stimulated had been the one stimulated first or more strongly (Hergenröder and Barth 1983b). Even with a time difference of only 2 ms most spiders turn toward the leg stimulated first (Wirth 1984). In the central nervous system (subesophageal ganglia) we found vibration-sensitive interneurons that receive sensory inputs from several legs and respond to Δd and Δt just as the whole spider does in its behavior (Speck-Hergenröder 1984; Speck-Hergenröder and Barth 1987).

What happens under natural conditions on the plant? Δt values are frequency-dependent, because signal propagation is dispersive. Consider the female courtship signal, which shows the male the way to a partner (see Chapter XX) and in which the main frequency component is at about 30 Hz; in this case the largest Δt values are about 10 ms, whereas the value for the two forelegs is 2 to 3 ms. For the male courtship signal (ca. 80 Hz) these values are reduced to about half, but are still in the behaviorally effective range. Prey signals, with still higher frequencies, give Δt values below 1 ms. For young and hence small spiders, in which the leg tarsi are considerably closer together than the 10 cm leg span of an adult *Cupiennius*, Δt values can also be this small. In the behavioral experiments such slight time differences elicited no responses. However, the scorpion *Paruroctonus* still responds with a Δt of 0.3 ms (Brownell and Farley 1979).

And Δd? The values found to be effective in the two-part-platform experiments are also, according to our measurements of the damping of vibration during propagation, to be expected on *Cupiennius* plants. However, the propagation of vibration is relatively heterogeneous even in the monocotyledonous plants preferred by *Cupiennius*, although these are quite homogeneous mechanically in comparison with many dicotyledons; therefore it would probably not be easy for the spiders to use Δd, or Δt, as directional indicators. In agaves, for example, the propagation velocity is higher in the basal region and along the middle of the leaf than it is in the distal region and along the edges. And the amplitude of the vibration often does not decrease uniformly with increasing distance from the stimulus source. Instead, the spider has to cope with a frequency-dependent spatial pattern of stimulus intensities (Wirth 1984). We undoubtedly need another round of experiments in order to learn more about how the spider handles Δt and Δd. Theoretically it has available not only time and amplitude differences but also phase differences and shifts in the frequency composition of the signal (on account of its dispersive propagation), which does not simplify matters.

Web and Water Surface

What can be said about angular orientation in web spiders, and in semiaquatic spiders? For comparison, these will now be considered briefly.

Zygiella x-notata, called in German the *Sektorspinne* (sector spider), and also *Nephila clavipes* turn their bodies when a prey vibration reaches the hub of their orb web; they aim themselves very precisely toward the stimulus source, with error angles (stimulus angle minus turning angle) of only 3.6° (±7.7°) and 7° (±8.2°), respectively (Klärner and Barth 1982). The spiders rotate with hardly any translational movement out of the resting position and then run straight to their target. Web vibrations not only spread along the radial threads, which are

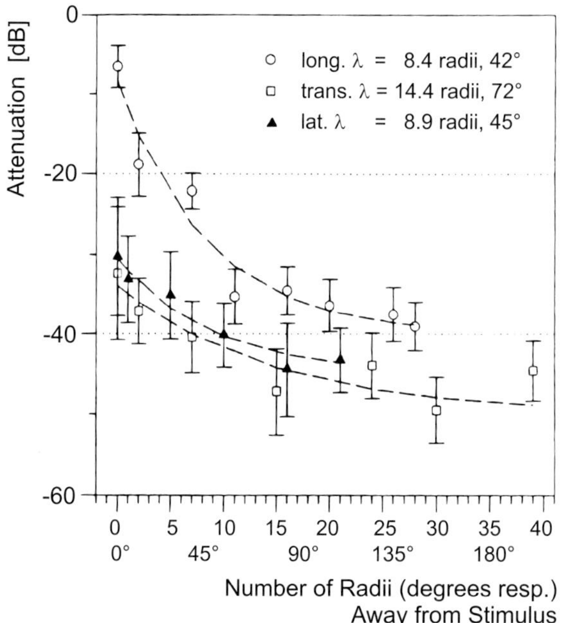

Fig. 11. Vibration transmission in the orb web of *Nephila clavipes* to neighboring radial threads. Attenuation values (dB) for longitudinal (○), transverse (□), and lateral (▲) vibrations. λ Space constant given in degrees and number of radii away from stimulated radius over which the levels of vibration fell to 1/e or 37% of original values. (Landolfa and Barth 1996)

initially set into oscillation, but also proceed from the radials throughout the web; therefore in a web, too, time and amplitude gradients arise that could show the spider the direction of the stimulus source (Masters 1984; Landolfa and Barth 1996). The amplitude gradients are largest for longitudinal oscillations, and about equal for transverse and lateral oscillations. A quantitative measure of this is the space constant (Fig. 11) (Landolfa and Barth 1996). When a *Nephila* is receiving longitudinal vibrations induced in a radius, the amplitude differences directly at the tarsi are 20 dB or more. Accordingly, amplitude differences are particularly good candidates as direction indicators for the web spiders as well. The metatarsal vibration sense organs can certainly resolve such gradients.

On the other hand, it is uncertain whether adequate time differences are available in the spider web; the question remains open. In comparison with a plant and a water surface, the propagation velocity of vibrations in an orb web is very high. Our measurements of the speed of a rectangular pulse in the *Nephila* web gave $986 \pm 390\ \mathrm{ms}^{-1}$ for longitudinal introduction, $129 \pm 7\ \mathrm{ms}^{-1}$ for transverse and $207 \pm 56\ \mathrm{ms}^{-1}$ for lateral (Landolfa and Barth 1996). Assuming that *Nephila* has an intertarsal separation of up to 10 cm, the time-of-arrival differences found for the three forms of stimulation would be 0.1 ms, 0.78 ms and 0.48 ms. It seems questionable, although not impossible, that *Nephila* would be able to resolve such small Δt. For young, small specimens of *Nephila* or for *Zygiella*, which is very small even as an adult, the Δt values are correspondingly lower and the resolution problem is therefore even greater.

On the *water surface*, in contrast to an orb web, time differences are potentially very good indicators of the direction of the stimulus source. The reason: propagation of water surface waves is relatively very slow. That *Dolomedes* actually does use Δt in this way has already been demonstrated by Bleckmann et al. (1994). The same applies to *Notonecta,* the backswimmer mentioned above, which is a typical inhabitant of water surfaces (Murphey 1973; Wiese 1974).

Amplitude differences are also large enough on the water surface to be used for orientation. However, according to Bleckmann et al. (1994) they play no role here, or at most a subordinate one.

Distance

In order to find a stimulus source an animal must determine both its distance and its direction. It must be admitted that we know very little about orientation to distance in spiders, including *Cupiennius.* Let that be an incentive for the next generation of sensory biologists to advance into this blank spot on our spider map! The problem is likely to be smallest for orb-web spiders. After they have turned to face the source of a disturbance, they need only run along the radius as though it were a rail and they are certain eventually to reach the trapped prey. On the plants of *Cupiennius* there is no such "track" to follow, and still less on the water surface. It may be that the frequency spectrum of the prey signal (or of a courtship or other signal) can be used to estimate distance. Because of the dispersive propagation, the proportion of high frequencies is greater near the stimulus source than further away. Surface-feeding fishes can make use of this difference (Bleckmann 1988). Whether *Cupiennius* or any other hunting spider also has and exploits this capability is at present uncertain and not experimentally documented.

4 Stimulation from Below and from Above

Even though one sensory channel may predominate in triggering and controlling a particular form of behavior, other sensory systems are usually also involved. And prey capture in *Cupiennius* is no exception. Substrate vibrations are quite sufficient to release this behavior. Experiments in which the trichobothria were inactivated (so that no air movements could be sensed) have put that beyond doubt. However: we shall see in the following section and in Chapter XIX that prey capture can also be induced by way of the air. A substantial part of the complication mentioned at the beginning of this chapter becomes evident when we try to understand the interaction of information from the two sensory systems. Our experiments on this question were designed from the viewpoint of substrate vibration; that is, how do air-movement stimuli affect the response to vibrations of the substrate (Hergenröder and Barth 1983 a)? The first surprise was that the trichobothria were activated even when we thought we were stimulating only by way of the substrate, as we discovered by comparing the threshold vibrations, the frequencies of the various behavioral responses, and the response times before and after elimination of the trichobothria.

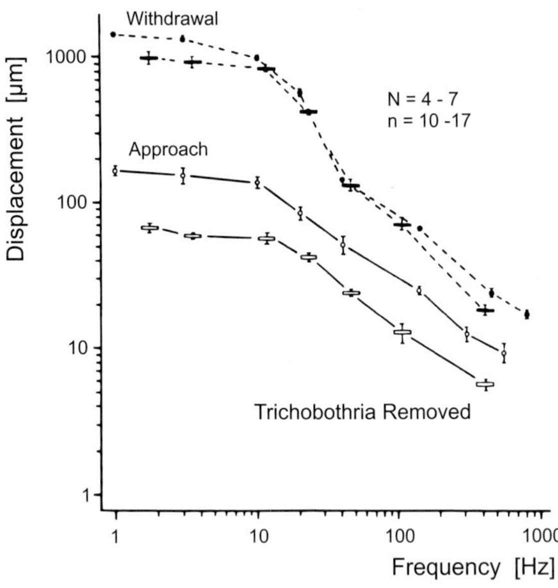

Fig. 12. Behavioral thresholds for approach and withdrawal reaction in response to vibrational stimulus (*circles* sinusoidal stimulus; *bars* band-limited noise) in animals with their trichobothria removed. (Hergenröder and Barth 1983a)

Thresholds

A reminder: *Cupiennius* turns either toward a stimulus (as the first phase of prey capture) or away from it, depending on the (displacement) amplitude of the substrate vibration (Fig. 12). The thresholds for turning away are distinctly higher than those for turning toward the stimulus. After elimination of the trichobothria the thresholds for the negative (away) response rise by about 10 dB, whereas those for the positive response do not change significantly.

Response Probability

Although the threshold curves for the positive response remain the same, removal of the trichobothria has a considerable and significant influence on the probability of a response to suprathreshold vibration stimuli. Figure 13 shows what happens when the frequency band is shifted while the deflection amplitude is kept constant at 140 μm. Differences in the outcome of this experiment at the different stimulus frequencies are closely correlated with the position of the 140 μm stimulus

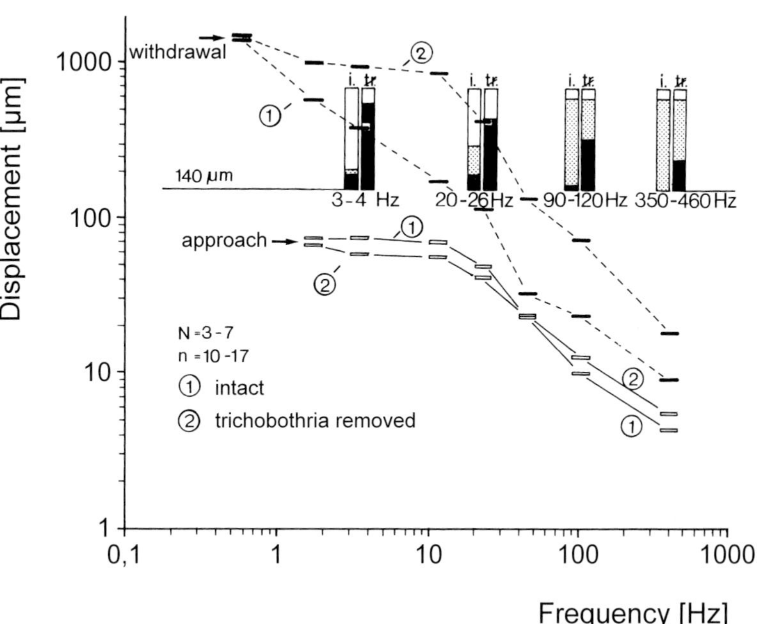

Fig. 13. Behavioral thresholds for approach (*continuous line*) and withdrawal reaction (*broken line*) to stimulation with band-limited noise. *1* Intact spiders; *2* spiders with trichobothria removed. Histograms show the frequency (in percent of all reactions, corresponding to 100% or entire column) for approach (*black areas*), withdrawal (*shaded area*), and no reaction (*white areas*) for intact animals (*i*) and animals without trichobothria (*tr.*) (stimulus displacement amplitude constant at 140 µm. SD within symbol size. (Hergenröder and Barth 1983 a)

amplitude relative to the threshold curve. As the frequency increases, this 140 µm value shifts progressively further above the threshold for the negative response; therefore the ratio of negative to positive responses rises at higher stimulus frequencies. Spiders without trichobothria always give a greater proportion of positive responses than do intact spiders. An example (Fig. 13): at 190–120 Hz and 350–460 Hz stimulus frequency intact animals almost always turn away, because at these frequencies the selected stimulus amplitude of 140 µm is above the threshold for this response by 15.5 dB and 24 dB, respectively. Elimination of the trichobothria significantly reduces the frequency of the negative response (more for stimulation with 90–120 Hz than with 350–460 Hz; in the second case the stimulus amplitude of 140 µm is very far above the associated threshold value for inducing the negative response).

Response Time

Whereas removal of the trichobothria does not significantly alter the response time for turning toward the stimulus, that for turning away is significantly increased. Both this result and the higher threshold of intact animals for the negative response indicate a close relationship between the presence of trichobothria and the negative response in particular.

All these behavioral experiments already imply that the trichobothria are deflected by substrate vibration, and this has proved to be the case. Because the trichobothrium is deflected proximally when the tarsus moves upward (and not distally, as we had wrongly supposed because in its resting position it points distalward), the effective stimulus presumably resides in the airborne sound (near-field) emitted by the vibrating substrate. When the tarsus is vibrated according to the threshold curve for turning toward a stimulus (prey capture), then a 400-µm-long tarsal trichobothrium is deflected by about 1°; the corresponding value for the threshold of the turning-away behavior is distinctly higher, 5–6°. In both cases the deflections are suprathreshold for the trichobothrium, as can be seen in Figure 14.

Back to the real question: how do the effects of stimuli from above (in the air) and below (from the substrate) interact? As the next chapter (XIX) will describe, *Cupiennius* can also be induced to perform prey-catching behavior by

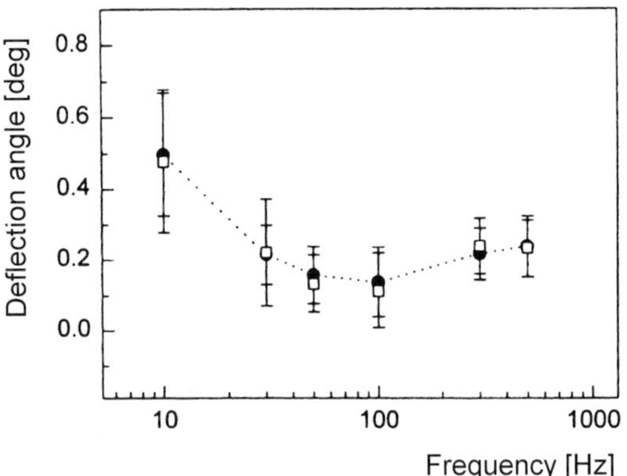

Fig. 14. Electrophysiologically determined threshold curve of a trichobothrium of metatarsal group D1 of *Cupiennius salei*. (Barth and Höller 1999)

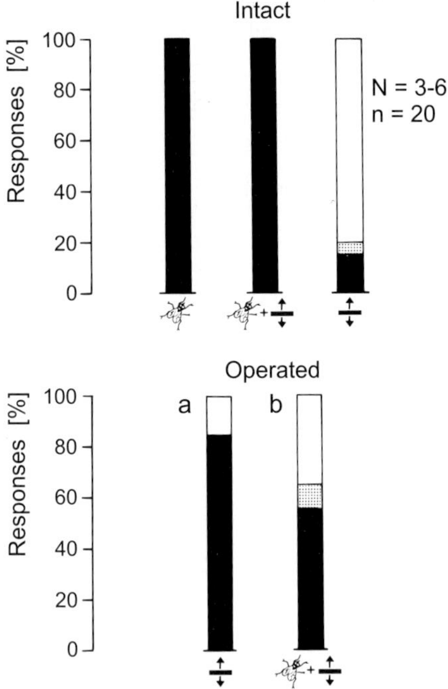

Fig. 15. Occurrence (in percent) of approach (*black areas*), withdrawal (*shaded areas*), and no reaction (*white areas*) in intact spiders (*top*) and spiders with various organs ablated (*bottom*). Stimulation with substrate vibrations (3–4 Hz, 140 μm) and/or with the airflow produced by a tethered flying fly. *a* All legs and pedipalps without trichobothria; *b* 3rd and 4th leg intact, all other legs without trichobothria. (Hergenröder and Barth 1983 a)

the air currents generated by a flying fly. How does such an aerodynamic stimulus influence the response to substrate vibration?

When a buzzing fly is held 2 cm away from a tarsus of *Cupiennius*, the spider willingly turns toward this stimulus if it is hungry. After removal of the trichobothria this response to stimuli "from above" is completely eliminated. If an intact animal is stimulated only by vibration "from below" (vibrating platform), the simultaneous stimulation of the trichobothria inhibits the release of a prey-capture or positive turning response (Hergenröder and Barth 1983 a). However, if there is an extra stimulation of the trichobothria by a real prey stimulus from above (buzzing fly), the response to this stimulus is additively combined with the turning angle induced by the stimulation from below. This effect can be illustrated by two experimental situations, in both of which all trichobothria were removed except for those on the right legs 3 and 4: (i) fly between leg 3 and leg 4, vibratory stimulation of legs 3 and 4; the spider turns more strongly than without the fly stimulus; (ii) fly as previously, but now vibratory stimulation of the forelegs as well (for instance, R_1, R_2 and R_3); the turning angle is smaller than in the situation without the fly stimulus. Before we summarize our observations in a circuit diagram for the interaction between trichobothria and substrate-vibration receptors – which should be regarded as a working hypothesis – I must mention one more experimental result, which at first was very surprising. Figure 15 shows a graph of this result.

- First column: intact animals, fly stimulus, 100% responses (positive turning)
- Second column: intact animals, fly stimulus plus vibration stimulus; 100% responses
- Third column: intact animals, vibration stimulus alone, only 15% responses to this stimulus, the same one used previously.

Ergo: this vibration stimulus is ineffective because the trichobothria are not stimulated. Wrong! It could also be that the trichobothria are being stimulated along with the leg but that this inhibits the response to the stimulus "from below". Experiments in which the trichobothria are eliminated support this latter interpretation.

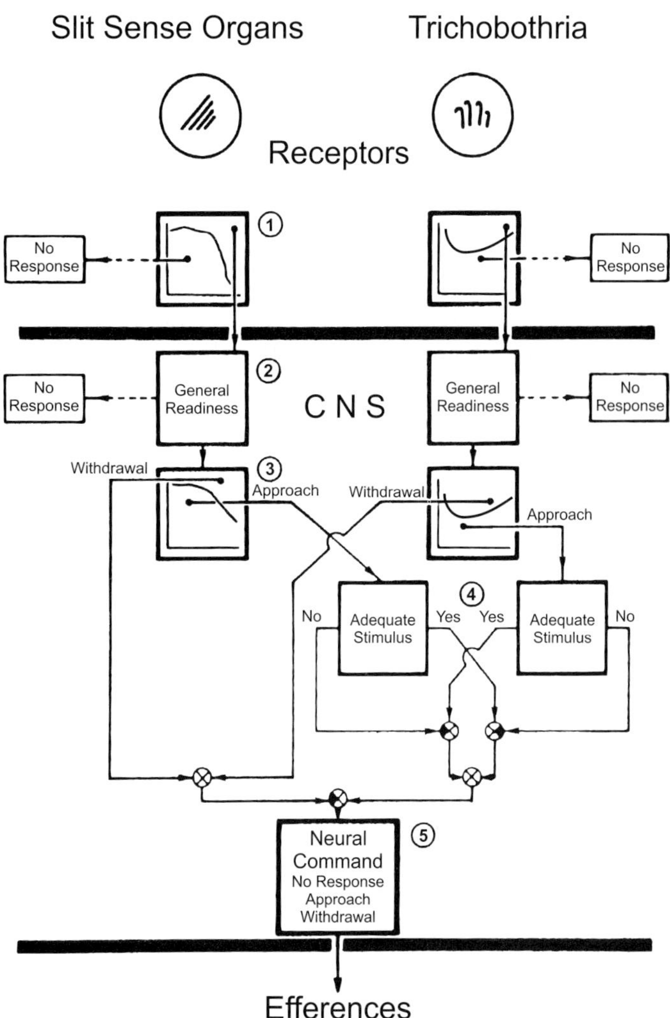

Fig. 16. Simple model of interaction of signals from slit sense organs and trichobothria in the central nervous system, mainly according to behavioral experiments. ⊗ addition, ⊗ subtraction. See text for details. (Hergenröder and Barth 1983 a)

- Fourth column (Fig. 15): all trichobothria removed, vibration stimulus as previously, 85% responses!
- Fifth column: trichobothria only partly removed, i.e. two legs (R_3, R_4) intact, fly plus vibration stimulus, 55% responses.

All these and other experiments confirm a significant and behaviorally relevant interaction of the two sensory systems in the central nervous system. Although this interaction is complex and by no means entirely understood, from what we do know we can draw some preliminary conclusions, which are represented in a simplified interaction diagram (Fig. 16). We hope this will provide an incentive for further analyses on this subject, which can have a significance that goes beyond the special case and the spiders: for animals in general, an interaction of several senses in generating particular forms of behavior is likely to be the rule rather than the exception.

We distinguish five consecutive steps, as numbered in the diagram.

- Step 1: Receptor thresholds; only suprathreshold stimuli elicit signals to the CNS.

- Step 2: Readiness to respond; degree of hunger, time of day, temperature and other factors determine an animal's momentary readiness to give a behavioral response; this central nervous threshold must be exceeded, just like the thresholds of the peripheral sensilla, if behavior (for example, prey capture) is to result.
- Step 3: Stimulus above the threshold for turning away; now if the stimulus is applied only through the substrate ("from below"), the effects of the responses of vibration receptors and trichobothria are added. Recall that the thresholds for turning away (not for turning toward a stimulus) are lower in intact animals than in those without trichobothria (except for very low frequencies of substrate vibration, ca. 0.6 Hz). Evidently the trichobothria play a role in detecting the frequency content of stimuli that elicit turning away. In contrast to intact animals, those without trichobothria do not have lower turning-away thresholds when noise bands (instead of sine waves) are used for the vibration stimulus.
- Step 4: Stimulus below the threshold for turning away; now two situations must be distinguished. a. Both trichobothria and vibration receptors signal "prey" (output "yes"); their effects are added. b. Trichobothria signal "not prey" (output "no"; for instance, in the presence of abiotic stimuli such as the wind, or when stimulated appropriately by substrate vibration, along with vibration receptors) and the stimulation is below the threshold for turning away; then the trichobothria inhibit the effect of prey signals sent out by the vibration receptors.
- Step 5: The final decision. The behavior ultimately observed in response to substrate vibration depends on the relative strengths of the stimuli "from below" and "from above". The signals to turn away sent out by the trichobothria inhibit the positive response otherwise elicited by the slit-sensillum signals, and conversely. It may be that the positive response is elicited when the prey signals from the slit sensilla dominate, whereas these signals themselves are dominated by the "turn away or not prey" signals from the trichobothria when the spider turns away. A failure of the spider to respond at all could then be interpreted as the result of equivalent inputs from trichobothria and slit sensilla.

There is a lot left for future neuroethologists to do!

XIX Signposts to the Prey: Airflow Stimuli

At the end of Chapter XVIII we discussed the complex interaction between the detection of substrate vibrations on one hand and of air movement on the other. Now I should like to consider more closely the movement of air as a stimulus in its own right, and its significance for prey capture. Remember that *Cupiennius* can be induced to perform capture behavior by air currents alone (for instance, those produced by a fly's wingbeats). There is thus biological justification for focusing in this chapter solely on the trichobothria, the sensilla that detect airflow.

1 Natural Signals

What kinds of airflow fields does a spider encounter in its habitat? Because *Cupiennius* must be able to distinguish relevant from irrelevant stimuli by means of its trichobothria, this is one of the first questions we have to ask.

In the initial study (Barth et al. 1995) we concentrated on three aspects in particular: *a.* the way air flows around a *Cupiennius* positioned in a laminar airstream in a wind tunnel; *b.* the background airflow to which the spider is exposed in its biotope, and from which it must

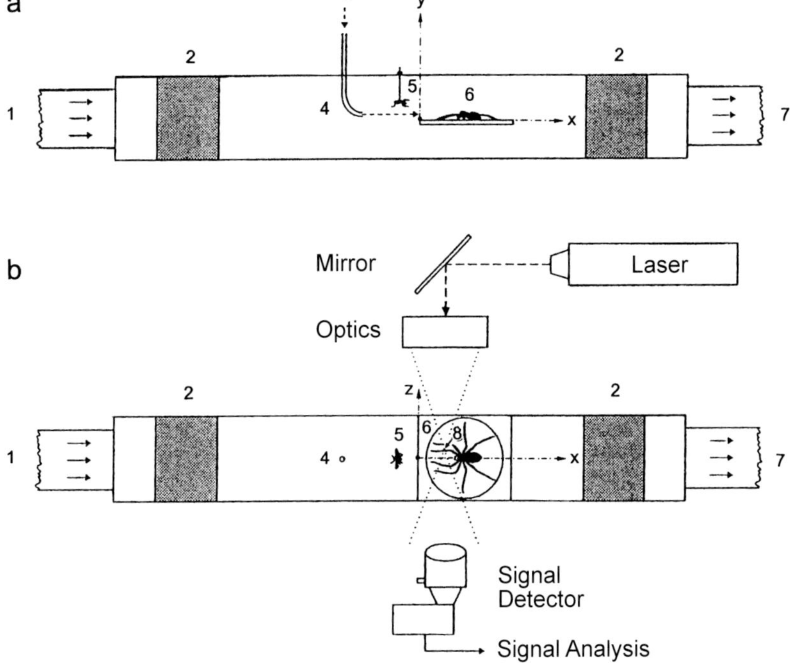

Fig. 1 a, b. Lateral (**a**) and dorsal (**b**) view of test section for the analysis of airflow above a spider in a wind tunnel of square cross-section (10×10 cm^2). *1* Entry to duct; *2* hexagonal mesh honey comb (diameter 0.5 cm, length 7.5 cm); *3* incense smoke for flow visualization; *4* smoke rake; *5* fly tethered to hypodermic syringe; *6* platform with spider; *7* exit of duct; *8* optical probe volume of laser anemometer defined by the crossing of two laser beams. (Barth et al. 1995)

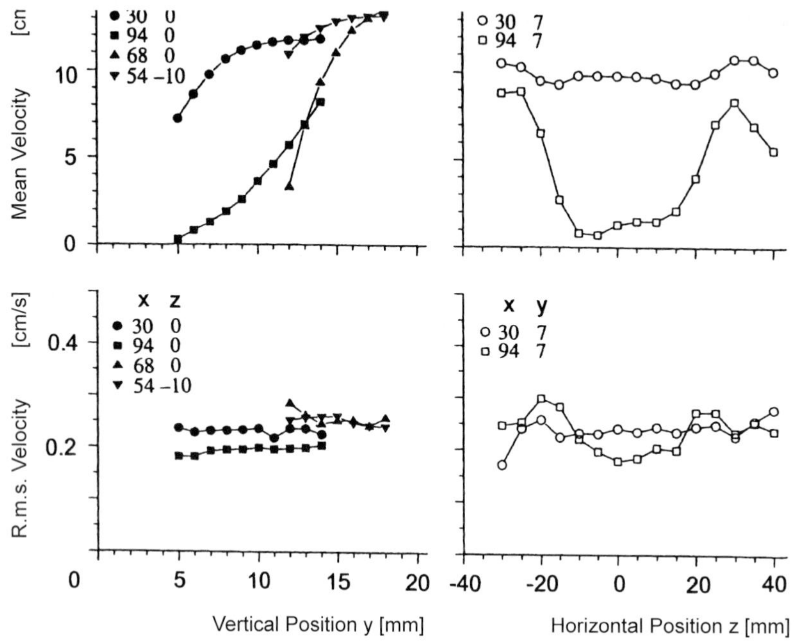

Fig. 2. Mean and r.m.s. streamwise velocity component profiles corresponding to the undisturbed airflow around an adult female of *Cupiennius salei*. $V_0 = 12$ cm/s. The leading edge of the spider platform corresponds to $x = 0$ mm; the spider prosoma begins at $x = 47$ mm; the opisthosoma ends at $x = 77$ mm; the spider is 10 mm wide at its widest point (opisthosoma) and 10 mm high at the highest point (prosoma). A horizontal position of $z = 0$ mm indicates a position of the measuring probe above the long axis of the spider. The positions of measuring profiles relative to spider are the following. *Vertical profiles* ● 17 mm in front of spider prosoma; ■ 17 mm behind opisthosoma; ▲ above opisthosoma; ▼ at level of prosoma but 10 mm away from body long axis above leg. *Horizontal profiles* ○ 17 mm in front of prosoma, and 7 mm above platform; □ 17 mm behind opisthosoma and 7 mm above platform. (Barth et al. 1995)

discriminate the biologically relevant signals; *c.* the airflow generated by a fly.

To tackle these three sub-problems, again we needed a certain amount of technology. Our apparatus included a wind tunnel, a laser Doppler anemometer, an anemometer suitable for field work and a lot of processing electronics (Fig. 1). As we were dealing with physics, it was once again important to collaborate with Pepe Humphrey, the fluid-mechanics expert in Berkeley. So what came of all this effort?

Fluid Flow Around Spider. It is immediately apparent from the *pattern of flow around the spider* in a laminar airstream (velocity in the range of the natural "airflow background") that we are dealing with a geometrically complex structure. Even under the simplest flow conditions, we must expect nonuniform distribution around the spider's body, with gradients in flow velocity. If the airstream is coming from the front, both the mean velocities and the r.m.s. (effective) values are higher above the legs than above the pro- and opisthosoma, and in the wake the velocities are appreciably reduced (Figs. 2 and 3). When the wind is not aligned with the spider's long axis, the velocity distribution becomes still more complex and, for example, causes increasingly discontinuous flow on the leeward side and a loss of the symmetry of the flow pattern. The details of this velocity distribution may be too variable to allow the spider to identify and localize a stimulus source. However, general pattern properties such as symmetry or asymmetry, the degree of

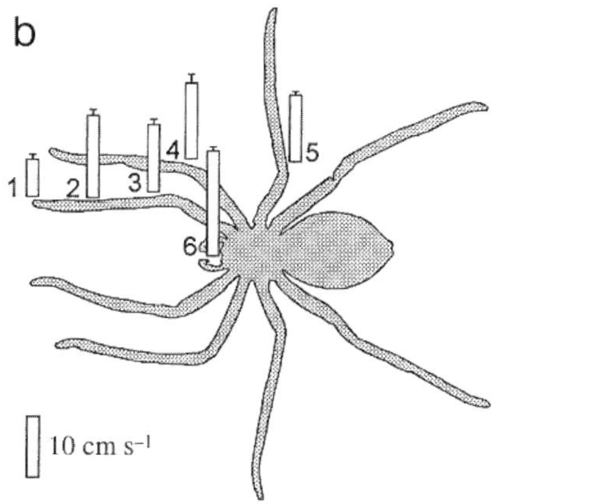

Fig. 3a,b. Undisturbed airflow around *Cupiennius*. $V_0 = 12$ cm/s, spider oriented horizontally and with long body axis parallel to airflow. **a** Note region of asymmetric recirculating flow in the wake of the opisthosoma and strong curvature of streaklines in and around the spaces defined by the spider legs. **b** Flow velocities measured immediately above various areas (1–7) of the spider body and behind it. (Barth et al. 1995)

discontinuity and the sequence in which the trichobothria are stimulated might let the spider know where the wind is coming from. The stereotyped leg postures could be helpful in this regard.

Background Flow. One of the big problems every sensory system must solve is to "filter out" the biologically important signals from a *background of noises and interfering signals.* To learn more about this background in the habitat of *Cupiennius*, we travelled to Las Cruces near San Vito Coto Brus in the far south of Costa Rica and, without disturbing the animals, made many measurements of the wind above spiders that were waiting for prey on their customary plants, in the normal nighttime activity period (Plate 12). We found that flow velocities below 0.1 m/s and r.m.s. values <15% are the most characteristic. Higher velocities, up to 0.4 m/s, were observed in only two out of 16 relatively long measurement periods. The background flow is dominated by low frequencies, <10 Hz, and is characterized by a narrow frequency spectrum (Fig. 4). Soon after sunset the wind speeds in the field and the likelihood of gusts almost always become noticeably lower. This is very probably one of the advantages that *Cupiennius* obtains by being active at night, beginning shortly after the sun has set.

The advantage becomes still more obvious when we consider a typical prey signal.

Prey Signal. Having found that we could so easily elicit prey-catching behavior of *Cupiennius* by holding a buzzing fly nearby, we decided to analyze the air currents a fly generates during stationary flight, in the hope of discovering which features are *typical of prey signals in general.* The difference from the background airflow is remarkable (Fig. 5). The fly signal is a spatially concentrated, downward-directed airstream that reaches velocities close to 1 m/s at a distance of 4 to 7.5 cm behind and below the fly; the fluctuations in local mean velocity in this airstream are of the order of 25% (to 56%), in comparison with 2–3% for the undisturbed background. The body parts of the spider can themselves introduce oscillations, even though the surrounding airflow is basically laminar, but their frequencies are only a few hertz. The fly, in contrast, produces oscillations in the region of 100 Hz. We conclude that the interference with the background airstream caused by the spider's body does not appreciably change the fly signal. That the airflow in the fly's wake is fluctuating and turbulent is not surprising; the wingbeat introduces a periodic pulsation, and vortices also break free from the edges of the wings (Maxworthy 1981). However, the proportion of high frequencies in the spectrum rapidly declines with increasing distance from the fly (Fig. 6). As in the case of substrate vibrations, therefore, the frequency content of a prey signal could be used as an indicator of the distance of the prey.

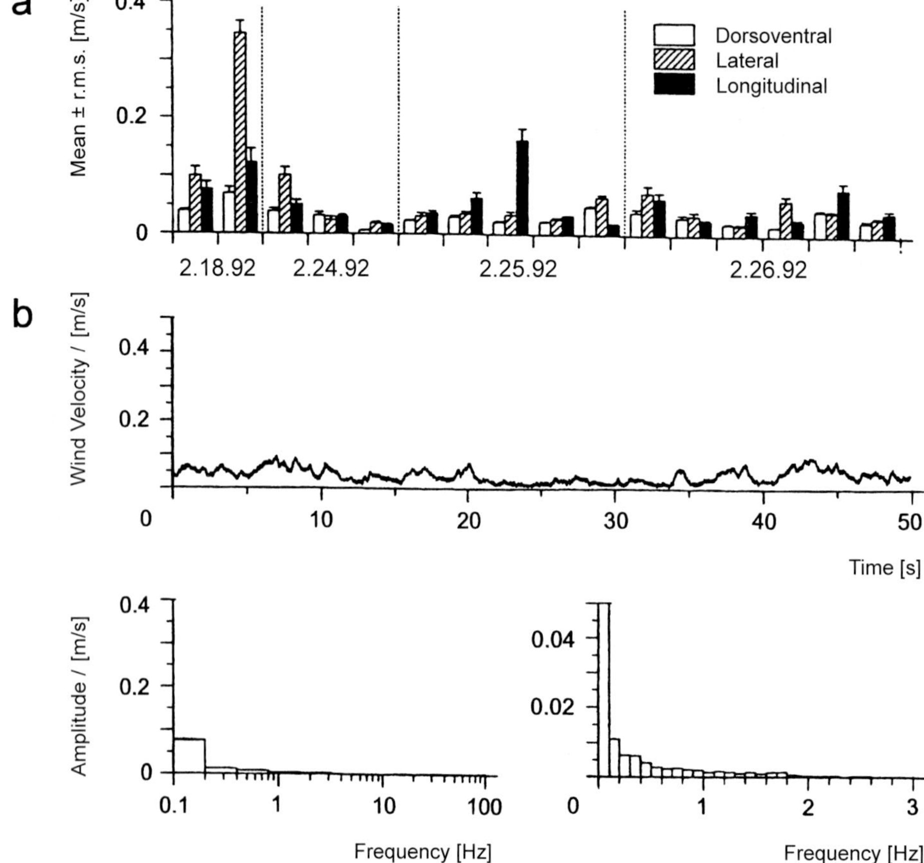

Fig. 4 a, b. Air movement in the natural habitat of *Cupiennius*. Wind velocities were measured roughly 1 cm above the prosoma in the longitudinal, lateral, and dorso-ventral direction relative to the spider. Spiders (N = 11) sat on their dwelling plants and all measurements were taken after sundown. **a** Summary of 16 measurements. **b** Typical real-time recording of air velocity for 50 s (see **a** 2.24.92, longitudinal movement); *below* spectra of the same signals shown with different resolution. (Barth et al. 1995)

For a spider like *Cupiennius*, which sits on a solid substrate such as a bromeliad leaf, the velocity of the airflow will be greater than it is in the open and presumably also greater than the values encountered in an orb web. The result is increased sensitivity, as can be seen by comparing the deflection of the trichobothria in the two situations (Fig. 7).

How dramatically the velocity profiles over a solid substrate can change, depending on whether the flow is laminar or has the complexity of a fly signal, is shown in Fig. 8. The particle velocity above the surface rises considerably, and the high r.m.s. values indicate a marked degree of fluctuation in the airstream (Barth et al. 1995).

When the airflow generated by a buzzing fly is measured at the site of the trichobothria on the various legs of *Cupiennius*, it becomes apparent that the velocities at the individual measurement sites depend on both the orientation of the spider and the velocity of the background current. The stimulus patterns to which the ensemble of trichobothria on all legs and the pedipalps is exposed in each case have directional characteristics that enable the spider to identify the direction of the stimulus source (Barth et al. 1995).

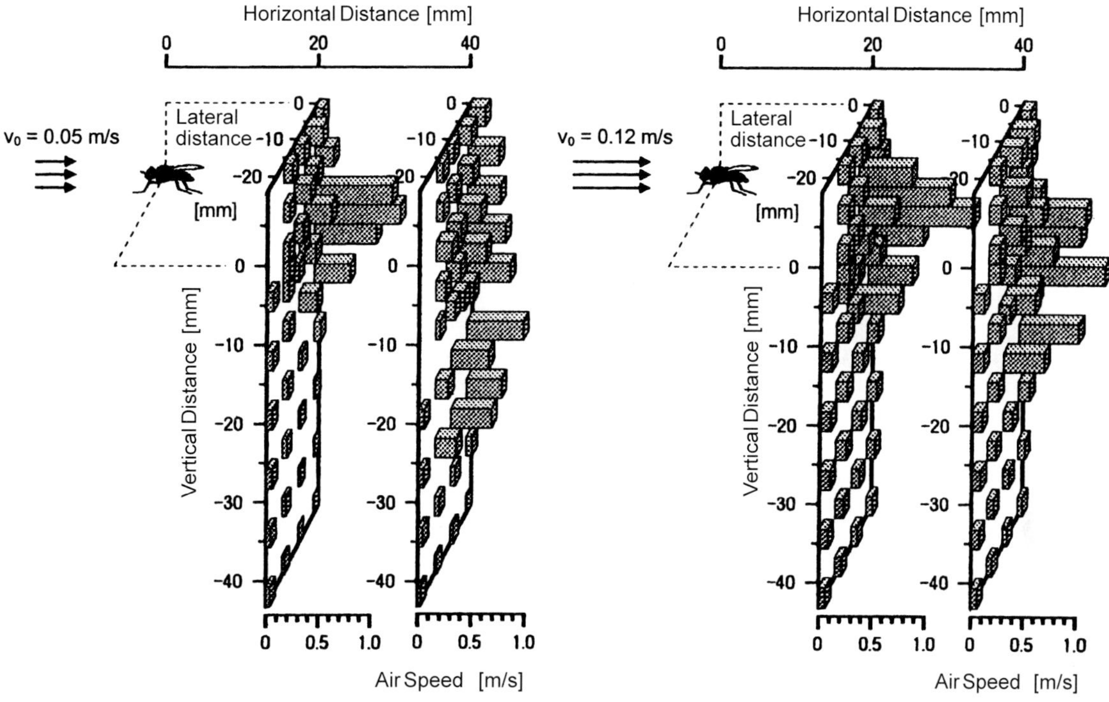

Fig. 5. Airflow produced in wind tunnel by tethered flying fly with laminar background at 0.05 m/s (*left*) and 0.12 m/s (*right*). (Barth et al. 1995)

Cupiennius snatches flying prey out of the air, but can more often be seen catching insects that are crawling over the plant and announce their presence from a considerable distance away, by the substrate vibrations they generate. As soon as these potential prey animals have come to within 1 to 3 cm of a spider leg, their locomotion generates air movements that can add to the effect of the vibrations conducted through the plant (Den Otter 1974; Reißland and Görner 1978; Hergenröder and Barth 1983b). The signal a spider receives in such a case is spatially confined and characterized by low frequencies (<20 Hz) and low particle velocities (≤2 cm/s) (Camhi et al. 1978; Gnatzy and Kämper 1990). Such a signal can be distinguished from the background wind by its pulsed nature.

All these measurements and observations of *Cupiennius* together revealed an interesting parallel between the detection of air movements and that of substrate vibrations. An effective stimulus is characterized by relatively broad

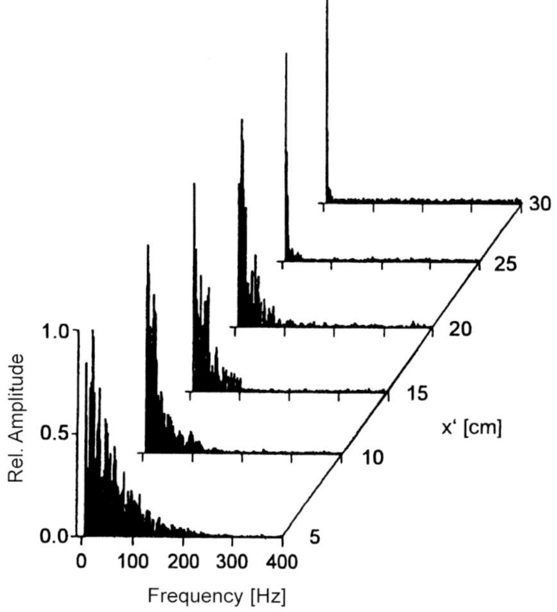

Fig. 6. Spectra of airflow behind tethered flying fly at increasing distance from fly. (Barth and Höller 1999)

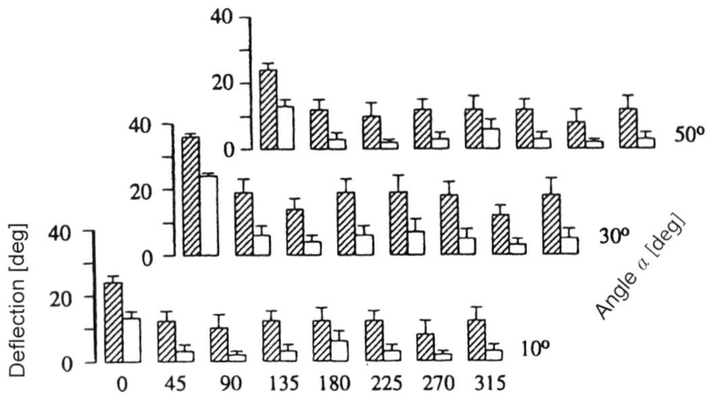

Fig. 7. Deflection (average and SD) of trichobothria on tarsus of *Cupiennius salei* by air movement due to a tethered flying fly (distance from prosoma constant at 8 cm) as a function of (1) the azimuth of the fly's position, (2) the fly's elevation above the spider given as angle α, and (3) the presence (*shaded bars*) or absence (*open bars*) of a platform under the spider. An azimuth of 0° refers to a case in which the long axis of the fly is aligned with that of the spider leg under study. N=5–6. (Barth et al. 1995)

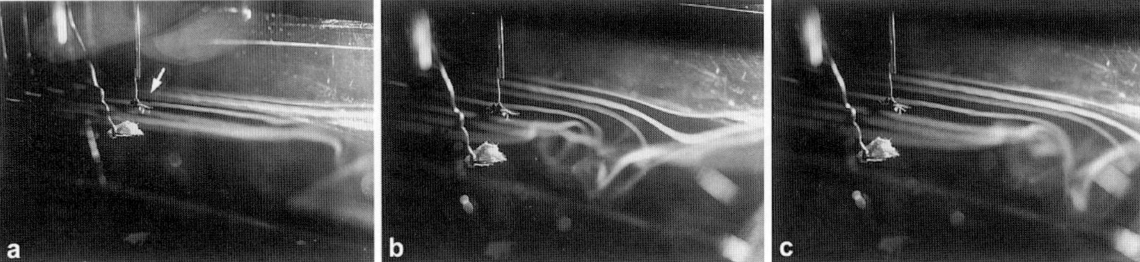

Fig. 8a–c. Visualization of airflow streaklines behind a stationarily flying fly in a wind tunnel. Laminar background flow velocity V_0=0.12 m/s; interval between photographs 0.25 s. (Barth et al. 1995)

bandwidth, by frequencies above those of the background noise, and by an irregular time course. For distances greater than about 25 cm, however, the fly signal progressively loses its characteristic properties. Frequency components above 50 Hz disappear and, interestingly, stimuli at about this distance also no longer elicit a behavioral response (see below).

2 The Response of Individual Trichobothria

General Characteristics

From the description of the natural stimulus situation in Section 1 – admittedly a rather pathetic attempt to describe something with such complex spatial and temporal dynamics – it is already clear that we really want to know what each of the nearly 1000 flow sensors is doing at each point in time when a natural stimulus is moving past the ensemble. Before this wish can be granted, considerably more analysis of the stimulus will be needed. For the present we must concentrate on smaller elements of the whole complex. How faithfully does the impulse discharge of a single trichobothrium signal the time course of the stimulus to the central nervous system?

The results of electrophysiological experiments to answer this question (Barth and Höller 1999) are grist to our mills. They support the hypothesis that important properties of natural and biologically relevant stimuli are reflected in the properties of the sensilla. This is not trivial, though it may seem so at first glance.

Figure 9 shows a typical response of a sensory cell of a trichobothrium to the air movements caused by a stationary, flying fly 5 cm away. After a first phasic peak of excitation, the physiological response clearly follows the time course of the stimulus. The response is proportional to the logarithm of both the mean flow velocity and the degree of turbulence (r.m.s.) of the current, over a broad range (from 1 mm/s to 1 m/s). That is, because of the logarithmic representation the sensor has a large working range – but only for frequencies up to about 150 Hz. As experiments with sinusoidal stimulation have shown, the responses to higher-frequency stimuli always consist of a single impulse per oscillation cycle, regardless of the stimulus intensity. Because the fly signal typically contains frequencies below 200 Hz, once again we have found a match between sender and receiver. And it should be recalled that the receptors have a distinct frequency tuning, with best frequencies between 50 and 120 Hz (see Chapter IX).

The lack of spontaneous activity and the marked phasic nature of the responses of trichobothria, which react mainly to the speed with which they are deflected, make these sensilla ideally suited to monitor the typically very inconstant stimuli encountered in the habitat. In figure 10 it can be seen that the response of the trichobothria to an irregular, noisy stimulus persists unchanged for a long time, whereas in the presence of a sinusoidal stimulus at 100 Hz the receptor responds for only a few seconds.

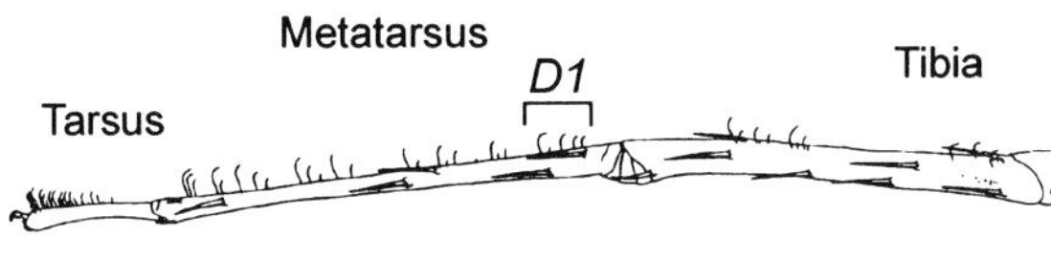

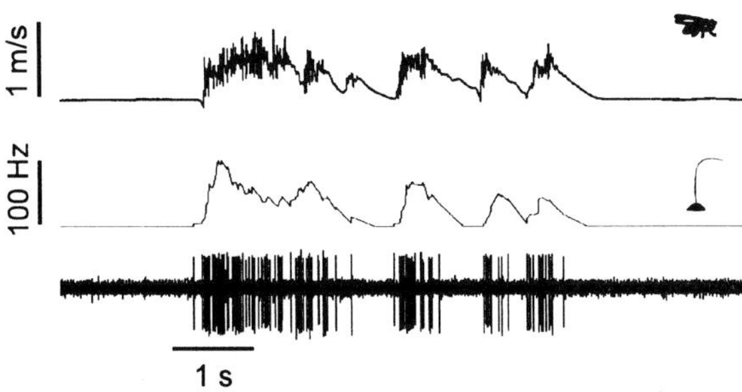

Fig. 9. Action potentials recorded from a trichobothrium of metatarsal group D1 in response to the airflow from a tethered flying fly (5 cm above substrate and with abdomen pointing toward the trichobothrium). *Above action-potential trace* the corresponding momentary frequency is given as well as the real-time values of the airflow velocity. Note similarity. (Barth and Höller 1999)

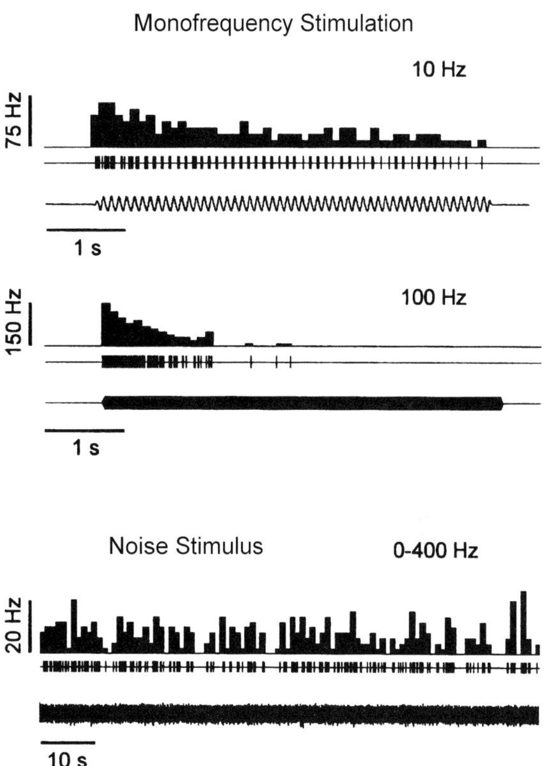

Fig. 10. Action-potential response of trichobothrium (metatarsal group D1) directly coupled to a vibrator to single-frequency stimulation at 10 and 100 Hz and to a noisy signal containing frequencies between 0 and 400 Hz. Note continuing response to noisy stimulus even after 90 s. (Barth and Höller 1999)

The Detection Range of the Sensors

Now that we know about some important physical properties of the stimulus, which can be taken as an example of many similar stimuli, we can also investigate the distance over which

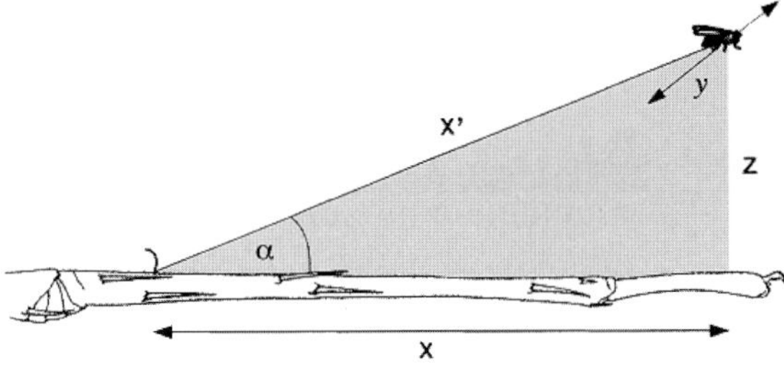

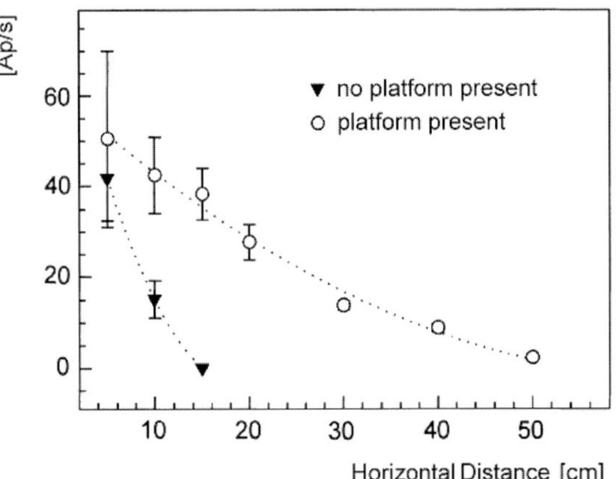

Fig. 11. The action-potential response of a trichobothrium (metatarsal group D1) to airflow due to tethered flying fly; note strong influence of presence or absence of platform beneath the spider leg on the range of the trichobothrium. *AP/s* Action potentials per s. Vertical distance $z = 5$ cm, horizontal distance x variable between 5 and 50 cm. (Barth and Höller 1999)

such stimuli can be detected by the trichobothria. Later we shall use a behavioral test for this purpose, but first we are concerned only with single hairs.

By measuring the distance of a stimulus at which it still causes a suprathreshold deflection of a trichobothrium, we obtain ranges of 50 to 70 cm (Fig. 11). These values agree closely with the results of electrophysiological experiments, in which the impulse responses of individual trichobothria to a buzzing fly at increasing distances were recorded (Barth and Höller 1999). The greatest distance from which a fly elicited such a response was 55 cm. To achieve this range, the spider's leg had to be resting on a plate; in the absence of such a solid, flat substrate the distance was reduced to 15 cm. As reported at the beginning of this chapter, the flow velocity over the leg is considerably greater in the presence of the plate (Barth et al. 1995). Because *Cupiennius* sits on flat parts of plants, for it this is the normal situation.

A complication arises in that the flow field generated by the fly (or other insects) is extremely directional (see this chapter, Section 1; Barth et al. 1995; Dickinson and Götz 1997). Therefore the response of a trichobothrium is not determined solely by the absolute distance from the stimulus source but also by the orientation of the spider relative to the source. The trichobothrium responds most strongly when it is in the center of the flow cone, which (from the fly's point of view) is directed downward at an angle of 25–45°. If the trichobothrium emerges from the flow cone, its response is greatly reduced. Presumably the resulting uneven distribution of the excitation of many trichobothria helps the spider become oriented to the stimulus source.

The following more detailed report on the effects of trichobothrial position relative to the flow cone will be easily understood by referring to figure 12a–c (Barth and Höller 1999).

Figure 12a: Variation of the horizontal distance (x) between 5 and 50 cm for two vertical distances (z) from the substrate, 2 cm and 9 cm. When $z = 2$ cm the response is largest when the horizontal distance is least, whereas with $z = 9$ cm the maximum coincides with a horizontal distance of 10 cm. The explanation: when the fly moves from position A ($x = 5$ cm, $z = 9$ cm) into position B ($x = 10$ cm, $z = 9$ cm), the trichobothrium is further inside the flow cone, so its response is stronger.

Figure 12b: Variation of the absolute distance of the fly (x′) between 5 cm and 35 cm with a constant angle α, which ensures that the trichobothrium is within the cone for all distances. Now the discharge rate decreases continuously with distance, following the decrease in average airflow velocity (0.56 m/s ± 0.08 SD at $x' = 5$ cm and only 0.08 m/s ± 0.14 SD at $x' = 30$ cm).

Figure 12c: When the distance (y) between the long axes of the fly (α constant at 40°) and the spider leg is increased from 0 cm to 11 cm, then the response of the trichobothrium again becomes much less strong as soon as it leaves the flow cone.

The fact that the decline in discharge rate with increasing distance from the fly is monotonic, as shown in figure 12b, might give the false impression that only the mean flow velocity is

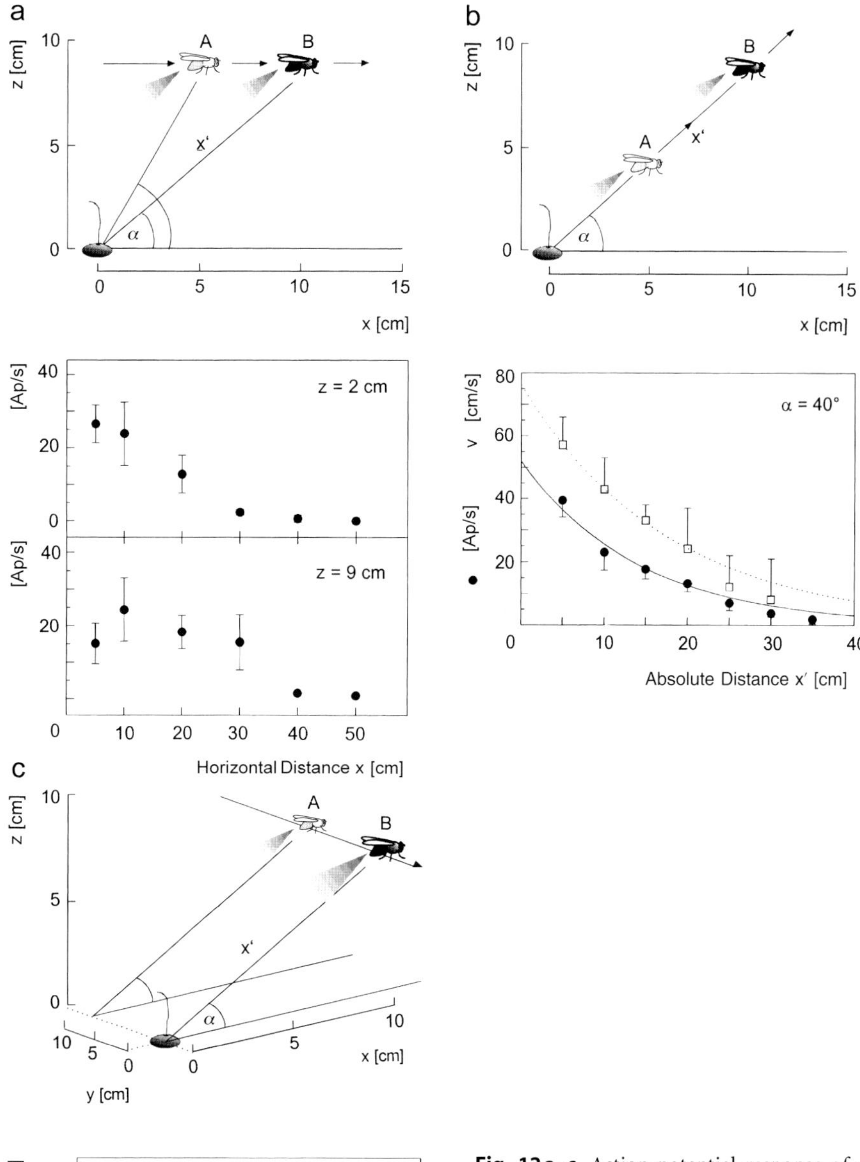

Fig. 12 a–c. Action-potential response of a metatarsal trichobothrium (group D1) to a fly signal (airflow) from different positions in space. **a** Constant elevation at z = 2 cm and z = 9 cm and with increasing horizontal distance x. **b** Constant angle $\alpha = 40°$, but increasing absolute distance x′ of fly; *lower graph* shows impulse response and in addition the mean velocity of the fly signal at increasing distances x′. **c** Constant elevation and constant angle $\alpha = 40°$, but increasing lateral shift (*y*) of fly. (Barth and Höller 1999)

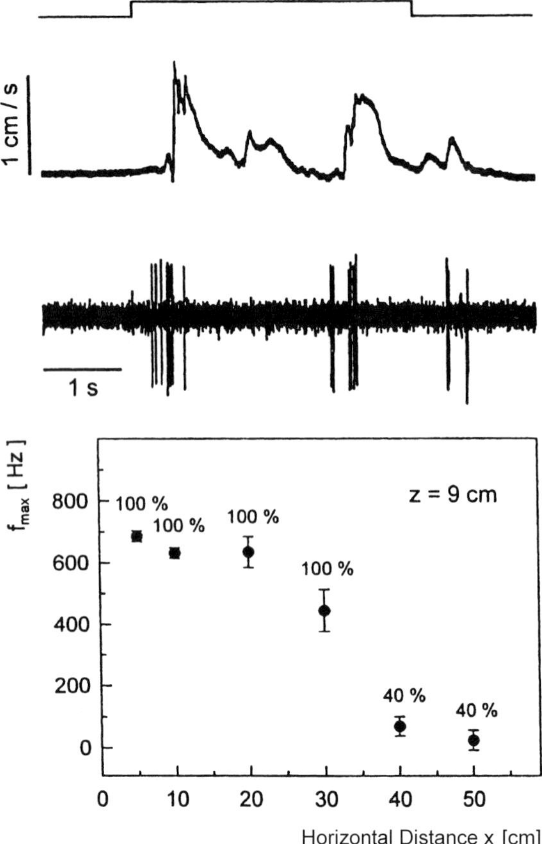

Fig. 13. *Above* Response of a trichobothrium (length 0.5 mm) to the airflow produced by a fly at a horizontal distance of x = 30 cm. *Top trace* Duration of flight; *middle trace* airflow velocity; *bottom trace* action potentials. *Below* Maximum frequency of action potentials with increasing horizontal distance *x* of fly. Percentages given above values denote the probability with which a stimulus (duration 1s) elicits a response (N = 1; n = 9). (Barth and Höller 1999)

changing. In fact, there is a simultaneous, dramatic change in the frequency spectrum of the signal. The high frequencies disappear so rapidly that beyond a distance of ca. 25 cm there are practically no components above 50 Hz (Fig. 6). That is, the components characteristic of prey insects are eliminated.

We noticed a similar complication when increasing the horizontal distance (x) of the trichobothrium from the fly (see Fig. 12 a). The signal becomes progressively more discontinuous so that finally, when the fly is only 30 cm away, it consists of individual pulses of air movement; as a result, the probability that a trichobothrium will be exposed to the stimulus is much lower, as is the discharge rate of the response (at *x* >30 cm the response probability is only 40%; see Fig. 13) (Barth and Höller 1999).

We shall soon return to these findings, because they are among the factors that explain the range of stimulus detection by trichobothria found in the behavioral experiment.

3 The Behavioral Response

Finally we come to the behavior itself. Although it was always in our sights, the path along which we have been approaching may seem overly winding and full of obstructions to the impatient reader. But I hope that it will become quite clear how important the questions we have been discussing are, if we are to understand the role of detection of airflow stimuli in a neuroethological and sensory-ecological context. Furthermore, for science too there is much truth in the religiously intended message of Buddhism, that the way is the goal. Tortuous paths usually offer the traveller more views than a straight, direct road that slices like an uncaring alien through the landscape.

The Jump Toward the Fly

Hungry spiders can time and again be observed jumping into the air at flies that fly close past them.

To reproduce this situation in the laboratory is difficult, so figure 14 (Brittinger 1998) shows the phenomenon in a slightly different situation, with the fly on a leash. The drama lasts less than 1.5 s. At the first attempt *Cupiennius* misses the target, but the second time it succeeds; the task is now simpler, because the fly is a shorter distance away. The next logical step makes it possible to elicit a spider's response in the laboratory with sufficient reliability: the fly

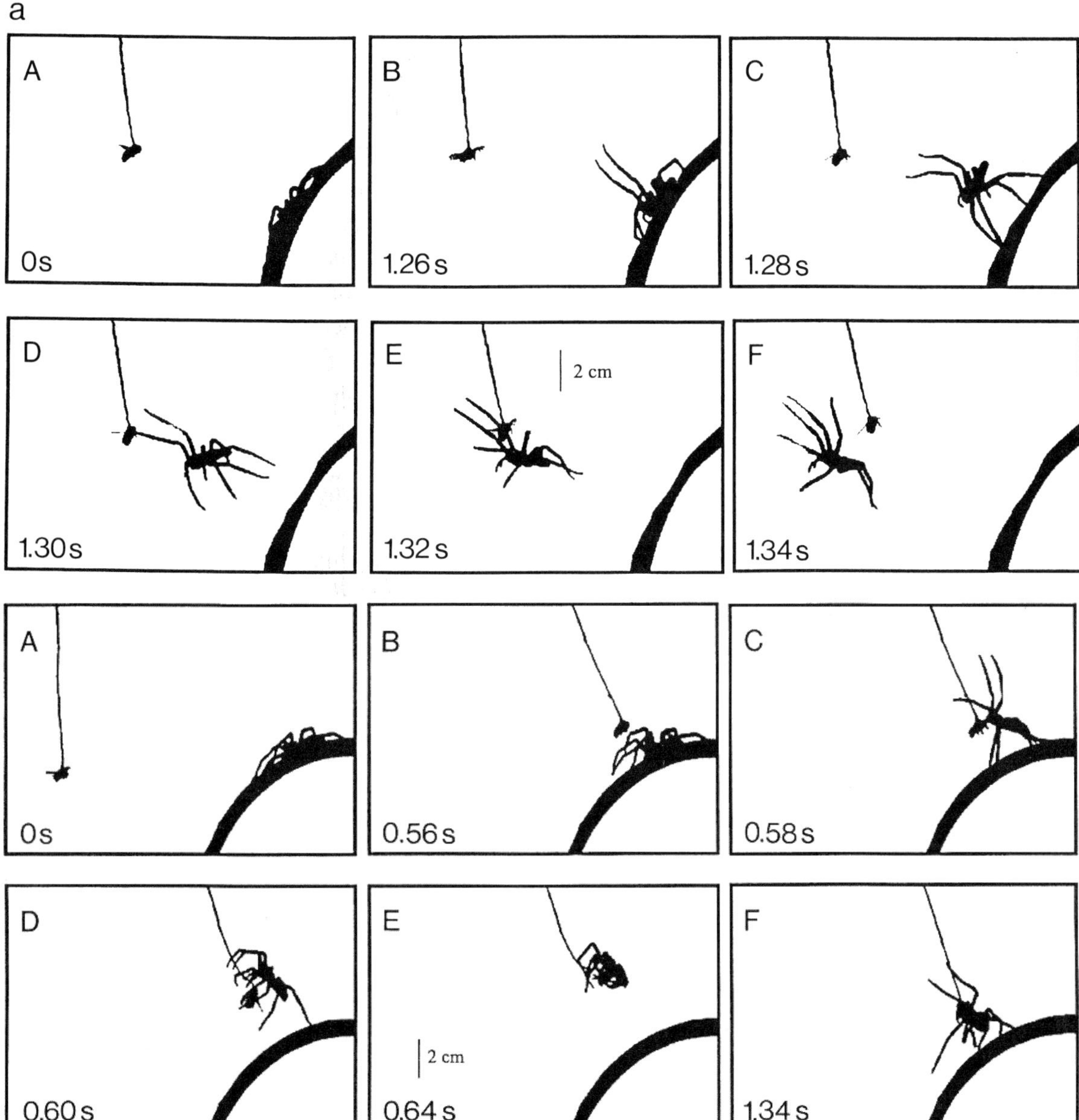

Fig. 14 a, b. *Cupiennius salei* jumps toward a fly flying on a leash (*thin thread*). Sequences of video frames; time between frames is indicated. Whereas the speedy jump shown in **a** is unsuccessful (presumably because of the fly's unnatural flight path), the spider catches the fly in the case shown in **b**. (Brittinger 1998)

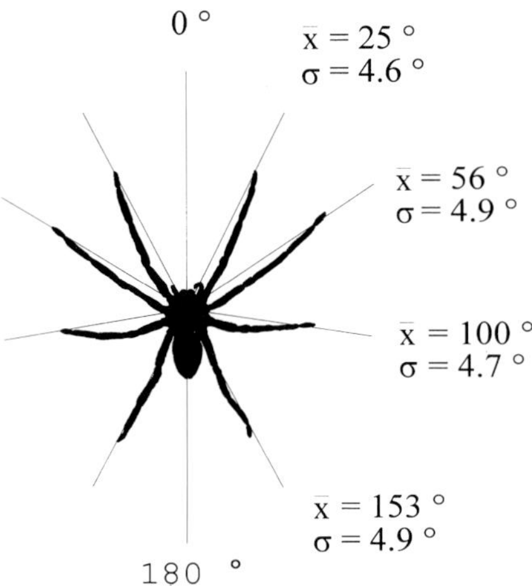

Fig. 15. Constancy of leg position in a resting *Cupiennius salei* ready for prey capture, given as the means of the angles between the long body axis and the respective leg (σ standard deviation). (Brittinger 1998)

is tethered so that it flies in one spot. Before the experiment the fly is glued by its thorax to a narrow strip of paper 1.5 cm long. A small piece of foam material is placed between its legs and the fly holds onto it. When the piece is taken away, the tarsal reflex is triggered, so that the fly begins to beat its wings. It is now flying while stationary, hanging from the strip of paper, and in this state it can be put into a precise position relative to the spider.

The spider responds to the airflow thus generated by turning toward the fly. This response is abolished when the trichobothria have been removed, even though the motionless fly is held only 5 mm away from the chelicerae. A spider without trichobothria "snaps" at the fly only when the fly touches it, or if the fly causes vibration by walking on the substrate (Brittinger 1998).

One way an initiate can tell that spiders are ready to respond is by their posture; a spider in wait for prey raises its body slightly above the substrate and puts its legs in a typical posture that varies only slightly (Hergenröder and Barth 1983b). The most recent data on this subject come from the dissertation by Waltraud Brittinger and refer to 7-month-old spiders, which still have to undergo a molt and hence are voracious (Fig. 15).

A similarly stereotyped leg position is known not only for other spiders (*Dolomedes*, Bleckmann et al. 1994; *Nephila*, Klärner and Barth 1982), but also for scorpions (*Paruroctonus*, Brownell and Farley 1979) and for insects with similar behavior (*Notonecta*, Wiese 1974).

As far as we know to date, web spiders behave differently from hunting spiders such as *Cupiennius*. At least orb-weavers such as *Zygiella x-notata* and *Nephila clavipes*, which because of its yellow silk is called the golden web spider, cannot be induced to exhibit prey-capture behavior by airflow stimuli (such as the "fly stimulus" so effective for *Cupiennius*). In 70% of the experiments (N=12 spiders, n=53 trials) *Nephila clavipes* ran up to a buzzing fly 20 cm away as soon as it touched the web. But if the fly was only 3 cm above the spider and not touching the web, only in 8% of cases (N=16, n=75) did the spider show elements of prey-capture behavior such as shaking the web, and none ever attempted to actually catch the prey. On the other hand, in 38% of the experiments the spider abruptly raised one or several of its first and second legs. We interpret this as defensive behavior. For *Nephila* airflow stimuli alone are not a sufficient prey stimulus. The same applies to *Zygiella*, which like *Nephila* is very willing to respond to thread vibrations (Klärner and Barth 1982).

Our hypothesis is that the difference in behavior with respect to airflow stimuli is correlated with the structure of the substrate on which the particular kind of spider lives. For a spider seated on relatively firm substrates such as banana leaves or bromeliads (*Cupiennius*), or on densely woven carpet webs such as those of *Agelena* and *Tegenaria* (Görner and Andrews 1969), it seems sensible to try to catch prey detected only by way of the air. Leaping up from an orb web is a different matter, and we have never seen *Nephila* jump into the air after an insect that was flying by, unlike *Cupiennius*. Evidently orb-web spiders need the web vibration as an indicator that an insect has become caught in the web and hence is approachable.

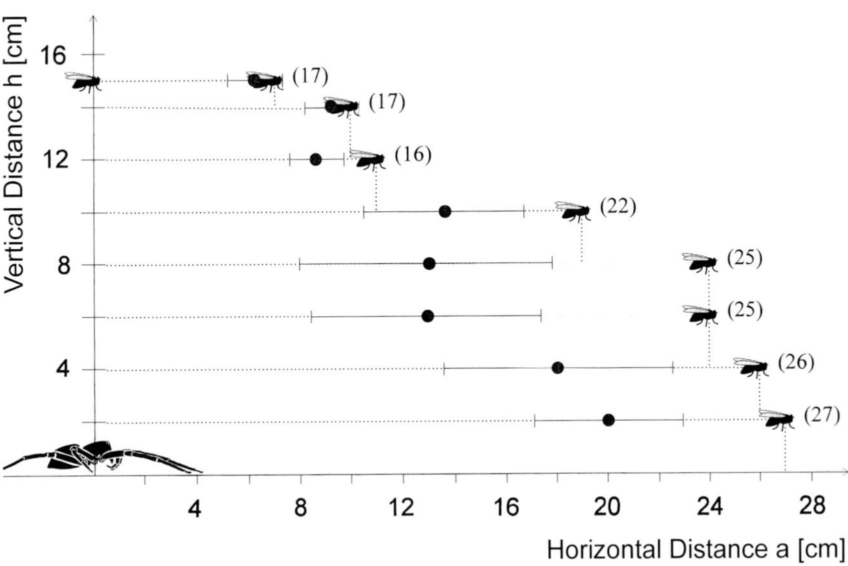

Fig. 16. Mean and maximal distances at which an intact *Cupiennius salei* still reacts with approach behavior to the airflow signal of a fly flying tethered at different elevations. Position of fly in horizontal plane (relative to the long axis of the spider) makes no difference, implying that in the given context the trichobothria work equally well all around the spider. (Brittinger 1998)

Once Again: The Range of the Trichobothria

Since *Cupiennius* proved so eager to respond to airflow stimuli by turning toward the source, we were able to get a better idea about the sensory space it occupies. Having previously estimated this only by the deflection of the hair shaft of individual trichobothria and the excitation of single sensory cells, we can now use the behavior to measure the range of the trichobothrial orientation system and find out how direction-dependent it is.

The answer is summarized in figure 16 (Barth et al. 1995; Brittinger 1998). Depending on the altitude of the fly above the substrate, the maximal distances from which a reaction can be elicited vary between 27 cm (fly height 2 cm) and 17 cm (height 15 cm). The angle α between fly, spider and substrate in effective stimulating positions varied from almost 0° to 90°. This shows that the airflow signal expands considerably (in the imagined plane extending downward from the fly's long axis), even though at large angles α it becomes progressively more difficult to trigger the behavior. Some spiders respond rarely or not at all when the stimulus reaches them from behind. If all the data from all animals are thrown into the same pot, this turns out not to be significant. In other words: the spider "sees" with its trichobothria equally well in all horizontal directions (Barth et al. 1995).

In the second part of this chapter I reported that the "mechanical" range of single trichobothria, for the same fly signal used in the behavioral experiments, is 50 to 70 cm and the "physiological" range agrees well with this result. Is it surprising, then, that the ranges found from the behavior do not quite reach 30 cm? It might have been expected that the ensemble of trichobothria would have a greater range than any individual hair. That this is not the case may be explicable on the basis of what we already know, as follows. The spider might be able to recognize the changes in the signal associated with increasing distance from the stimulus source (see Figs. 3 and 13) and "know" that a signal originating far away has taken several seconds to arrive, so that an attempted capture now has no chance of success – quite apart from the long time it would take for the spider to approach the signal source.

The ranges approaching 30 cm found for *Cupiennius* are remarkably high, given that Görner and Andrews (1969) were able to elicit prey-capture behavior in agelenids only when the stimulus source was less than 1 cm away from the spider. As a result, they classified the sense mediated by the trichobothria as "touch at a distance", in analogy to the term coined by Dijkgraaf (1947) for the lateral lines of fishes and amphibians. A major factor in this definition is the importance of distinguishing between pressure receptors in the far field of a stimulus source and movement receptors in its near field. That the trichobothria of *Cupiennius* should be counted as movement receptors in this sense is beyond doubt. But it is confusing to speak of "touch" in view of the range of these sensilla, in particular the deflection of the trichobothria by a fly signal originating at distances of 70 cm or more. By the way, this value agrees well with the results of experiments in which filiform hairs on the caterpillars of *Barathra* were stimulated with the air current generated by a flying parasitic wasp, which the caterpillars try to escape (Tautz and Markl 1978). It is likely that *Cupiennius salei*, too, uses its trichobothria not only to detect prey but also as a warning system against predatory wasps (Pompilidae) and parasitic Neuroptera (Mantispidae). In any case, we have repeatedly observed the approach of such enemies in the field: *Cupiennius* responds by raising its forelegs, which looks like a defensive reaction.

4 Details of the Turning Movement

The apparatus used for the experiments on range also enabled a detailed analysis of the orienting movement itself (Fig. 17) (Brittinger 1998). The most important findings can be summarized as follows.

- As has been mentioned, the range is equally large in all horizontal directions. The residual stimulus angle α' (the angle remaining after the turn) is always very small, but for stimulation in the direction of a first leg it is significantly smaller ($11 \pm 8.4°$) than for stimulation

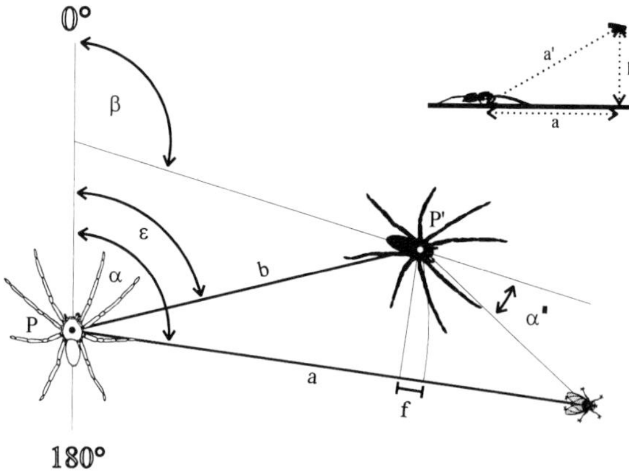

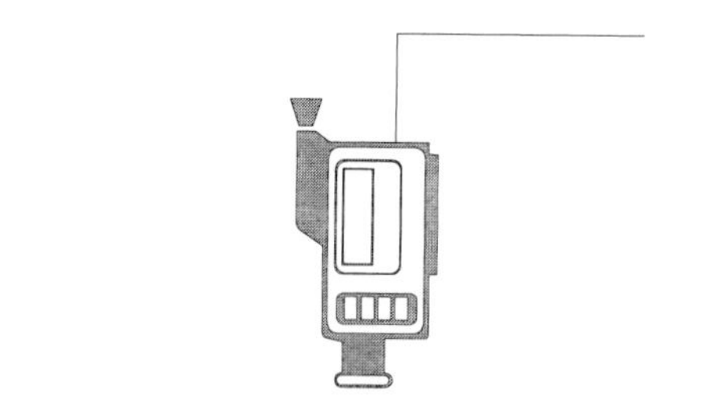

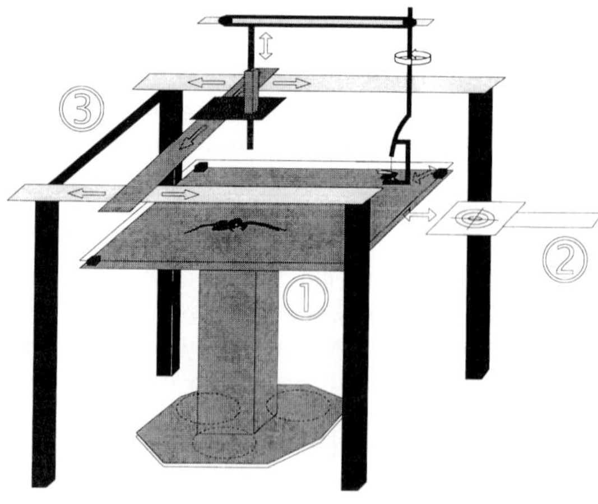

Fig. 17. Apparatus used to record the various geometrical parameters involved in prey-capture jump. *Above* Position of spider before (*white*) and after (*black*) the first turn toward the prey; *P, P'* center of prosoma before and after jump; *a* distance to stimulus source; *b* translatory component; *f* walking error; α stimulus angle; β turning angle; ε turning angle of prosoma; α' remaining stimulus angle. *Inset a'* distance to stimulus source; *h* elevation of stimulus. *Below* Experimental table (*1*) with glass plate about 1 cm above surface; *2* Plexiglas disk with concentrically arranged circles for the measurement of *a*; *3* device for the exact and easy positioning of the fly in three-dimensional space. (Brittinger 1998)

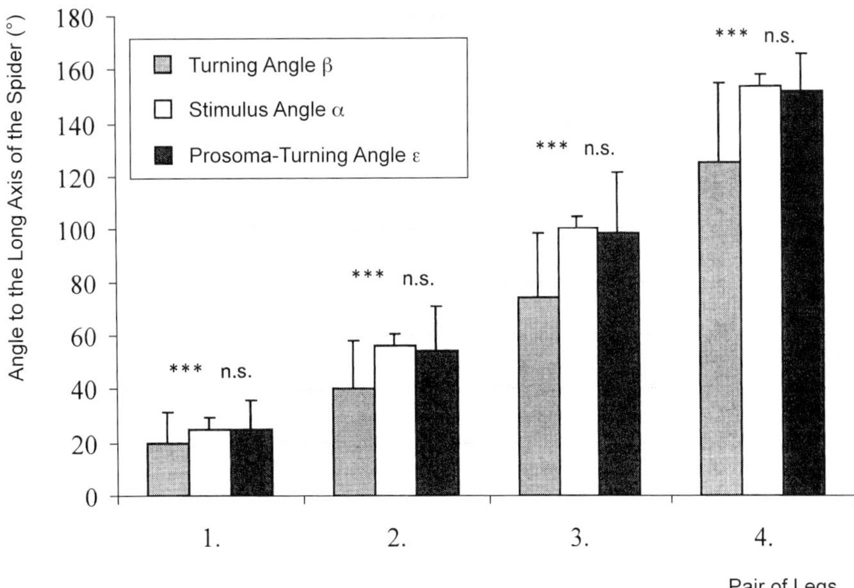

Fig. 18. Comparison of stimulus angle α, turning angle β, and prosoma turning angle ε in situations where the stimulus comes from directions identical with the orientation of one of the legs (*1–4*). Stimulus: stationary fly, flying 2 cm above platform with spider, and at a distance from it which is most effective in eliciting a behavioral response; stimulus angle and turning angle always differ from each other significantly (***), whereas this is not the case for stimulus angle and prosoma turning angle (n.s.). N=53, n=101–106. (Brittinger 1998)

in the direction of a second, third or fourth leg (ca. 23–29°) (Figs. 18, 19). The scatter of the values is also least for stimulation in the direction of a first leg and increases as the stimulus comes from progressively more lateral or more posterior directions. This implies that *Cupiennius* turns most precisely toward stimuli from the front.

The same applies to the response to vibratory substrate stimuli in the orb-web spider *Nephila clavipes* and to a considerably greater degree in *Cupiennius* (Klärner and Barth 1982, Hergenröder and Barth 1983a) and in the semiaquatic spider *Dolomedes okefinokensis* (Bleckmann et al. 1994). However, after all these spiders have turned toward the prey in response to vibration, they are standing so that the prey is always within the angular region bounded by their forelegs.

- Values of the translation component b are independent of the stimulus direction, but the scatter increases with the distance of the stimulus source (for example, with a = 15 cm: b = 5.1 ± 1.76 cm; with a = 25 cm, in contrast: b = 6.7 ± 3.59 cm; fly 2 cm above the substrate). That is, the localization becomes somewhat less accurate at greater distances from the stimulus.
- The most interesting result: both the latency l (time between stimulus onset and beginning of spider movement) and the response duration d (time between the beginning and the end of the spider's movement) are independent of stimulus angle α (Table 1). Therefore the spider's turning speed rises with increasing stimulus angle α. This seems entirely reasonable: the spider will have a chance of catching even prey that is approaching from behind. However, the significant increase in latency l and response duration d with stimulus distance indicates that a greater distance makes it harder for the spider (for example, with a = 15 cm: l = 0.3 s, d = 0.4 s; with a = 25 cm: l = 0.7 s, d = 1.8 s).

As figure 19 shows, in the great majority of cases (77%) the spiders do not turn quite far enough to face the stimulus source directly. This is

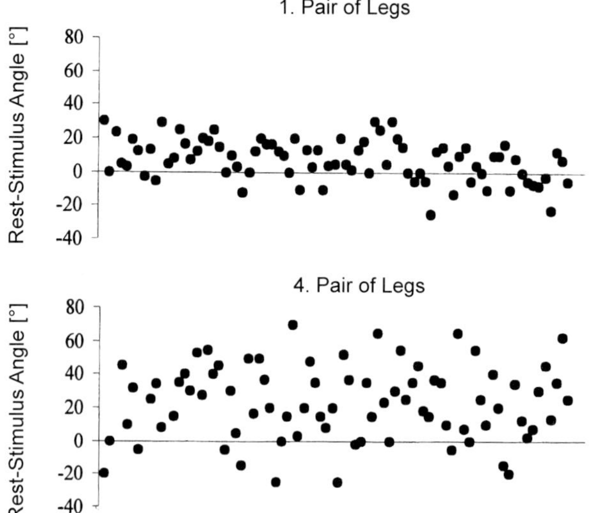

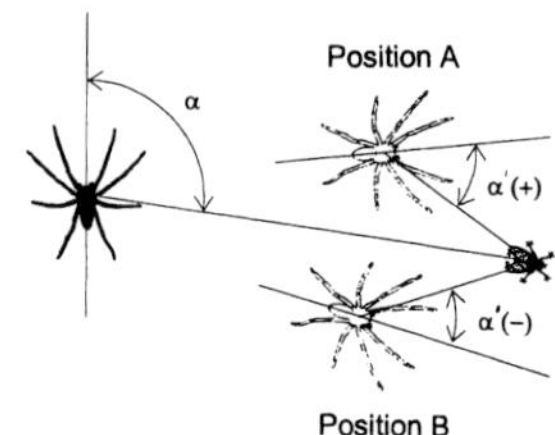

Fig. 19. Stimulus angle α' remaining after *Cupiennius salei* has turned toward stimulus source directed along long axis of a first or a fourth leg. X axis corresponds to position of long axis of spider after turning. Positive values denote that the spider did not turn far enough; negative values, that it turned beyond the direction leading to its prey (see *inset*). N=41, n=79–82. (Brittinger 1998)

Table 1. Latency *l* and reaction time *d* of the behavioral response of *Cupiennius* to stimulation from different directions. Stimulus: stationarily flying fly, elevation 2 cm, distance from spider prosoma 20 cm. *M* median, *V* range of variation. (Brittinger 1998)

	1st leg N=17; N=23	**2nd leg** N=20; n=27	**3rd leg** N=10; n=11	**4th leg** N=20; n=25
Latency l M, V (in s)	0.5 (0.2–4.0)	0.5 (0.1–2.0)	0.6 (0.2–3.2)	0.7 (0.2–2.5)
Reaction time d M, V (in s)	0.7 (0.3–6.3)	0.8 (0.5–4.3)	0.8 (0.3–2.7)	1.1 (0.3–7.7)

presumably a matter of saving time, in view of the fact that the prey is initially seized not with the chelicerae but with forelegs. As these have a relatively long reach, there is no need for the long axis of the spider's body to point precisely toward the stimulus source. Unfortunately, no one has yet checked whether there is a quantitative relation between error angle and actual prey-catching success.

5 The Interplay of the Trichobothria: First Glimpses

For her dissertation Waltraud Brittinger (1998) also removed trichobothria in various patterns, in order to find out how many of these sensors are really needed to trigger turning-toward-stimulus behavior and whether and how the trichobothria of the eight legs collaborate. The first insights are a good beginning, no less but also no more.

Minimal Number. Given that animals with no trichobothria at all can never be induced by airflow stimuli to give a behavioral response (but still respond quite normally to substrate vibrations and if they are male also court and copulate normally), the question naturally arises: how many trichobothria must be intact to ensure orientation to a stimulus source?

When up to 50% of all trichobothria are removed from the same segments of all eight legs, the individual geometric parameters of the movement (see Fig. 17, a, b, α', β, ε) are all the same as in the intact animal, and the same applies to the latency and the response time (l, d). This is true regardless of whether the trichobothria were removed from the tarsus, metatarsus

Table 2. Behavioral reaction of *Cupiennius salei* to tethered flying fly after removal of the trichobothria of various segments of all of its legs. + trichobothria intact, – trichobothria removed. N = 12. (Brittinger 1998)

Ablation				Behavior
Tarsus (ca. 30 trichobothria per leg)	Metatarsus (ca. 30 trichobothria per leg)	Tibia (ca. 50 trichobothria per leg and pedipalp)	Percentage of all trichobothria	
–	+	+	ca. 25	normal
+	–	+	ca. 25	normal
+	+	–	ca. 50	normal
–	–	+	ca. 50	normal
–	+	–	ca. 75	none
+	–	–	ca. 75	none

or tibia (t-test; N = 8), which implies that there is no differential allocation of tasks. In contrast, the reactions are abolished when 75% of the trichobothria are removed, in the same regular distribution (Table 2, Fig. 20a).

Pattern of Ablation. Figure 20 shows what happens when the removal of trichobothria is not distributed uniformly over the eight legs.

1. *Removal on one side* (see Fig. 20b). When all the trichobothria on the same side of the body were gone, in 60% of the cases (N = 7, n = 59) the animals turned through the correct absolute angle but in the wrong direction; that is, they turned toward the intact side of the body even though the stimulus was directed toward a leg lacking trichobothria. In intact animals this wrong-way turning occurred in only 0.8% of the cases, with a sample of 496 response-eliciting stimuli. The only error the animals without trichobothria made was in the sign of the response; otherwise the movement was "normal", with a, b, α', β, ε (see Fig. 17) and the latency and response time all the same as in intact animals (t-test). What can we conclude? *a. Cupiennius* makes use not only of the information from the side stimulated but also of that coming from the opposite, intact side. *b.* The spider turns correctly in the sense that normally the stimulus is not on the unstimulated side. *c.* Regardless of whether the stimulus passes over the trichobothria in the direction of the prosoma or away from it, evidently nothing changes in the way the angle is determined – otherwise

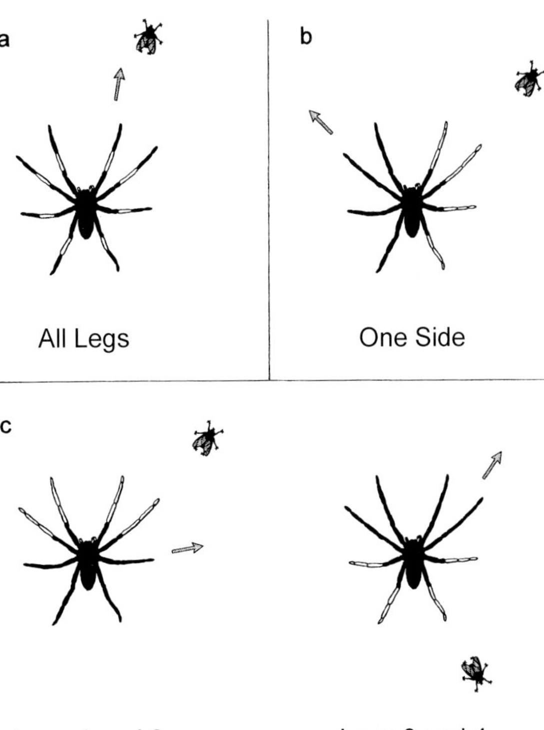

Fig. 20a–c. Behavioral reaction of *Cupiennius salei* to tethered flying fly after ablation of 50% of all trichobothria in different patterns. *White* Segments of legs and pedipalps without trichobothria. When stimulated with the fly from the direction indicated in the drawings, the spider turns into a direction indicated by the *open arrow*. N = 7–8. (Brittinger 1998)

Cupiennius would also have chosen the "correct" body side after the trichobothria had been removed.

2. *Front vs. back* (see Fig. 20c). To check whether the signals from the more anterior trichobothria are analyzed differently from those of the more posterior ones, the responses were compared after removal of all trichobothria from the first two pairs of legs plus the pedipalps or from leg pairs 3 and 4. When the stimulus was directed toward a leg lacking trichobothria, the animals turned as though the nearest intact leg had been stimulated. Hence it appears that the spider turns toward the leg stimulated earliest or most strongly, and it is this that determines the direction and angle of turning. This simple postulate could also explain the response to unilateral removal as described in (i) above. Neither anterior nor posterior legs have any discernible special role.
3. *Single intact legs.* If the conclusions drawn for cases (i) and (ii) are correct, will a single intact leg suffice for "correct" turning when the stimulus is directed toward that leg? First, under such conditions the spiders are not very eager to respond at all. Out of 10 animals, only 6 could be induced to give a response, three of them not until 3–4 weeks had passed since the operation. The translation component (b) of the movement and also the latency (l) and the response duration (d) were

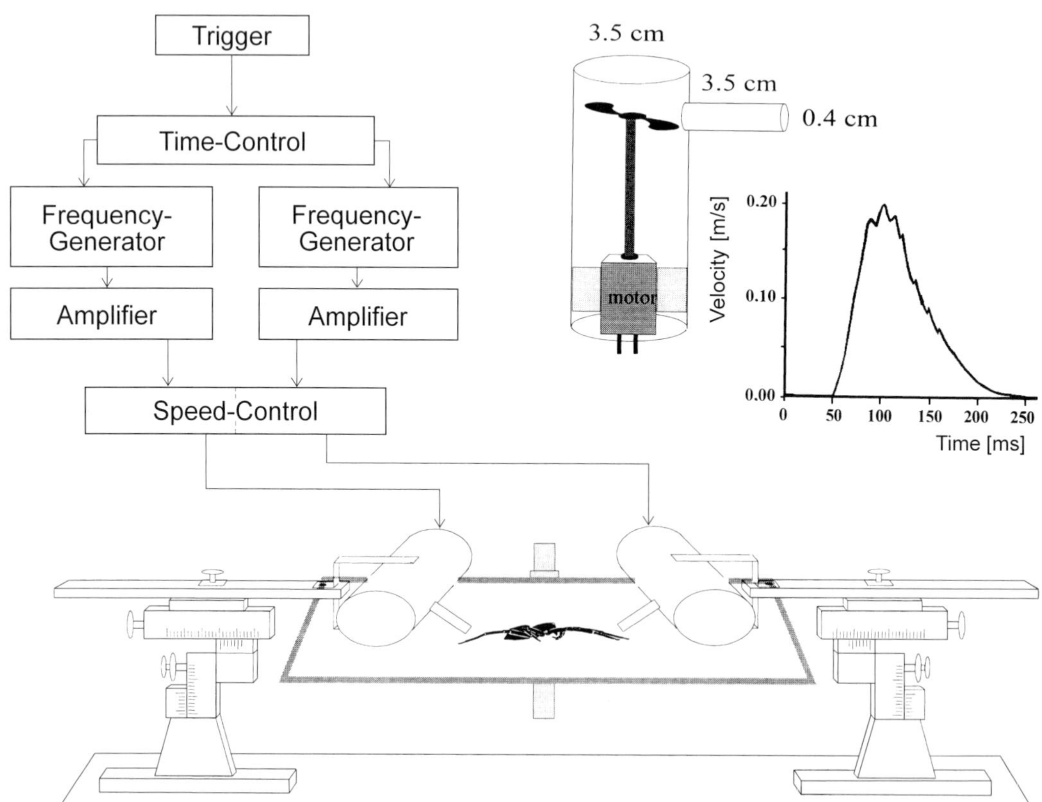

Fig. 21. Experimental arrangement used to study the elicitation of prey-capture behavior in *Cupiennius* by artificially produced air pulses and independent stimulation of two legs. *Inset* The screw used to produce a stimulus pulse of 200 ms duration and 0.233 m/s airflow velocity; time course of stimulus given. Distance of nozzle outlet to spider metatarsus 5 mm. (Brittinger 1998)

significantly ($p<0.001$) smaller (b) or longer (l, d) than in intact animals. The results nevertheless support the hypothesis. When stimulated in the direction of the intact leg, the spider almost always responded by turning toward the stimulus. When the stimulus was directed toward one of the legs without trichobothria, the spider often made mistakes, which should not come as a surprise. Altogether, in no less than 79% of the cases the spiders turned toward the body side with the single intact leg, and they always did so when the stimulus source was on that side of the body. This, too, is in favor of the hypothesis.

Selective Stimulation. It is also possible to take the opposite approach and present the stimulus on such a small scale that it affects only one leg, even though the animal is intact. To do this, Waltraud Brittinger developed the apparatus shown in figure 21. The key element here is a motor-driven propeller from a toy ship. This generates an airstream that is very effective as a stimulus, reproducible and also accurately controllable in time. Because the airstream emerges toward the spider from a nozzle that is only 0.4 cm wide and 3.5 cm long, the stimulation can be spatially confined. If the trichobothria are removed from alternate legs (that is, first and third, or second and fourth), then stimulation of an intact leg elicits prey-capture behavior: *Cupiennius* jumps toward the nozzle, grasps it and usually tries to bite. There is no response to stimulation of a leg that lacks trichobothria. Having thus shown that the "propeller apparatus" can indeed selectively stimulate trichobothria on only one leg, she then used it to test intact animals. Again it turned out that in 86% of the cases (N=9, n=63), regardless of the leg chosen for the stimulus, *Cupiennius* attacked the nozzle as though it were prey.

By duplicating the propeller apparatus and adding some electronics it is possible to present the same stimulus separately to two legs, either simultaneously or with a slight delay. This experiment shows that simultaneous stimulation of the trichobothria on two non-adjacent legs (otherwise in any desired combination) is somehow confusing or irrelevant to the spiders. In 89% of the cases (N=8, n=28) they do not respond. The simultaneous occurrence of the same stimulus at such different places is decidedly unnatural; the behavioral response is omitted.

This blockage can be released in two ways (Fig. 22). *First*, two adjacent legs can be stimulated. Then, in 67% of the cases (N=7, n=15) *Cupiennius* jumps in a direction between the two legs. From the viewpoint of sensory ecology

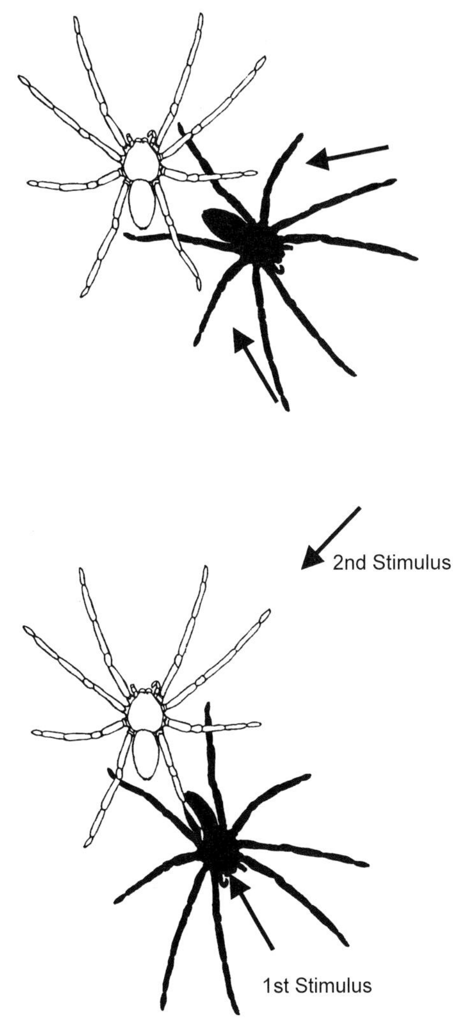

Fig. 22. The turning movement of *Cupiennius salei* when its trichobothria on two neighboring legs are stimulated simultaneously (*top*) or with a time delay of 60 ms or more (*bottom*). Spider is drawn in *white* before reaction and in *black* afterward. See Fig. 21 regarding the stimulus. (Brittinger 1998)

this seems understandable. Such a stimulus situation is certainly more natural than the preceding one; if an air current is broad enough, it would be expected to strike the trichobothria of directly neighboring legs. A *second* means of triggering the prey-capture response is to introduce a time difference. To put it simply, if one of the two stimuli begins at least 60 ms after the other, in 80% of the cases *Cupiennius* turns toward the first stimulus. This corresponds to the response frequency observed when only one leg is stimulated and is also reminiscent of the result of similar experiments using substrate-vibration stimuli. By changing the time difference systematically, it was shown that a delay ≥50 ms was needed in order for there to be significantly ($p<0.001$) more turns toward the first than toward the second stimulus, although a tendency in this direction becomes apparent with as little as 10 ms delay. As in the case of the vibration-sensitive slit sense organs (see Chapter XVIII), it can be inferred that the trichobothria on the leg first stimulated have a strong inhibitory influence on the information coming from those on the other legs.

One more number in conclusion: Waltraud Brittinger found out that in the best case at least three trichobothria on one leg must be intact if stimulation of this leg with an air pulse (from distal to proximal) is to trigger a directed response! With two trichobothria all that can be produced is a listless twitch.

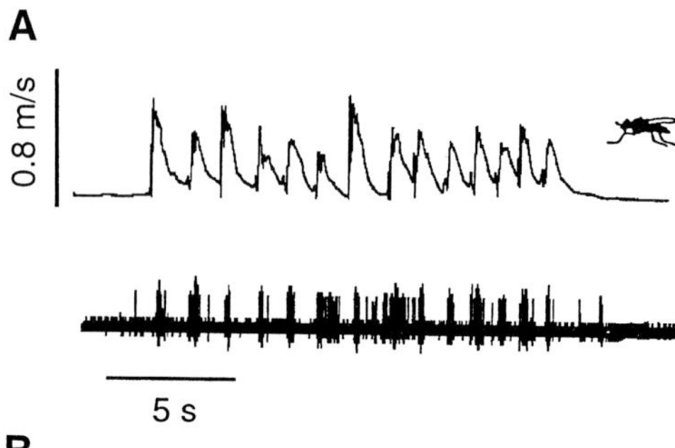

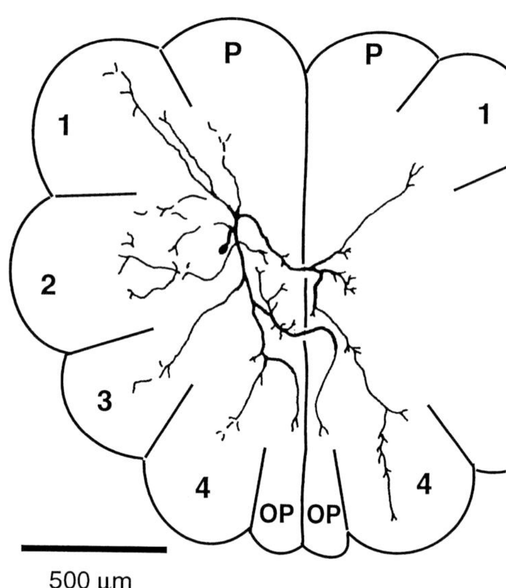

Fig. 23 A, B. Response of an interneuron in the subesophageal mass to airflow produced by a tethered flying fly. **A** Velocity of airflow (*top trace*) and action potentials (*bottom*). **B** Structure and topography of an interneuron sensitive to such airflows. *1–4* Leg neuromeres; *P* pedipalpal neuromere; *OP* opisthosomal neuromere.(Friedel and Barth 1997)

6 Airflow-Sensitive Interneurons

Only a little is known about the actual processing of the information from the trichobothria by interneurons in the central nervous system. The reason lies mainly in the technical problems involved in electrophysiological recording. We know considerably more about the analogous neurons belonging to the cercal system of cockroaches (Camhi and Levy 1989, Ritzmann and Pollack 1990, Ritzmann 1993) and crickets (Bodnar et al. 1991, Miller et al. 1991, Theunissen and Miller 1991, Baba et al. 1995), the directional sensitivity of which is particularly interesting.

Nevertheless, we can say the following. Gronenberg (1989, 1990) and Gramoll (1990) were the first to demonstrate, by stimulating various legs of *Cupiennius salei*, that inputs from several legs converge onto plurisegmental wind-sensitive interneurons in the subesophageal ganglion mass. Another feature that had already been inferred from behavioral experiments was also confirmed: the presence of interneurons that re-

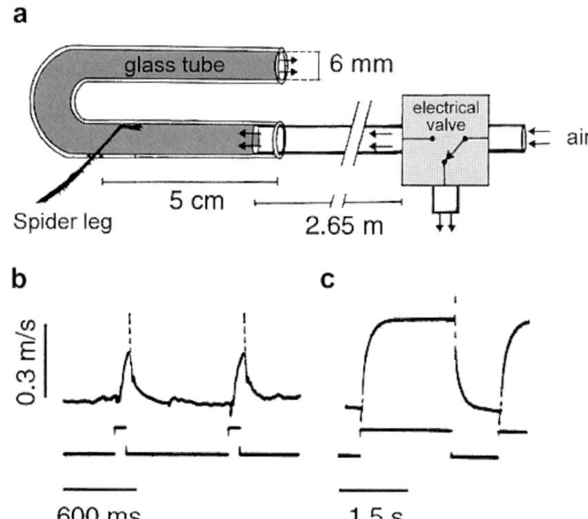

Fig. 24 a–c. Apparatus used to stimulate single legs or a combination of legs with artificial air pulses. **a** The distal leg protrudes into a glass tube through a tiny hole; the air stream flowing through the glass tube is controlled by an electric valve. **b, c** Two air pulses differing in their time course (measured with a hot wire anemometer). (Friedel and Barth 1997)

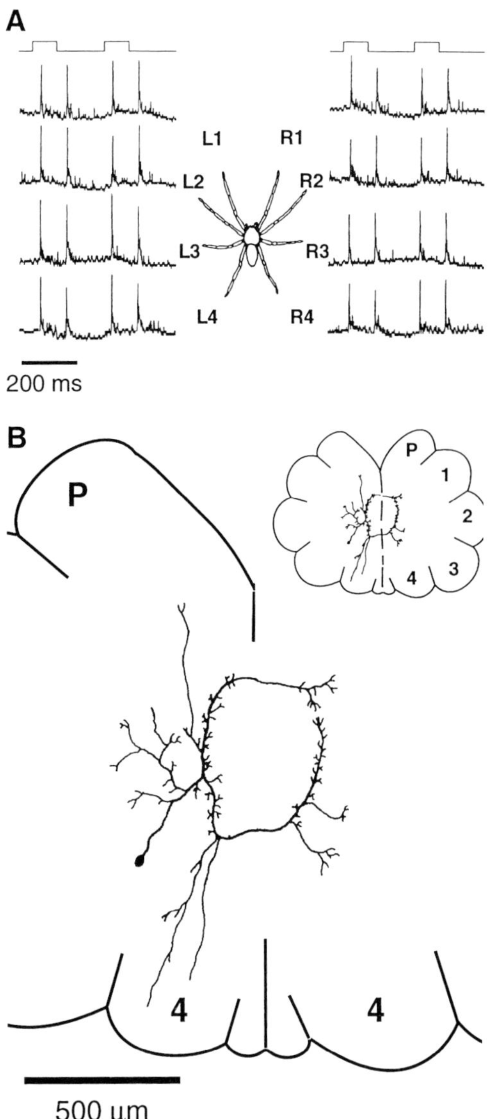

Fig. 25 A, B. A Response of an interneuron to spatially and temporally separate stimulation of the eight legs of *Cupiennius salei* (stimulus: duration 100 ms, velocity 1.0 m/s). The interneuron responds to the stimulation of all legs. **B** Structure and topography of the same interneuron. *1–4* Leg neuromeres; *P* pedipalpal neuromere. (Friedel and Barth 1997)

ceive inputs from both trichobothria and receptors sensitive to substrate vibration (Gronenberg 1990, Friedel and Barth 1995, 1997).

Furthermore, recordings are available of interneuronal responses to the same fly stimulus that was used for the behavioral experiments and the tests of individual trichobothria (Fig. 23). All nine of these neurons responded with short latency and predominantly to the sudden increases in flow velocity. Accordingly, they are excellently suited to encode the highly fluctuating time course of prey signals. The responses shown in figure 23 are from one such neuron. It is a bilaterally projecting plurisegmental interneuron with its soma in the ventral region of the neuromere of the second leg. The main neurite ascends to the central tract (see Fig. 7, Chapter XIV), within which it projects toward both anterior and posterior, sending out branches into the lateral neuropils of the neuromeres. In these the branches terminate ventrolaterally, together with the primary afferents of metatarsal and tarsal trichobothria (Anton and Barth 1993). An anterior and a posterior projection end contralaterally in the first and fourth, respectively, sensory longitudinal tract (SLT 4, see Fig. 3, Chapter XV), together with the afferents of metatarsal trichobothria (Anton and Barth 1993).

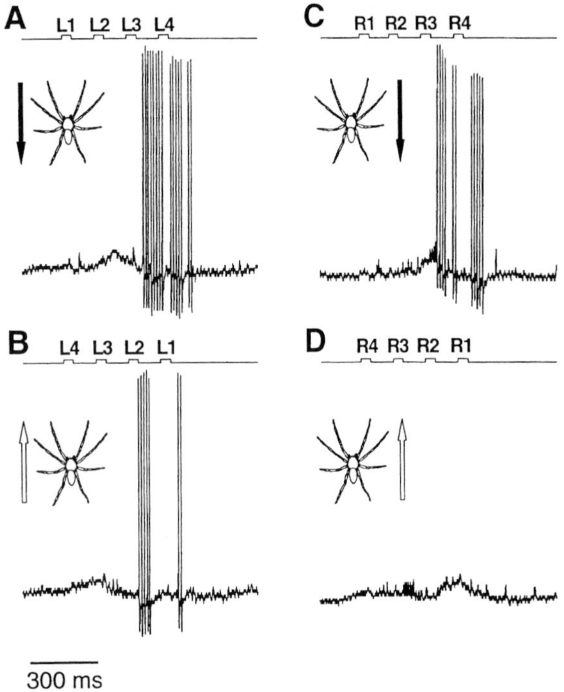

Fig. 26 A–D. Response of an interneuron in the subesophageal ganglionic mass to the successive stimulation of neighboring legs (stimulus: duration 50 ms, velocity 1.0 m/s). The response is larger when the stimulus moves from anterior to posterior (**A, C**) than when it moves in the opposite direction (**B, D**). (Friedel and Barth 1997)

Figure 24 shows how the very difficult feat of stimulating while recording from neurons in the central nervous system is accomplished. The ends of the legs are stuck into U-shaped glass tubes, the walls of which are provided with a hole for this purpose. One end of the tube is connected to a stimulus source, and the stimulating air current, after it has flowed over the trichobothria of the tarsus and metatarsus, is guided away so that it cannot stimulate adjacent legs. By means of a microprocessor and electrical valves, air pulses with different flow velocities (0.4–2 ms^{-1}), durations (50–100 ms) and time offsets (60–200 ms) can be applied to any desired single legs or leg combinations (Gramoll 1990, Friedel and Barth 1997). Thomas Friedel used this apparatus to obtain some initial data on the mechanisms by which information about stimulus direction is processed in the CNS.

All 23 of the central neurons successfully investigated with the artificial air pulses received inputs from the trichobothria on at least two legs: all eight legs, the first three leg pairs or all legs on one side of the body (Fig. 25). Nineteen of the 23 interneurons gave markedly phasic responses to the onset and/or the end of the air pulse. Two especially remarkable neurons responded only to the successive stimulation of the legs on one side of the body, and not to the stimulation of single legs (Fig. 26).

Both of these neurons were distinctly more strongly activated by stimulation of the legs from front to back than in the opposite order. It may be that these neurons are involved in the localization of moving objects and can identify the direction of movement.

When examined closely, fly catching proves to be a very complicated affair. Simple as it may seem to the casual observer, we are still far from a complete understanding of this behavior – even though it has evidently fascinated people for millennia. Humans in the Mesolithic, about 10,000 years ago, were so moved by it that they referred to it in their hunting sorcery (Fig. 27).

Fig. 27. Spider with flies; drawing from a cave in the Gasula gorge in Spain dated 5000 to 10,000 B.C. (Schimitschek 1968)

XX Courtship and Vibratory Communication

That males and females of *Cupiennius* are each capable of independent life but as a species, like all animals with two sexes, are conceivable only in partnership is in itself a trivial statement. It becomes interesting when the community of two individuals is considered in detail. In *Cupiennius* the time they spend together is less than an hour, after which the sexes go their own separate ways as before. The brief period of their meeting is full of ingenious biological mechanisms to ensure that the genetic equipment of the parental animals will be passed on to the next generation in a new, unique combination. Sociobiology would put it very bluntly: all the splendor of our spiders is a product of the spider genes, created by them for the sake of their own continued existence; the egg makes the chicken so that the egg will have a future!

In fact, the precopulatory behavior of *Cupiennius* (and other spiders) is a complicated sequence of motor activities. On the way to the goal of copulation, the partners must master various tasks. One of them is to avoid wasting energy on a potential partner that belongs to the wrong species or wrong sex, or is sexually incompetent (for instance, is not old enough). Individuals that are suitable partners have the problem of finding one another and managing actually to come together. Because adult spiders are generally intolerant and aggressive, nonreproductive behavior – in particular, prey-catching – must somehow be suppressed. Finally, the activities of the two partners must be synchronized in order to enable pair formation and copulation. Their sensory systems are certainly not involved merely in achieving mutual orientation; they are also responsible for guaranteeing sexual readiness, species recognition and the suppression of inappropriate behavior. The selection pressures under which the courtship behavior of *Cupiennius* and the role of the sensory systems evolved must therefore have been many and diverse.

This chapter is concerned not with copulation *per se* but with the preceding courtship and its crucial element of vibratory communication. A special aim will be to show how important and rewarding it is to approach such a complex of questions at the broad level of the whole organism, and to see how the various aspects of spider biology interact. If the following account begins to seem confusing, as I try to give at least some idea of the complexity and multilayered structure of an important form of behavior, please keep this ultimate aim in mind. A particularly appealing part of the story is the participation of plants in the encounter between these sexual partners, as well as the fact that communication by means of substrate vibrations is highly developed in *Cupiennius* and can serve as an example of what vibratory communication in general can accomplish.

Our path will lead us from below upward, from the signals through the neuronal responses of the vibration receptors and vibration-sensitive interneurons to the problems of species identification and reproductive isolation (Barth 1997).

First, though, a few points, formulated as working hypotheses, regarding the complex of adaptations that play a role in the vibratory courtship of spiders.

If the sensory system of the receiver has a relation to the signalling behavior of the sender, if the signalling behavior to some degree reflects the properties of the signals, and if the signals have a relation to the habitat and the physical properties of the communication channel, then we can expect the following.

1. The *vibratory signals* emitted by the courting spiders should be such as to minimize losses during transmission under the special physical conditions that prevail.
2. The *sensory systems* of the receiver will be such that under natural conditions they detect and identify just these vibratory signals, and can localize the site of their origin.
3. *Adaptive compromises* are to be expected, because for example (i) the vibratory sense of spiders must deal not only with courtship signals but also with other vibrations (see Chapters VIII and XVIII), such as the signals generated by prey animals; (ii) although the courtship signals should be conspicuous to the sexual partner, they should escape the notice of a potential enemy; (iii) the same signals are likely to be used in different behavioral contexts, such as species identification and male competition; (iv) and finally, the place and time of the signalling and reception should be chosen so as to enable adequately undisturbed communication.

We are concerned here with an abundantly interlinked system with many components, within which the vibration sense must fulfill its function.

1 Survey

The courtship behavior of *Cupiennius* can be subdivided into three sections (Fig. 1) (Rovner and Barth 1981, Barth 1993, 1997). With reference to the type of stimulus most important in each, we call these the chemical, the vibratory and the tactile phase of courtship. In the case of our night-active spider there is no evidence for an exchange of visual signals during courtship.

Chemical Phase

Sexual communication is initiated by the female, as is known of other spiders as well (Tietjen and Rovner 1982, Pollard et al. 1987). Her dragline,

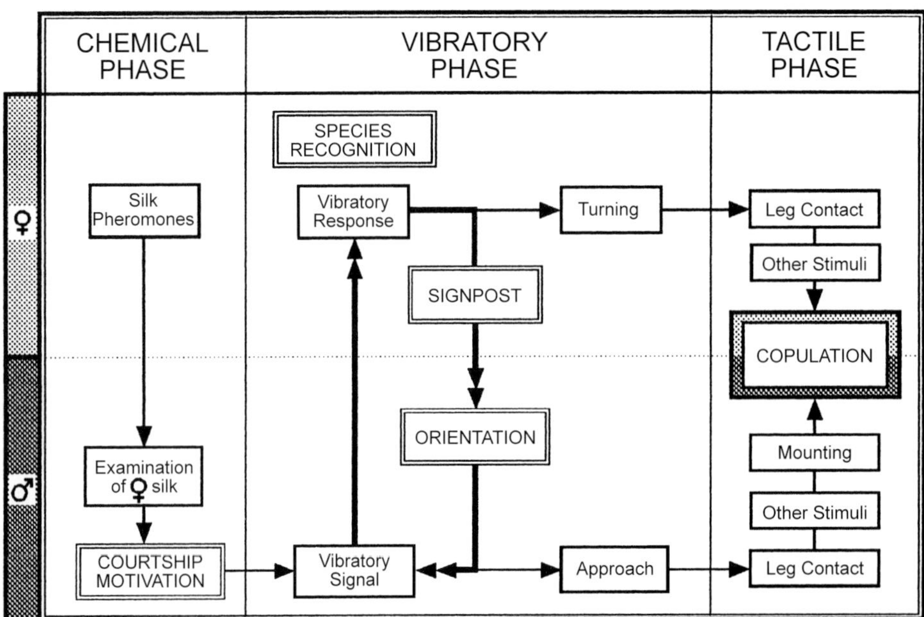

Fig. 1. The different phases of the courtship behavior in *Cupiennius salei*. *Heavy lines* and *double arrowheads* in vibratory phase indicate that the respective behaviors may be repeated several times before the next stage of courtship. (Barth 1994)

with which she is always connected, contains a pheromone, as do the threads at the attachment disks and in the webs spun to cover the shelter. This has recently been identified as the S-dimethyl ester of citric acid (see Chapter XII). When the male on his nocturnal wanderings encounters a female dragline, he examines it with his pedipalps, which bear chemoreceptive tip-pore sensilla (see Chapter XII); then he searches in the immediate vicinity of the thread and assumes the courtship posture, recognizable by a slight raising of the body (Plate 4).

There is a 25-minute color film (also available as video) on the courtship of *Cupiennius salei*, with German or English text, which shows these and other details (Barth FG, 1993, Scientific Film C 2318 of the Österr Inst f d wiss Film, ÖWF, Wien).

Once the male has become sexually excited by the female pheromone, the vibratory phase follows.

Vibratory Phase

Now the male begins to make scratching and drumming movements of the pedipalps and oscillatory up-and-down movements of his opisthosoma, which induce vibrations in the plant. These are propagated through the plant as bending waves with a mean attenuation of about 0.3 dB/cm (see Chapter XVIII) and reach the female, if she is still present. The female responds to the male's vibrations by sending out her own signal, within a narrow time window of 0.9 ± 0.5 s (*C. salei*, Rovner and Barth 1981). Having received her signal, the male starts out toward the female, which at least in *C. salei* remains largely stationary. During the approach this reciprocal vibratory "flirting" is repeated several times.

Tactile Phase

The third or tactile phase begins as soon as the male has come within reach of the female and lasts into the period of copulation (Plate 9). Although the most conspicuous stimuli in this phase are touch stimuli, it could be that both airflow stimuli in the near field (trichobothria, see Chapter XIX) and chemical stimuli are also significant here.

2 The Vibration Receptors

All of Chapter VIII is dedicated to the vibration receptors. Here I can be brief and select only a few aspects that seem important in the present context.

Among the various vibration-sensitive sense organs of spiders, the metatarsal organs stand out because of their sensitivity and structural adaptation to their special function. These are in the category of slit sensilla, which illustrate impressively how much potential there is in a structure as simple as a hole in the cuticle (see Chapter VII). The most obvious morphological adaptation of the metatarsal organs is their dorsal position at the distal end of the metatarsus of the walking legs and the orientation of their 21 slits, perpendicular to the long axis of the leg. The organ is arranged so that it bridges a groove in the metatarsus near the joint with the tarsus. When the tarsus is moved either up and down or sideways by substrate vibrations, the slits of the metatarsal organ are stimulated by compression. The position of the metatarsal organ in the groove increases its mechanical sensitivity.

The threshold curves of the slits of the metatarsal organ are not maximum curves but have the same shape as the frequency-response curves of high-pass filters. That is, their sensitivity with respect to deflection of the tarsus is low up to about 10–40 Hz but increases at higher frequencies, by up to 40 dB/decade; in this part of the curve, the threshold deflection corresponds to constant acceleration. Hence the slits of the metatarsal organ evidently operate as displacement detectors at low frequencies but as acceleration detectors at higher frequencies (see Chapter VIII).

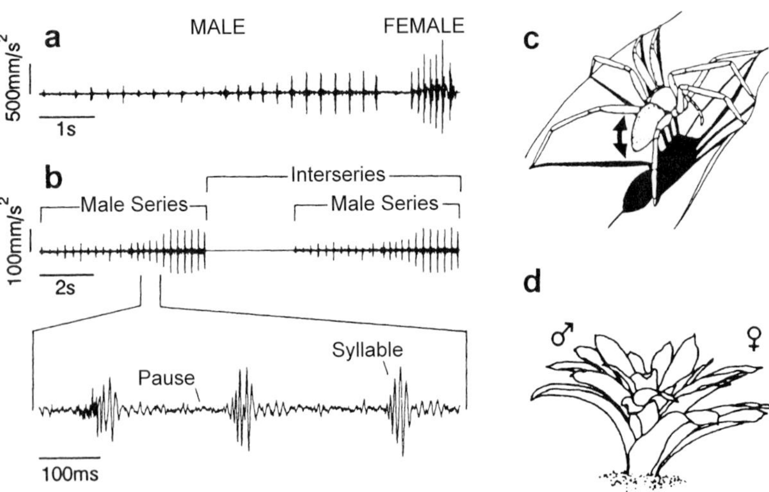

Fig. 2a–d. Vibratory signals of *Cupiennius salei* used during its courtship on a plant (bromeliad) and recorded by an accelerometer. **a** Male signal and female response. **b** Temporal structure of male courtship vibration. **c** Movements of the opisthosoma used by the male to produce courtship vibrations. **d** Courtship situation on a typical "*Cupiennius* plant". (Barth 1997)

3 Vibratory Courtship Signals on the Plant

The courtship of *Cupiennius* takes place on the same plant where the animal hunts. The plant serves as signal-transmitting channel and is the most important environmental factor in vibratory communication.

Male and Female Vibrations

The *male vibrations* consist of series of up to 50 syllables, each of which comprises as many as 12 pulses. They are produced by oscillatory movements of the opisthosoma (see below) and transmitted into the plant through the legs. The individual packets of syllables are separated by "silent" pauses lasting up to 10 s (Fig. 2). As a rule of thumb, the syllable duration of *Cupiennius salei* can be taken to be 100 ms and that of the pause between syllables, 250 ms.

The frequency spectrum of the syllables has its most prominent peak between ca. 75 Hz and 100 Hz. The pedipalp signals last only a few milliseconds and contain high frequencies, up to more than 1 kHz, in particular at the beginning of the very brief signal (Baurecht and Barth 1992) (Fig. 3). However, these signals are often omitted, whereas the opisthosomal signals are necessary and sufficient for normal courtship; therefore for the moment we shall ignore the pedipalp signals.

The *female vibrations* are sent out shortly after the end of the male's series of syllables and have widely varying durations, lasting ca. 0.1 s to 1.8 s. The temporal pattern so obvious in the male signal is lacking here. The main frequency components are at ca. 20 Hz to 40 Hz and thus are lower than those in the male signal (Plate 9).

Functions and Adaptive Properties

When asking what is advantageous and adaptive about these courtship signals, we soon find ourselves immersed in a motley mixture of problems.

Fig. 3a–d. Spectral analysis of a typical male courtship signal of *Cupiennius salei* and its transmission through the plant (bromeliad). *Inset* in **a** shows experimental situation with the male spider and the two accelerometers (●) on the ipsilateral and a contralateral leaf. **a, b** Oscillogram and "sonagram" (vibrogram) of the signals recorded on the two leaves. Signals produced with pedipalps (*) contain the high frequencies typical of them only on the ipsilateral leaf. Likewise, the increase in amplitude of the two final opisthosomal signals is not transmitted to the contralateral leaf. **c, d** FFT (see p. 278) of a pedipalpal signal and of an opisthosomal signal recorded on the ipsilateral and the contralateral leaf. Note the much stronger attenuation of the high-frequency components as compared to the low frequencies. 0 dB corresponds to highest amplitude in **d**. (Baurecht and Barth 1992)

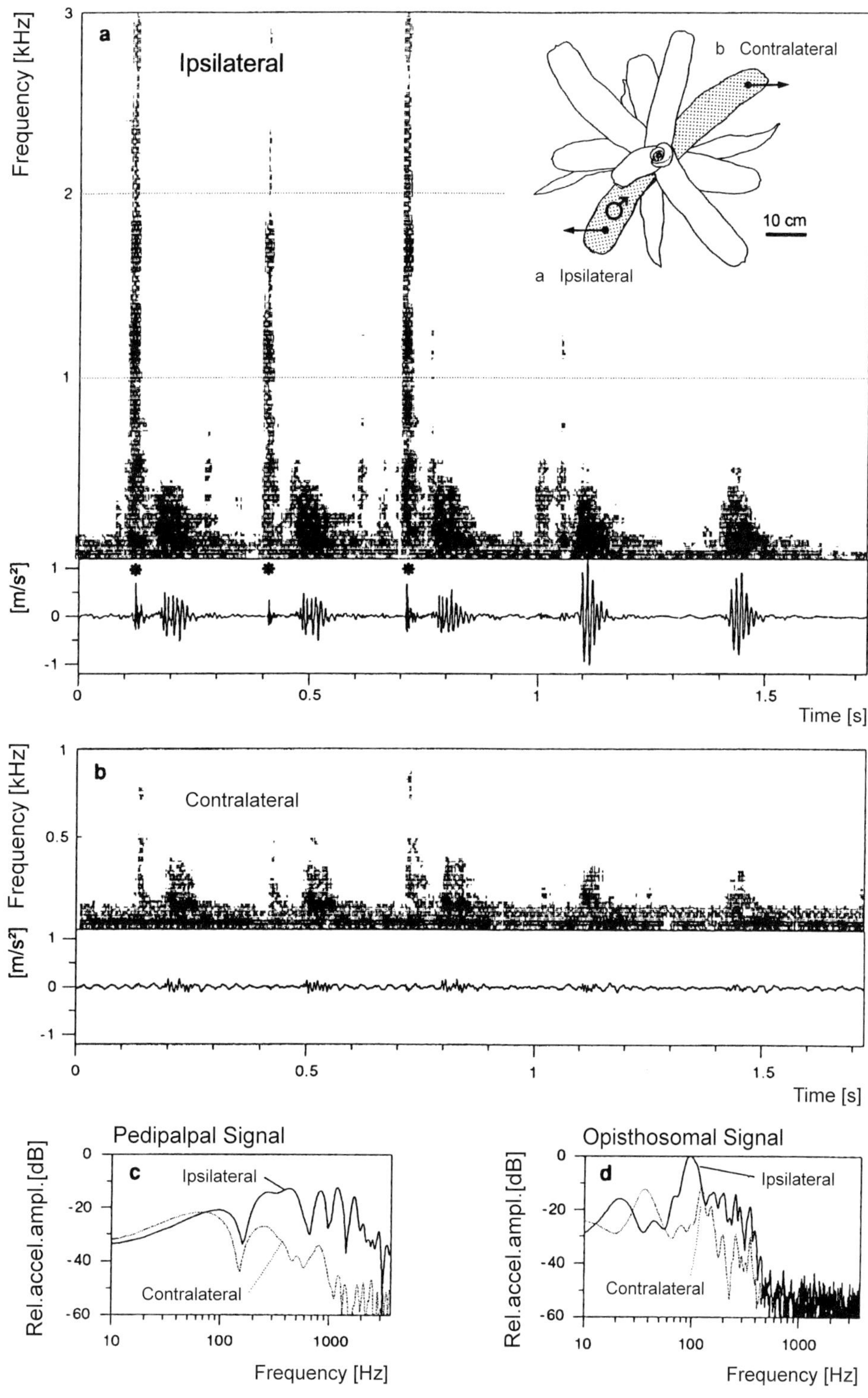
a
Ipsilateral
b Contralateral
10 cm
a Ipsilateral
Frequency [kHz]
[m/s²]
Time [s]
b
Contralateral
Frequency [kHz]
[m/s²]
Time [s]
Pedipalpal Signal
c
Ipsilateral
Contralateral
Rel.accel.ampl.[dB]
Frequency [Hz]
Opisthosomal Signal
d
Ipsilateral
Contralateral
Rel.accel.ampl.[dB]
Frequency [Hz]

Why Substrate Vibrations? *Cupiennius* is a genus of night-active spiders. Vibratory communication continues to function at night. Airborne signals, a potential alternative, can be effectively produced only by animals that are large with respect to the wavelength of the emitted sound (for a dipole-like sound source $r_0 > \lambda/2\pi$, where r_0 is the radius of the sound source). It is for this reason that so many insects use the ultrasonic region, which in turn is why the mechanisms of frequency multiplication, in particular stridulation, are so important in insects. Substrate vibration is different: even small animals can effectively send out low frequencies. Because the transmission range of vibration signals is relatively short and they persist only briefly, they permit an intimate private conversation, which reduces the danger that competitors or enemies will take notice. With airborne sound, as well as olfactory and visual stimuli, this is usually not the case. Furthermore, the latter signals propagate more diffusely than do vibratory signals through the plants preferred by *Cupiennius*.

Vibratory signals also have little tendency to drift off track and are not severely affected by obstacles on the transmission route (unlike chemical or visual stimuli), which must be advantageous to a receiver searching for the location of the sender.

The Differential Functions of the Courtship Signals of *Cupiennius*. As figure 1 shows, the vibratory signals exchanged by males and females have entirely different functions. The male signals are species specific (Fig. 4). They differ from one another most prominently in their temporal pattern, less in frequency content, and serve mainly to identify the male's species. It follows that the female must make the first decision whether the prospective partner is a conspecific male. Then her signal informs him that a conspecific female is present. If the male had belonged to some other species, she would not have responded, or her response would be much less reliable (see Section 6). The female signal also shows the male the way to the female.

It also seems likely, although we have no experimental evidence for this as we do for the preceding functions, that the male signals reduce the female's aggressiveness, and that the vibrations of both partners help to synchronize the partners. According to our experiments with *Cupiennius* sexual selection cannot be distinguished from signal recognition. Modulatory long-term effects of the courtship signals on the motivation and the behavior of the receiver at each end have not yet been analyzed experimentally.

The Suitability of the Signals. What could we say if asked whether and how the vibrations are suitable to perform the functions just listed? Well, at least we have some partial answers.

The low frequencies. From the viewpoint of the bioacoustician, at least, the vibrations of *Cupiennius* are of very low frequency. α. As mentioned above, communication by airborne signals would be extremely ineffective if similarly low frequencies were used, with wavelengths of several meters – unless a minimal transmission range were acceptable and the receiver were to measure the near-field air movement (rather than the sound pressure). In fact, it is conceivable that *Cupiennius* might use the trichobothria (see Chapter IX) during the tactile phase of courtship to detect such airborne near-field vibrations. In an earlier stage of courtship, however, this certainly would not work. The male searching around in the darkness must be able to send signals over distances of a meter or more in order to learn whether a female ready for courtship is sitting on the plant. β. Frequencies around 30 Hz (female) or 90 Hz (male) propagate very well, with little attenuation (ca. 0.3 dB/cm) in the plants on which *Cupiennius* sits. Frequencies only a little higher, over a few hundred Hertz, are considerably more severely damped. For example, the pedipalp signals with their comparatively high frequencies are considerably more strongly damped than the opisthosomal signals and, accordingly, have a much shorter range (Fig. 3). γ. Finally, it is notable that the vibrations produced by the female are of particularly low frequency. Because the female signals show the male the way to a partner, we can postulate an additional adaptive value of the low frequencies: the lower the frequency, the smaller is not only the attenuation

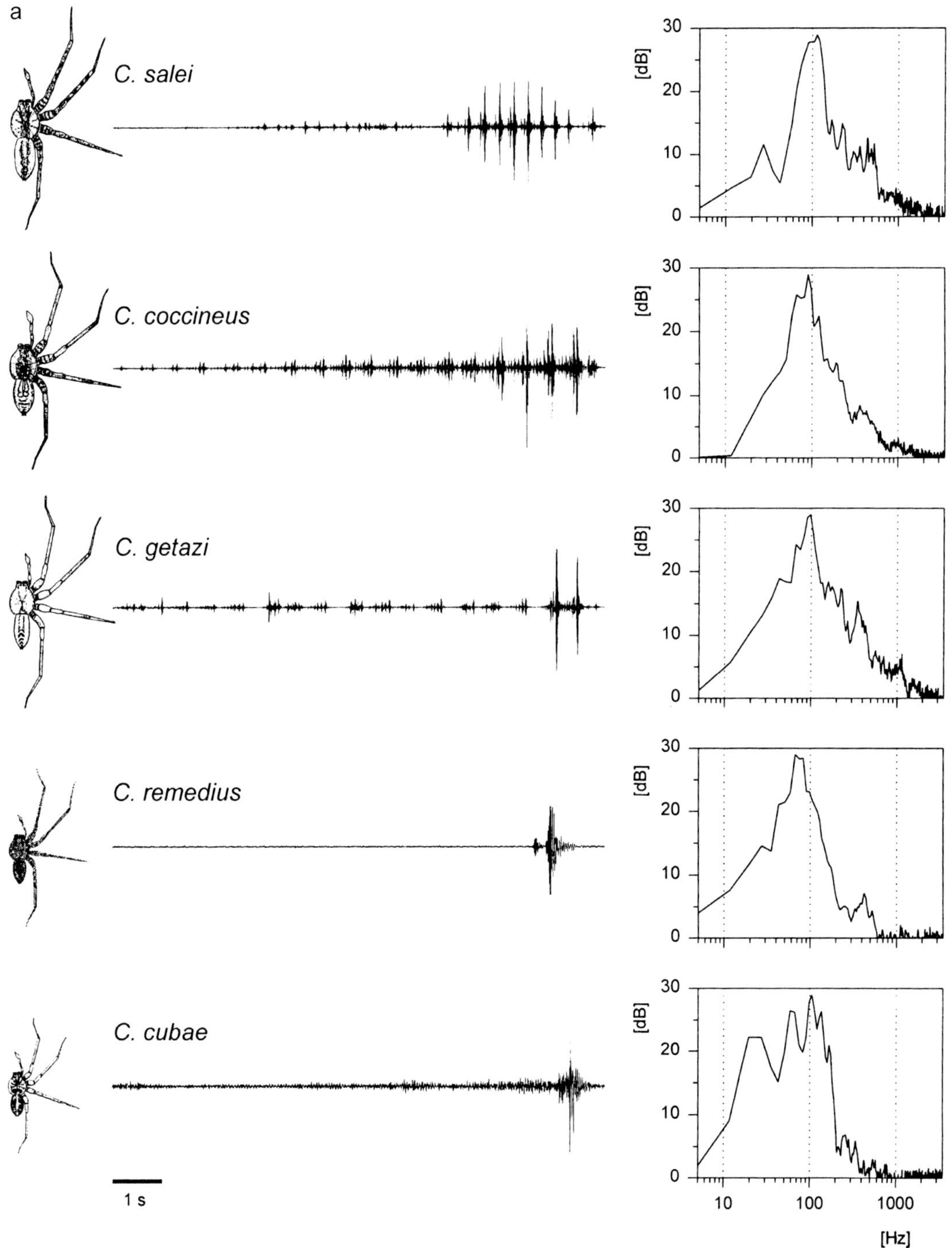

Fig. 4 a, b. The vibratory courtship signals of the males of five different species of the genus *Cupiennius*. **a** Oscillograms (recorded with accelerometer) and spectra. **b** Vibrograms together with the corresponding oscillograms (Fig. 4 b see page 276)

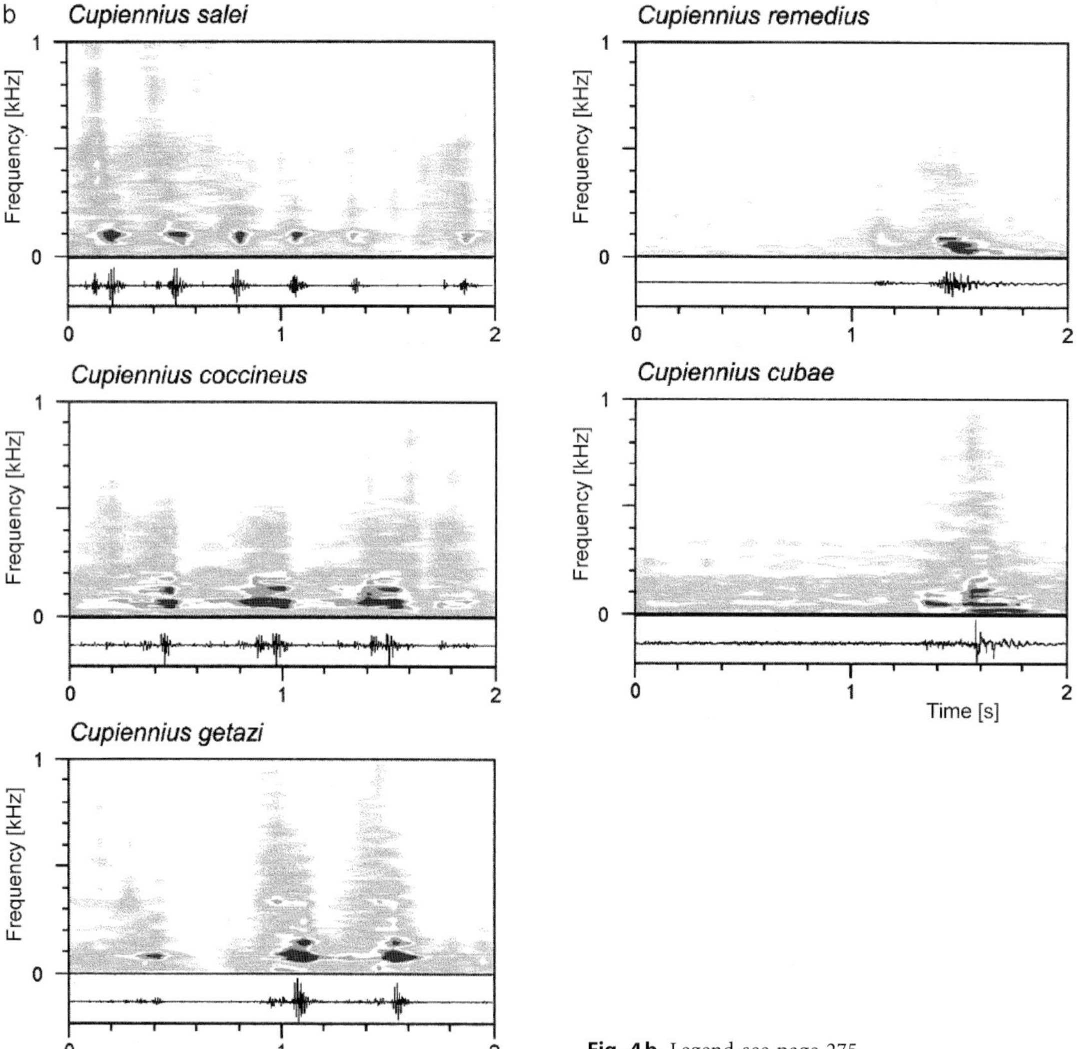

Fig. 4 b. Legend see page 275

but also the propagation velocity. This in turn results in an increase in the time-of-arrival differences (Δt) of the vibration signal at the various legs of the male. Because Δt is used for orientation to a source of vibration, the lower frequencies must be advantageous in this respect as well. This point was discussed at greater length in Chapter XVIII.

The temporal pattern. The most striking characteristic of the male courtship vibrations in *Cupiennius* is their temporal arrangement – like cricket song! There are several advantages to this. α. First, it allows the courtship signals to be discriminated from typical background vibrations and from vibrations caused by prey animals (see Chapter XVIII; Barth 1985 a). Also, the courtship vibrations of males of various species of *Cupiennius* can be distinguished from one another, as they differ most conspicuously in temporal pattern. That temporal parameters are of outstanding significance in species identification and reproductive isolation has been demonstrated experimentally and is discussed in detail in Section 6 of this chapter. β. The frequency content of the syllables in a male series hardly varies. Therefore all the syllables propagate in the plant with the same velocity, and the temporal pattern of the syllables is retained. γ. The rapidity of the female response, which is produced barely a second after the end of the male's series, presumably helps him to recognize the vibration as an answer to his own. In noisy surroundings a frequently observed virtual synchronization (Shimizu and Barth 1996) also contributes in this regard: the female often responds at a time that would have been occupied by a male syllable, if the male had continued his series.

The amplitude. As described in Chapter XVIII, the range of the courtship signals of *Cupiennius*, which depends on the original signal intensity, the damping during transmission and the sensitivity of the receiver, is remarkably long at ca. 2 m. In the field in Costa Rica a female of *C. coccineus* sitting on a banana plant gave a courtship signal in response to the signal of a male 3.8 m away (Barth 1993). Behavioral experiments in the laboratory, using synthetic male signals (Schüch and Barth 1990), indicate that the male series must exceed an acceleration-amplitude threshold of about 8 mm/s^2 at the site of the female in order to induce the female to respond. In addition to this lower threshold, there is also an upper one. In the laboratory responses could be elicited in a few cases (< 10%) with signals as strong as 8000 mm/s^2. The natural amplitude range extends up to about 1000 mm/s^2. The reason for the marked decline in response frequency at higher amplitudes, instead of a further increase, will become clear in Section 5 below (Barth 1993, Baurecht and Barth 1993). The large working range of the female reflects the large variability of signal strength under natural conditions. In contrast, the other signal parameters may vary only within substantially narrower limits in order to pass the receiver's filter (see Section 6). From the viewpoint of sensory ecology, this is entirely reasonable.

The signal amplitudes found under natural conditions must be a compromise, to satisfy many different selection pressures. Among these, particularly in relation to the lower threshold, are the receptor-cell and neuronal sensitivities of the receiver; also the distance between sender and receiver over which communication has to function and which depends on, among other things, the structure of the spider population and the plant community it occupies; and the attenuation of the signal by the conducting substrate as well as the noise level. The upper limit of the vibration intensity is determined by the energetic constraints on the sender; by the risk of alerting predators and other enemies; and by the response properties of the neuronal apparatus of the receiver (see below).

4 How is the Courtship Signal Produced?

Although the widespread occurrence of vibratory communication in arthropods has long been known (for review see, for example, Markl 1969, 1983), there are still many open questions regarding the production of vibratory signals. This initially applied also to the mechanism by which *Cupiennius* generates "opisthosomal" signals. It moves the opisthosoma up and down without even touching the substrate, to say nothing of beating against it. The oscillations are conducted into the substrate through the legs.

How do male spiders generate the vibrations so important for species identification, and how does it happen that their main frequency is between ca. 80 and 100 Hz? Do the muscles move the opisthosoma directly at this high frequency? Or are passive mechanisms operating? For instance, the spider might produce a brief, pulsatile movement of the opisthosoma, the main frequency of which is amplified by the resonance of one of the components of the oscillatory system composed of the spider plus the plant (such as the legs or the leaf).

Stefan Dierkes took up this subject as part of his doctoral work. He studied *Cupiennius getazi*, the males of which are particularly suitable for the purpose because their courtship behavior can be so readily elicited. We have no doubt that the basic conclusions to which he came also apply to the other species of *Cupiennius* (Dierkes and Barth 1995).

For a precise analysis of the opisthosoma movements that produce the signal, both high-speed videography (400 frames/s) and laser Doppler vibrometry were employed. By means of

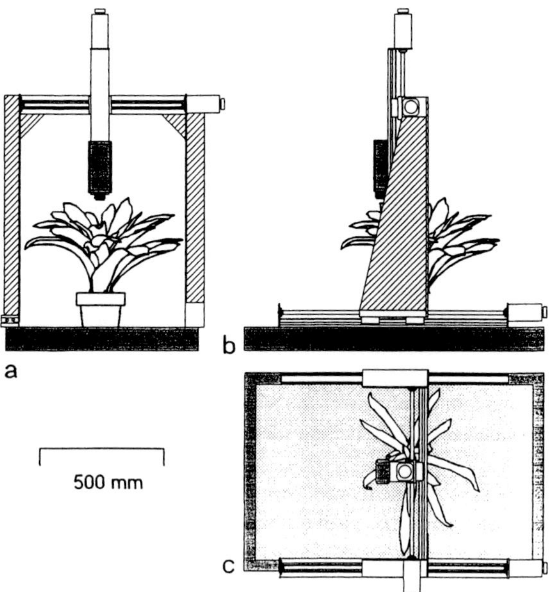

Fig. 5 a–c. Apparatus used to position the laser Doppler vibrometer precisely and quickly on the spider opisthosoma and on the plant by means of stepping motors, in order to record vibrations produced by a freely walking male. **a, b,** and **c** frontal, lateral and top view. (Dierkes 1992)

stepping motors (step size 10 μm; movement velocity up to 10 cm/s) the laser could be made to follow the spider accurately as it courted freely on a bromeliad, so as to provide a non-contact measurement of the oscillations of the opisthosoma (Fig. 5) (Dierkes and Barth 1995). In order to determine the transmission characteristics of the plant, broad-band noise was introduced to it by means of an electrodynamic vibrator and the resulting oscillations in the plant were measured with the laser vibrometer at various distances from the site of introduction; the frequency spectra at these measurement sites were found with the fast Fourier transformation (FFT) and finally compared. The transmission characteristics of the spider body were investigated similarly; in this case the vibrations were introduced in the "reverse" direction, that is, through the plant to the walking-leg tarsi.

Finally, Stefan Dierkes took a close look at the complicated anatomy in the region of the muscles responsible for the opisthosoma movement, and by chronic implantation of electrodes recorded the activity of as many as 6 muscles while the animal was courting – and simultaneously obtaind a laser-vibrometric measurement of the movement of the opisthosoma.

The Courtship Movements

To generate the syllables male *C. getazi* move their opisthosoma dorsoventrally about an axis situated in the petiolus (the spider's "waist"), with only slight lateral deviations (Fig. 6). Whereas at the beginning of a series the deflec-

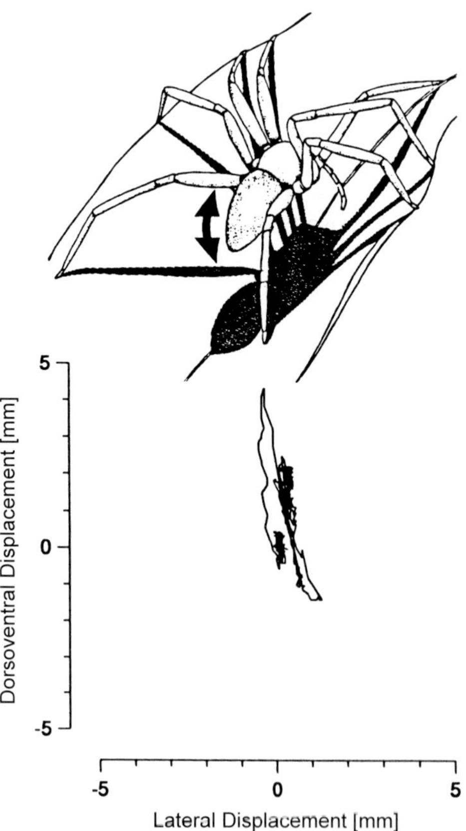

Fig. 6. Displacement of the opisthosoma of a male *Cupiennius getazi* during vibratory courtship, recorded with a high-speed video camera (400 frames/s). Coordinate origin (0) set to the spinnerets' position in the first of a total of 1600 analyzed frames. The position of the substrate was at ca. –8.5 cm on the ordinate. (Dierkes and Barth 1995)

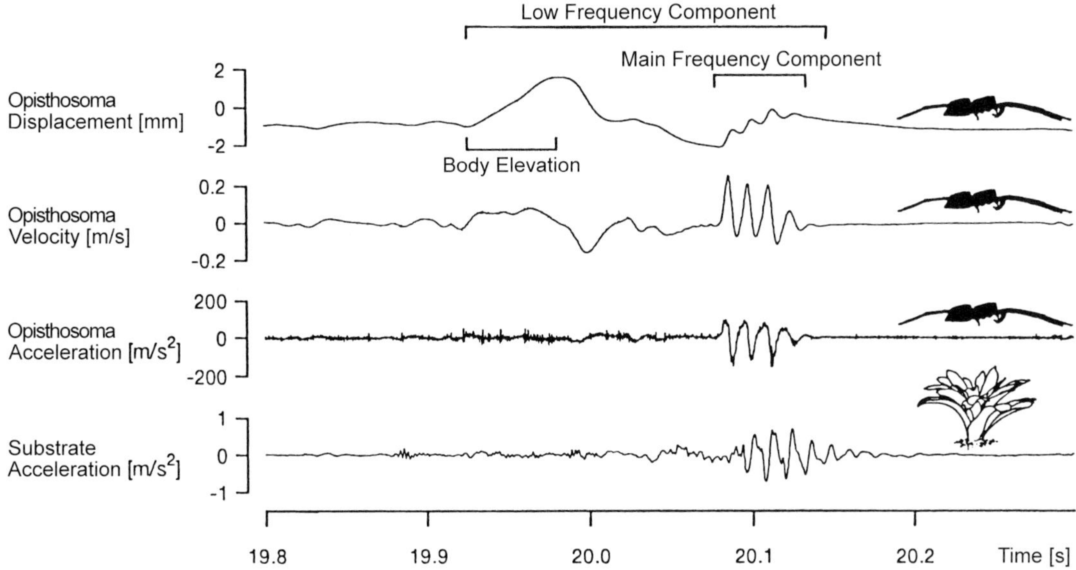

Fig. 7. The displacement, velocity and acceleration of the male opisthosoma during the production of a vibratory signal (*Cupiennius getazi*) and acceleration recorded on the plant at a distance of 15 cm from the spider. (Dierkes and Barth 1995)

tion of the opisthosoma (measured at the spinnerets) amounts to only ca. 0.4 mm, or ca. 2°, it is typical of *C. getazi* that its amplitude becomes much larger in the three or four final syllables of the series, increasing to ca. 6 mm or 30°. Because these final syllables in themselves are just as effective as the complete signal in eliciting a female response (Schmitt et al. 1992), all further analyses and explanations will refer to these.

In figure 7 a nice detail of this movement can be seen. Even during the powerful oscillations associated with the final syllables, the opisthosoma does not strike the substrate. The spiders manage this by raising the body to an appropriate height just before the final syllables.

The movement of the opisthosoma contains a low-frequency component at 10 to 20 Hz. Upon this is superimposed a higher-frequency component, which appears during the last upswing at the end of each syllable and corresponds to the main frequency component of the male vibration signal, around 80 Hz. The acceleration in the simultaneously measured substrate vibration is remarkably closely correlated in time with the rapid movement component of the opisthosoma (Fig. 7). Much the same applies to the frequency spectra (Fig. 8). Both the spectrum of the opisthoso-

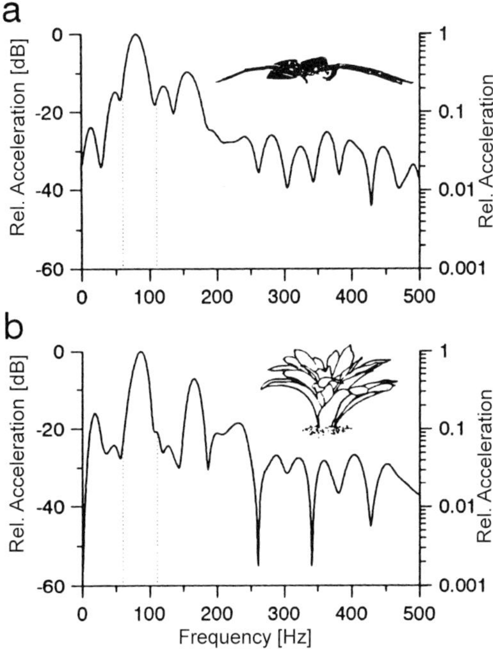

Fig. 8a,b. Frequency spectra of the acceleration of the opisthosoma (**a**) and of the plant (**b**) during the syllable of the male courtship vibration shown in Fig. 7. (Dierkes and Barth 1995)

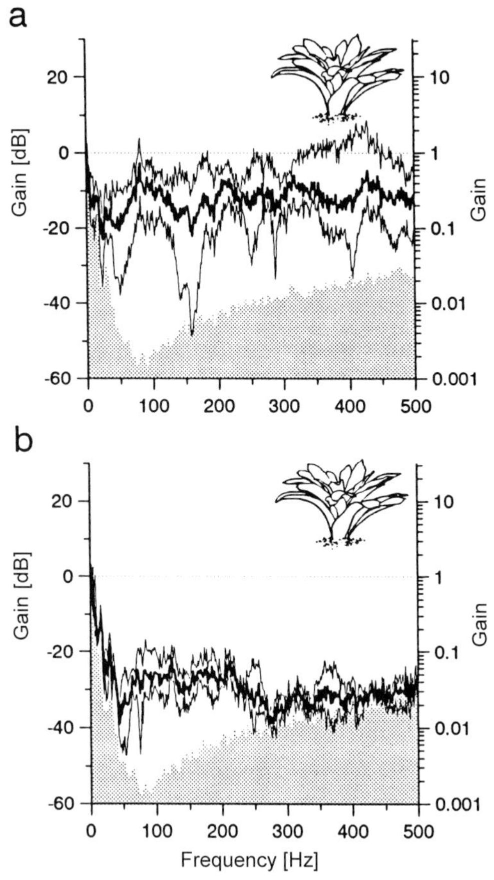

◀ **Fig. 9 a, b.** Mean (*bold line*) and range (*thin lines*) of the transfer functions of two leaves of a bromeliad (*Aechmea fasciata*) for substrate vibrations. *Shaded area* indicates mean background noise. The reference amplitude (gain = 1 = 0 dB) is the vibration amplitude of the plant at the attachment point of the vibrator (input). **a** Ipsilateral leaf (that is, the leaf coupled to the vibrator); mean and range of transfer functions measured at distances of 5, 10, 15, 20, and 25 cm from site of input. **b** Contralateral leaf; mean and range of transfer functions measured at distances of 5, 10, 15, and 20 cm from the leaf base. (Dierkes and Barth 1995)

mal movement and that of the plant vibration have a large peak at ca. 80 Hz, a peak at ca. 15 Hz (the lower-frequency component) and additional peaks near the first harmonic of the main frequency, at around 150 Hz.

The Transmission of the Vibrations

The most essential things to know about the propagation of vibrations (bending waves) in the plants relevant to *Cupiennius* can be found in Chapter XVIII. For the present purposes the finding of chief importance is that in the frequency range of the signals there were no resonances of the plant (bromeliad) even though frequencies in the relevant range, between 50 and 150 Hz, are less attenuated than the adjacent fre-

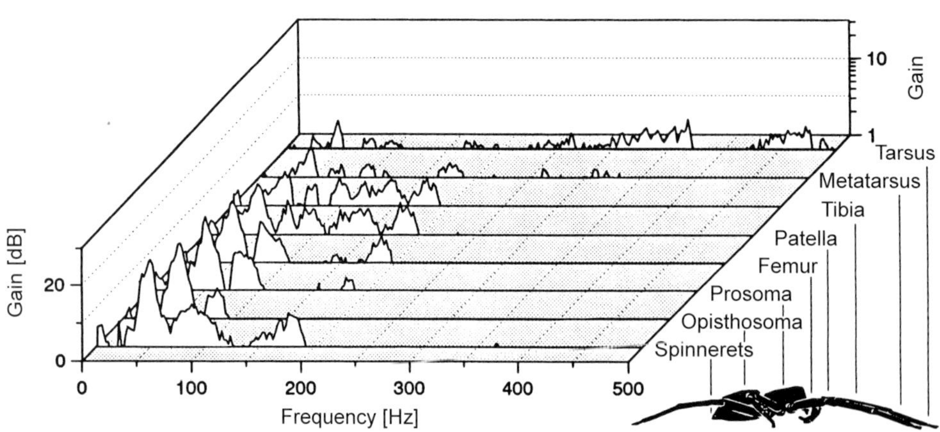

Fig. 10. Transfer functions of several parts of the body of a male *Cupiennius getazi* for substrate vibrations. The reference amplitude (gain = 1 = 0 dB) is the vibration amplitude of the plant 1 cm in front of the tarsus of the spider. (Dierkes and Barth 1995)

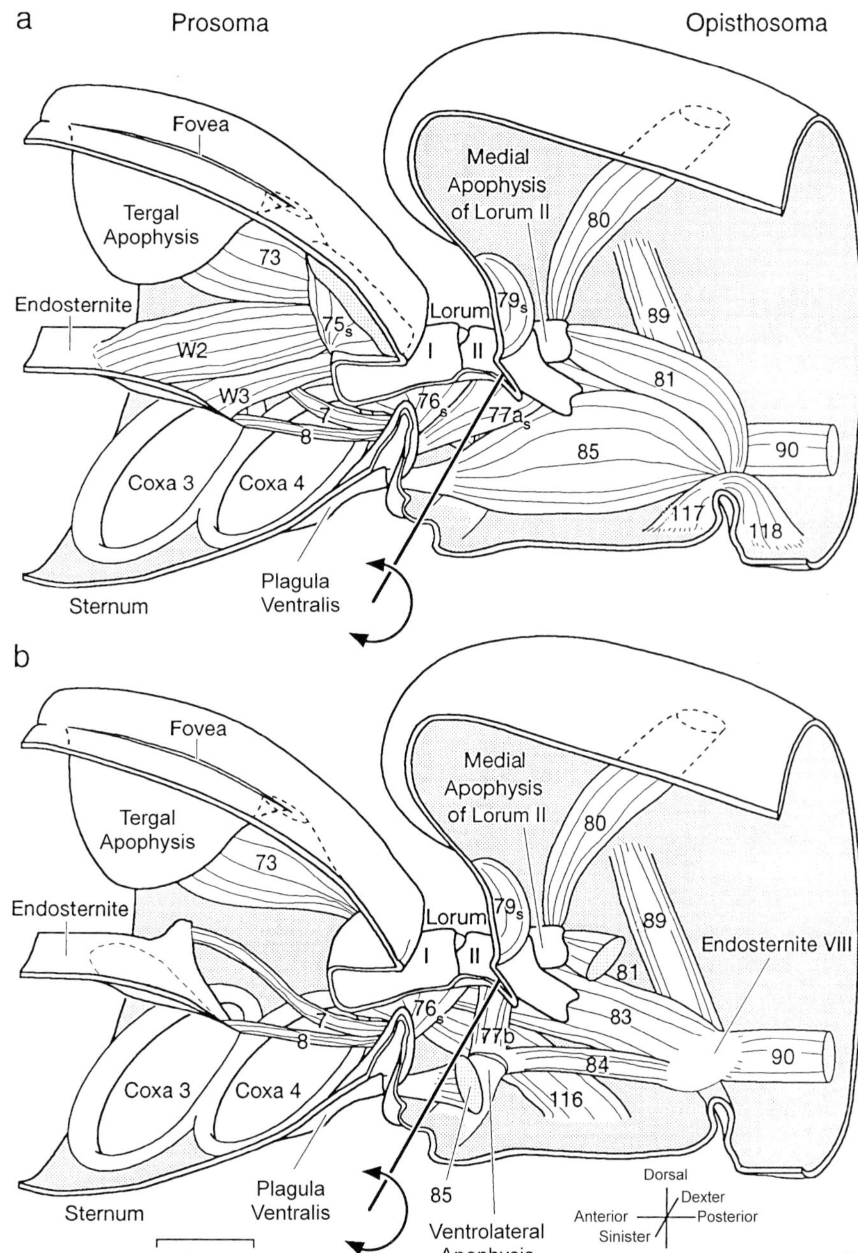

Fig. 11a, b. Musculature in the region of the petiolus of *Cupiennius getazi* (parasagittal section). *Double arrow* marks axis of dorso-ventral rotation of opisthosoma. Muscles W_2, W_3, 75_s, $77a_s$, 85, 81, 117, and 118 have been removed from preparation shown in **b** in order to show more laterally located muscles. Subscript "s" indicates muscles on left body side. (Dierkes and Barth 1995)

quency ranges, and on "contralateral" leaves (relative to the site at which the vibrations were introduced) only frequencies below 250 Hz can be distinguished from the background noise (Fig. 9).

The transfer properties of the spider's body look different from those of its plant: there are resonances (that is, amplification of the input signal) between 0 and about 250 Hz. In the example of figure 10 the largest such peaks are at

50 Hz and 80 Hz, whereas frequencies above 200 Hz are reduced in amplitude as they pass through the spider.

Such experiments are difficult. Not only do they involve a whole, living animal; the animal must stay in the "courtship position" without actually courting, and although it is receiving vibration through the plant on which it stands, it should keep its leg and body posture unchanged for a long enough time that the measurements can be completed. Under these conditions exact reproduction of results cannot be expected from measurements on different animals or even on the same animal at different times. Nevertheless, a typical pattern does emerge, of which the example in figure 11 is representative.

The Muscles and Their Activity

To a non-specialist it may be surprising that in the region of the petiolus responsible for movement of the opisthosoma there are no less than 17 paired muscles and 2 unpaired ones, for a total of 36 muscles. Figure 11 shows their bewildering complexity.

By the criteria of position and insertion points, several of them might conceivably serve as courtship muscles. In particular muscle 85, a large paired muscle attached anteriorly to the sternite of the petiolus and posteriorly to the endosternite above the interpulmonary fold, is a good candidate for the opisthosoma depressor. The muscles 83, 84, 79 and 80 might also function to bend the opisthosoma down. The only one that looks like an elevator is muscle 81, which (like muscle 85) inserts at lorum II and at endosternite VIII (Fig. 12). What the contraction of these muscles actually contributes to the courtship movements of the opisthosoma can be discovered only by recording their activity electrophysiologically during courtship. Fortunately, Stefan Dierkes was able to do such experiments in the laboratory and with the aid of Norbert Elsner in Göttingen.

The most important finding: muscle 85 is active only during courtship; it is not involved in locomotion or in attaching silk to the substrate. Its activity pattern corresponds to that of the opisthosomal movements. Figure 13 shows an example of its contraction pattern during a syllable of the male signal. The first muscle action potential (1) is followed by a downward movement of the opisthosoma. In the subsequent phase (2), which lasts about 40 ms, the muscle discharges 9 action potentials with a frequency of about 200 Hz. These are accompanied by a tetanic contraction, and the opisthosoma remains in its ventrally deflected position. In phase 3 an unidentified elevator muscle (for technical reasons the attempts to obtain clear recordings from muscle 81 failed) moves the opisthosoma back upward. As it rises, muscle 85 again contracts three to five times at intervals of 11 to 12 ms. This corresponds to a frequency of 80 to 90 Hz – the main frequency component in the male courtship signal, the origin of which is thus explained! Figure 13 shows this very clearly: upon the upward movement of the opisthosoma are superimposed, in the rhythm of the action potentials of muscle 85,

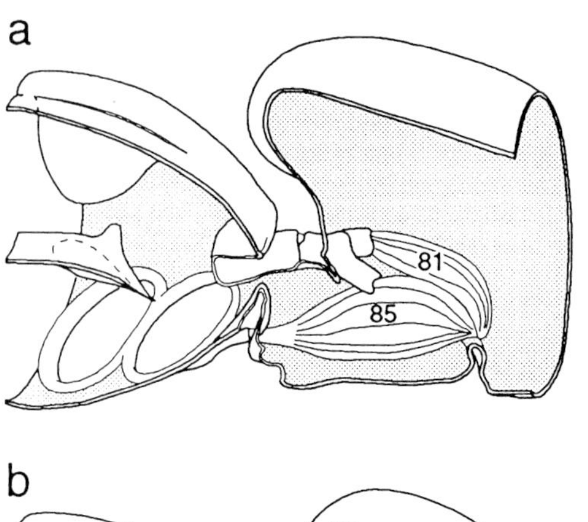

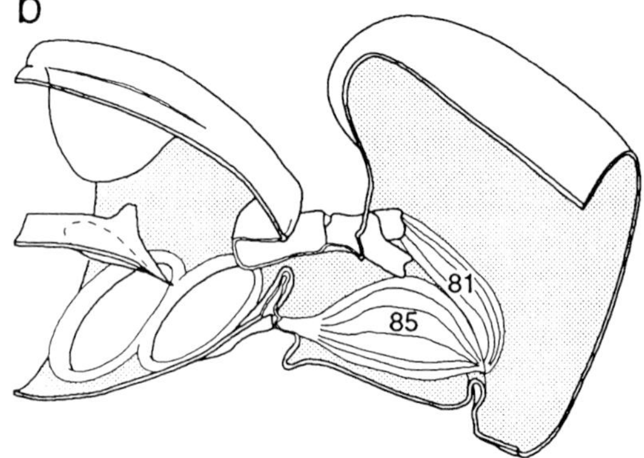

Fig. 12a, b. Roles of muscles *81* and *85* for the dorso-ventral movement of the opisthosoma. **a** Upper, dorsal position due to contraction of muscle 81. **b** Lower, ventral position due to contraction of muscle 85. (Dierkes and Barth 1995)

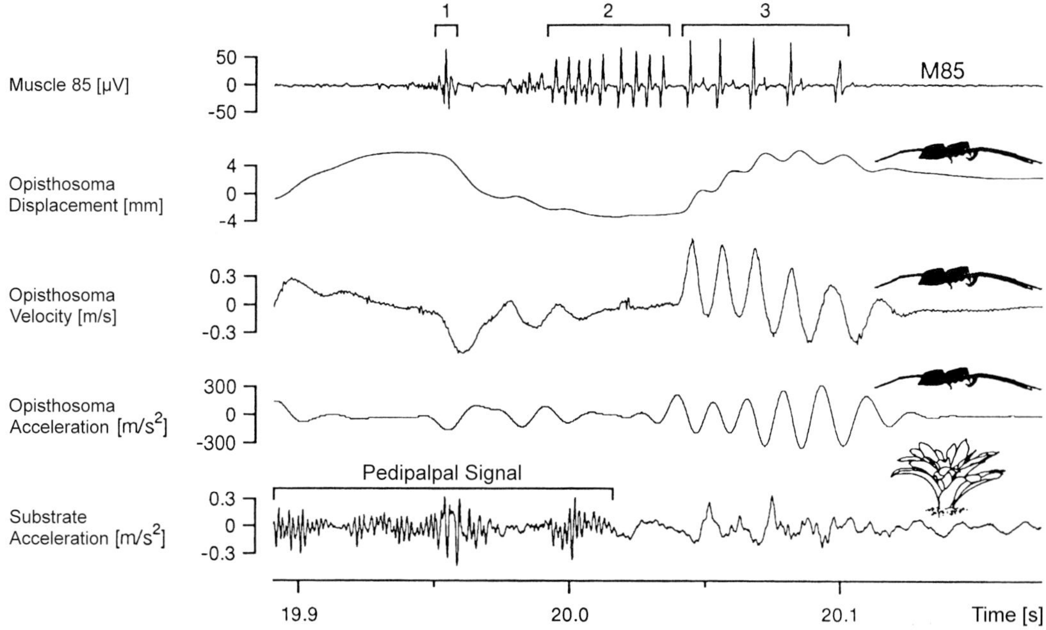

Fig. 13. Simultaneous recordings of the activity of depressor muscle 85, of the movement of the opisthosoma (displacement, velocity, acceleration) and of the plant vibrations during a courtship period of a male *Cupiennius getazi* lasting 21 s. *1* Single action potential; *2* frequency of action potentials ca. 200 Hz; *3* frequency of action potentials 80 to 90 Hz. (Dierkes and Barth 1995)

small downward movements with high velocity and acceleration.

So it turns out that the main component of the signal is not introduced by resonances in the plant. Instead, it is actively and directly produced by the spider. Because the contraction frequencies of the crucial muscle 85 are in the range of resonant frequencies of the spider's body, the vibration is amplified as it passes through the legs into the plant.

What is the significance of the low-frequency components in the opisthosomal movement? In behavioral experiments with *C. getazi* 70% of the females responded to natural male courtship signals, only 7% to the low-frequency components alone (0–60 Hz, filtered signal), but again 80% to a low-pass-filtered signal (0–110 Hz) containing both components (Schmitt et al. 1994). Therefore both components of the signal contribute to its efficacy.

In a physiological rather than ethological context, it is particularly striking that *Cupiennius* uses such a simple mechanism to produce the opisthosomal signals. Only one pair of rapidly contracting muscles (No. 85) is needed for the downward movement, while the antagonist (for the upward movement) can be a slow muscle. Hence no special devices for frequency multiplication are required.

5 Responses in the Peripheral and the Central Nervous System

The Level of the Vibration Receptors

The threshold curves for the slits of the metatarsal organ were shown in Chapter VIII. As is the rule for determining threshold curves, sinusoidal stimuli at different frequencies were used. That curves so constructed do not in themselves suffice to predict the responses to natural signals, such as the temporally highly structured

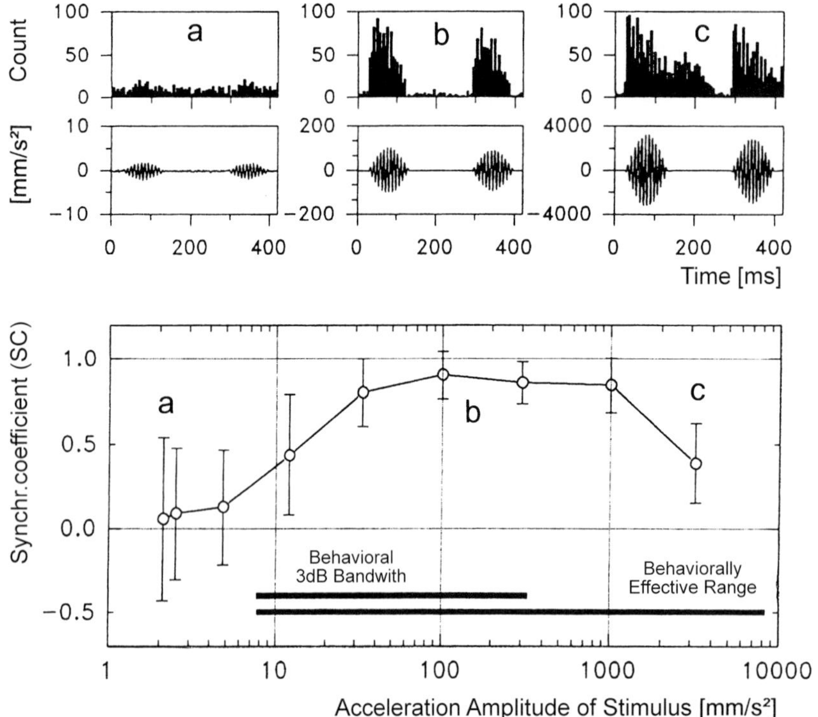

Fig. 14. Responses of the female metatarsal organ (*Cupiennius salei*) to male courtship vibrations. Copying properties of the long slit 2 of the organ (see Fig. 2, Chap. VIII) as a function of the acceleration amplitude of the synthetic signals used in this experiment (carrier frequency 114 Hz, syllable duration 105 ms, pause duration 160 ms). Counts of action potentials refer to 10 series of 10 syllables each (bin width 10 ms). *Top* Peri-stimulus time (PST) histograms of the receptor response to stimuli of increasing amplitude (note different calibration of ordinates in *a*, *b*, and *c*). *Bottom* Synchronization coefficient *SC* as a function of stimulus amplitude (acceleration). *SC*=1 indicates perfect copying of syllable and pause durations; *SC*=0 indicates no copying. (Barth 1993, after Baurecht and Barth 1991)

male courtship signal, became very clear ten years later. Using suprathreshold natural and synthetic male signals (*C. salei*) as stimuli, Dieter Baurecht discovered that the metatarsal organ of the female is extremely well tuned for reception of certain parameters of the male signal in particular, and that considerable filtering goes on even at the sensory periphery. In certain respects, at least, this result matches remarkably closely the results of behavioral experiments (Baurecht and Barth 1992, 1993).

(i) Amplitude

From behavioral experiments (see Section 6) we know that the amplitude of the male signal is one of its less influential parameters, when the effectiveness of the signal is measured by the number of female responses. The receptors, too, respond over a broad range of acceleration amplitudes. The relationship between their discharge rate and the acceleration amplitude of the syllable is logarithmic, so that one receptor can cover the entire natural amplitude range of about 3–1000 mm/s^2 (which is necessarily accompanied by a reduced ability to discriminate amplitudes). For both the animal's behavior and the individual slits of the metatarsal organ, the threshold for eliciting a response is ca. 8 mm/s^2. It is particularly interesting, however, that with amplitudes above ca. 1000 mm/s^2, greater than the highest naturally occurring amplitudes, the synchronization between the vibration signal and the receptor response rapidly decreases (Fig. 14). The reason is that more impulses are discharged between the syllables. In the behaviorally most effective amplitude range, in contrast, the receptor copies the temporal pattern of syllables very accurately. The sensory misinterpretation at too-high amplitudes presumably contributes to (and may even be the most important cause of) the rapid decline in the vibratory responses of the female in the same amplitude range. The fact that the response rate of the female to synthetic male signals passes through a maximum indicates the female's preference for the male to be separated from her by a particular range of distances.

(ii) Frequency
In the light of the difference in their ethological significance, it is remarkable that the male pedipalp and opisthosoma signals are to a great extent processed separately by the metatarsal organ of the female. The series of syllables generated by the opisthosoma excite predominantly the long, distally situated slits of the metatarsal organ, whereas the higher frequencies contained in the pedipalp signals elicit responses of all the slits so far investigated, including the short proximal ones. Furthermore, it is the long slits in particular (for instance, slit 2) that exhibit the logarithmic stimulus-excitation relationship mentioned in (i). Their functional significance is also clear in the fact that the response curves of the small slits (for instance, slit 11) are not logarithmic but rather linear between 10 and 10,000 mm/s^2 (Baurecht and Barth 1992, 1993) (Fig. 15). The main frequency component of the male syllables is represented as a distinct peak in the distribution of the spike frequencies.

(iii) The temporal pattern
A third group of filter properties, which cannot be derived from a simple threshold curve but nevertheless indicates important adaptations or matches of the metatarsal organ to characteristics of the male courtship signal, brings us back to its temporal patterns. Their special significance for species identification is treated in Section 6.

- The synchronization coefficient: although the number of nerve impulses per syllable decreases as the series proceeds, the quality of the representation of the temporal pattern of syllables and pauses is maintained (largely constant synchronization coefficient; Baurecht and Barth 1993).

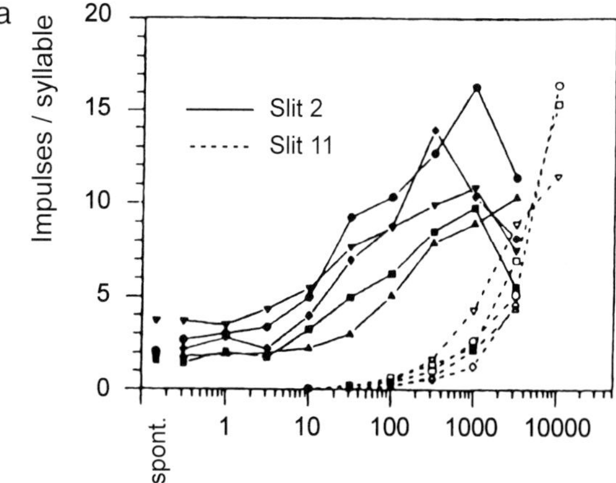

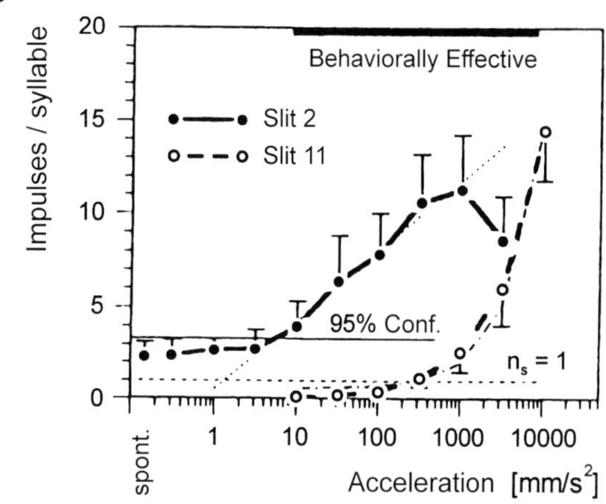

Fig. 15 a, b. Response characteristics of the long slit 2 and the short slit 11 of the metatarsal lyriform organ to vibrational stimuli (syllables) of increasing acceleration amplitude. **a** Number of action potentials per syllable (n_s); means for five animals (*Cupiennius salei*, female). *spont.* Spontaneous activity of slit 2. **b** Means of n_s of all animals examined (SD shown in one direction only). *95% confidence interval* of spontaneous activity indicates the threshold of slit 2, whereas the mean value of one action potential per syllable is regarded as the threshold for slit 11. Note the logarithmic relationship between the response of slit 2 and stimulus acceleration between 3 and 1000 mm/s^2. In contrast, the same relationship is linear in the case of slit 11 over the entire range tested. (Baurecht and Barth 1993)

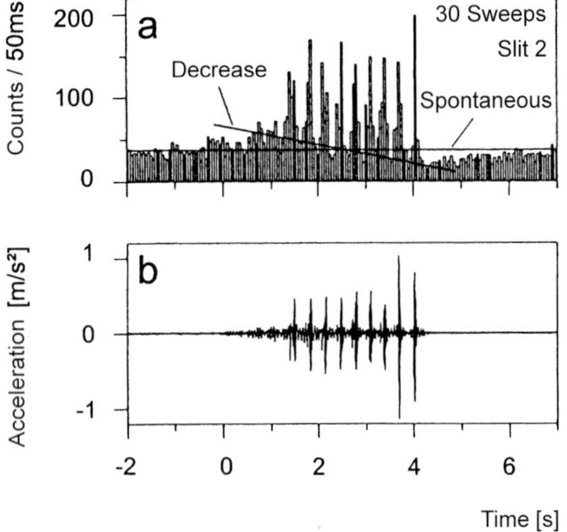

Fig. 16a,b. Response of slit 2 of the metatarsal lyriform organ of a female (*Cupiennius salei*) to an entire male courtship series consisting of nine syllables (three animals, ten runs per slit). **a** Peri-stimulus time histogram of response. Note the decrease of impulse frequency between successive syllables during the course of the series. As a consequence, the signal-to-noise ratio increases. After the end of the stimulus the activity of the slit falls below the spontaneous level for up to about 2 s. **b** Oscillogram of stimulus. (Baurecht and Barth 1992)

- The number of syllables: the behavioral experiments with synthetic male courtship signals (see below and Schüch and Barth 1990) had shown that a series must consist of at least three syllables in order to cause a female to respond in 50% of cases. The response rate increases until the series contains 12 syllables. What does the receptor do when stimulated with series of various lengths? Dieter Baurecht tested this as part of his dissertation study. The result: with increasing syllable number the signal-to-noise ratio becomes larger, because the synchronization of stimulus and excitation improves progressively and the number of impulses discharged between the syllables is progressively reduced (Fig. 16).
- The series: the end of the packet of syllables that constitutes a series is specially marked by a poststimulatory depression of the receptor response: the discharge rate falls below the spontaneous rate. Previous behavioral experiments had shown that only syllables grouped to form series are effective; an evenly spaced sequence of 1000 syllables never elicited more than two responses of the female (Schüch and Barth 1990).

Vibration-Sensitive Interneurons

Not much is known about the central nervous processing of the sensory information spiders receive about vibrations. Once the specializations of the metatarsal organ for the reception of courtship signals had become evident, Thomas Friedel obtained electrophysiological recordings from vibration-sensitive interneurons in female *Cupiennius salei*, to examine their responses to male courtship signals (Friedel and Barth 1995). Because his recordings were intracellular, the neurons could be stained after the experiment and then reconstructed from serial sections.

Of the 30 neurons that responded to natural courtship vibrations, 20 could be morphologically identified in this way. Nineteen of the identified neurons were plurisegmental neurons, having either uni- or bilateral projections within the subesophageal ganglion mass (Fig. 17). Their cell bodies were all in the ventral somata layer of the subesophageal mass (see Chapter XIV) and typically were near the midline, as had already been discovered during previous extracellular recording (Speck-Hergenröder and Barth 1987). The main neurites (into which the recording electrode was inserted) of all vibration-sensitive plurisegmental interneurons run in one of the large longitudinal tracts (central tract, centrolateral tract or sensory longitudinal tract 4; see Figs. 8 and 9 in Chapter XIV). From there they send branches into the lateral neuropils of the various neuromeres and/or into other longitudinal tracts. The bilaterally projecting plurisegmental interneurons pass through one of the segmentally arranged commissures to the contralateral side. Remarkably, none of the vibration-sensitive plurisegmental interneurons were found to send a projection into the brain. This finding underlines the significance of the subesophageal region for the processing of vibratory inputs. Interneurons ascending into the brain, as far as we know at present, are rare but not completely lacking. Wulfila Gronenberg found that out of 32 mechanosensitive plurisegmental interneurons in *Cupiennius salei*, 9 had branches in the subesophageal mass as well as branches that ran into the brain (Gronenberg 1990) (Fig. 18).

Thirty neurons are certainly only a small proportion of the interneurons involved in the reception of male courtship signals. Therefore the following classification into functional groups should be regarded as preliminary and presumably incomplete.

Two types of interneuron give responses that represent the microstructure of the courtship vibration, while the four others represent its macrostructure. No corresponding differences in cell morphology could be discerned. On the basis of characteristic differences in latency, however, primary interneurons can be distinguished from higher-order interneurons. Whereas the former (latency 20–26 ms) extend into ventrolateral sensory neuropils and/or sensory longitudinal tracts (see Fig. 17) and there very likely make direct connection with the sensory affer-

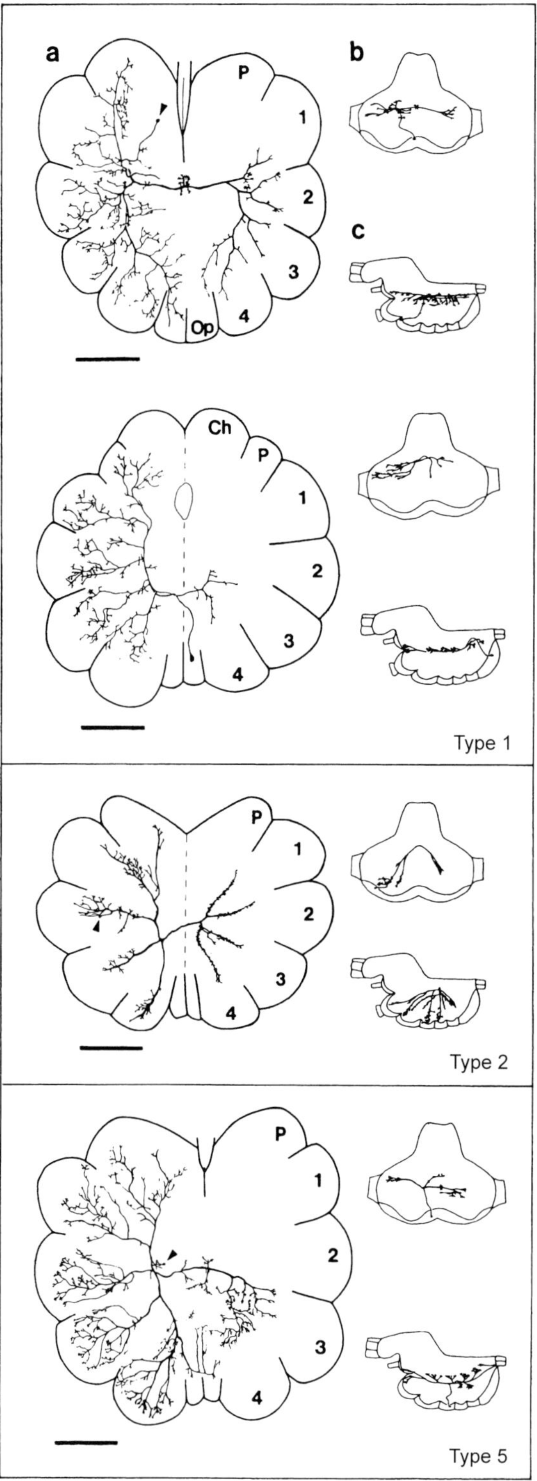

Fig. 17 a–c. Vibration-sensitive interneurons in the subesophageal nervous mass of the CNS of *Cupiennius salei*. The figure shows four neurons after cobalt-filling (**a** dorsal; **b** frontal; **c** sagittal view), which are attributed to four different classes according to their response to male courtship vibrations. *1–4* Leg neuromeres; *Ch* cheliceral neuromere; *OP* opisthosomal neuromere; *P* pedipalpal neuromere. *Bar* 500 μm. *Type 1* Bilateral, plurisegmental interneuron which copies the microstructure of the signal as far as the low-frequency opisthosomal syllables are concerned. High-frequency pedipalpal signals are not reliably represented by action potentials. *Top* Arrowhead points to cell soma in the cellular rind of the pedipalpal neuromere. *Type 2* Bilateral, plurisegmental interneuron, a spectral filter like type 1 interneuron, but mainly representing the high-frequency pedipalpal signals instead of the low-frequency opisthosoma signals. *Type 5* Bilateral, plurisegmental interneuron that is special with respect to the modulation of its response to ten consecutively applied male courtship series. Whereas it behaved like a neuron of Type 1 in the beginning, starting with stimulus series 3 it responded to the two first and the two terminal syllables only. After a pause of 2 min it again behaved like a Type 1 neuron. (Friedel and Barth 1995)

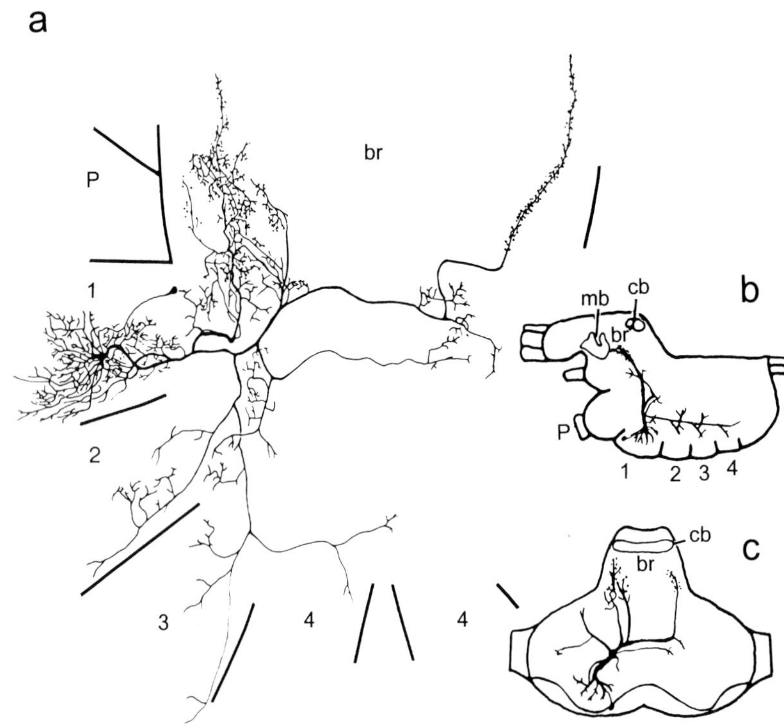

Fig. 18a–c. Interneuron (Lucifer Yellow stain; **a, b,** and **c** dorsal, sagittal, and cross-sectional view) responsive to low-frequency substrate vibrations and to airflow stimuli. It sends ascending collaterals to the protocerebrum. *br* brain; *cb* "central body"; *mb* "mushroom body"; *1–4* leg neuromeres; *P* pedipalpal neuromere. (Gronenberg 1990)

ents, the neurons with distinctly longer latencies do not branch at all in any of the sensory neuropils.

Neurons that Represent the Microstructure. More than half of all neurons (16) copied the syllabic structure of natural signals faithfully, but did not respond reliably to the pedipalp signals (Fig. 19a). Accordingly, these interneurons are spectral filters for the low-frequency opisthosomal signals. Their responses do not adapt, as long as the male series are presented in the natural rhythm, and they are remarkably independent of the intensity of individual syllables. Surprisingly, most of these neurons do not respond to synthetic male signals (amplitude-modulated single-frequency syllables), although these very effectively elicited behavioral responses. We must assume that these neurons require the greater spectral complexity of the natural signal. They also accurately reflect the characteristics of heterospecific natural signals (of *C. coccineus* and *C. getazi*). Responsiveness exclusively to conspecific signals was never observed.

Other neurons that represent the microstructure of the male courtship vibrations respond to the pedipalp but not the opisthosoma signals. In these neurons action potentials were triggered almost entirely by the higher frequencies in the pedipalp vibrations.

The parallel processing of low- and higher-frequency vibrations that begins in the metatarsal organ is thus continued in the central nervous system.

Neurons that Represent the Macrostructure. These interneurons do not copy the syllables but rather signal the beginning and the end of a series, or only its end. An example is shown in figure 19b. The response is evidently elicited by the substantial changes in amplitude of the syllables in a series, which occur at the beginning and at the transition to the large terminal syllables. Other neurons in this category were characterized by adaptation (decreasing discharge rate) during the presentation of several series in succession, or were spontaneous "bursters", which changed their discharge rhythm when the spider

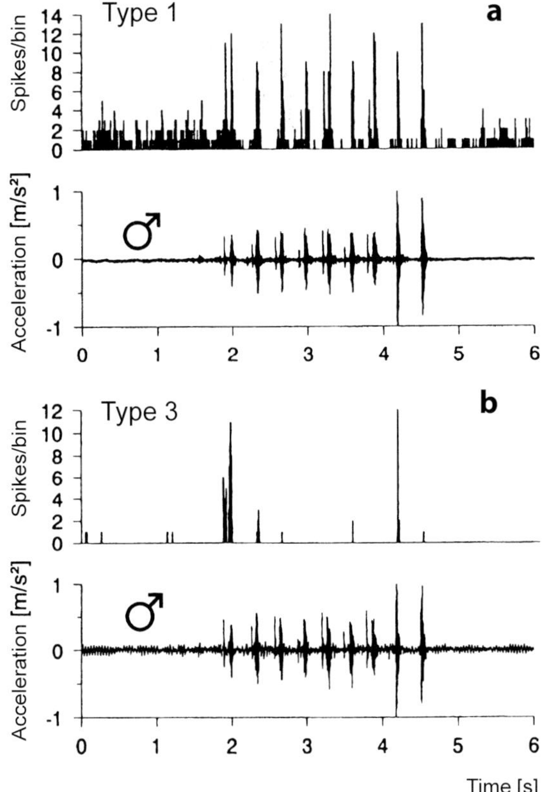

Fig. 19 a, b. Responses of vibration-sensitive interneurons in the subesophageal nervous mass of a female (*Cupiennius salei*) to the vibratory courtship signal of a conspecific male. **a** Interneuron (see Fig. 17, Type 1), which mainly responds to low-frequency opisthosomal signals but does not copy high-frequency pedipalpal signals. **b** Interneuron, mainly responding to large-amplitude changes at the beginning and shortly before the end of the series. Histograms summarize responses to ten presentations of the same courtship series; bin width 10 ms. (Friedel and Barth 1995)

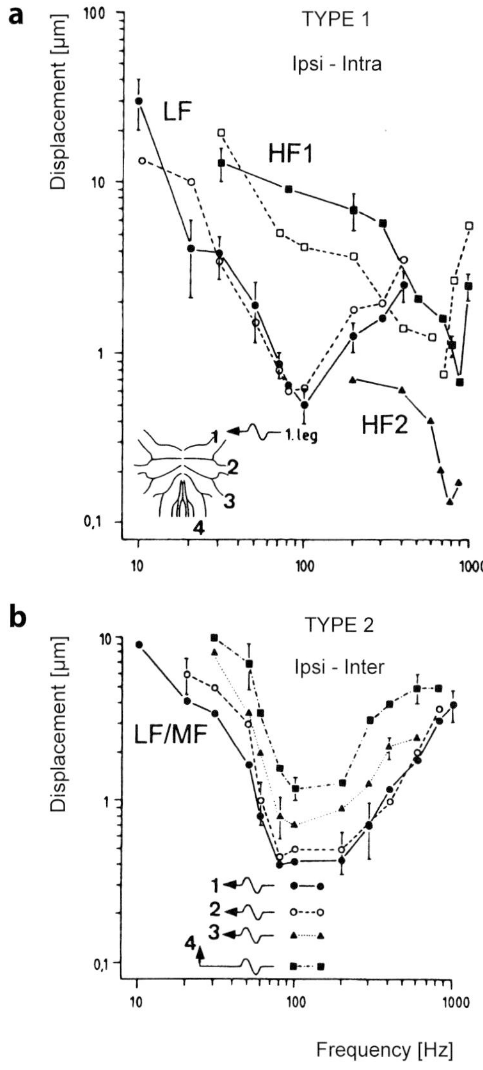

Fig. 20 a, b. Tuning curves (at threshold) of vibration-sensitive subesophageal interneurons of two different types. **a** *Type 1* Ipsi-intraganglionic; responds only to stimuli applied to leg corresponding to same neuromere from which recording was made. Figure shows examples for leg 1. Depending on the best frequency range low-frequency units (*LF*) can be distinguished from high-frequency units (*HF*). **b** *Type 2* Ipsi-interganglionic; neurons of this type were only found in recordings from 1st and 4th neuromere. They respond to stimulation of all ipsilateral legs but not contralateral legs. Tuning broader than that of Type 1, with best frequencies between 80 and 200 Hz (*LF/MF* low/medium frequency neurons). Thresholds in this frequency range are lowest with stimulation of leg 1 and increase with stimulation of legs 2, 3, and 4. (Speck-Hergenröder and Barth 1987)

was stimulated with courtship signals and continued this new pattern of discharge for as long as several minutes after the end of stimulation.

Threshold Curves. Here we must return to the threshold curves of vibration-sensitive interneurons briefly mentioned in Chapter XVIII. These interneurons are situated ventromedially in the subesophageal mass and differ from one another in the convergence of their sensory inputs. In all cases, however, the threshold curves measured from 10 Hz to 1 kHz had a band-pass character-

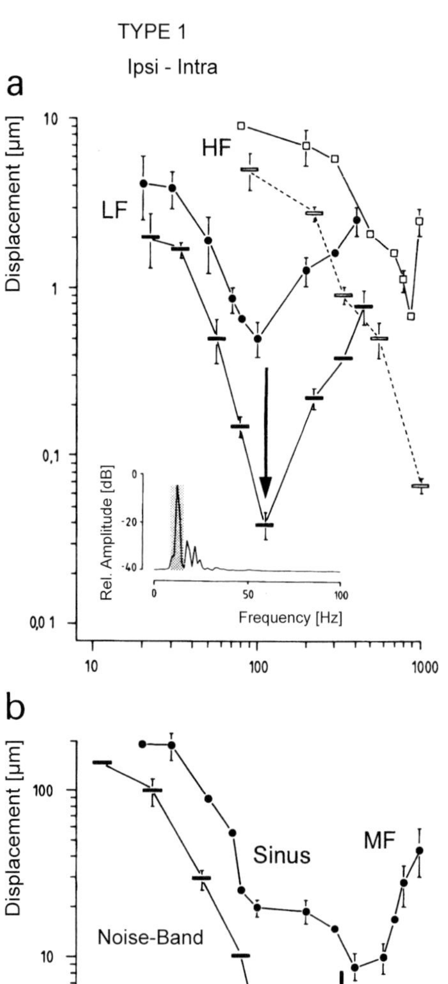

Fig. 21 a, b. The effect of band-limited noise on the threshold curve of an interneuron of Type 1 shown in Fig. 20. Stimulation with narrow frequency bands (*bars* range 1/3 octave) instead of sinusoidal vibrations (●) lowers the thresholds significantly. *LF, HF* (**a**), and *MF* (**b**) low-, high-, and medium-frequency interneurons. *Inset* in **a**: spectrum of band-limited noise stimulus (10–13 Hz) used in these experiments; *inset* in **b** recording site (●) and stimulation of leg 1. (Speck-Hergenröder and Barth 1987)

istic (Speck-Hergenröder and Barth 1987). This is remarkable, first, because no such characteristic was found for the slits of the metatarsal organ, and second, because the best frequencies are in the region of the courtship signals and the vibrations produced by prey animals. Interneurons which respond only to vibrations of the leg associated with the same neuromere are tuned to either low (ca. 80–100 Hz), intermediate (ca. 400 Hz) or high (ca. 800 Hz) frequencies. The frequency tuning of neurons that can be excited by the vibration of several legs is distinctly broader (ca. 80–200 Hz) (Fig. 20) (Speck-Hergenröder and Barth 1987). It springs immediately to mind that tuning to frequencies near 100 Hz is related to processing of the male courtship signal. This inference is corroborated by several other findings: the large dynamic range of these neurons, more than 46 dB; the decrease in threshold, by as much as 20 dB, when band-limited noise (1/3 octave) is used instead of pure sinewave stimuli; and the low sensitivity at the very low frequencies of vibrations of abiotic origin (Fig. 21).

6 Species Identification and Reproductive Isolation

In its natural habitat in Central America *Cupiennius* is exposed to all sorts of vibrations, which it must identify or discriminate if they are to be useful as a source of information to guide behavior (see Chapter XVIII). Courtship signals of the animal's own species must be distinguished not only from prey and background vibrations but also from the courtship signals of other *Cupiennius* species. The question of adaptedness thus goes beyond the boundaries of a single species. It follows that a physiologist must take an interest in taxonomic and ecological questions in order to find out what species there are and which of them are likely to meet one another – to enter into the situation of needing to differentiate during courtship.

The three large species, *Cupiennius salei, Cupiennius getazi* and *Cupiennius coccineus*, are an ideal team for experiments on reproductive isolation. They are not only easy to handle, to breed and to induce to perform courtship in the laboratory; they also differ in the sharing of habitat. Whereas *C. coccineus* and *C. getazi* are sympatric, *C. salei* is allopatric with respect to both of them (see Chapters II and III).

The Three Filters

The three phases of courtship (see Fig. 1) together constitute a reproductive barrier between the species. The efficacies of the chemical, vibratory and tactile phases as filters can be quantified by letting individuals of all three species court one another and noting the probability of passing through the filter in each stage. The probability of attaining copulation equals the product of the three probabilities of passing the individual filters. The result of such experiments, in which one male and one female, of the same or different species, are put together on the same bromeliad, is shown in Table 1 (Barth and Schmitt 1991).

First, it should be noted that ethological barriers in themselves effectively keep the species apart. This implies that the differences in structure of the sex organs, which play such a great role in spider taxonomy as species-defining characters, are irrelevant in the present context. Their significance as "lock-and-key" structures must be rethought (Eberhard 1985). In conspecific pairings the overall probability of passing successfully through the entire courtship is between 0.44 and 0.88. In heterospecific pairings, by contrast, it was 0.00 in 4 of the 6 cases and only 0.07 and 0.15 in the two remaining combinations (with a 100% effective postcopulatory barrier).

What do each of the three phases of courtship contribute to this obvious preference for the conspecific sexual partner?

The Chemical Phase. The silk spun by the female is the smallest hurdle between the species. In all three species the probability (p_m) that a male will begin courting when he makes contact with conspecific female silk is higher than that for contact with heterospecific female silk (except for *C. getazi* contacting *C. coccineus* silk); however, on average they do begin to court in no less than 58% of the latter cases. It follows that the silk pheromones are largely anonymous (Fig. 22a) (Barth and Schmitt 1991) and by themselves are certainly not capable of keeping the species apart. Heterospecific silk reduced p_m by an amount ranging from 35% (*C. getazi* × *C. salei* female silk) to 77% (*C. salei* × *C. coccineus* female silk). The males of *C. salei* discriminate most strongly and those of *C. getazi* least. This is also evident at the level of the individual animal, as can be seen from the following numbers: 30% of the males of *C. salei*, 20% of those of *C. coccineus* and 50% of those of *C. getazi* respond less often to conspecific than to heterospecific silk; only 3 out of 10 *C. getazi* responded more frequently to conspecific silk, and 7 out of 10 *C. coccineus*. Experiments on the taxis behavior of males in a Y-maze point in the same direction: the decision to follow the thread laid along one branch of the Y does not depend critically on its "same-" or "other-"specificity (Barth and Schmitt 1991).

The pheromones bound to female silk have an obvious releaser function, in that they induce the male promptly to perform courtship behavior. Thus one of the three functions of signals in sexual communication has been fulfilled. Their role in mate assessment has not been experimentally documented and, as we have seen, species identification depends only to a very limited extent on these pheromones. The near-anonymity of the chemical female signal does not interfere with the most important function, as was indicated in figure 1: to announce the presence of a female and arouse the male.

The Vibratory Phase. Neither silk nor pheromone is necessary for the male to pinpoint the female's location. The female's ability to identify her caller's species and the male's discovery of the way to the female rely on the vibration signals of male and female, respectively.

To judge by her responses to the male signals, the female can discriminate very well between con- and heterospecific male courtship vibrations (Table 1, Fig. 22b). The females of *C. geta-*

Table 1. Combined probability p(*) with which con- and heterospecific pairings of females and males of different species of *Cupiennius* proceed successfully through all phases of courtship. p(*) is the product of the probabilities of passing through the chemical phase (p_m), the vibratory phase (p_f), and the tactile phase (p_c). (Barth and Schmitt 1991)

Pairings Female	× male	Probabilities p_m ×	p_f ×	p_c =	$p(*)$
C. cocc ♀	× *cocc* ♂	0.67	0.73	0.90	0.44
	× *get* ♂	0.92	0.13	0.00	0.00
	× *sal* ♂	0.50	0.33	0.00	0.00
C. get ♀	× *get* ♂	0.92	0.84	0.90	0.70
	× *cocc* ♂	0.42	0.13	0.00	0.00
	× *sal* ♂	0.20	0.00	–	0.00
C. sal ♀	× *sal* ♂	0.88	1.00	1.00	0.88
	× *cocc* ♂	0.34	0.64	0.33	0.07
	× *get* ♂	0.60	0.25	1.00	0.15

zi are the most selective and those of *C. salei*, least so. In all cases their own males are clearly preferred to "foreign" ones; the female responds to the former about 4 to 5 times more frequently than to the latter. The responses to the heterospecific signals are not only fewer, they also come with greater delay and after a larger number of male signals (at least in *C. getazi* and *C. coccineus*). The differences in significance of the male and the female vibrations for the courtship events, previously defined for figure 1, now become still clearer. The males do not discriminate between the signals of different female species. As soon as a male has received a first answer, he raises his own signalling rate (significance demonstrated for *C. salei* and *C. coccineus*), turns toward the female and approaches her, regardless of whether she belongs to his own species or not (Barth and Schmitt 1991). One observation made during our early experiments on the courtship of *Cupiennius salei* shows very impressively that the male also uses the female's signal for orientation (Rovner and Barth 1981; see also Chapter XVIII). A male was climbing down a banana plant, and when he reached the point at which the petioles of five leaves diverged radially, he had to decide which leaf to choose. In this difficult situation the male

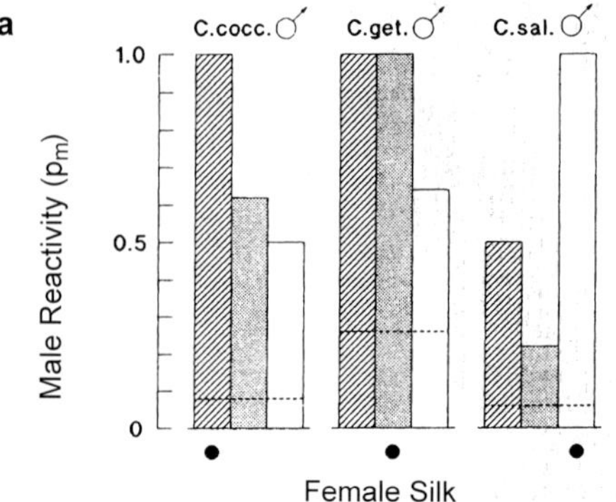

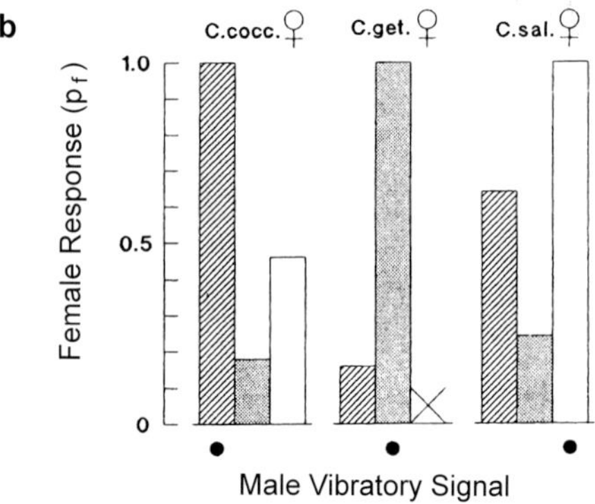

Fig. 22 a, b. Behavioral responses to conspecific courtship signals and those of closely related species: ▨ *Cupiennius coccineus*, ▩ *C. getazi*, and □ *C. salei*. **a** Probability (p_m) of the initiation of vibratory courtship by males upon exposure to conspecific (●) and heterospecific female silk. Values for conspecific silk always taken as 1. *Dashed horizontal lines* indicate "spontaneous" courtships in control situations without silk. On average, all males courted significantly less often in the heterospecific than in the conspecific situation, with the exception of *C. getazi* exposed to silk of *C. coccineus* (Friedman test, all $P<0.05$). In each species, the two heterospecific situations differed significantly from each other. For each situation N (animals) = 10, n (experiments) = 94–160. **b** Probability (p_f) of female vibratory courtship response to conspecific (●) and heterospecific male vibratory courtship signals. Values for conspecific case always taken as 1. (Barth and Schmitt 1991)

signalled several times while standing in the middle and then felt several petioles in succession, as though he were looking for the leaf with the female signal. In seven out of ten such cases the male finally chose the right one, the one with the female. In the other cases the female presumably did not answer.

The Contact Phase. This last phase of precopulatory behavior is an insuperable obstacle to heterospecific males (if they have managed to get this far through the courtship) when the females are either *C. getazi* or *C. coccineus*. In the case of *C. salei* females, their chances are better (see p_c in Table 1). In contrast, the probability of moving on from the vibratory phase to copulation is excellent for conspecific males: 0.9 (*C. coccineus*, *C. getazi*) or even 1.0 (*C. salei*).

As has been mentioned (see also Chapter III), *Cupiennius coccineus* and *Cupiennius getazi* are sympatric species, and as far as we know they do not live within the range of *Cupiennius salei*. What could sympatry and allopatry have to do with our courtship story? Higher reproduction barriers are expected between sympatric species than between allopatric species, which will not normally encounter one another. According to our behavioral experiments, the reproductive isolation between *Cupiennius getazi* and *Cupiennius coccineus* is indeed greater than when either is paired with *Cupiennius salei*. However, when the sexes are considered separately, the matter becomes more complicated. As expected, the females of *Cupiennius salei* are least selective regarding the males of the other species. Among the males *Cupiennius getazi* is the least selective species, while *C. salei* is just as selective as *C. coccineus*. This seems not to contradict the theory, given that female selection during the vibratory phase is more important for the whole discrimination process than male selection.

Why are the females of *C. getazi* and *C. coccineus* so especially unfriendly to the males of the other sympatric species? And why do *C. salei* females respond more often to males of *C. coccineus* than to those of *C. getazi*? Why do females of *C. coccineus* respond to males of *C. salei* significantly more often than to those of *C. getazi*?

The "mutual dislike" in heterospecific pairings of sympatric species could derive from "character displacement" by competition in the same habitat. On the other hand, it could indicate a low degree of relatedness. Sequencing of DNA fragments of the mitochondrial 12 S + 16 S sr DNA genes (Huber et al. 1993, Felber 1994) by means of the polymerase chain reaction (ca. 684 nucleotides) and phylogenetic evaluation of the findings support the idea of a close phylogenetic relationship between *C. salei* , *C. coccineus* and *C. getazi* as well as the monophyly of the genus *Cupiennius*. In addition, and this is more important in the present context, they point to a particularly close kinship between *C. salei* and *C. coccineus*. This implies that our results in the courtship experiments could indeed reflect differences in the kinship relations. However, this does not rule out "character displacement" as an alternative explanation.

Regarding the possible significance of different diurnal activity rhythms for the isolation of the sympatric species *Cupiennius coccineus* and *C. getazi*, see Chapter IV.

The Filter of the Female (Innate Releasing Mechanism)

Once behavioral experiments had demonstrated the special significance of male courtship vibration for species identification and hence also the predominant role of the female in dealing with this problem, the next question was: how closely is the female's releasing mechanism tuned to the various parameters of the male's signal? Are some of these parameters more important than others, when the male is trying to induce his partner to respond?

The experiments on this question were carried out with *Cupiennius salei* (Schüch and Barth 1990). They exploited the fortunate circumstance that the females respond even when they are standing, free to move, on a nearly uniformly vibrating platform coupled to an electrodynamic vibrator so that it can be driven by synthetically generated signals, producing "male oscillations" modified in various ways (Fig. 23). The number of responses by the female serves as a measure of the attractiveness of the various signals and also provides information about the tuning of the female filter for the parameter of interest. In all cases groups of 5 adult females were tested. As in the natural signal, the acceleration amplitude of the syllables increased continuously during a series. Each synthetic series comprised 24 syllables.

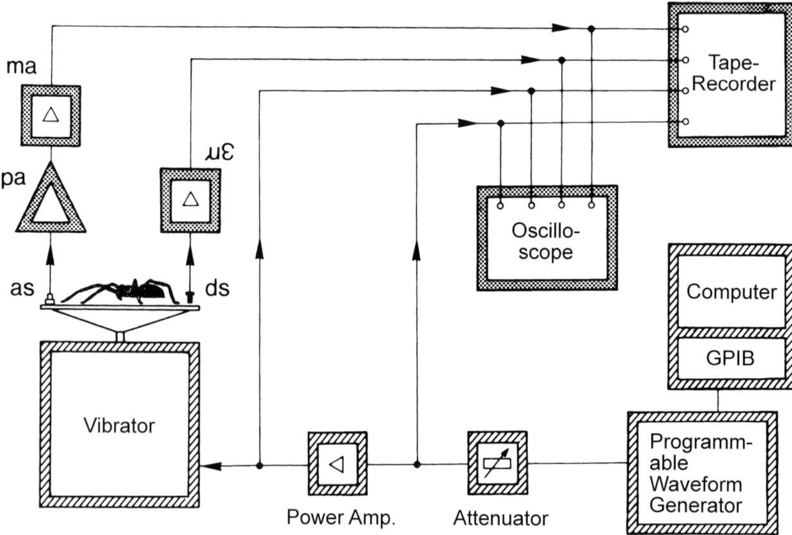

Fig. 23. Experimental setup used to expose adult females of *Cupiennius salei* to synthetic male vibrations. *as* Accelerometer; *ds* displacement receiver; *GPIP* general-purpose interface bus; *ma* measuring amplifier; *pa* preamplifier; *με* movement detector. (Schüch and Barth 1990)

The varied parameters of this mock male opisthosomal signal were as follows: the carrier frequency of the syllables (CF) as the "chief component" of the natural signal; the syllable duration (SD); the duration of the silent pause between two consecutive syllables (PD); the "duty cycle" (DC), the percentage of SD in SD+PD; the shape of the rising and falling flanks of the syllables; and the rate at which the amplitudes of the syllables rise within a series.

The measure for comparing the importance of the different parameters: the 3-dB bandwidth and the Q value.

These are quantities used in technology to describe the sharpness of tuning of electronic filters. In the present context the 3-dB bandwidth describes the range of parameter values on the X axis obtained by drawing a horizontal line 3 dB away from the peak of the response curve (see Fig. 24). The Q value is obtained by dividing the maximal effective value of the parameter in question by the 3-dB bandwidth; hence it is dimensionless.

The Result. Figure 24 summarizes the experiments on carrier frequency, syllable duration, pause duration and syllable shape. The response curves of the females do indeed exhibit distinct maxima.

In other words: the female tolerates variation in the male signal only to a very limited extent; the slopes of the curves indicate narrow spectral and temporal filtering in the female's releasing mechanism. As measured by the Q values, the parameters to which the female is particularly sensitive are (in descending order): the "correct" carrier frequency, syllable duration and pause duration. Females also tolerate only a small range of syllable repetition rates, as would be expected given that this parameter is determined by the syllable and pause durations.

Carrier frequency: No response was elicited by frequencies below 57 Hz or above 228 Hz. The so-called effective region (defined as the region in which at least one of the five females responded in more than 50% of the tests) was between 60 Hz and 200 Hz, and the maximum was at 133.3 Hz.

Syllable duration: Hardly any responses below 53 ms and above 210 ms (effective range), maximum at 105 ms.

Pause duration: Effective range between 60 ms and 350 ms, maximum at 169 ms.

Syllable repetition rate (SRR): With a constant duty cycle of 40%, a repetition rate below 2.3/s never elicited a response; the maximum was at 3.8/s. The change in duration and energy content of the syllables associated with changing SRR plays little or no role here.

It should come as no surprise that the effective regions found for these parameters in the synthetic signals are very close to and include the values characteristic of the male signals received by a female sitting on a plant in her natural

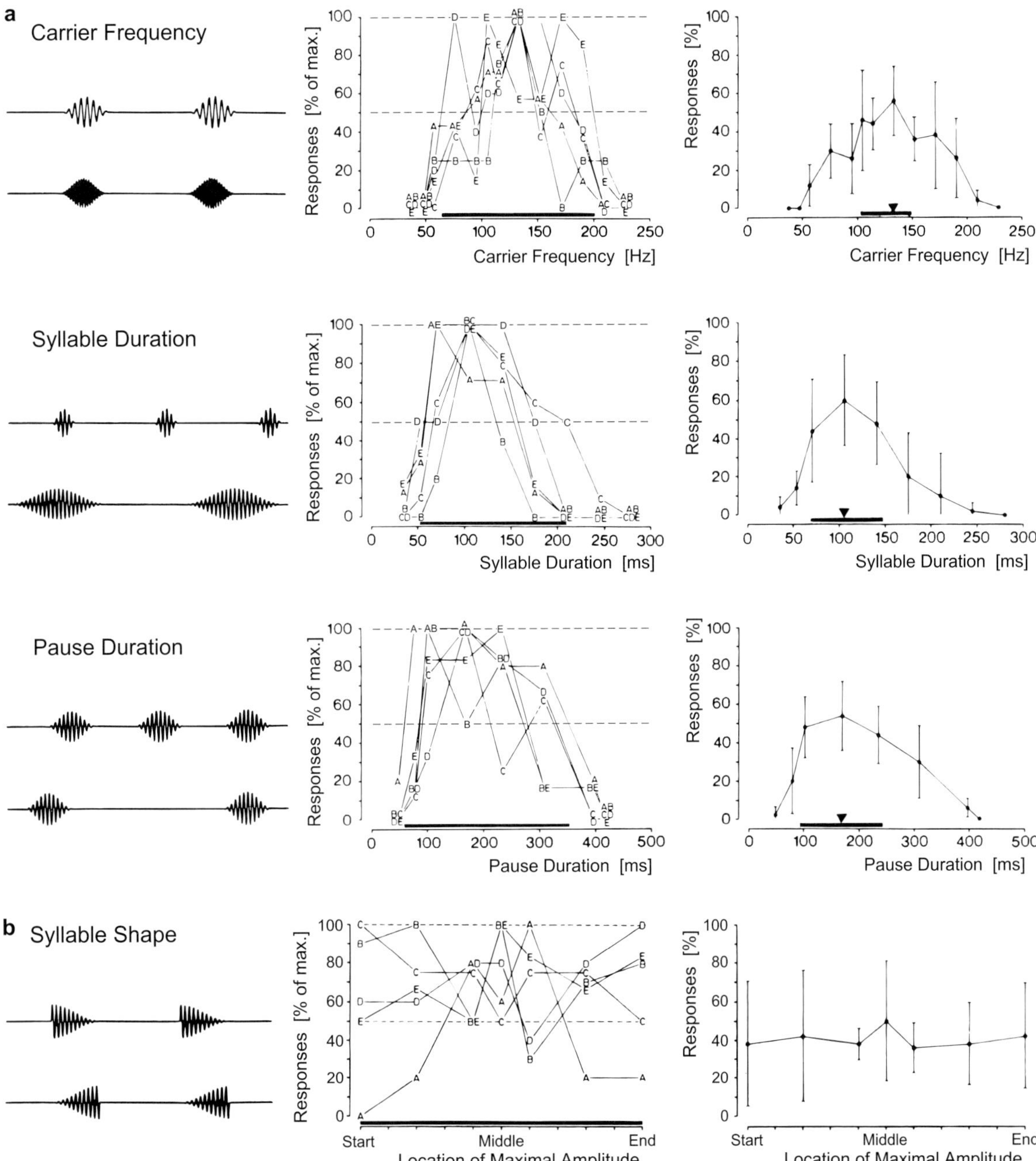

Fig. 24 a, b. a Responses of females (*Cupiennius salei*) to synthetic male courtship vibrations as a function of various signal parameters (carrier frequency, syllable duration, pause duration). *Left* The parameters varied; center: responses of 5 females (*A, B, C, D, E*) as percentage of maximum response frequency; *right* same responses given as percentage of total number of trials. *Bars above x-axes* indicate effective ranges (*center*) and 3-dB bandwidth (*right; arrowhead* points to most effective value). **b** Syllable shape, one of the parameters without influence on response frequency. (Schüch and Barth 1990)

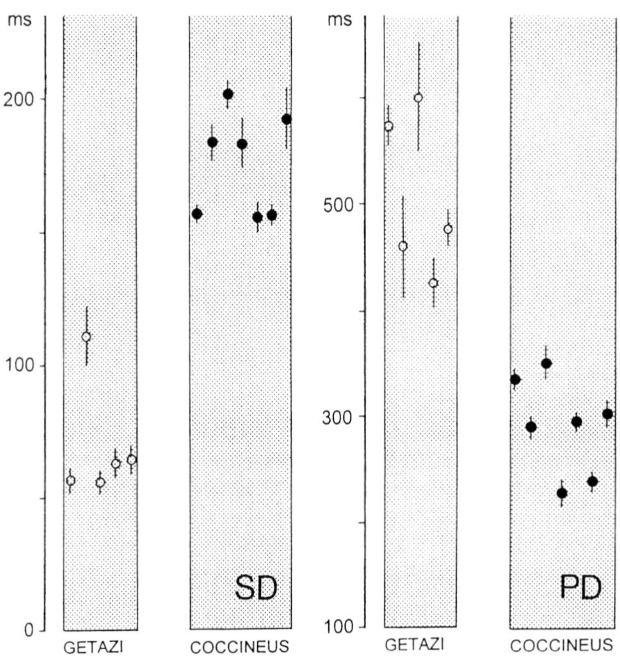

Fig. 25. Syllable duration (*SD*) and pause duration (*PD*) of male courtship vibrations in *Cupiennius coccineus* and *C. getazi*. Each value represents the mean (± SD) of five (*C. g.*) and seven (*C. c.*) courtships of four animals. The number of values used to calculate the means is 4–53 in the case of *C. g.* and 23–115 in the case of *C. c.* (Barth 1993)

habitat (Rovner and Barth 1981; Schüch and Barth 1985, 1990). Nor that SD and PD differ in the different species. Figure 25 shows this for the sympatric species *C. getazi* and *C. coccineus.*

Both the syllable repetition rate and the duty cycle are composite quantities, always associated with a particular ratio of syllable duration to pause duration. By systematically changing this ratio we were able to synthesize a signal that was more effective than the natural one – at least, the natural one with which we were familiar. Figure 26 demonstrates how influential SD and PD are in combination, as well as independently. In this graph regions of decreasing response frequency (100–70.7%; 70.7–50%; 50–10%) surround a center, the maximally effective combination. The naturally occurring range of SD-PD combinations is also plotted. From this graph it must be concluded that the natural male signals are already optimized with respect to SD but not PD. This implies that syllable duration is more important than pause duration. Will the evolution of the signals proceed in the direction of shorter pause duration?

We shall return to the temporal parameters of the male signal shortly. First, the considerably less critical properties: the signal amplitude (but see Section 5), the amplitude change within a series, the shape of the syllables and the number of syllables per series.

Acceleration amplitude: Effective range between 8 mm/s^2 and 8000 mm/s^2, but above only 300 mm/s^2 only one female responded once to more than 50% of the series; between 8 mm/s^2 and 316 mm/s^2 all females responded with a rate of at least 50%. 3-dB bandwidth 10–168 mm/s^2, maximum at 42 mm/s^2.

Amplitude change within the series: With an amplitude change of 0 to 3 mm/s^2 between consecutive syllables, the response frequency increased with increasing (but not decreasing) amplitude, although the differences in all cases were small with respect to the effects of the influential parameters CF, SD, PD, DC.

Syllable shape: Natural syllables are largely symmetrical, with the acceleration peak in the middle of the syllable; when the position of this maximum is shifted between the beginning (0%) and end (100%) of the syllable, the females' response behavior hardly changes. Only one of the five females preferred the symmetrical syllables; the other four responded regardless of syllable shape.

Number of syllables per series: At least 3 syllables per series were required to elicit responses to more than 50% of the series; with 6 or more syllables 3 to 5 females reached the 50% criterion; the response frequency increases up to 12 syllables/series.

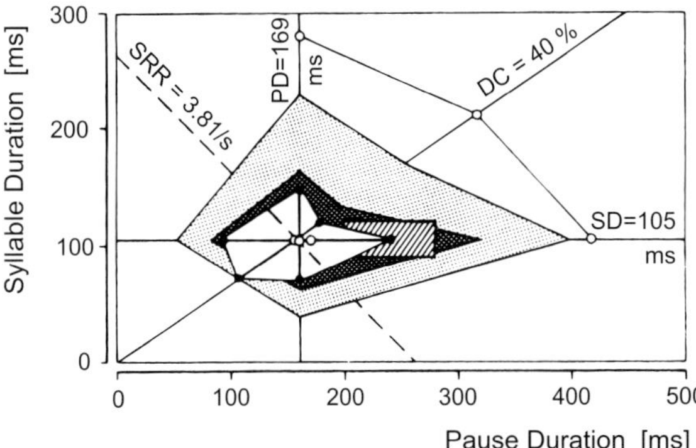

Fig. 26. Response of the female (*Cupiennius salei*) to different combinations of values for syllable duration (*SD*) and pause duration (*PD*) in synthetic male courtship vibrations. *Innermost white area* 70.7–100% responses; *intermediate dark area* 50–70.7% responses; *dotted area* 10–50% responses. *Rectangular hatched area* gives range of values typically found in natural male signals. *SRR* Syllable repetition rate, *DC* duty cycle. (Schüch and Barth 1990)

In not one case did synthetic pedipalp signals elicit a vibratory response of the female. This underscores the conclusion, already drawn from other observations, that pedipalp signals are neither sufficient nor necessary for the purpose.

Now back to the temporal pattern of the male signals. Because the female is evidently tuned to both temporal and spectral properties of the male's vibration, there is no single parameter that in itself is necessary and sufficient to trigger a female response. Instead, a whole group of parameters together determine the attractiveness of the signal. However, the Q values reveal differences in the weighting of the parameters and hence also in the female's selectivity with respect to them.

Many arguments document the special importance of the temporal pattern of the male signal in species identification and reproductive isolation, and it has recently received further support from a quite different direction (Shimizu and Barth 1996). An investigation of the influence of ambient temperature on vibratory communication during the courtship of *Cupiennius salei* has shown that the male signals change with temperature. As would be expected of a poikilothermic animal, the durations of most elements in the male signal (syllable, pause, series and so on) decrease with rising temperature. The duration of the female response also becomes shorter (Fig. 27). How does the female deal with this situation? Theoretically, her receiver system could vary correspondingly with temperature. But there might also be temperature-invariant features of the male signal, which would make this unnecessary. In fact, the duty cycle of the male vibrations (like the ratio of series duration to interseries duration, the number of syllables per series and the number of pulses per syllable) does not change between 13 °C and 34 °C! Nevertheless, females kept at different temperatures prefer male signals recorded at the same temperature to others. It follows that the female could also make use of the first of the two possibilities.

If we assume that clock mechanisms in the spiders measure the intervals between series, sequences (SD+PD) and pulses, then these could very well run faster at higher temperatures. And one might well speculate that only one such mechanism accomplishes all this. There are three reasons: (i) all these parameters are correlated with one another; (ii) the Q_{10} values between 15 °C and 21 °C are practically the same for the repetition rates of these parameters (series 2.3; sequence 2.2; pulse 2.5); (iii) the regression lines of the various repetition rates or frequencies all pass through zero at ca. 7 °C. A notable aspect of the same subject is that the female often responds at a time when the male would have emitted a syllable if it had continued its signal beyond the end of the series (Fig. 28). That is, the female responds not only within a brief time window of 0.9 s following the last male syllable, but even preferentially in synchrony with "projected" additional syllables. The temperature-dependence of the inferred clock is presumably the same for males and females, given that this synchronization (which presumably assists the males in recognizing the female's response in vibrationally noisy surroundings) was observed at all the temperatures tested.

The annual mean temperature in the habitat of *Cupiennius salei* in Central America is somewhat below 20 °C and changes only slightly over the year. The highest temperature used in our experiments, 34 °C, does occur during the day

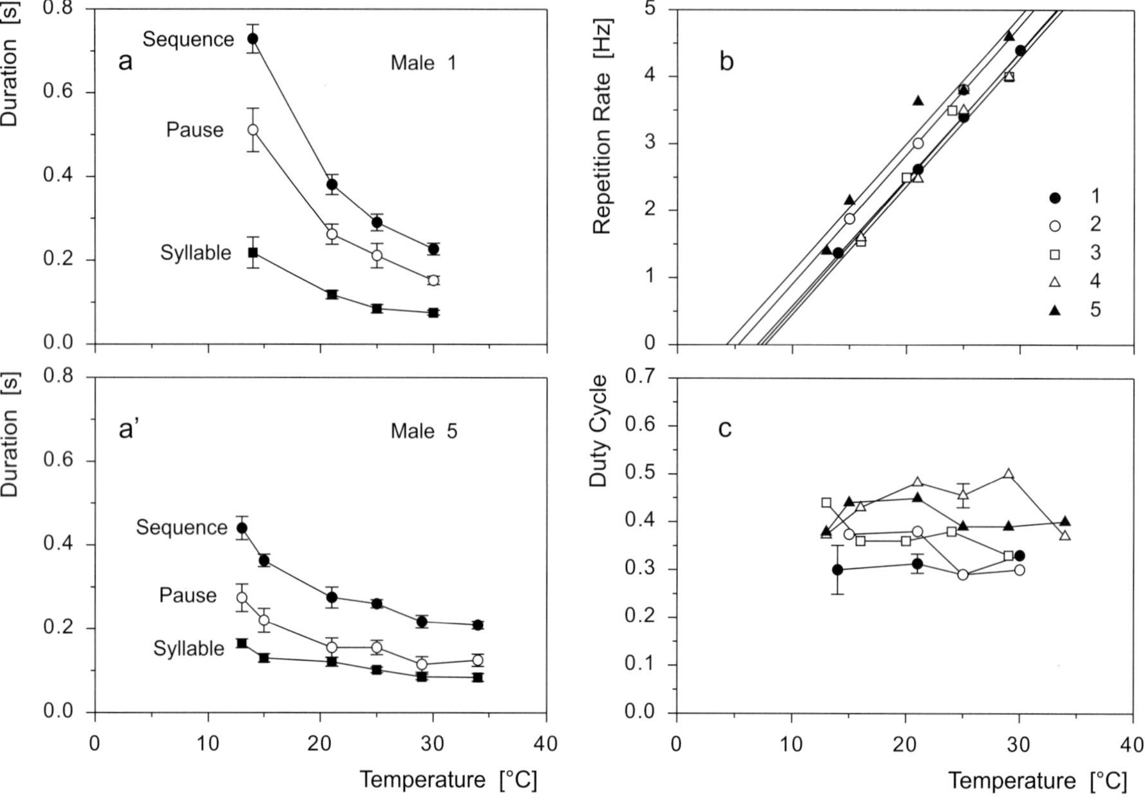

Fig. 27 a–c. The effect of ambient temperature on the microstructure of the male vibratory signals (*Cupiennius salei*). **a, a'** Duration of sequence, syllable, and pause, for two males (1 and 5; n = 10–15; SD indicated). **b** Repetition rate of the male sequence calculated from mean values of sequence duration. **c** Noninfluence of temperature on the duty cycle; symbols as in **b**. (Shimizu and Barth 1996)

but is unlikely during the nocturnal activity period (see Chapter III). At this high temperature only two out of five males gave courtship signals and only two females responded to a male signal. The courting activity is higher at lower temperatures; in experiments, it was in fact highest at or near 20 °C in four out of five females (Shimizu and Barth 1996). In our field observations, at a time of year in which the temperature sometimes fell to 12 °C at night, we repeatedly found females with newly spun egg sacs. This implies that even at such low temperatures courtship occurs, which presumably reflects adaptation to the climate of the habitat just as does the spiders' unwillingness to court at high temperatures.

A Brief Comparison with Other Arthropods

In *Cupiennius* the carrier frequency (main frequency component) of the male signal is one of its most important characteristics. This distinguishes the "vibrational recognizer" of the female *C. salei* from the "phonotactic recognizer" of the cricket *Gryllus campestris*. In the cricket, although the frequency has to be within the audible range, the feature of overriding significance proved to be the syllable repetition rate (Thorson et al. 1982). That the main frequency in the male signal is highly significant for the female *C. salei* is underlined by the fact that the female is certainly sensitive over a broader frequency range than that

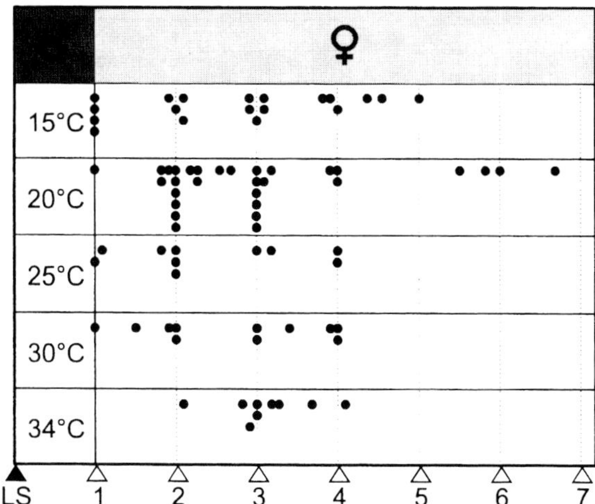

Fig. 28. The temporal relationship between male signals and female responses recorded at various ambient temperatures. *Black triangle* (*LS*) indicates the onset of the last syllable of the male series. *Open triangles* (*1–7*) represent the supposed time of the onset of male syllables if the train of male syllables had not stopped. *Filled circles* denote the actual onset of the female response. (Shimizu and Barth 1996)

found to be effective in the synthetic-signal experiments, and the metatarsal organ is even more sensitive to frequencies outside this effective region (see Chapters VIII and XVIII). These facts also show that the response curves obtained by means of synthetic signals result less from peripheral than from central filtering.

The characteristic frequency alone, however, is not a sufficient condition for the attractiveness of the male signal. It also does not suffice to categorize a vibration source as "male sex partner". Without the specific temporal pattern frequencies in the "effective region" can elicit both turning away (escape) responses and turning toward the source in preparation for prey capture, depending on the frequency and the signal amplitude (Hergenröder and Barth 1983a; see Chapter XVIII).

As in *Cupiennius*, the temporal parameters syllable duration, pause duration and syllable repetition rate are of major significance in crickets (Thorson et al. 1982, Stout et al. 1983, Doherty 1985, Stout and McGhee 1988). Accordingly, it is in the syllable-pause structure that the courtship signals of the males of various *Cupiennius* species differ most conspicuously (see Fig. 5). Much the same applies to representatives of many other arthropod groups, including stoneflies, Plecoptera (Rupprecht 1968, Zeigler and Stewart 1986), cicadas, Delphacidae (Bieman 1986, de Vrijer 1986), alder flies, Megaloptera (Rupprecht 1975), ghost and fiddler crabs, Ocypodidae (Salmon 1965, Altevogt 1970, Horch and Salmon 1971), wolf spiders, Lycosidae (Stratton and Uetz 1983) and cribellate spiders (*Amaurobius*) (Krafft 1978).

The shape of the syllable (position of the maximum) in *Cupiennius salei* has no influence at all on the number of female responses, although in the natural signal the maximum departs from the middle only slightly, by up to ca. ± 10% of the syllable duration. In field crickets (Acrididae), however, the von Helversens at the University of Erlangen (von Helversen and von Helversen 1983) found that the efficacy of the signal used by the male for auditory communication depends on the slope of the rising flank of the syllables and the position of the maximum at its onset.

With respect to the difficulties that temperature-related changes in the signal structure present to communication between transmitter and receiver, the solution that has been adopted by crickets (Walker 1957, Oecanthidae, flower tree crickets; Doherty 1985, *Gryllus bimaculatus*) and grasshoppers (von Helversen and von Helversen 1987, 1990, Acrididae, *Chorthippus montanus*, *Ch. parallelus*) is temperature coupling between males and females. The same thing has been described for the acoustic communication of frogs (*Hyla*) (Gerhardt 1978). The other strategy, to utilize temperature-independent signal parameters, has been found in *Chorthippus biguttulus*, another acridid grasshopper (von Helversen and von Helversen 1987).

As stated above, in *Cupiennius* it seems that both ways of solving the problem are used.

7 Parental Investment

The courtship of *Cupiennius*, in particular the vibratory communication, can be viewed in a still higher biological context. Then we shall better understand both the obvious and the subtle differences in behavior between the two sexes (Schmitt et al. 1994). In general, given that a male's investment in the next generation is small in comparison to that of the female, the parental investment theory (Trivers 1972) predicts certain aspects of male behavior:

1. The males compete for females,
2. they have a low threshold for sexual excitation, and
3. they are not very selective in choosing a sexual partner. The females would be expected to be more selective and better able to reject heterospecific or unsuitable conspecific partners (Thornhill 1979, Alcock 1989).

In *Cupiennius* the situation is clear: the males fertilize the females and that's it – nothing else is done to provide for the offspring! The females, on the other hand, produce a large number of eggs. In *Cupiennius salei* there are about 1500 eggs in the first sac and altogether as many as five egg sacs may be filled after one copulation, for a total of up to ca. 5000 eggs (Melchers 1963a, b). The female carries the sac around with her for several weeks, loosening the envelope from time to time, and finally releases the young by biting small holes in its wall shortly before they emerge. Prior to emergence she hangs the egg sac up and spins a tangled mass of threads around it, in which the spiderlings can stay and molt as long as their yolk supply lasts (Plate 10). After a few days the mother abandons the empty sac, and after about three weeks she begins to construct the next one.

Obviously, there is a big difference in the amounts males and females invest in their progeny. Alain Schmitt investigated the theoretically predicted consequences in detail for *Cupiennius getazi* and confirmed them all (Schmitt et al. 1994).

1. The males of this species in particular can very easily be put into a state of sexual excitation; once they have made contact with female silk there is a probability of 0.98 that they will begin vibratory courtship. Quite unlike the females, they sometimes begin to signal spontaneously in the absence of any other spider (Barth and Schmitt 1991).
2. Males approach other males that are courting and compete with them by sending out vibratory signals of their own (Schmitt et al. 1992). The fights between males are ritualized – unlike the attacks females make against males – and their result is not correlated with the age of the combatants, their leg length or body weight, or the frequency of signalling. But if a female is nearby, then the larger and heavier male wins.
3. Like the males of the other large species of *Cupiennius*, those of *C. getazi* are less selective than the females. In the behavioral experiments described above (see Table 1 and Fig. 22) the females to a great extent decided "against" heterospecific males, whereas the males by no means rejected heterospecific female vibrations (Barth and Schmitt 1991).

8 The Releasing Mechanisms of Males and Females

In the case of *Cupiennius getazi* it is possible to compare the selectivity of the releasers of males and females directly by using the same vibratory signal. This remarkable opportunity derives from the fact that the courtship vibrations males generate with the opisthosoma elicit not only the female's courtship response but also vibratory behavior of competing males (Fig. 29). To fit the parental investment theory, the releasing

Table 2. Comparison of male and female selectivity toward various properties of the male courtship vibration (*Cupiennius getazi*). (Schmitt et al. 1994)

Male vibratory signal	Choosiness*		
	Male		Female
1 Maximum amplitude		<	
2 Amplitude change within series		=	
3 Number of syllables		<	
4 Syllable duration		<	
5 Sequence duration		≪	
6 Inter-series duration		<	
7 Model, synthesized		=	

* Selectivity of innate releasing mechanism

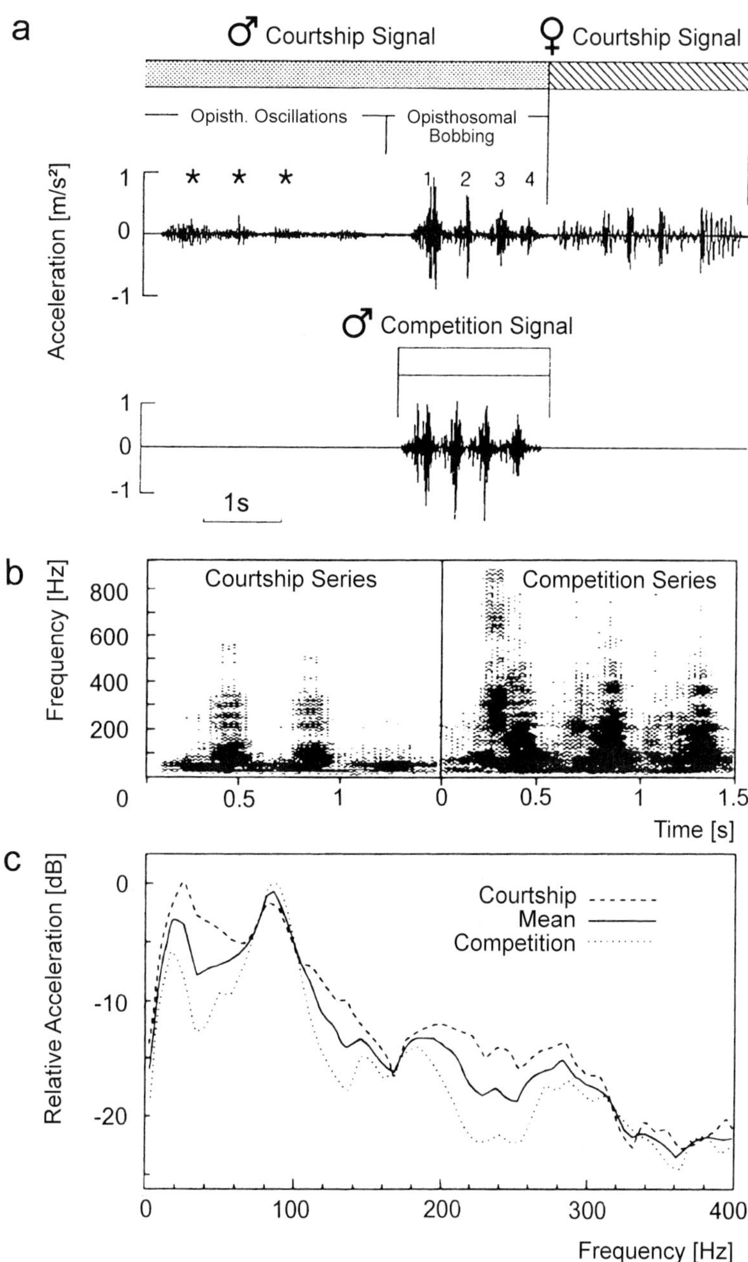

Fig. 29 a–c. Oscillograms (**a**), vibrograms (**b**), and spectra (**c**) of vibratory courtship signals and vibratory signals that the male produces during male competition (*Cupiennius getazi*). **a** The final 5 s of a courtship signal; * pedipalpal signals; *1–4* the four final syllables of the series, followed by the female response. *Below* Competition signal. **c** Frequency spectra of male courtship signal (mean FFT of first eight syllables of eight series) and the competition signal (n = 7) as well as the mean of both signal types (n = 15). (Schmitt et al. 1994)

mechanism by means of which the males recognize the signals of other males would have to be less selective than that by which the females categorize a male as conspecific.

Alain Schmitt (Schmitt et al. 1994) pursued this problem in our laboratory in Vienna, together with Martin Schuster. Again the experimental procedure was to expose females and males to systematically altered male signals and synthetic imitations of these signals, and to test their readiness to respond in each case. The result: males are the less selective sex (Table 2). Although this agrees well with the predictions of the parental investment theory, it can still be interpreted in two ways, which are not mutually exclusive:

1. The "vibratory" innate releasing mechanism of the males is in fact less selective than that of the females;
2. the males are simply easier to arouse sexually than the females are.

XXI Kinesthetic Orientation

1 An Experiment

The scenario: a hungry *Cupiennius* notices a buzzing fly; alerted by its trichobothria, it runs promptly towards the fly and bites into it. Immediately it is separated from the fly and gently but firmly driven away from the scene of the deed – 20, 30, as far as 70 cm away. The presumably frustrated spider usually stays where it has been driven for a few minutes, then turns around and runs back to the place where it has caught its prey.

What is special about this behavior?

- All the eyes have been blinded with black paint. Therefore it cannot be orienting visually to the capture site.
- After the spider has been driven away from the fly, which has been immobilized by the bite, the fly is either removed altogether or put somewhere else in the arena, usually closer to the spider than it was originally and often only 1 cm away from the spider's nearest leg. Nevertheless, the spider runs back to the original capture site. This can mean only one thing: olfactory orientation also plays no role in this experiment.
- The experimental arena is a smooth, round plastic plate with no "landmarks" or gradients; that is, it has been precisely levelled. Hence orientation by tactile stimuli or gravity is also ruled out.
- All experiments in which the spider had spun a dragline were discarded. In the cases that were retained, then, the spider did not orient with reference to a dragline.

So how does the spider find its way back to the capture site? It orients kinesthetically. That means that it uses no external references but only stored information about its own previous movement sequences. Formulated somewhat more practically, the spider "knows" exactly how it got from the capture site to the place to which it was banished, and with this information it finds the way back.

Under the natural nighttime conditions this amazing ability of *Cupiennius* is likely to be involved in returning not only to prey animals but also to its shelter. The first thing *Cupiennius* does after catching a large animal is to spin threads that enclose the prey and attach it to the plant (Melchers 1963b). If the spider is disturbed in the process, for instance by a predator dangerous to itself, it immediately leaves the prey packet and comes back to it later. It is also likely that when a female leaves her egg sac for some reason, she finds her way back by kinesthetic orientation, as has been demonstrated for the wolf spider *Pardosa amentata* (Lycosidae) (Görner and Zeppenfeld 1980).

Use of the term "kinesthetic orientation" in this context is not entirely uncontroversial. For instance, it has been proposed to replace it by "endokinetic orientation" (in contrast to exokinetic), because the content of the individual's memory is the crucial thing (Jander 1970). Later Mittelstaedt and Mittelstaedt (1973) introduced the broader term "idiothetic orientation" (in contrast to allothetic), to emphasize that the stored information used for the backtracking need not derive solely from proprioreceptors; it can also be available in the form of efference copies (that is, copies of motor signals). In our own work on this subject we were primarily interested in the sensory bases of idiothetic orientation and we did indeed identify, for the first time, sense organs that provide information about movement sequences for idiothetic purposes: the lyriform organs. Therefore the old term "kinesthetic" orientation still seems appropriate for the present chapter.

2 The Involvement of Lyriform Organs

The first goal we pursued with *Cupiennius* was to demonstrate the involvement of identified sense organs in the return walk, in order to show that there is in fact a movement sense (kinesthesia). To do this, we inactivated specific sensors from spiders tested in the arena described above (Barth and Seyfarth 1971; Seyfarth and Barth 1972). At that time we knew very little about the slit sense organs, but there was no doubt that they had something to do with locomotor behavior, if only because they are most commonly situated on the walking legs. Surprisingly, it became clear that the inactivation of lyriform organs on femur and tibia does *not* impair normal walking and the associated coordination of the leg movement, but it does affect the ability to orient (Seyfarth and Barth 1972; Görner and Zeppenfeld 1980; Seyfarth and Bohnenberger 1980).

At the beginning of the experiment a fly (*Calliphora erythrocephala*) was impaled on two insect pins attached to a hook and connected to a source of electric current. A cover plate made of transparent plastic was mounted 3 cm above the actual experimental arena (80 cm diam.), mechanically unconnected to the arena. In the cover plate holes (1.25 cm diam.) were drilled at regular distances of 4 cm from one another. The holder with the fly was stuck into one of these holes in such a way that the fly could not touch the floor of the arena. When the fly began to buzz, hungry spiders (3–4 months old) approached it from as far away as about 30 cm. As soon as the spider bites the fly, a short current pulse is applied to the fly, whereupon the spider releases its prey and can be urged away from the site with a fine brush.

The lyriform organs are best inactivated by piercing them with an electrolytically sharpened tungsten needle. Preliminary trials had shown that the mechanical intervention can be well controlled with respect to both depth and area. This does not apply to the other method initially tested, inactivation by the application of heat with a cautery. The control animals had intact lyriform organs but holes were made in the nearby cuticle.

The impairment of orientation is manifest in the frequency of arrival at the capture site, the starting direction and the route taken to return to the capture site.

Frequency of Arrival

A simple measure of the success of an orientation task, and the most important one, is the proportion of trials in which the animal actually reaches its goal. On average, intact animals will approach the goal to within 5 cm. In the case of the spiders used for these experiments, this means that the tarsus of one foreleg would have touched the prey if it had still been present at the site. The criterion we chose for successful arrival was a distance of not more than 10 cm between middle of the prosoma and capture site; in this situation one foreleg tarsus is no further than 5 cm from the goal and the spider reacts promptly to a moving object (see Chapters IX, XIX). According to this definition, the approach has failed when the distance between middle of prosoma and goal is greater than 10 cm.

As figure 1 (Seyfarth and Barth 1972) shows, intact animals are successful in 95% of all trials, whereas the animals with nonfunctional lyriform organs reach the goal only 20% to 33% of the time (depending on the pattern of organ inactivation). The operation thus reduces the arrival frequency dramatically. The elimination of a total of only 8 lyriform organs on the femora has the same effect as elimination of 32 organs on the tibiae, which underlines the prominent role of the femoral lyriform organs in kinesthetic orientation. The high success rate of the control animals corresponds exactly to that of intact animals, confirming the significance of the inactivation effect.

Failure occurs in two ways: either the spider starts out but stops too soon, or the return walk covers a long enough distance but the spider does not reach the goal because its walking error (see below), which also takes into account the angular deviation, is too great.

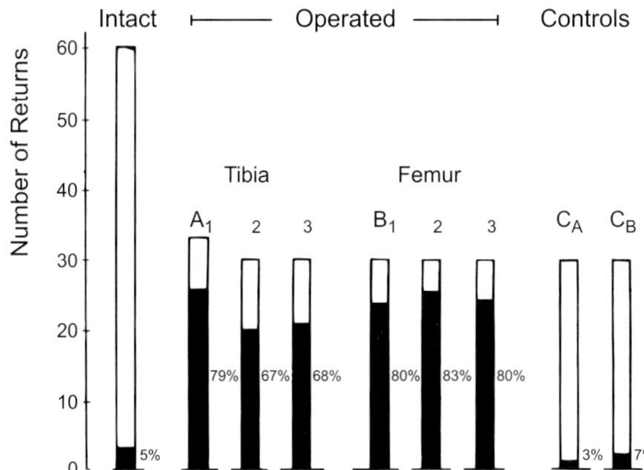

Fig. 1. Successful and unsuccessful returns of *Cupiennius salei* to the site of preceding prey capture. In the case of intact and control animals 95% of the returns are successful, with the spiders coming closer than 10 cm (mean 5 cm) to the prey-capture site. Of the animals with impaired lyriform organs at least 65% miss the goal by more than 10 cm. *Black parts of the bars* indicate unsuccessful fraction of returns. *A*, *B* Groups of animals with lyriform organs destroyed on the tibiae and femora, respectively; *C* control animals; A_1 all organs on tibia destroyed; A_2 (A_3) all organs on the anterior side of the right (left) tibiae and on the posterior side of the left (right) tibiae destroyed; B_1 all organs on all femora; B_2 (B_3) organs on anterior side of all right (left) and posterior side of left (right) femora; C_A, C_B control animals with holes pierced in the cuticle on both sides of all tibiae and in all femora, respectively. (Seyfarth and Barth 1972)

Starting Direction

The starting direction certainly does not have the significance often ascribed to it in homing experiments with all kinds of animals. But it is easy to measure, as the angle between the ideal direction and the direction in which the animal actually begins walking (Fig. 2) (Seyfarth and Barth 1972). Plots of the results as circular diagrams, evaluated with the appropriate circular statistics, showed that the starting directions of intact animals come very close to the ideal; their distribution is unimodal and the mean vector departs insignificantly, by only 2°, from the ideal direction. In all groups of animals with inactivated lyriform organs the values are significantly more scattered than in the intact animals, and this applies particularly to the groups in which the affected organs were on the femur. The mean starting direction departs significantly from that of intact animals in only three inactivation groups (all tibial organs; organs on the anterior surface of the right-leg tibiae and the posterior surface of the left-leg tibiae; all organs on the femora) (Fig. 3). The values for the control groups correspond to those for the intact animals.

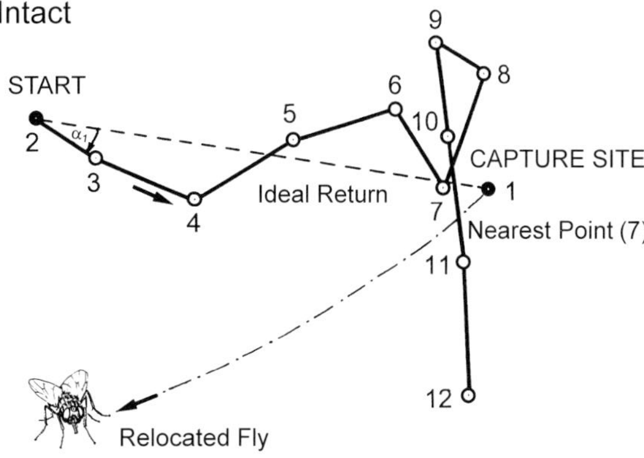

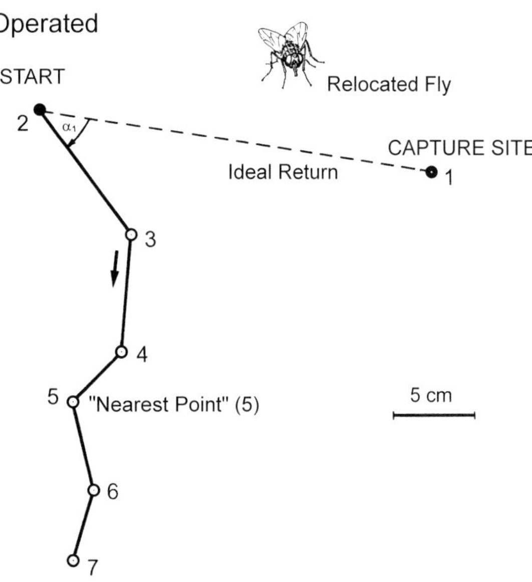

Fig. 2. Examples of the return path of an intact spider (*top*) and one with all lyriform organs on all tibiae destroyed (*bottom*), again showing a successful and an unsuccessful return, respectively. *1–12* Points where spiders pause and/or turn. (Seyfarth and Barth 1972)

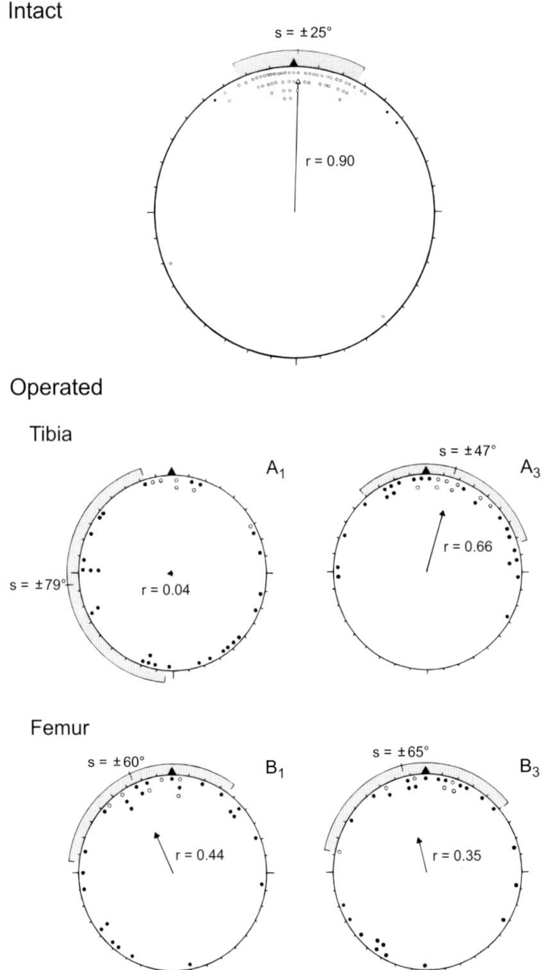

Fig. 3. Circular distribution of starting angles of the return paths of intact and operated animals. *Intact* 22 animals, 60 runs; the direction of the mean vector *r* deviates from the ideal return direction (0°) only insignificantly. *Operated* Values for two animals and 30 runs in each group; A_1, A_3 and B_1, B_3 experimental groups as explained in Fig. 1. *s* Mean angular deviation about *r*; ○ successful returns; ● unsuccessful returns. (Seyfarth and Barth 1972)

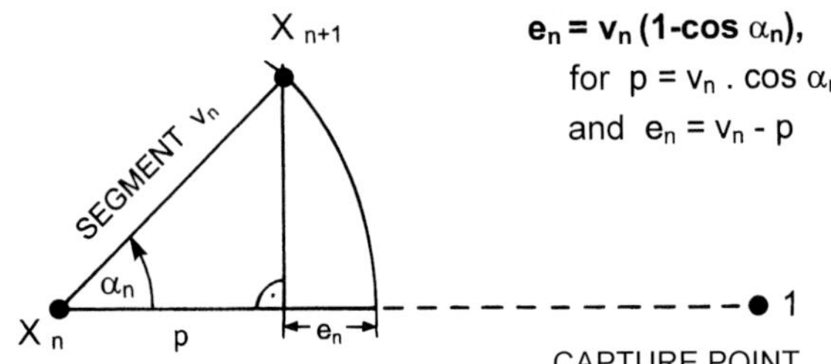

Fig. 4. Definition of walking error e_n. For details see text. (Seyfarth and Barth 1972)

Walking Error

A "good" starting angle alone does not guarantee a successful arrival, and a "poor" one does not rule it out. The route taken to the goal is decisive – "the Way is the Goal". *Cupiennius* travels in several straight-line segments, each a vector with specific length and direction (Fig. 2). The spider always makes a walking error when at a turning point X_n it does not walk straight toward the goal but instead toward a point X_{n+1} that is not on the ideal route. The magnitude of the error e_n is determined not only by the angle of deviation but also by the length of the segment in the wrong direction. Figure 4 shows that the error in this sense is given by

$$e_n = V_n(1 - \cos \alpha_n)$$

where V_n is the length of the path segment and α_n is the angle of deviation.

Comparison between the mean walking errors e_n of intact animals, animals that were operated on and control animals shows certain interesting results.

First, it is notable that in intact animals the walking error is considerably smaller during the approach to the point nearest to the goal than at the nearest point itself (by the factor 1/5). This finding reflects the fact that the spider begins walking in search loops as soon as it is close to the goal. Evidently it has information not only about the angle with respect to the goal but also about its distance (Figs. 2 and 5). In principle a precise choice of direction in itself would suffice to bring the spider back to the prey (or whatever its goal is). By additionally keeping track of the distance covered, and then walking in search loops when near the goal, the spider can increase its success rate even if, for example, the prey was not completely immobilized and has moved slightly away from the original site. Furthermore, the search loops also allow for a

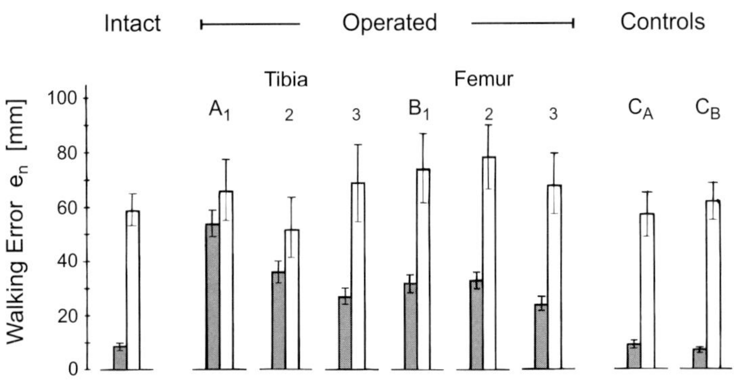

Fig. 5. Mean walking error e_n for all experimental groups (names as in Fig. 1). *Shaded bars* denote mean walking error (±SE) for all walking segments up to the nearest point (see Fig. 2) whereas *white bars* show values of walking error at the nearest point. Note significant increase of walking error at nearest point, which is interpreted as an indication of the spiders' distance orientation. (Seyfarth and Barth 1972)

degree of inaccuracy, and it may have been considerably simpler for evolution to find an adequately good solution rather than the most accurate one conceivable.

In animals with inactivated organs, the mean walking error before the nearest point is 4- to 5-fold larger than in intact animals. Organ inactivation has thus altered the length and direction of the course segments. Interestingly, e_n at the nearest point is about as large as in the intact spiders and the control animals (Fig. 5). It is tempting to conclude immediately that the distance measurement is unaffected by the operation. However, if the successful and unsuccessful walks are considered separately, it is only in the unsuccessful ones that the walking error is significantly greater than in intact animals both before and at the nearest point. Whether the distance orientation, like directional orientation, depends on intactness of the lyriform organs of femur and tibia must for the present remain an open question.

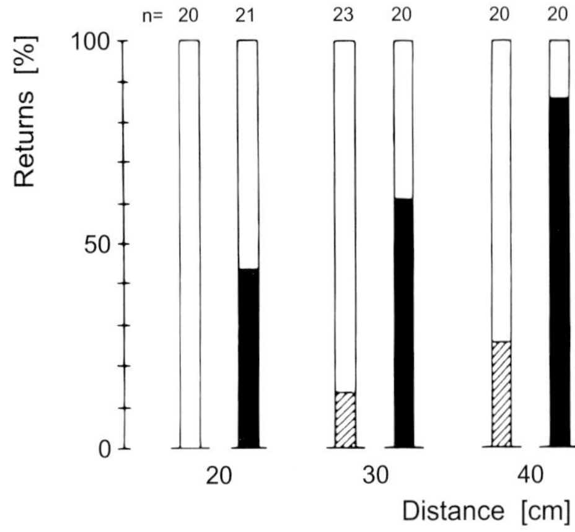

Fig. 6. Successful and unsuccessful returns of *Cupiennius salei* after rectilinear chases from the prey-capture site. *Shaded part of left bars* gives percentage of unsuccessful returns of intact spiders; *black part of right bars* gives percentage of unsuccessful returns of operated animals. *Numbers above bars* indicate number of returns observed. (Seyfarth et al. 1982)

3 More About Distance Orientation

The position at which the search loop is begun can be taken as a measure of the ability of *Cupiennius* to estimate correctly the distance from the starting point to the prey-capture site (Seyfarth et al. 1982). The number of successful return walks (that is, the walks that bring the middle of the prosoma to within at least 5 cm from the goal) decreases as the distance to be covered increases: after they have been driven 20 cm away from the capture site, all intact spiders return successfully; from 40 cm about two-thirds do so, and from still greater distances ca. 60% reach the goal. The longest successful return was achieved by a spider that had been driven 77 cm away from the capture site, the longest distance possible in the arena. After all lyriform organs on the femora had been inactivated, the success rate fell considerably (Fig. 6). With the exception of the 20-cm group, the failure rate was then ca. 75%.

The starting angle is hardly related at all to the distance by which the spider was displaced. In intact animals the departure from the ideal direction is not statistically significant, and even the animals with missing organs have average starting directions not significantly different from those of the corresponding group (same displacement distance) of intact animals.

The standard deviations of starting angle, however, for displacement distances ≥25 cm are significantly greater following an operation than in intact animals. The exception: displacement distances > 41 cm, for which even the intact animals exhibit a relatively large scatter. The net result is thus that the distribution of starting angles is affected both by the displacement distance and the elimination of the organs on the femora.

Back once again to the search loop: in intact animals the walking error e_n, even for displacement distances of over 40 cm, increases by 3- to 4-fold at the "nearest point"; that is, the distance from the capture site is evidently correctly estimated. In animals with all organs on the femora inactivated the walking error is greater than in intact animals even before the nearest point. Therefore it increases less at the nearest point than in the latter animals, but there is still a clear increase in most of the operation groups. Apart from a degree of weakening, therefore, distance estimation still functions after elimination of the femoral organs.

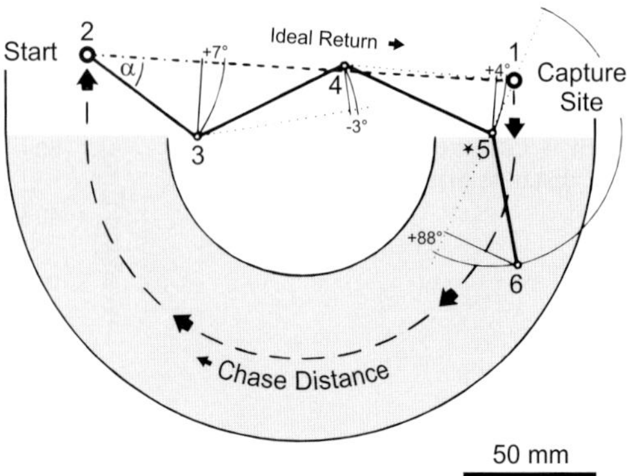

Fig. 7. Examples of return paths to prey capture site (*1*) after chases through a semicircular corridor (*arrows*) to site 2, from where the spider starts its return. Intact animals take a short cut and approach the capture site successfully (distance smaller than 5 cm). *Small numbers* and *thin lines* refer to walking error at each turning point. *Positive and negative numbers* Errors to the right and left side of the ideal return path (*dotted line*), respectively. (Seyfarth et al. 1982)

4 Compensation for Detours and the Question of the Precise Role of Lyriform Organs

The experiments described so far show that lyriform organs do participate in bringing about correct orientation, but they still do not reveal just how the proprioreceptive information they provide is used. Is it that information is collected during the spider's outward trip, away from the capture site, and stored in the CNS for later use? Or does the spider orient by means of stored efference copies, copies of the motor programs that controlled the spider's locomotion along the outward route? In the latter case the function of the lyriform organs would be to provide information about the *return route* to the capture site, and the spider would have to compare this with the stored data representing the outward route.

At least partial answers are obtained by driving the spider away from the capture site not in a straight line, as in all experiments previously described, but instead along enforced detours; for instance, by driving it through a semicircular corridor. Then it turns out that intact animals do not rigorously follow semicircular routes back to the goal, but save time and energy by taking the straight short-cut (Fig. 7). At the point nearest the goal they abruptly change direction, evidently "knowing" that they have gone the right distance (beginning of the search loops). The length of the path travelled is not appreciably longer than the ideal way back, and is about 50% shorter than the route through the semicircular corridor. And 85% of all return walks by intact animals are successful.

The animals with missing lyriform organs arrive at the goal in less than 50% of the trials. When the deficit is unilateral, an interesting directional effect appears, which demonstrates the importance of side-specific information from the lyriform organs. For instance, if the organs on the femora of the right (or left) legs had been inactivated, the spiders were considerably more successful after being driven away toward the left (right) than after being driven to the right (left) (Fig. 8). That is, the return trip was better directed when the organs still intact were on the side toward the center of curvature of the corridor through which they were driven. When the operations were carried out on both sides, this direction-dependence was no longer observed.

That even intact animals do not compensate fully for a detour is evident from the distinct tendency of the starting angle to be directed toward the curved corridor. In animals with missing lyriform organs on both sides (for instance, all organs on the femora inactivated) this preference disappears (Fig. 9).

The walking error e_n shows the same tendency as the starting angle: in intact animals it is always greater on the "corridor side", corroborating the tendency to walk back toward this side. After operations on both sides, the spiders are largely free of such a tendency, and may even be slightly inclined to walk toward the opposite side. When the operations are unilateral, the right/left distribution of the walking errors is the same as that for the intact animals, and hence the success rate is equally high.

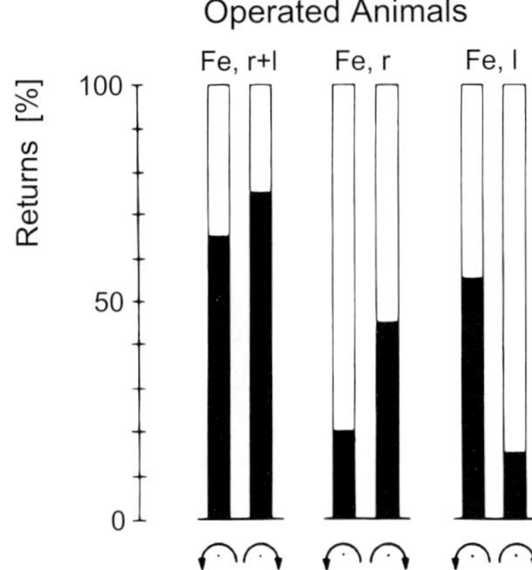

Fig. 8. Returns after curvilinear chases from the prey-capture site. *Black areas.* Fraction of unsuccessful returns. *Fe,r+l*; *Fe,r*, *Fe,l* Ablation of all lyriform organs on all femora, and on the right and left femora, respectively. Animals with unilateral ablations have high success rate only if affected legs were on the outer corridor perimeter during the chase. (Seyfarth et al. 1982)

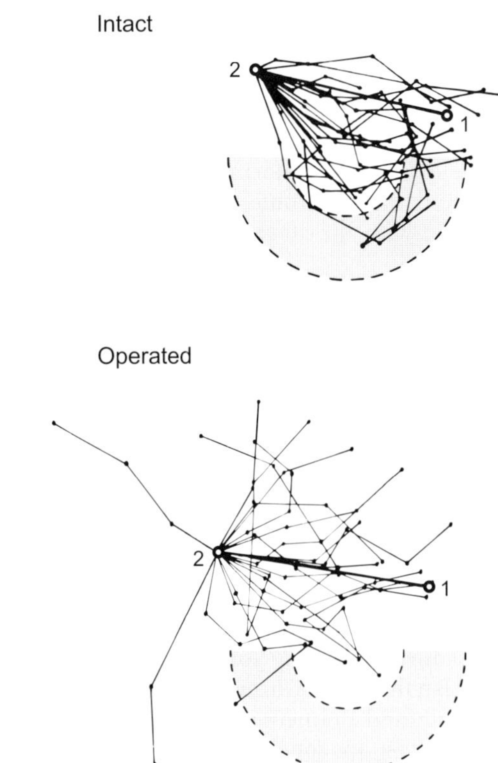

Fig. 9. Comparison of returns of intact spiders and those in which all lyriform organs on all femora were destroyed, after curvilinear chases. Unlike the intact spiders, the operated ones show no bias toward the corridor. (Seyfarth et al. 1982)

What Conclusions Can Be Drawn?

The side-specific effects of this orientation behavior are clearly correlated with the inactivation patterns and the asymmetric situation of being driven through the corridor. It follows that the kinesthetic (or idiothetic) memory does indeed depend at least partially on proprioreceptive information provided by the lyriform organs of tibia and femur while the animal is walking away from the capture site. However, this does not exclude the possibility that the same organs also inform the spider, step by step, about its return route and thus enable a comparison with the stored information (Seyfarth et al. 1982).

Cupiennius is so far the only instance, apart from the wolf spider (Lycosidae) *Pardosa amentata* (Görner and Zeppenfeld 1980), of a demonstrable participation of morphologically identified proprioreceptors in idiothetic orientation – and in such a way that we can justifiably continue to use the word "kinesthetic". A particularly remarkable aspect is that after inactivation of the lyriform organs, the spiders have no obvious problems in maintaining their normal locomotor coordination. Evidently they do not use the information mediated by the lyriform organs during walking for this purpose (Seyfarth and Barth 1972; Görner and Zeppenfeld 1980; Seyfarth and Bohnenberger 1980).

However, in conclusion it should be pointed out once again that this rather peculiar mode of orientation is not a specialty of *Cupiennius*. It has often been described previously, and presumably exists in many more animals. Detour compensation has been reported, for instance, for the spider *Agelena labyrinthica* (Görner 1958, 1973), for fiddler crabs *Uca rapax* (von Hagen 1967) and for gerbils (Mittelstaedt and Mittelstaedt 1980). Among the invertebrates, spiders have been particularly thoroughly investigated with regard to idiothetic behavior patterns (Görner and Claas 1985; Mittelstaedt 1985).

When saying spiders, I mainly refer to the funnel-web spider *Agelena labyrinthica* and its ability to find its way back from the sheet web to its retreat. *Agelena* does not lose its bearings even in total darkness, when it relies on idiothetic orientation exclusively. Under biologically relevant conditions, however, external sources of spatial information are involved as well. The directional cues then used are gravity, the mechanical properties of the web, and the distribution of the ambient light.

We must conclude with an even broader excursion, in which the focus is not so much on the sensory inputs *per se* as on what the central nervous system does with them. Thanks in particular to the many excellent analyses by Rüdiger Wehner and his coworkers at the University of Zurich on the desert ant *Cataglyphis* (see, for example, Wehner and Wehner 1986, 1990; Wehner 1992; Müller and Wehner 1994; Hartmann and Wehner 1995), it has become clear that "homing" or return orientation in arthropods – whether they are spiders, ants, bees or crustaceans – always conforms to the same principle. By continuous integration of information about the outward route and its angle, a mean homeward vector is obtained. This amounts to an Ariadne's thread, by way of which the animals are constantly connected to their starting point (Wehner and Wehner 1990). The term for this widespread type of navigation, in which the animal itself is always at the center of the spatial reference system, is "route integration" or "dead reckoning". It does not matter where the information about distance and angle of the outward path comes from. It may be kinesthetic information, as in the case of *Cupiennius* and other spiders, or the information provided by a sky compass in ants and bees (Wehner 1992). And, of course, dead reckoning does not exclude the possibility that landmarks are also used for orientation. At least when relatively long distances are covered, this seems a useful means of fine regulation, especially because once errors are made during route integration, these add up and must be compensated.

Readers interested in neuronal models of route integration are referred to recent publications by Hartmann and Wehner (1995) as well as Wittmann and Schwegler (1995), and to the older papers by Mittelstaedt (1985) and Mittelstaedt and Eggert (1989).

XXII Visual Targets

Chapter XI, dealing with the eyes, leaves no doubt: even in *Cupiennius*, the night-active hunting spider, vision is a highly developed sensory system. Our findings on the neuroanatomy of the visual system in the brain reinforce this impression (Chapter XVI). But why and how does *Cupiennius* make use of its visual capacities in behavior? Both the electrophysiologically measured sensitivity of the eyes and the sensitivity mainly estimated from their fine structure suggest that *Cupiennius* can see even in extremely dim light – that is, not only shortly after sunset but also under considerably worse conditions, as in moonlight (see Chapter XI).

This chapter reports the first steps toward an understanding of the role of this sensory potential in behavior. We consider first the spiders' ability to walk to a visible target, then a light-controlled change in gait, and finally the eye muscles, which in *Cupiennius*, as in other spiders, can move the retinas of the principal eyes behind the rigid lens. There is nothing in *Cupiennius* like the abundance of visually guided behaviors shown by day-active jumping spiders and also some wolf spiders. We thus have nothing to report about visually guided prey capture, recognition of sexual partners or the astounding ability to visually conceive detours to a prey animal (Hill 1979; Jackson and Blest 1982; Forster 1985; Clark and Uetz 1993; Tarsitano and Jackson 1993; McClintock and Uetz 1996; Hebets and Uetz 1999).

1 The Discrimination of Visual Stimuli: AM Eyes or PM Eyes?

The experiment: an adult *Cupiennius* male is cautiously released into an arena, the floor (a square 2.5 m on a side) and side walls of which are uniformly illuminated. On the wall opposite the spider, 2 m away from it, visual stimuli cut out of black cardboard are mounted. Usually there are two such stimuli, 1.5 m apart from one another. The simple question: how will the spider walk and toward what, especially when the cardboard shapes are different? With such experiments Axel Schmid (1997) made some observations that are important in the initial stage of this analysis.

First, it was essential to establish that a response to visual stimuli can be elicited in the laboratory at all. *Cupiennius* does walk straight toward the targets! If the eyes are covered with an opaque mixture of wax and soot, its walks are undirected – so the response is indeed visual. We had reason to hope that *Cupiennius* would approach visual patterns even before Axel Schmid's experiments, because during field work at night we repeatedly saw that males (these are the true vagabonds, wandering about considerably more than the females; see Chapter IV) would walk straight toward the pseudostem of a banana plant, or else toward one of our trouser legs. In the arena, when two vertical bars are presented simultaneously, spiders with eight intact eyes approach one or the other with equal frequency. But when simultaneously presented stimuli have different shapes, the spiders prefer the vertical bar to a tilted bar (22° away from the vertical) or an upside-down V; in both cases, the difference in response is highly significant.

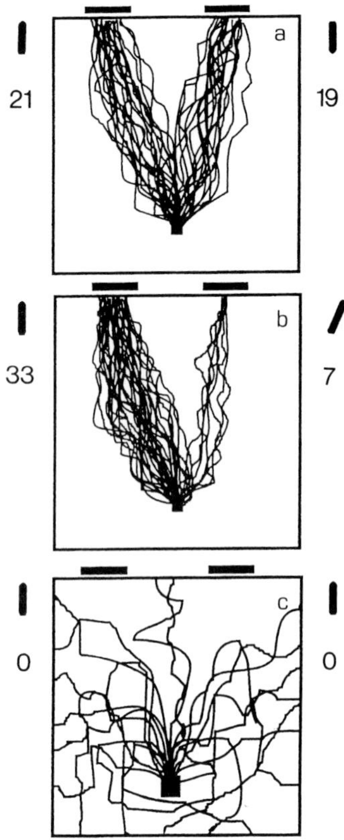

Fig. 1 a–c. Walking paths of *Cupiennius salei* toward two different visual targets. **a** Visual stimuli are *vertical bars* (height 50 cm, width 24 cm). **b** The same bars as in **a** but the *right one* tilted by 22°. **c** Visual stimuli as in **a** but spiders with all eyes covered. *Black square* indicates area at which spiders were introduced into the arena. Ten animals, four runs each. (Schmid 1998)

There is also a significant preference for an upright rather than an inverted V (Fig. 1, Tables 1 and 2). However, there is no significant difference between the number of walks toward a vertical bar or an upright V.

The elimination of various eyes by painting the lens in some cases had drastic effects, depending on the combination of eyes so treated (see Table 2). This result suggests that the AM (anterior-median) and PM (posterior-median) eyes play different roles. It is clear that the spiders can detect the visual patterns with either set of eyes. The number of undirected walks is

Table 1. Combinations of visual stimuli presented to *Cupiennius salei.* (Schmid 1998)

Target Combinations	Shape of Targets A	Shape of Targets B	Width/Height (mm) and Orientation (°)
Type 1	|	|	240/500 versus 240/500 at 0°
Type 2	|	/	240/500 versus 240/500 at 22°
Type 3	|	Λ	240/500 versus 2×120/500 at 22°
Type 4	|	V	240/500 versus 2×120/500 at 22°
Type 5	V	Λ	2×120/500 versus 2×120/500 at 22°

the same as when the animal is intact if only the two AM eyes are available. However, it rises from ca. 20% to over 40% when only one AM eye, or either one or two PM eyes, remains unpainted. Still more important: as long as the AM eyes are intact, the differential attractiveness of the various targets persists, whereas the spiders cannot distinguish the visual stimuli with their PM eyes alone. All this indicates that the AM eyes are used for discrimination, and the PM eyes merely detect that a visual stimulus is present.

That *Cupiennius* would see differently with its AM and PM eyes had been expected, given that although the visual fields of these eyes largely overlap, they differ considerably in their structure and their central-nervous connections (see Chapters XI and XVI). Given the results just described, we are more confident in our supposition that *Cupiennius* males in particular, while wandering around at night, use their visual sense to locate not trouser legs, necessarily, but plants on which they can count on finding females or prey and/or a retreat.

It is no longer entirely correct that, as Grenacher wrote in 1879: "We must accept dimorphism as a given fact and its physiological significance as totally incomprehensible to us, and leave it at that." There is still much to be done, however. The next generation of spider researchers should now concentrate especially on the electrophysiological characterization of the interneurons in the visual pathways of the two types of eyes.

Table 2. Number of choices spiders make when presented with the five stimulus pairs explained in Table 1. Control animals had untreated eyes only. In the other experimental groups the untreated eyes are the ones indicated, the other eyes being covered with a mixture of beeswax and charcoal. (Schmid 1998)

Type	Control		2 AM		2 PM		1 AM		1 PM	
	A	B	A	B	A	B	A	B	A	B
1	21	19	15	17	18	14	13	19	16	16
2	33	7***	21	7**	14	14	22	10*	17	15
3	33	7***	24	4**	14	14	20	12	18	14
4	25	15	20	12	15	17	16	16	18	14
5	35	5***	23	9**	17	15	23	9**	19	13

AM Anterior median eye, *PM* posterior median eye. Seven to eight animals and 28 to 32 runs per experimental group. Control group ten animals, 40 runs. * $p<0.05$; ** $p<0.01$; *** $p<0.001$

2 Change of Gait

A second experiment on the participation of the eyes of *Cupiennius* in behavior is not unlike the first. Again the spider walks toward a visible target. But now it does this not on solid ground but on a spherical walking compensator, which is controlled by a motor so that as the spider walks freely, the ball rotates under its legs in such a way to keep the spider more or less in the same place. Axel Schmid had the opportunity to use such an apparatus in Gernot Wendler's laboratory in Cologne. In his experiments, *Cupiennius* performed a "stationary" walk toward a black cardboard strip 40 cm away. When in total darkness (that is, in infrared light at 950 nm, which it cannot see) the spider not only walks undirectedly but converts from a gait using all eight legs to one using only six. In the dark it employs the first legs as antennae – you might almost say, like a blind person's cane. The extended forelegs are moved up and down and also sideways, feeling out the immediate surroundings. The conversion from the normal walking movement to this sensor function occurs as soon as the normal room light is switched to the invisible (to the spider) infrared (Fig. 2). If the spider has no forelegs, the antenna function is not taken over by the second pair of legs. The influence of the visual input on the movement pattern is evidently specific to the first leg pair.

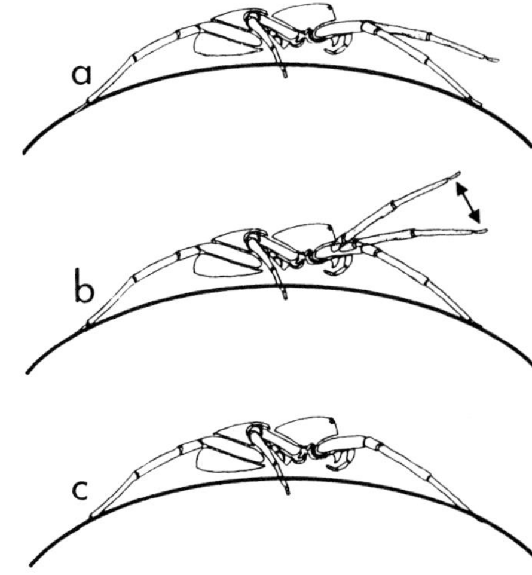

Fig. 2 a–c. Leg posture of *Cupiennius salei* when walking on a sphere in complete darkness (**b**). **a** Spider walking with all eight legs with lights on. **c** Spider without front legs; second pair of legs do not take on the feeler-like function of the first legs as shown in **b**. (Schmid 1997)

3 The Movement of the Retina

The third story about visually controlled behavior relates to the behavior of the eyes themselves! Like other spiders, *Cupiennius* has muscles with which it can move the retinas of the AM (anterior-median or principal) eyes behind the rigid lenses. This specialty of spiders has been known for 160 years. The Dutch scholar A.

Brants discovered it as early as 1838 in *Mygale*, a bird spider, but it is still insufficiently understood.

Dr. Franz Leydig of the University of Würzburg wrote about this in 1855 in the journal Müller's Archiv (Archiv für Anatomie, Physiologie und Wissenschaftliche Medicin), saying that unlike Johannes Müller (the famous Berlin physiologist and editor of that journal), he himself was very well able to make a statement about the mobility of the visual tools of the spiders: "in both *Mygale* and other spiders I see the muscles of the choroidea very clearly ...". And in order easily to observe the movements in the living animal "one should cripple a living spider by cutting off the legs, to make it motionless, and focus on the eyes with slight magnification and incident light, which can be intensified by convex lenses. Then one can see that the back of the eye, which either is dark or glows due to a tapetum, pulls together, not slowly, gradually, but quite powerfully, it makes jerky contractions."

Whereas Franz Leydig (1855) was still quite erroneously regarding these muscles as a sphincter, to constrict and expand the pupil, Grenacher (1879) recognized clearly that their basic function was rather different:
"Each contraction of these will necessarily result in a change in position of the axis of the retina, a shifting of the latter with respect to the lens, and these movements have nothing in common with those we associate with the notion of accommodation ... A necessary consequence of shifting the retina in a plane perpendicular to the axis of the eye, however, is that other regions of the outside world are projected onto it, and if one is determined to make a comparison, such shifts could better be put in parallel with the rotations of our own bulbi mediated by the external eye muscles."

The only group investigated in detail has been the jumping spiders (Salticidae), and for this information we have to thank in particular Michael Land (1969a,b). Jumping spiders have six eye muscles, with which they can produce not only translatory but also rotatory movement of the retinas of the AM eyes. Land (1969a,b), in what still counts as a pathbreaking paper, found both spontaneous retinal movements and saccades that serve to fixate the fovea on a visual target. He also describes "tracking", by means of which the image of a moving object is fixed on the fovea, and "scanning", which presumably serves for pattern recognition.

Cupiennius has only four eye muscles, two dorsal and two ventral (Fig. 3). The ventral muscle in each of the two AM eyes is 650 μm long and contains about 33 striated tubular muscle fibers; it is thus somewhat larger than the dorsal muscle (600 μm, ca. 28 cross-striated tubular fibers). In this respect *Cupiennius* closely resembles the wolf spiders *Lycosa poliostoma* and *L. travassosi* (Melamed and Trujillo-Cenóz 1971). This muscle originates at the inner surface of the clypeus (see Fig. 3) and inserts at the ventro-lateral surface of the eye cylinder. The dorsal muscle runs from its origin at the exoskeleton between the two PM eyes to the dorso-lateral surface of the eye cylinder. The contractions of both muscles are opposed by the elastic forces of the eye cylinder (Kaps and Schmid 1996).

Each of the two pairs of eye muscles (comprising a dorsal and a ventral muscle) is innervated by a motor nerve. It comprises four axons and splits into two branches shortly before reaching the eye cup. A dorsal branch with only one thick (diameter ca. 12 μm) axon runs to the dorsal eye muscle. The ventral branch contains the remaining three axons (one ca. 10 μm in diameter and two ca. 8 μm) and runs along the wall of the eye cup, to the ventral eye muscle.

Looking back toward the brain, the two branches join to form a nerve strand that runs dorsalward above the optic nerve of the AL eye and then along the optic nerves to the venom glands. Its course beyond this point is complicated: the oculomotor nerve passes through the narrow gap between venom glands and pharyngeal muscles (musculus dilator pharyngis anterior and m. d. p. posterior, according to Palmgren 1978) and joins the ca. 150-μm-thick cheliceral nerve (Babu and Barth 1984), which ultimately enters the brain below the optic neuropils (Kaps 1998).

The similarity of not only the eye muscles themselves but also their innervation to the situation in *Lycosa* is remarkably close (Melamed and Trujillo-Cenóz 1971).

Florian Kaps and Axel Schmid (1996) recorded the muscle activity (electromyogram) at the same time as the retinal movements (video recordings) in order to understand how the movements are produced. First the spider was fixed to a holder so that it could not move. It turned out that the dorsal muscle is spontaneously active, unlike the ventral one. Its activity, 12±1 Hz at the resting level, increases sharply (80 Hz) when an air puff is directed through a pipette onto the tarsus of the second walking leg, stimulating the trichobothria there. The ventral muscles also respond to such a stim-

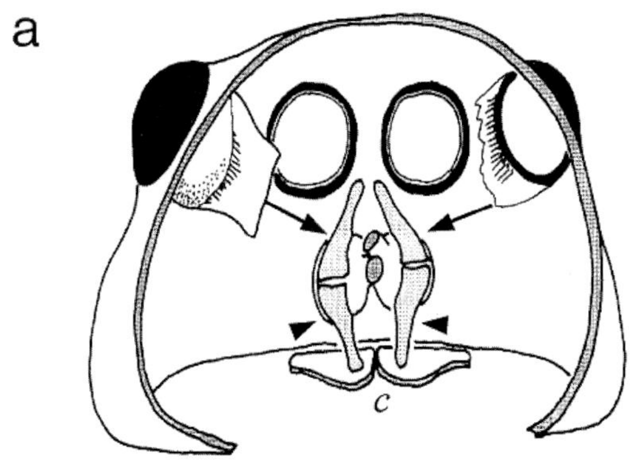

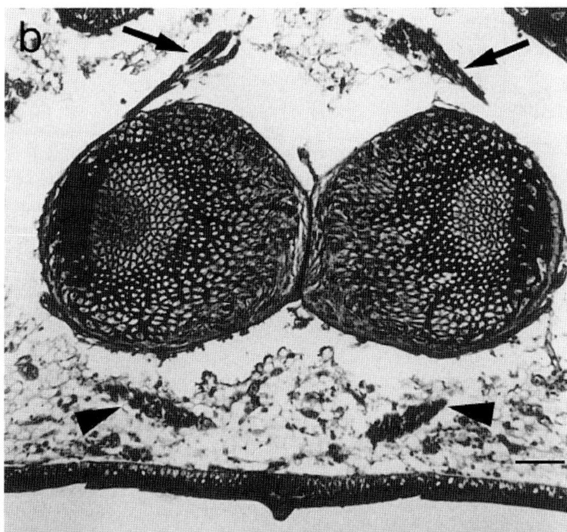

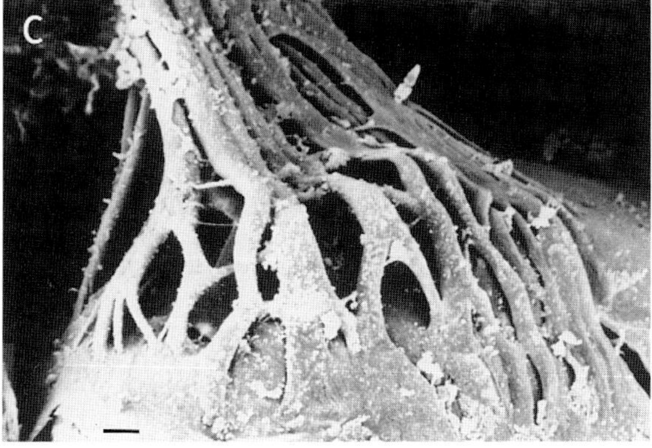

ulus. During active retinal movements the dorsal muscles, like the ventral muscles, of the two principal eyes are never activated simultaneously, and the video analysis showed that there is no correlation between the two eyes in either the occurrence of retinal movements or their direction.

According to Kaps and Schmid (1996), *Cupiennius* makes spontaneous retinal movements, so-called microsaccades, as well as the retinal movements induced by mechanical stimuli as described above.

The microsaccades consist of continuously recurring twitch-like retinal movements of 2° to 4° in the dorso-median direction. They last ca. 80 ms and are produced by the activity of only the dorsal muscle. Because each microsaccade corresponds to one action potential, their frequency is identical to that of the spontaneous discharges of the dorsal muscle. Accordingly, with each action potential the visual field is shifted ventrally (in the direction opposite to that of the retinal movement) and is then returned to the original position by the elastic restoring forces of the eye.

In contrast, the retinal movements induced by mechanical stimulation are brought about by both dorsal *and* ventral muscles and can be considerably larger, with deflection as great as 15°, than the spontaneous retinal movements. Because both muscles are involved, the direction of movement can also vary, depending on their relative levels of activity (Fig. 4). If the activity increases (above the spontaneous level) as much in the dorsal as in the ventral muscle, then the retina moves in the medial direction, and if the

Fig. 3 a–c. Muscles of the anterior median eyes (principal eyes) of *Cupiennius salei*. **a** Inside view of anterior region of prosoma; *arrows* point to two dorsal eye muscles, *arrowheads* to two ventral eye muscles. *Bar* 500 μm; *c* clypei. **b** Frontal section through principal eyes, also showing the eye muscles. *Bar* 200 μm. **c** Scanning electron micrograph of dorsal eye muscle where it attaches to the principal eye. *Bar* 20 μm. (Kaps and Schmid 1996)

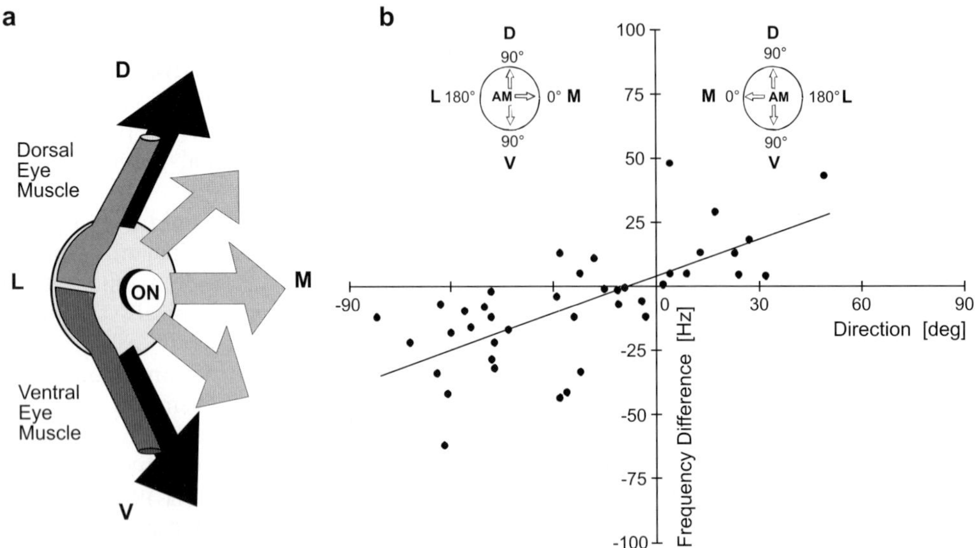

Fig. 4 a, b. Arrangement of the eye muscles and direction of pull on the eye by their contraction. **a** Posterior view of eye cylinder (*ON* optic nerve). Contraction of the dorsal or ventral muscle only (*black arrows*) displaces the eye tube in the direction of the respective black arrows. When both eye muscles contract simultaneously, the shift of the retina results from the vector sum of the forces generated. *Gray arrows* indicate resulting deflection directions. *D* Dorsal; *V* ventral; *M* medial; *L* lateral. **b** Correlation between the direction of the retinal shift (x-axis) and the difference in activity of the two eye muscles (electromyogram *EMG*, impulse frequency). 0° corresponds to medially directed movement, 90° to dorsalward movement (see *inset*). (Kaps and Schmid 1996)

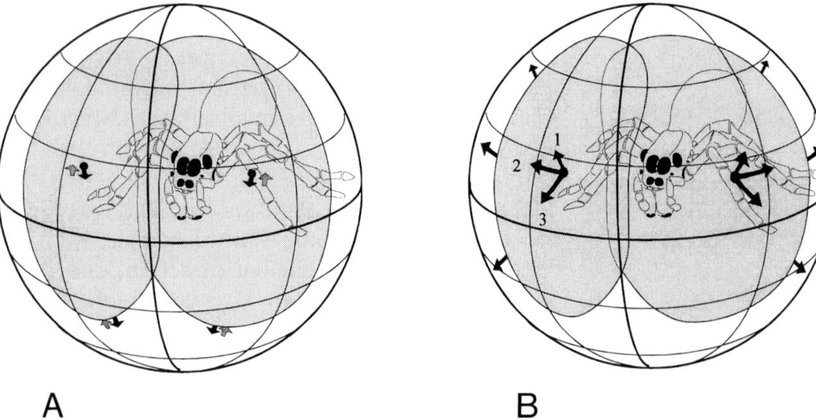

Fig. 5 A, B. The visual fields of the principal (AM) eyes of *Cupiennius salei* (after Land and Barth 1992) and their shift (*black arrows*) due to active eye movement. *Gray arrows* indicate passive return movement. **A** Spontaneous ventrally directed microsaccades. **B** Stimulus-induced shift of visual fields in dorsolateral (*1*), lateral (*2*), or ventro-lateral (*3*) direction. (Kaps and Schmid 1996)

"dorsal" or "ventral" activity predominates, the movement will be more toward dorsal or ventral, respectively. Both muscles insert on the eye cylinder in a somewhat lateral position, so that each movement comprises a median component. The associated visual-field displacements of the principal eyes are illustrated in figure 5. When a leg is stimulated mechanically with an air puff, the visual field of only the ipsilateral AM eye shifts toward the stimulus site, while the contralateral eye is unaffected. An enlargement of the binocular overlap region of the visual fields of the two principal eyes is not possible, because of the lateral insertion of the muscles. On the other hand, it is only this special feature of the arrangement of the muscles that allows them to collaborate synergistically.

Some time ago we interpreted the microsaccades of *Cupiennius salei* as a mechanism to prevent the photoreceptor cells of the AM eyes from adapting to a motionless visual stimulus (Land and Barth 1992). This idea is appealing in view of what is known about jumping spiders, and also vertebrates (Land 1969b). The new findings by Kaps and Schmid (1996) provide further support: the microsaccadic movements of 2–4° match closely the angle of about 3° between the receptor cells of the AM eyes (Land and Barth 1992) and should therefore work excellently to prevent adaptation. This mechanism then probably also plays a role in the spider's capacity for visual discrimination of stationary objects, discussed in Part 1 of this chapter, and its dependence on the intactness of the AM eyes (and not the PM eyes). The adapted secondary eyes would then, as in the jumping spiders (Land 1985), serve as a second visual information channel with the function of providing information about the movements of objects. Once again, then, we have a hypothesis that is supported by the structure of the eyes as well as the neuroanatomical findings in the CNS (Chapter XVI) (Land 1985; Strausfeld and Barth 1993; Strausfeld et al. 1993).

In more recent experiments for his dissertation Florian Kaps (1998) gave the spider more freedom of movement, to the extent that the constraints of electrophysiological recording of the muscle activity allowed. For this purpose the animal was still connected to a rod by its sternum but was provided with an air-suspended, easily movable styrofoam ball on which it could walk in place (Fig. 6).

Under these conditions, the spontaneous microsaccades proved not to occur continuously as they did in the fixed spiders; the dorsal eye muscle does not fire permanently at about 12 Hz. Instead, the eye movements now depend on the current "activity state" or motivation of the spider. These states can be distinguished from one another by the body posture and leg positions. In the "fright posture" there is no eye-muscle activity at all. A spider in the "preparedness posture" exhibits repeated spontaneous contractions, in the absence of any mechanical or visual stimulus. In this situation microsaccades (contractions of only the dorsal eye muscle) are more common than a directed displacement of the visual fields (saccades) by the activity of the dorsal and the ventral muscles (Fig. 7). When in the preparedness posture *Cupiennius* is more likely to begin walking spontaneously than when in other states. Interestingly, an impending period of locomotion is signalled by an increase in eye-muscle activity. In this case the frequency rises continuously and more slowly than in the sudden spontaneous bursts of activity, beginning as long as 5 s before the onset of walking (Fig. 8). No relationship between the resulting visual-field displacement and the starting direction of the walk has been discerned.

The situation is different when the spider is actually underway. Then the direction of gaze is indeed shifted in the direction of the locomotion. That is, the spider looks where it is going and does it (under the experimental conditions) at each walking movement (Table 3).

When a spider standing on a styrofoam ball is presented with motionless visual stimuli in the form of black cardboard strips, nothing changes in the activity of the eye muscles but walking activity increases (although it is still very slight). The walking direction is in no way related to the position of the stimulus in the visual field. During walking the spider again shifts the field of view of the principal eyes in the current direction of locomotion, as though the visual stimulus were not present at all.

A moving visual stimulus is considerably more effective than a stationary one. It induced walking activity in 20 out of 22 spiders. Both slowly (36°/s) and rapidly (72°/s) moving stimuli are effective. But again the spider seems not to be concerned with the position of the stimulus in the visual field. It does not clearly direct its walk toward the moving stripes; in a typical experiment there were 313 instances of locomotion in the direction of the stimulus, and 311 in some other direction.

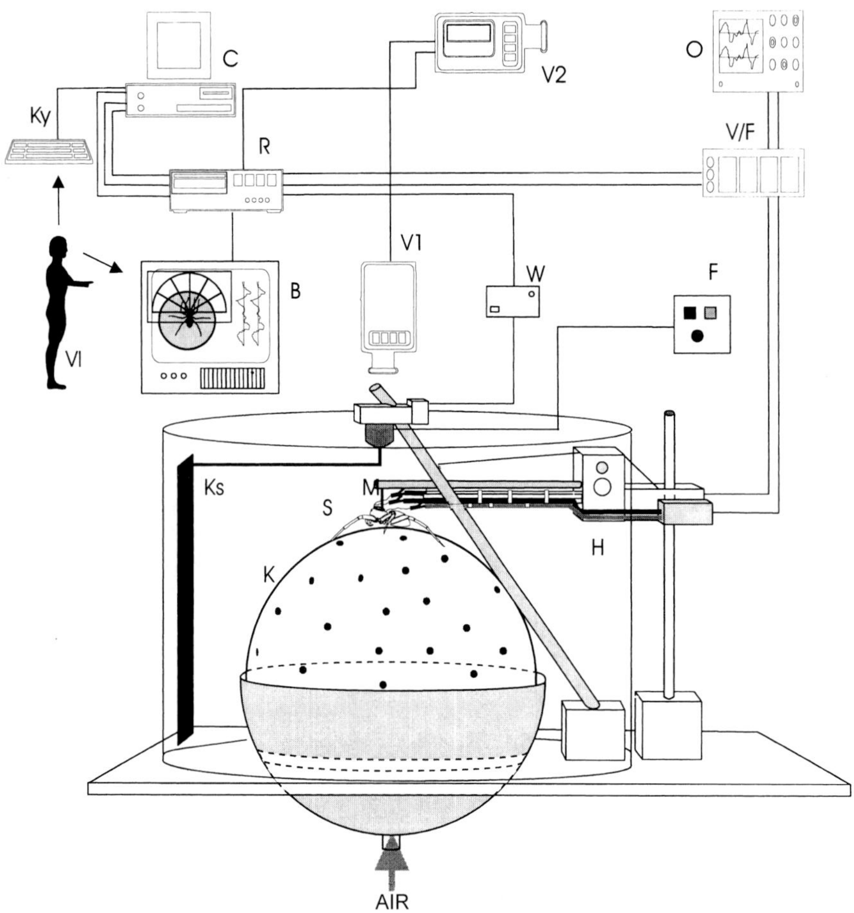

Fig. 6. Experimental setup used to study retinal movements in *Cupiennius salei* by synchronously recording the shift of the visual field of the principal eyes (AM), the body movement of the spider, and the visual stimulus presented. (1) Eye-muscle activity: recording with manganin wires (*M*); *V/F* amplification and filtering; recording on the two audio tracks of a video recorder (*R*). (2) Stimulus: position of cardboard strip (*Ks*) recorded on audio track of video recorder (*R*). (3) Movement of the sphere (*K*) resting on an air cushion and moved by the stationarily walking spider: recorded on video track of *R* using two video cameras (*V1*, *V2*; image-splitter) simultaneously with the image (*O*) of the electrophysiologically monitored eye-muscle activity. After the experiment the video tape is played (*screen B*) and the signals recorded on the three audio tracks of *R* are fed into a PC (*C*). The movements of the walking sphere are observed on the monitor and assigned to different directions. (Kaps 1998)

In addition to the fright and preparedness postures, *Cupiennius* has what can be termed a "resting" posture. As in the preparedness posture, the activities of the dorsal and the ventral eye muscles are correlated with the position of the moving visual stimulus in the field of view. This underlines the two main findings: a resting spider shifts the field of view of its principal eyes in the direction of a moving stimulus, whereas a walking spider shifts it in the direction of locomotion. The saccades of the resting spider on a styrofoam ball last up to three seconds and presumably help the spider in its attempt to keep the moving stimulus in the field of view of the principal eyes as long as possible. Remarkably, such saccades can also be elicited when the stimulus is seen only by the secondary eyes. This is an indication of the connection be-

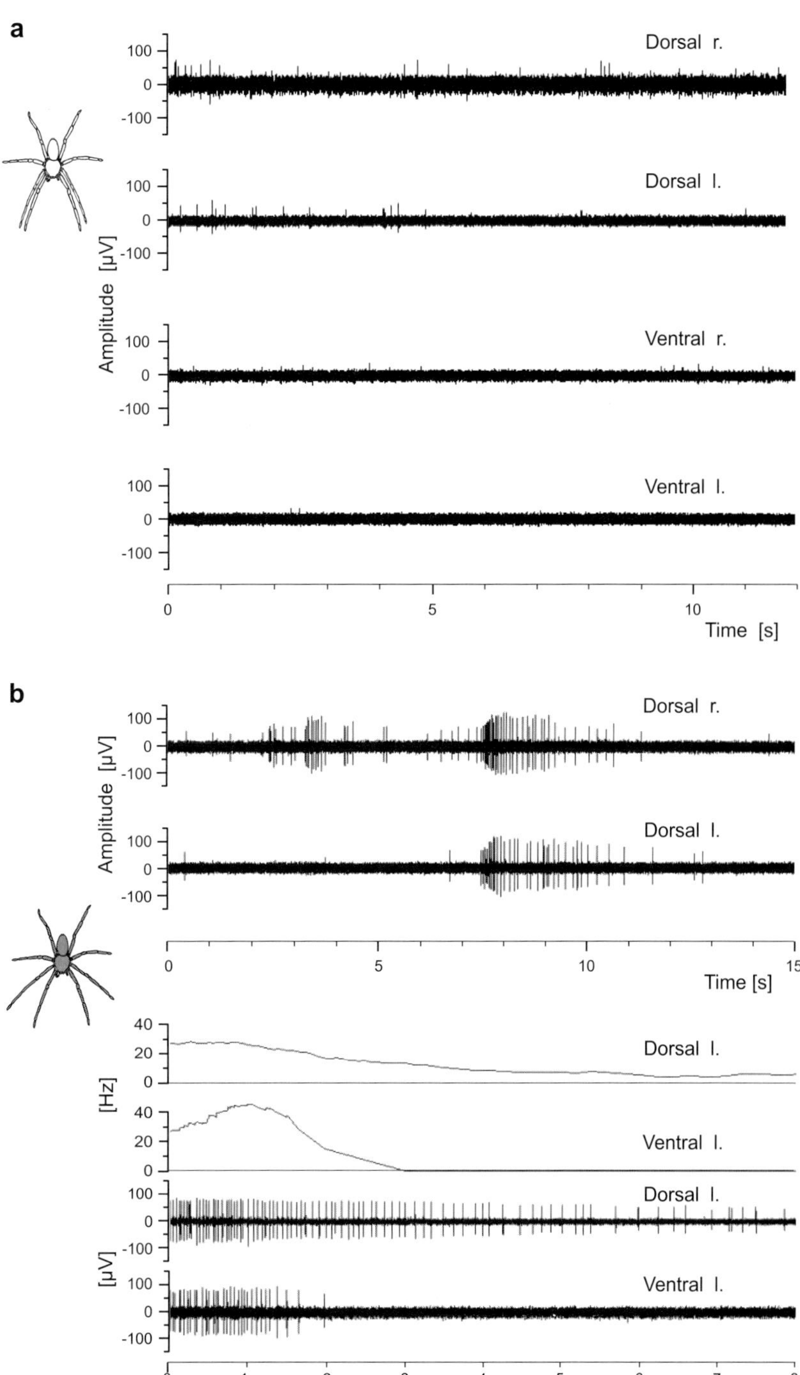

Fig. 7 a, b. Eye-muscle activity in *Cupiennius salei*. **a** Spider passive and in defensive posture, without mechanical or visual stimulation. Only sporadic activity of right and left dorsal (traces *1* and *2*) and right and left ventral (traces *3* and *4*) muscles. **b** Spider assuming readiness posture, again without mechanical or visual stimulation. Note typical phases of activity. Activity of ventral muscles usually is shorter than that of dorsal muscles. (Kaps 1998)

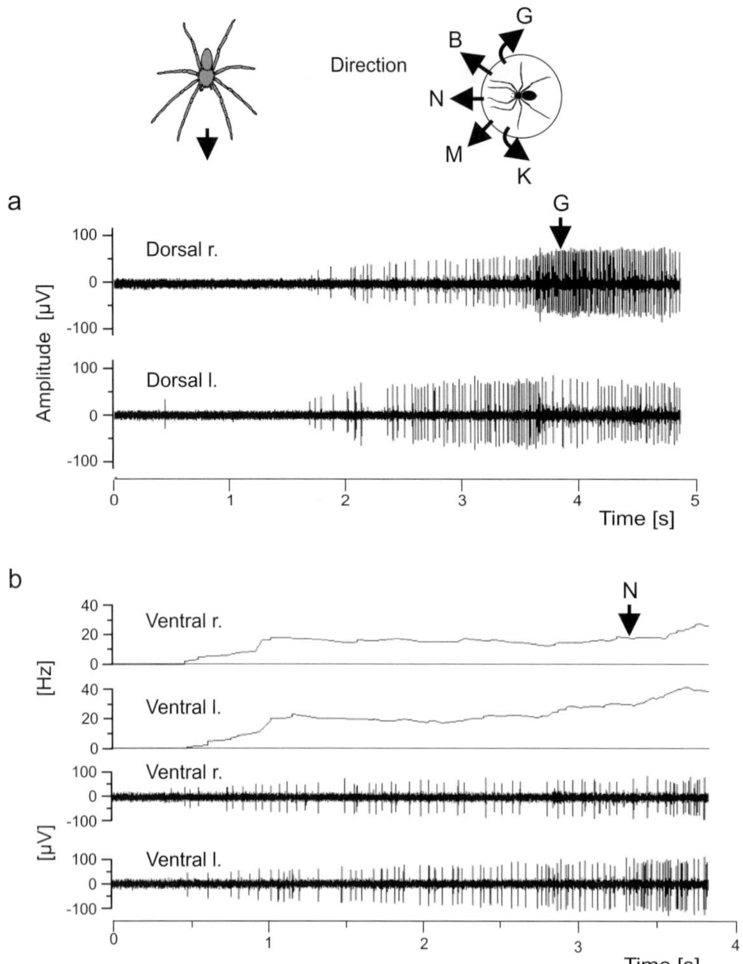

Fig. 8 a, b. Eye-muscle activity of *Cupiennius salei* before and during spontaneous locomotion. Both the right and left dorsal (**a** traces *1* and *2*) and the right and left ventral (**b** traces *3* and *4*) muscle are active for as much as 3 s before the onset of the walking movements (*arrow*; note walking direction; see *inset*). **b** shows original recordings and action potential frequencies. (Kaps 1998)

tween principal and secondary eyes. In contrast, the two sides of the body are evidently separated. The moving cardboard strip never elicits saccades in both principal eyes; instead, it is effective only when it is in the field of view of the eyes on the corresponding side of the body.

The story is obviously complicated. And we still haven't come to the truly natural situation in which the spider obtains information about the consequences of its visually released behavior. In contrast to such a "closed-loop" situation, the spider on the styrofoam ball is in an "open loop": the visual stimulus does not change position in the visual field, despite the fact that the spider is walking! The open-loop results are therefore presumably only seemingly inconsistent with Schmid's (1998) finding that genuinely free-walking spiders proceed very directly to visual stimuli.

If we now, keeping the styrofoam ball experiments in mind, consider once again the significance of microsaccades and saccades, the most

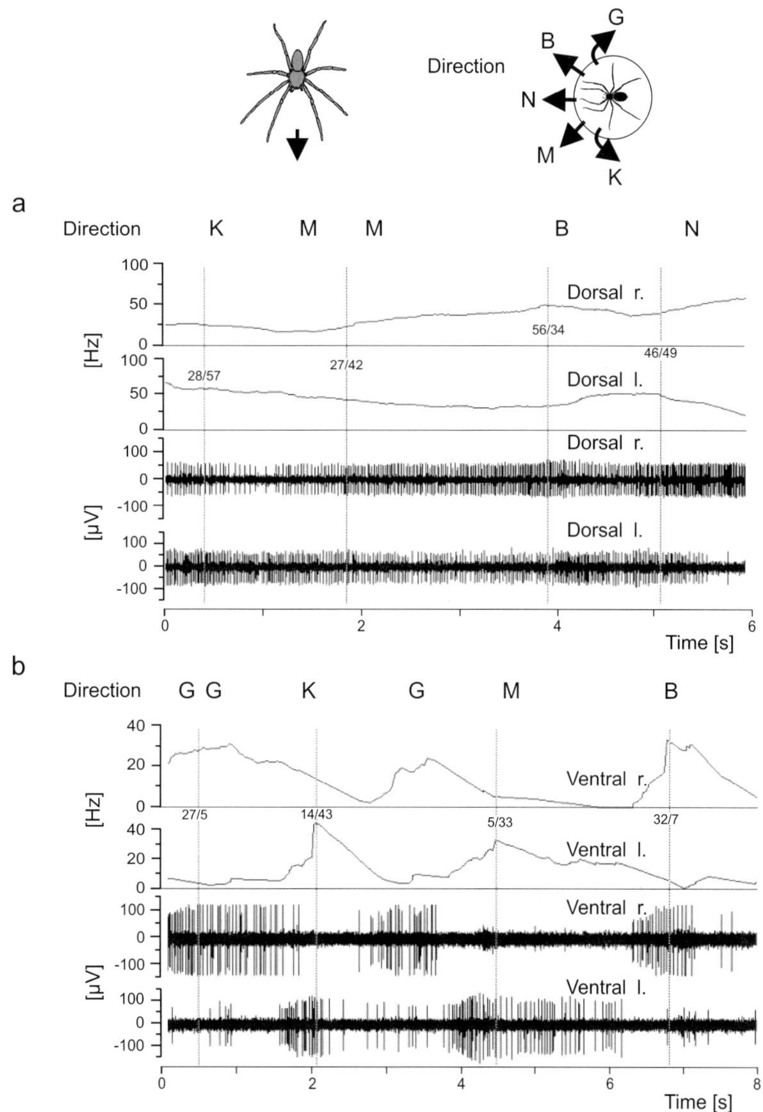

Fig. 9 a, b. Eye-muscle activity and direction of locomotion during spontaneous walking. **a** Dorsal muscles, **b** Ventral muscles. *Upper traces* Action potential frequency; *lower traces* original recording. As a rule the eye muscle moving the retina towards the direction of locomotion (*G, B, N, M, K*; see *inset*) is the only or most active one. (Kaps 1998)

important point turns out to be that *Cupiennius* sees a motionless stimulus only if its principal eyes are moving.

1. Microsaccades evidently allow the spider now and then to form an image of the stationary surroundings. Between retinal movements stationary stimuli probably disappear from the field of view because the visual cells adapt, as a result of which moving objects stand out all the more distinctly from the background. In order to test this hypothesis it will be necessary, for instance, to measure the actual adaptation characteristics of the visual cells.

Table 3. Eye-muscle activity during spontaneous locomotion and without the presentation of a stimulus. The activity phases of the dorsal and ventral eye muscles clearly correlate with the direction of locomotion. (Kaps 1996)

NO STIMULUS N=4 n=10	TOTAL NUMBER OF MEASURED ACTIVITY PHASES	ACTIVITY PHASES CORRELATED WITH WALKING DIRECTION	CORRELATED ACTIVITY PHASES DURING WHICH ONLY THE EYE MUSCLE ON THE CORRESPONDING BODY SIDE IS ACTIVE
DORSAL EYE MUSCLES N=2 n=6	24 100%	23 96% OF ALL MEASURED ACTIVITY PHASES	8 35% OF ALL CORRELATED ACTIVITY PHASES
N=2 n=4 VENTRAL EYE MUSCLES	33 100%	32 97% OF ALL MEASURED ACTIVITY PHASES	20 63% OF ALL CORRELATED ACTIVITY PHASES

2. The stimulus-induced saccades with which *Cupiennius* tracks the stimulus visually show that the moving stimulus is seen (regardless of its position in the visual field). In comparison with the horizontal extent of the visual field of the principal eyes, about 60°, the saccadic turning angle of up to 15° is small; furthermore, the saccadic movements are not reinforced by body movements.

XXIII Raising the Body When Walking over an Obstacle

One of the most urgent tasks of neuroethology is to discover functional connections in the nervous system that are related to specific behavior patterns and are responsible for the execution of these patterns. Whenever possible, the analysis should be based at the level of single cells. Of course we too wanted to achieve such an analysis, for the first time in a spider. A considerable amount of data has now accumulated that characterizes interneurons involved in the processing of sensory information on vibrations and air movements. Although in these cases (see Chapters VIII and IX) the sensory periphery and the afferents to the CNS are well known, we know far too little about the other half of the story, the motor system.

Ernst-August Seyfarth, my first doctoral student and subsequently long-term colleague in Frankfurt, had the first major success in this regard, studying a simple behavior of *Cupiennius salei* together with Wolfgang Eckweiler and drawing on the electrophysiological expertise of Jürgen Milde (of the University of Cologne). They described correlates of the reflex activity of a walking-leg muscle from the sensory input to the motor output (Eckweiler and Seyfarth 1988; Milde and Seyfarth 1988; Kadel 1992). This work laid a valuable foundation for the comprehensive neurobiological understanding of a simple behavior pattern in a spider.

A spider's walking is not preprogrammed, like the movement of a wind-up toy. Instead, the spider depends on a continual stream of sensory feedback, so that it can adjust to the conditions in the terrain it encounters and to the behaviorally relevant aspects of the situation, choosing whether to accelerate or slow down, climb up or down, change direction or crawl over an obstacle. Here we are concerned with the last case – and with the role of touch-sensitive hairs in this behavior.

When *Cupiennius salei* bumps against an object with its ventral surface, so that tactile hairs on the lower surface of the proximal leg segments are deflected and stimulated, it raises its body by means of corresponding leg reflexes. The same coordinated and sudden movement of the eight legs can also be elicited in the laboratory with no great difficulty, even when the animal is standing on an air-suspended styrofoam ball and – fixed by its back – is forced to walk on the spot. Even in an electrophysiological experiment to record the activity of central interneurons, during which the animal has to be fixed with its back down, it is possible to reflexly activate coxal muscles involved in the body-raising response by deflecting only a single hair.

In addition to *Cupiennius salei* and four other species of *Cupiennius* (*C. getazi*, *C. coccineus*, *C. panamensis* and *C. foliatus*), Eckweiler and Seyfarth (1988) stimulated ventral tactile hairs in jumping spiders (*Phidippus regius*) and bird spiders (*Brachypelma sp.*), which wander about on substrates similar to those chosen by *Cupiennius salei*. All of them responded with the typical raising of the body.

1 The Behavior

The part of the body closest to the substrate when a spider is walking is not the sternum, as might be expected. Below it are the ventral surfaces of the walking-leg coxae. These are kept 6 to 9 mm above ground, as long as no obstacle is

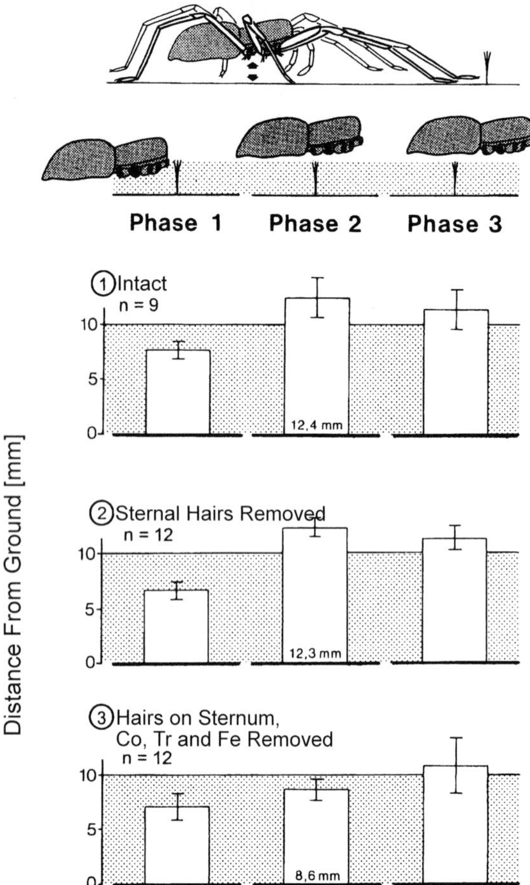

Fig. 1. Adjustment of body height and the effect of ablation of tactile hair sensilla in a *Cupiennius salei* passing freely over a wire obstacle. *Above* Body height is measured as distance from the ventral edge of the coxa of leg 4 to the ground. *Below* Mean values (±SD) of body height at three successive phases; *n* number of runs of the same animal. *Shaded area* indicates obstacle elevation of 10 mm. Intact animals (*1*) approaching the obstacle increase mean body height to 12.4 mm upon touching the wires, high enough to just pass over them. Removal of the sternal hairs (*2*) does not change this behavior. Only after additional removal of all hairs ventrally on the coxae (*Co*), trochanter (*Tr*), and femora (*Fe*) (*3*) does *Cupiennius* collide with the obstacle (*phase 2*), because body height is only 8.6 mm. In *phase 3* hairs on the opisthosoma touch the wires and induce lifting of the prosoma. (Eckweiler and Seyfarth 1988)

in the way. That the spider does not walk with its legs extended, as though on stilts, has a good mechanical reason: the stability of the body posture is increased when the center of gravity is lower. For the same reason we also feel more secure when proceeding through difficult terrain if we crouch somewhat instead of walking fully upright. When the spider is presented with an obstacle, for instance in the form of a fine wire brush (Fig. 1) (Eckweiler and Seyfarth 1988), one of two things can happen. If the spider first encounters the brush with its forelegs, pedipalps or chelicerae, then it stops, inspects the object, raises its body slowly and continues walking. In the other case the coxal hairs, many of which are up to 2.1 mm long, make the first contact. Now the spider does not stop but suddenly, within 100 to 160 ms, raises the pro- and opisthosoma far enough that it can pass over the obstacle without touching it again. After removal of the hairs on the ventral surfaces of the sternum, coxae, trochanters and proximal femora, the spider collides with the obstacle and does not raise the body until the brush touches the (also hairy) ventral surface of the opisthosoma. An intact animal goes through all the events leading to behavior in just over a tenth of a second: the detection of the tactile stimulus, the central nervous filtering and evaluation of the information provided by the sensilla, the activation of the motoneurons and finally the contraction of the relevant muscles and elevation of hemolymph pressure in order to move the legs.

In the laboratory *Cupiennius* can be made to walk in place by putting it on a lightweight ball suspended on an air jet; when the spider moves its legs, the ball rotates under it. Then individual tactile hairs can be selectively stimulated. Repeated stimulation of a given hair elicits the consecutive body-raising responses and also shows very clearly that the associated coordinated activity of all eight legs can be brought about by stimulation of a single hair (Fig. 2) (Eckweiler and Seyfarth 1988). The specificity of the response becomes clear when it is compared with the behavioral responses to stimulation of other tactile hairs. Dorsal surface of the proximal leg segments (coxa, trochanter, femur) or of the opisthosoma: escape response; dorsal surface of the distal leg segments (patella, tibia, metatarsus, tarsus): withdrawal of the leg; ventral surface of the distal leg segments: extension and lifting of the leg.

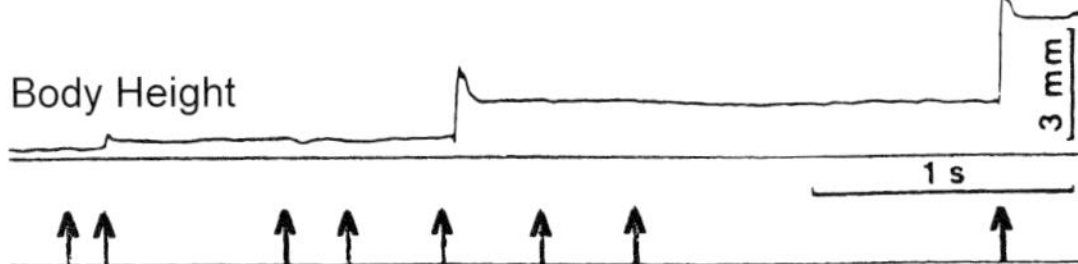

Fig. 2. Adjustment of body height following the repeated stimulation of a single hair ventrally on the trochanter of the right hindleg. Approximately every third stimulus (see *event marks in bottom trace*) induces body raising as recorded with capacitive transducer in upper trace. (Eckweiler and Seyfarth 1988)

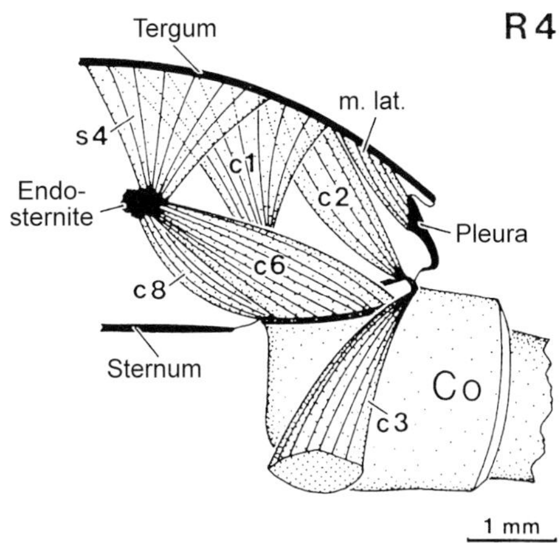

Fig. 3. Muscles within the prosoma and at the prosoma-coxa articulation in a hindleg of *Cupiennius salei* (vertical section through right half of prosoma; view from posterior). Nomenclature follows Palmgren (1978, 1981). *c2* Promotor/adductor coxae (*Co*); *c3* remotor/adductor coxae; *m.lat.* musculi laterales. *c1*, *c6*, and *c8* attach to the proximal inner edge of coxa. *s4* is attached at the cuticular endosternite in the center of the prosoma. *c3* has been dissected away and folded down to allow unobstructed view of *c2*. (Eckweiler and Seyfarth 1988)

2 Joints and Muscles

Mechanically, the raising of the body is produced mainly by pulling the coxae toward the prosoma and simultaneously extending the leg joints. Peripheral to the joint between trochanter and femur, the extension is brought about hydraulically, by hemolymph pressure (see Chapter XXIV; Blickhan and Barth 1985). The coxa is moved by muscles situated substantially in the prosoma (Fig. 3). When a tactile hair on a leg is touched, nearly simultaneous activity of many of these muscles is elicited in that leg within 1 to 3 ms, and in the other legs within 5 ms. A particularly clear response is given by the muscle C2, which originates at the tergum of the prosoma and inserts in the coxa, in the dorsal region halfway along its anterior surface (Fig. 3). As a result of the contraction of C2 the coxa is pulled closer to the prosoma and slightly forward. Thus the prosoma-coxa joint is stabilized. Presumably this muscle also contributes to the hydraulic extension of the distal sections of the leg. It is useful for further, more detailed analysis of the reflex response because of its low threshold for excitation, the reliability of its response and its good accessibility (at the tergum) for electromyographic recording by means of fine, carefully implanted copper wires. However: when the spider is lying on its back and immobilized (tethered) for an electrophysiological experiment, the response to tactile stimulation is restricted to the stimulated leg, even though many hairs have been contacted vigorously (local reaction). The muscle C2 is not spontaneously active, and its response to such stimuli is limited to slow motor units, whereas when the spider is struggling to get free (for instance, after being subjected to a sudden puff of air) fast units are recruited in the same muscle.

Although the electrical activity of the muscles during body raising is very brief, the spider remains in the new position for seconds and sometimes even minutes, with no recordable muscle activity. How it does this is unknown. Some possibilities have been discussed, in analogy to similar phenomena in insects, by Eckweiler and Seyfarth (1988) (see also Hoyle 1983).

3 The Hairs as Triggers

The hairs that trigger the body-raising reflex and activation of the muscle C2 are within a clearly circumscribed field extending along the ventrolateral surface of the coxa to a narrow strip on the femur, near the joint (Fig. 4). In adult animals this field measures around 12 mm^2 on each leg and contains approximately 3000 tactile hairs (for a total of 24,000 hairs on the eight walking legs). With ca. 400 tactile hairs/mm^2, the part on the coxa has the greatest hair density (Plate 15).

In small spiders the fields are not yet as large. The first tactile hairs develop while *Cupiennius* is still in the egg sac, as a so-called second incomplete stage. In the "sensory field" for the body-raising response shown in figure 4, there are then only 10 tactile hairs, and stimulation of these does not elicit coordinated raising of the body but merely an apparently uncoordinated stamping of the legs. This changes in the next stage, the first complete stage; this molt still occurs in the egg sac, but the tiny spiderling then leaves its shelter. Now both the number and the density of the tactile hairs have increased sharply. On the ventral coxa alone there are 32 hairs, and in the entire area (ventral surfaces of the proximal segments of each leg) 52. Furthermore, animals in this stage are already able to raise the body upon hair stimulation just as the adults do. The behavior is thus completely there (Höger 1994; Höger and Seyfarth 1995). Only a few days after emergence from the egg sac, the spiderling runs out of stored yolk. It must then

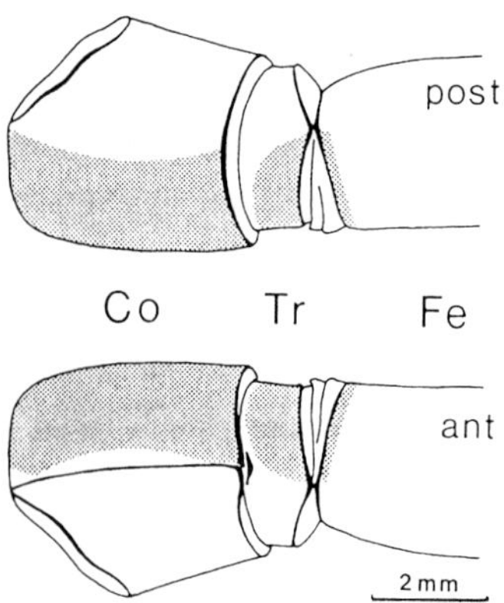

Fig. 4. Tactile reflex field for muscle c2 ventrolaterally on the coxa (Co), trochanter (Tr), and femur (Fe) of walking legs. Stimulation of tactile hairs *within shaded areas* elicits reflex activity in muscle c2 of the same leg. *post* Posterior; *ant* anterior. (Eckweiler and Seyfarth 1988)

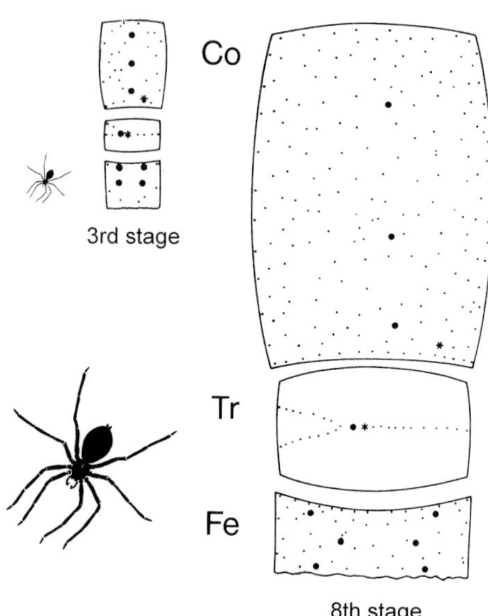

Fig. 5. Tactile hairs on the ventral aspects of the proximal leg segments in *Cupiennius salei* at 3rd and 8th developmental stage. *Dots* mark position of hairs. *Heavy dots* Individually identifiable long hairs are present from the beginning; *asterisks:* lyriform slit sense organs *HS*-1 (*Co* coxa) and *HS-2* (*Tr* trochanter). *Fe* femur. In their 3rd developmental stage spiderlings are just beyond their first molt outside the egg sac. In their 8th stage they are beyond their sixth molt. (Höger 1994)

really come to terms with its surroundings and hunt for prey itself. This is also the period during which the "drop and swing" behavior is easiest to induce in *Cupiennius* (see Chapter XXV).

That small spiders having such a meager array of tactile hairs do the same thing as the adults, with their ca. 24,000 tactile hairs (ventral surfaces of the proximal segments of all legs), may seem astonishing. Why bother to add so many more? Ulli Höger (1994) considered this problem while with Ernst-August Seyfarth in Frankfurt and came up with something interesting: although the total number of tactile hairs increases greatly at each molt, their density remains about the same from the first complete stage on. The increase in number is merely to keep the enlarged surface area equally well supplied with hairs. Presumably, therefore, the spatial resolution stays constant. The increase in area is enormous, as can easily be visualized by comparing the overall sizes: the first complete

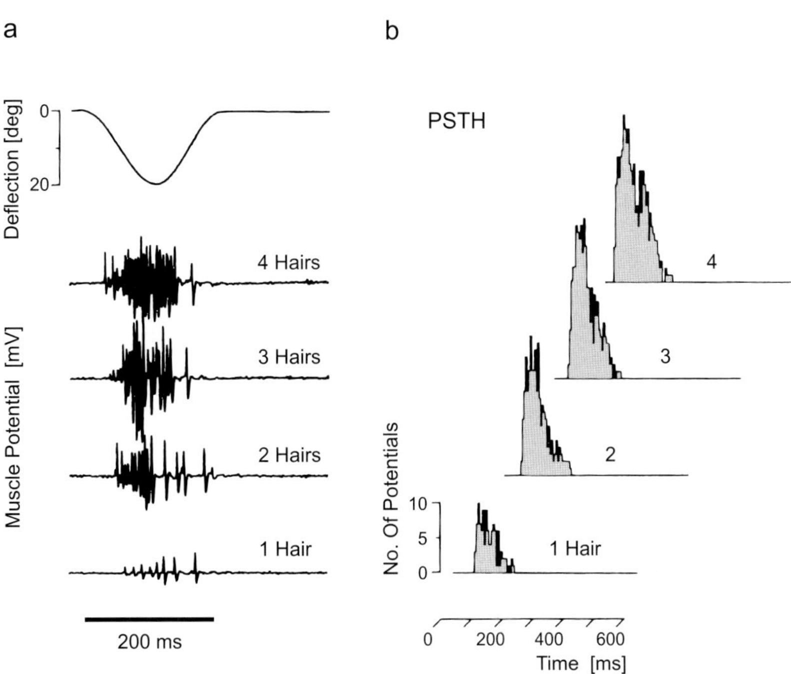

Fig. 6 a, b. Tactile reflex in muscle c2 (see Fig. 3) and the effect of successive ablation of an increasing number of trochanteral tactile hairs. **a** Muscle response to sinusoidal stimulation (5 Hz, deflection ca. 20°) initially of 4 intact hairs and reduction of response after successively plucking out three of them. **b** Peristimulus-time histograms of ten muscle responses for each of the four ablation cases. (Eckweiler and Seyfarth 1988)

stage has a leg span of about 8 mm, whereas that of the adult animal (12th complete stage) is over 100 mm (Fig. 5) (Höger 1994; Höger and Seyfarth 1995).

By the way, simply touching the surface of the cuticle or pressing gently on the lyriform organs are ineffective stimuli, eliciting no muscle activity (C2). In contrast, as in the case of the behavioral response, deflecting only a single hair is effective (Fig. 6). Slow deflections at 0.1 Hz are sufficient, if the deflection amplitude is 10° to 20°. When the hairs on the coxa are deflected with a rectangular time course (rise time 10 ms), the latency for the muscle response is only 23.1 ms (±2.7 SD) (Eckweiler and Seyfarth 1988).

The electrophysiological response of the hairs to such deflections, as seen in extracellular recordings, comprises action potentials of three sizes. Given that tactile hairs in spiders have a threefold innervation (Foelix and Chu-Wang 1973; Harris and Mill 1977; Eckweiler 1983), this

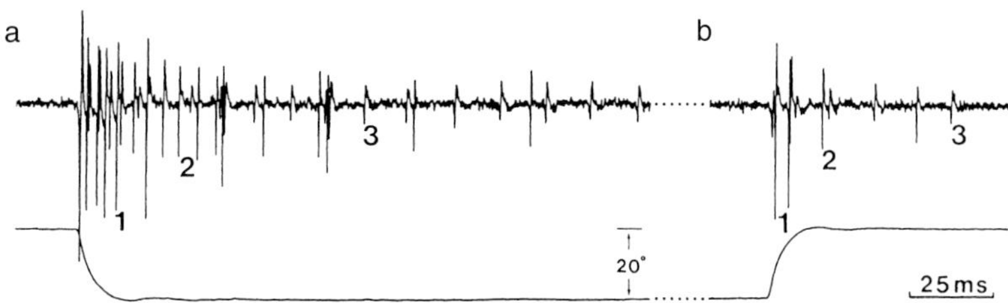

Fig. 7 a, b. Action-potential response of a single tactile hair on the trochanter of *Cupiennius salei* to step-like, maintained deflection by 20°. **a** Initial response to first 160 ms of a 3-s stimulus. **b** Activity upon the return of the hair shaft to its resting position. The three sensory cells (*1, 2, 3*) innervating a tactile hair adapt with different speed. While cell 1 has already ceased firing after 15 ms, cell 3 remains active for up to 2 s. (Eckweiler and Seyfarth 1988)

was to be expected: each sensory cell has its own spike amplitude. The three sensory cells also adapt at different rates: the cell with the largest impulses adapts to step stimuli within milliseconds, whereas the one with the smallest impulses continues to discharge for about 2 s (Eckweiler and Seyfarth 1988; Höger and Seyfarth 1995).

Which of the three cells triggers the reflex? Clearly the cell with the smallest impulses suffices. The evidence is as follows. (i) When stimulated with single sinusoidal deflections (amplitude 20°), the two other cells do not respond at frequencies below 2 Hz. However, the reflex is still elicited at 0.1 Hz, a frequency to which only the "small" cell responds. (ii) When several of these single sinusoidal stimuli are applied in succession, at intervals of 250 ms, then only the "small" cell remains active beyond the second such stimulus, and it continues to respond up to 20 or more hair deflections. This corresponds to the conditions under which the motor response is triggered (Fig. 7).

4 The Way into the Brain

The axons of anteriorly situated tactile coxal hairs run far anterior, into the corresponding neuromere. This contains a few fibers that pass ventrad along the anterior edge of the neuromere and a larger group of fibers, the branches of which run further dorsally in the rostro-caudal direction. Both groups remain on the same side of the body and end in the vicinity of intersegmental tracts (see Chapter XIV); that is, they do not terminate in the dorsal motor association areas (Babu and Barth 1984; Eckweiler and Seyfarth 1988). Individual sensory cells in hairs on the *ventral* surface of the coxae were stained with Lucifer Yellow during electrophysiological experiments, and a similar picture was obtained (Milde and Seyfarth 1988). The axons of these cells have only short collaterals with typical presynaptic "blebs" along the anterior neuromere boundary, and their terminal arborizations are near the midline of the subesophageal ganglion mass (Fig. 8). The sensory terminals and the C2 motoneurons are separated by several hundred μm. This anatomical feature, like the delay of at least 23 ms between hair stimulation and muscle response (see above), indicates the interposition of interneurons and implies that the sensory cells are not monosynaptically connected to the motoneurons. Milde and Seyfarth (1988) and Kadel (1992) found and described good candidates for such interneurons (see below).

5 Motor Neurons and Interneurons

When a microelectrode is inserted progressively further into a neuromere, beginning at its ventral surface near the front edge, and a tactile hair is repeatedly deflected, the responses of various units are recorded in sequence: first primary sensory afferents, then interneurons and – near the dorsal boundary of the neuromere – motoneurons. Figure 8 shows the anatomical arrangement schematically.

The activity of the muscle follows that of the motoneurons in a 1:1 ratio. To establish this, Milde and Seyfarth simultaneously obtained intracellular recordings from the neuron and extracellular recordings from the muscle. It turned out that the coxa muscle C2 is innervated by at least three excitatory, slow motor axons. Figure 9 (Milde and Seyfarth 1988) shows typical arborization patterns of such motor neurons, the somata of which are in the ventral cortex of the ipsilateral neuromere (as expected) and have a relatively large diameter of ca. 50 μm (see Chapter XIV, Fig. 9). Their branches, like those of the sensory neurons, are exclusively ipsilateral and are limited to the same neuromere. Fibers directed toward the midline presumably make contact with interganglionic fibers, as do the primary afferents (Fig. 8) (see Chapter XV).

The interneurons that clearly become active in correlation with the hair-sensillum discharge, and/or with the reflex activity of the muscle C2, can be categorized as either local or plurisegmental.

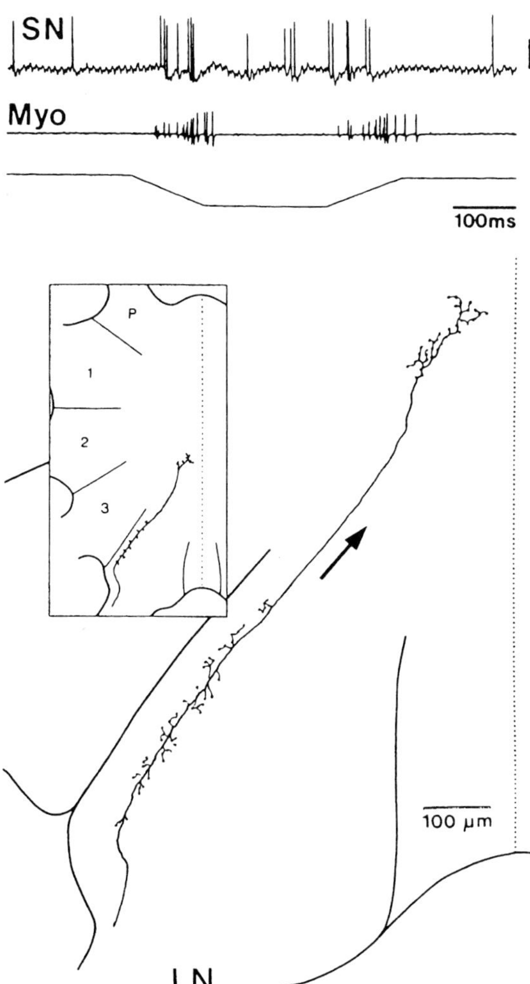

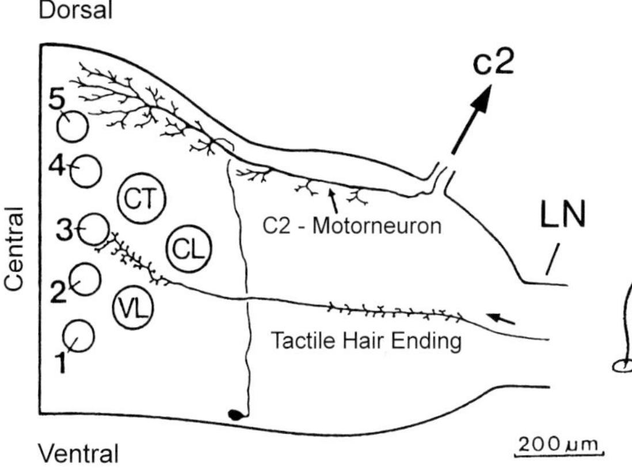

Fig. 8. Primary sensory endings of a tactile hair on the left hindleg of *Cupiennius salei*. *Above* Electrophysiologically recorded response of sensory cell (*SN*) and muscle c2 (*Myo*) to ramp-like deflection (*lower trace*) of about ten tactile hairs on hindleg coxa. *Center* Axon of sensory cell stained with Lucifer yellow. *Inset* Position of sensory endings in subesophageal central nervous system drawn on reduced scale. *1–4* Leg neuromeres; *P* pedipalpal ganglion. *Below* Posterior view of cross-sectioned 4th-leg neuromere; nomenclature of intersegmental tracts according to Babu and Barth (1984). *LN* Leg nerve; *1–5* sensory longitudinal tracts; *CT* central tract; *CL* centro-lateral tract; *VL* ventro-lateral tract. (Milde and Seyfarth 1988)

Local Interneurons

Their branches are restricted to an ipsilateral leg neuromere and the areas occupied by the main branches coincide with those of the afferents on one hand and those of the motoneurons on the other.

Plurisegmental Interneurons

As the name implies, these neurons project into several neuromeres, which complicates the situation. In the example shown in Fig. 10 (Milde and Seyfarth 1988), although the reflex response to hair deflection is limited to the stimulated leg, activity in another neuromere can be recorded simultaneously. Other plurisegmental interneurons arborize in all ipsilateral neuromeres and send collaterals to the contralateral side. With this arrangement, they could well play a role in coordinating the muscles of all eight legs (raising the body of a non-immobilized spider). Furthermore, it is well conceivable that they, like the local interneurons, are also involved in behaviors other than raising the body.

6 Internal Joint Receptors as Triggers of the Plurisegmental Response

When the behavioral response and the muscle potentials resulting from a tactile stimulus are examined with high temporal resolution, it be-

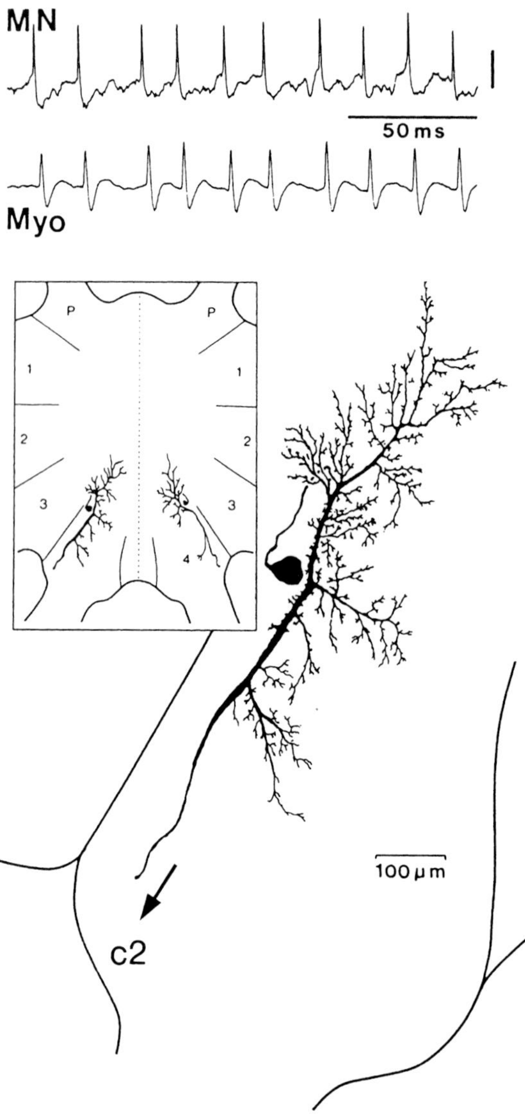

Fig. 9. Organization of two excitatory motor neurons innervating muscle c2. *Above MN* activity of motor neuron and corresponding myogram (*Myo*) of muscle c2. *Bar* 10 mV. *Below* Dorsal view of the same motor neuron in the left hindleg neuromere. The cell soma is located in the ventral cell rind and sends a process into the dorsal region of the neuropil whereas the axon leaves the neuromere dorsally on its way to muscle c2 in the prosoma (see Fig. 8). (Milde and Seyfarth 1988)

comes evident that about 30 ms after stimulus onset the muscles of the stimulated leg are active, and after another 30 ms (within a period of 2 to 4 ms) those in the other legs become active as well. The raising of the body, finally, occurs about 120 ms after the stimulus.

As M. Kadel and C. Bickeböller, in collaboration with E.-A. Seyfarth in Frankfurt, discovered by recording the muscle activities while selectively excluding sensory information (Kadel 1992; Seyfarth 1993), the second or "plurisegmental" response, in which the legs not directly affected by the tactile stimulus move so as to raise the body, is elicited not by this tactile stimulus but by the movement of the coxal joint (at which muscle C2 inserts) caused by local muscle contraction. Convincing evidence of this is provided by experimentally immobilizing the coxal joint: then, although the muscles of the touched leg contract, the other legs are not activated and the body is not raised. Conversely, they showed that the "plurisegmental" response can be elicited by movement of the coxal joint alone. After a joint receptor in the coxal joint has been surgically destroyed, this response is abolished. Ergo: the "plurisegmental" response is very probably induced by a muscle receptor organ (see Chapter X, 3) that itself is stimulated by the local muscle contraction resulting from the tactile stimulus.

Interestingly, E.-A. Seyfarth and his coworkers found plurisegmental interneurons in the subesophageal ganglion region of *Cupiennius salei* that are effectively stimulated by a movement stimulus at the coxal joint of another leg. Depolarization of such a neuron by current injection (+4 nA) triggers an electrical response of the muscle and thus a motor response that corresponds to raising the body. The projections of these plurisegmental interneurons are situated exclusively in the ventral region of the subesophageal ganglion complex. Contacts with the dorsally situated motoneurons are very probably made by identified, non-spiking local (that is, not plurisegmental) interneurons (Seyfarth 1993).

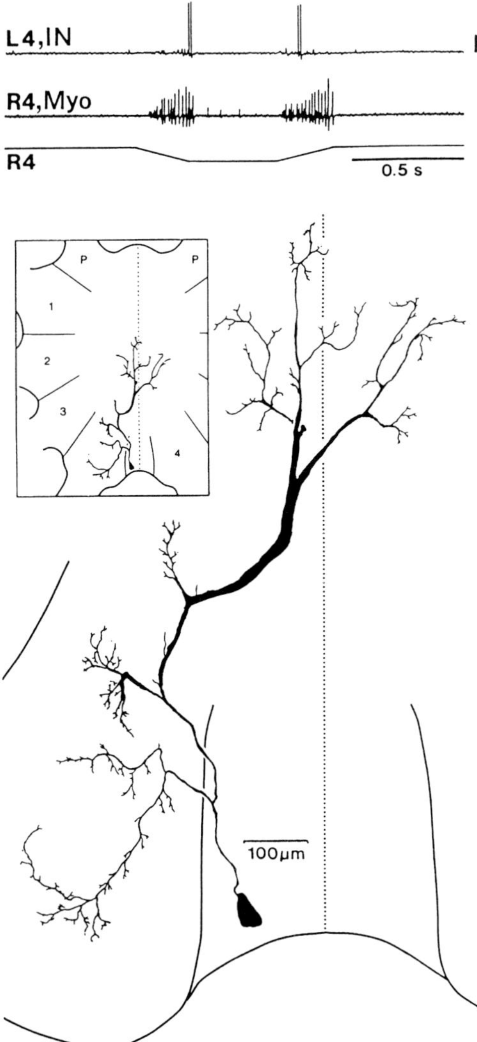

Fig. 10. Plurisegmental interneuron. *Above* Stimulation of tactile hair of right hindleg (*R4*), reflex activity in same leg (*R4, Myo*) and activity of interneuron (*L4, IN*) on contralateral side of same neuromere. *Bar* 10 mV. *Below* Dorsal view of the interneuron with its projections; soma in left opisthosomal neuromere; contralateral processes ending in intersegmental tracts. (Milde and Seyfarth 1988)

XXIV Locomotion and Leg Reflexes

1 Locomotion

First R_4, L_3, R_2, L_1, then L_4, R_3, L_2, R_1: this is the "alternating tetrapods" walking pattern, long familiar in many arachnids, in which the eight legs are moved as two groups of four. Here R and L stand for the right and left sides of the body, and the indices 1 to 4, for the first (fore-) to fourth (hind-) legs. As in insects, with their six legs, the locomotor pattern of spiders is a diagonal rhythm: the legs diagonally opposed to one another on the two sides of the body are moved synchronously – not truly in synchrony, but one shortly after the other. The result is that the legs on a given side step in a wavelike forward progression, which can easily be observed especially during slow walking; when the walk is quite fast, the synchronicity of the movements within one of the two leg pairs is greater. Andreas Brüssel (1987) found out, while working on his dissertation, that when *Cupiennius salei* is walking at a mean speed of 10 cm/s it uses the following patterns, in order of decreasing frequency: 4-2-3-1 (68.1%); 4-1-3-2 (14.8%); 4-3-1-2 (12.8%); 4-2-1-3 (4.3%). By far the most common ipsilateral step sequence was thus the 4-2-3-1 rhythm, and after the spider has been walking for a short time it is almost the only one (91%) (see also Seyfarth and Bohnenberger 1980). Figure 1 shows it very clearly. However, a number of observers have stressed that the exact time at which a leg is moved can depart considerably from the strict pattern of alternating tetrapods, and spiders can quickly adjust to the loss of one or even two legs (Kaestner 1924, Wilson 1967, Seyfarth and Bohnenberger 1980, Seyfarth 1985).

From these findings, the neurobiologist infers that sensory information plays an important role in configuring the walk, and would not expect locomotion to be explicable as the product of a rigid central nervous program.

We might suppose that slit sense organs would make a substantial contribution here. That they are stimulated during locomotion has been demonstrated directly, at least for certain lyriform organs (Blickhan and Barth 1985; see also Chapter X). Nevertheless: the inactivation of lyriform organs in *Cupiennius* by no means had the expected effect; instead, the treated legs continued to make the normal rhythmic movements, just as the untreated ones did (Seyfarth and Barth 1972). Even more drastic intervention, such as complete deafferentation of the

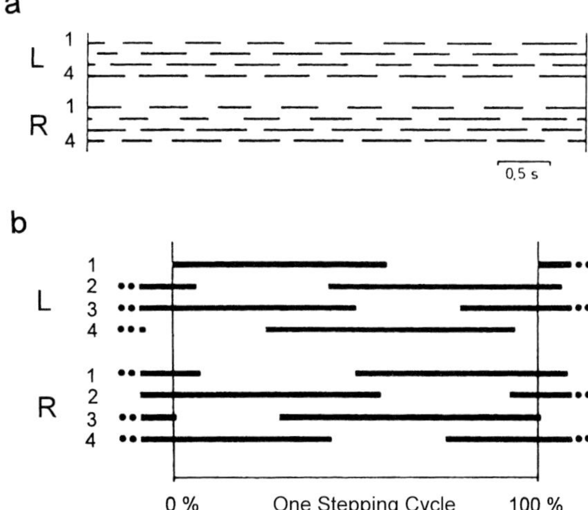

Fig. 1 a, b. Stepping pattern of *Cupiennius salei* walking at a mean speed of 10 cm/s. **a** Sequence of steps of right (*R1–4*) and left (*L1-4*) legs during a period of about 5 s. **b** Mean stepping pattern for same walking speed as in **a**. *Horizontal bars* indicate the stance phase of the stepping cycle. (Brüssel 1987)

joint between tibia and metatarsus by transection of both sensory nerves, in many cases had no influence on the normal use of the affected legs; occasionally an omission of some steps was observed, but even this had often disappeared within an hour after the operation (Seyfarth 1985).

We are thus left in an unsatisfactory position from a neurobiological viewpoint, because we cannot say nearly enough about the relative significance of central programs and of sensory information for the locomotion of *Cupiennius*. The lack of long-term effects, even after severe measures to remove sensory information, seems to indicate either a dominance of central programs or a considerable redundancy in the sensory input. But were the behavioral observations sufficiently differentiated? The participation of the lyriform organs in kinesthetic orientation may teach us something here: the clear effects of their elimination came as a great surprise, since walking in itself had not been affected (see Chapter XXI).

2 Leg Reflexes

Ernst-August Seyfarth, in the doctoral dissertation he subsequently published in the Journal of Comparative Physiology, investigated proprioceptive muscle reflexes in the walking legs of *Cupiennius salei*. Our motivation at that time was a desire to know more about the behavioral relevance of slit sense organs, since their participation in kinesthetic orientation had been demonstrated so convincingly. Can reflexes be elicited by stimulating identified proprioreceptors?

The response to this question was provided by experiments in which the activity of sensilla and that of leg muscles was recorded simultaneously. Seyfarth (1978a,b) was able, first, to elicit so-called resistance reflexes by stimulating internal proprioreceptors; second, he found that stimulation of lyriform organs triggered synergic reflexes. Both reflex responses are limited to the muscles of the stimulated leg.

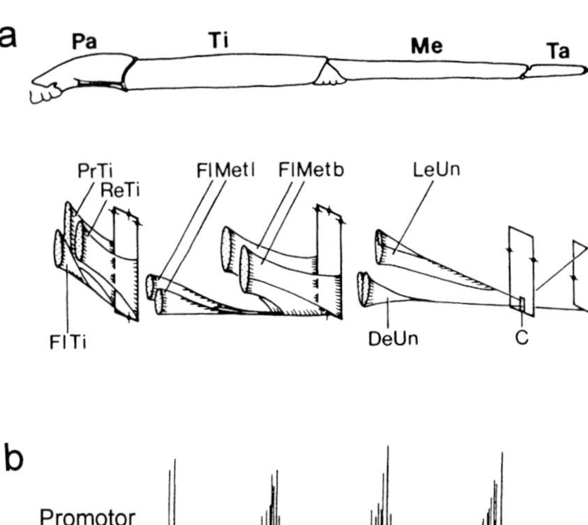

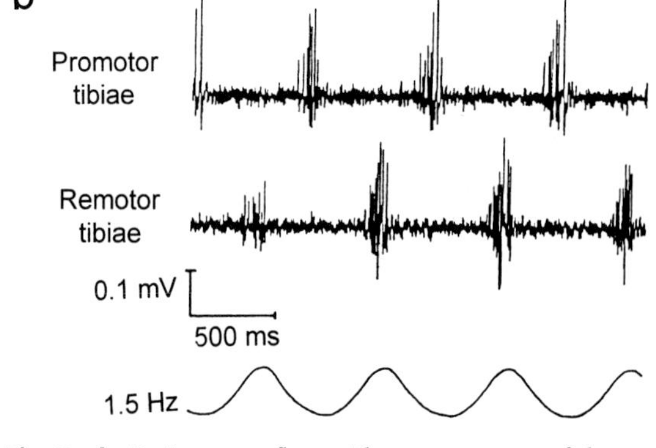

Fig. 2a,b. Resistance reflex. **a** The arrangement of the muscles in the distal leg segments of *Cupiennius salei*; schematized diagram of muscles and their distal insertions. Joints are drawn as *rectangles*, and the positions of the pivot axes are indicated by *dot-and-line* symbols. *PrTi/ReTi* Promotor/remotor tibiae; *FlTi* flexor tibiae; *FlMetl/FlMetb* flexor metatarsi longus/bilobatus; *LeUn/DeUn* levator/depressor unguium; *C* collar serving as a slide-guide for claw tendons. **b** Resistance reflex at the patella/tibia joint; simultaneous recording of the activity of the pro- and remotor tibiae muscles. Stimulus amplitude ±30° from resting position; upward movement of stimulus trace indicates passive promotion of tibia, downward movement passive remotion. (**a** Seyfarth 1985; **b** Seyfarth 1978a)

Resistance Reflexes

Resistance reflexes act to oppose forces that tend to rotate the joint in its main plane of movement, in that intrinsic muscles (muscles associated with the same joint) are activated. Resistance reflexes have been observed in muscles of all the leg segments investigated in this regard (patella, tibia, tarsus; see Fig. 2). They persist even after the removal of all external proprioreceptors at the affected joint – that is, all hairs and lyriform organs – but were abolished by transecting the sensory nerves containing the axons of the internal joint receptors (Seyfarth and Pflüger 1984). These are the multiterminal sensory cells described in Chapter X, the dendrites of which end between the hypodermis cells below the cuticular joint membrane. Muscle receptor organs, by which is meant sensory cells that are excited by the stretching of a specialized receptor muscle, are unknown in spiders. However, they have been found in the postabdomen of scorpions by Bowerman (1972a,b), who also demonstrated their involvement in resistance reflexes. The functional significance of resistance reflexes in spiders might be similar to that in other arthropods, where such reflexes have frequently been described. Resistance reflexes have been studied primarily in crustaceans; here they are important in maintaining the body posture and in the fine control of locomotion (Bush 1965, Barnes et al. 1972, Macmillan 1975). Elimination of the associated reflex-triggering sense organs causes the step length to become too long ("overstepping") or the joint to bend too far ("hyperflexion"), so that the animal has difficulty initiating the antagonistic phase of periodic walking movements (Bässler 1972, Fourtner and Evoy 1973, Wong and Pearson 1976, Seyfarth 1978b).

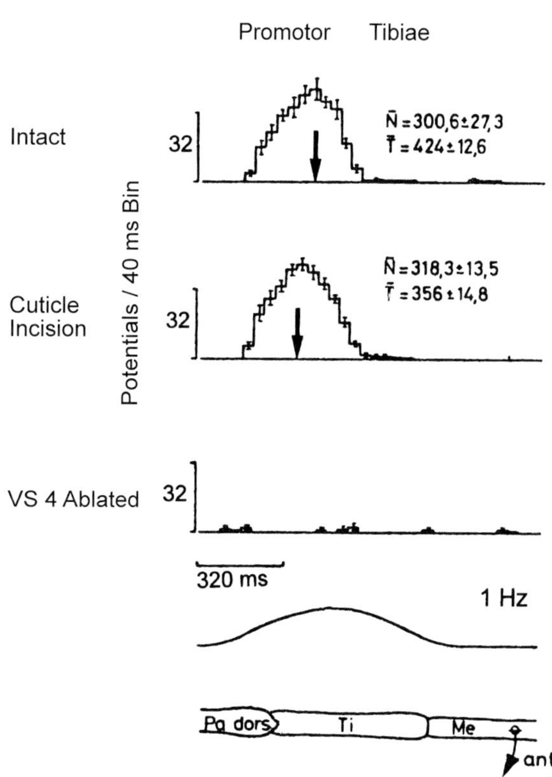

Fig. 3. Synergic reflex. Response of promotor tibiae to passive forward deflection of metatarsus from its resting position by ca. 5°. Small incision close to intact lyriform slit sense organ VS4 (see Chap. VII, Fig. 24) does not significantly modify reflex pattern. After ablation of VS4, however, the reflex fails. (Seyfarth 1978b)

Synergic reflexes

These reinforce an imposed movement of a joint in a direction perpendicular to the main plane of movement by the activity of extrinsic muscles (muscles at other joints in the same leg). For example, the metatarsus can be deflected by up to about ±5° in a lateral-anterior or lateral-posterior direction about a small dorsally situated joint tubercle, thereby selectively stimulating the lyriform organs on the two sides of the distal tibia (Barth 1972b). As a result, muscles in the patella are activated (promotor tibiae when the metatarsus moves forward, remotor tibiae when it moves backward). At the tibia-metatarsus joint there are only flexors. If the lyriform organs stimulated by these metatarsal deflections are eliminated, then the reflex response of the associated patella muscle is abolished (Fig. 3). Just as our expectations regarding the participation of the lyriform organs in the control of normal spider locomotion were disappointed, so here it turned out that their stimulation elicited not resistance reflexes but synergic reflexes. That the connectivity of the individual lyriform organs can be highly differentiated was shown by selectively eliminating either the tibial organ HS 8 or the organ HS 9, which is located only about 500 μm above the former and is also on the back surface of the leg (see Chapter VII). Although both organs are stimulated when the metatarsus moves passively backward, only elimination of HS 8 abolishes the re-

a Resistance Reflexes

SENSE ORGAN	TOPOGRAPHY	RECEPTOR STIMULATION BY	EFFECTS OF AFFERENT INPUT	ABLATION EFFECT IN FREE BEHAVIOR	AUTHORS
Stretch Receptor Locust	Meta. Wing	Wing Elevation	Activ. of Depressor Inhib. of Elevator	Reduction of Wing Beat Frequency	Wilson and Gettrup (1963) Pabst (1965) Burrows (1975)
Hair Plate Cockroach	Co Tr Fe	Femur Flexion	Activ. of Extensor Inhib. of Flexor	»Overstepping«	Wong and Pearson (1976)
Chordotonal Organ Stick Insect	Co Tr Fe Ti	Tibia Flexion	Activ. of Extensor	»Exaggerated Movements«	Bässler (1965, 1978)
Muscle Receptor Organ Scorpion	Postab. 3. 4.	Flexion of Postabdominal Segments	Activ. of Extensor +Lateral Muscles	?	Bowerman (1972)
Myochordotonal Organ Crab	Mer Ca	Carpopodite Flexion	Activ. of Extensor Inhib. of Flexor	»Hyperflexion«	Evoy and Cohen (1969) Fourtner and Evoy (1973)

b Synergic Reflexes

SENSE ORGAN	TOPOGRAPHY	RECEPTOR STIMULATION BY	EFFECTS OF AFFERENT INPUT	ABLATION EFFECT	AUTHORS
Campaniform Sensilla Locust	Meta. Wing	Wing Depression	Activation of Depressor	Impairment of Constant Lift Reaction	Gettrup (1965, 1966) Wendler (1978)
Campaniform Sensilla Cockroach	Co Tr Fe	Distortion of Trochanter; Loads on Animal	Activation of Depressor (=Extensor)	?	Pringle (1938) Pearson (1972)
Lyriform Slit Sense Organ Spider	Pa Ti Me	Lateral Deflection of Metatarsus	Activation of Patellar Muscles (Synergic Reflexes)	Failure of Reflexes	Barth (1972) Seyfarth (1978)

flex response of the remotor tibiae. This second type of reflex causes, for example, the leg to be pulled away from dangerously large forces. It may also initiate or reinforce the stance phase of walking. Remarkably, in insects it also appears that the "force receptors" (campaniform sensilla) in particular are the ones that elicit such stimulus-synergic responses (Pringle 1938; Gettrup 1965, 1966; Pearson 1972). In contrast, according to Seyfarth (1978b) the resistance reflexes are a domain of position- and deflection-detectors, such as "hair plates" (cockroach: Wong and Pearson 1976), stretch receptors (grasshopper wing: Pabst 1965, Burrows 1975) or chordotonal organs (stick insect: Bässler 1972). The dendrites of the mechanoreceptive cells in these cases are coupled to movable structures such as joint membranes, accessory muscles and connective tissue, which are readily deformed when parts of the body move with respect to one another. These are quite unlike the "force receptors" embedded in hard cuticle, which are only slightly strained and displaced by such movements (see Chapter VII). Figure 4 summarizes a few typical examples.

In conclusion, we turn to an entirely different aspect, an excellent illustration of how important it can be to know one's experimental animal very well and to find a good balance between reductionistic work and general biology. In pilot experiments Ernst-August Seyfarth had noted that the leg reflexes of *Cupiennius* are most easily elicited in the evening twilight and the synergic reflex, in particular, can often be triggered only when the laboratory is dark. At that time we did not fully realize what is now reported in Chapter IV about the daily activity rhythm: the normal locomotor activity of *Cupiennius* is restricted to the dark period of the day. Evidently the lighting conditions or the associated endogenous rhythm affect the spider's readiness to respond, even if it is "merely" a matter of simple muscle reflexes.

Fig. 4a,b. Types of reflexes in arthropods. **a** Displacement and position sensors are found to be components of reflex arcs activating muscles that oppose the induced movement (*resistance reflex*). *Solid arrow* indicates movement resulting in stimulation of sensor; *dashed arrow* indicates reflex movement. **b** Force detectors are found to be components of reflexes activating muscles which are synergic to the induced or active movement (*synergic reflex*). *Solid black symbols* mark location of sensors in the exoskeleton. *Ca* Carpopodite; *Co* coxa; *Da* dactylopodite; *Fe* femur; *Me* metatarsus; *Mer* meropodite; *Meta* metathorax; *Pa* patella; *Post-ab* post-abdomen; *Pr* propodite; *Ti* tibia; *Tr* trochanter. (Seyfarth 1978b)

3 Mechanical Stresses in the Skeleton

In Chapter VII on mechanoreception mention was made of biosensors that measure the most minute strains in the spider exoskeleton. And it is already clear that if we hope ever to understand these slit sense organs, we must have an accurate knowledge of the mechanical events to be expected in the exoskeleton under natural conditions. In its role as supporting apparatus and lever system this skeleton is exposed to many forces; it can tolerate these not only because of the material of which it is made and the structural organization of that material, but also because mechanical loads can be detected, and hence if necessary avoided, by the several thousand sensors incorporated into the skeleton. Being concerned here with the relationship between locomotion and leg reflexes, we now have an opportunity to take a closer look at the distribution of load in entire sections of the skeleton.

Andreas Brüssel in 1987 presented a comprehensive dissertation on this subject. He first determined the ground reaction forces present in a *Cupiennius salei* standing quietly, and then went on to use miniature strain gauges to measure the strains that arose in various behavioral situations (such as slow and rapid walking on differently sloping substrates as well as starting, stopping and jumping movements). His measurements were made mainly in the tibial skeleton of the walking legs, with a site on the femur for comparison. The results very usefully supplemented earlier measurements on the loading of the tibial lyriform organs during slow, calm locomotion (Blickhan and Barth 1985). They give us a considerably improved picture of the diversity of the natural loading states.

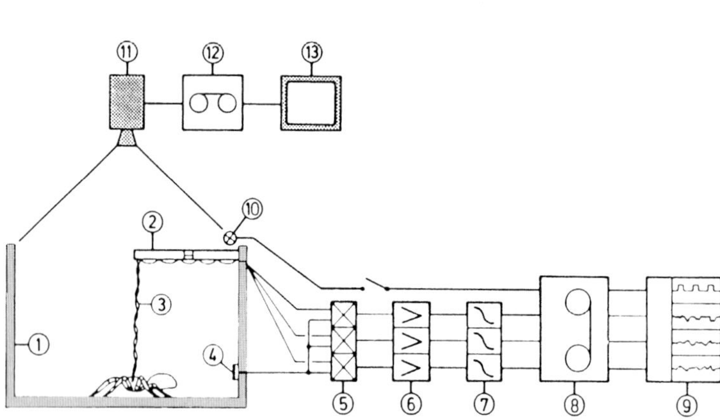

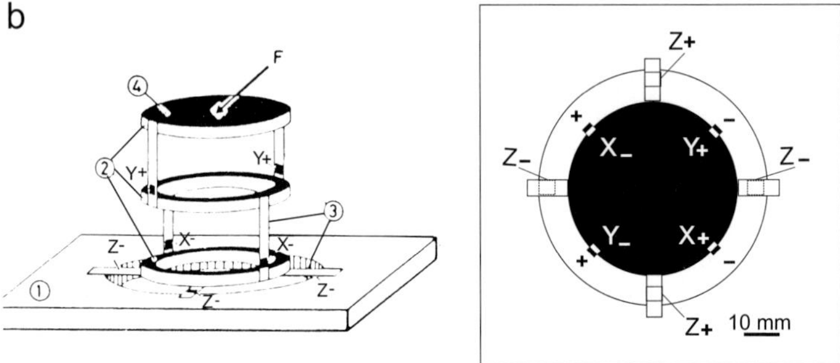

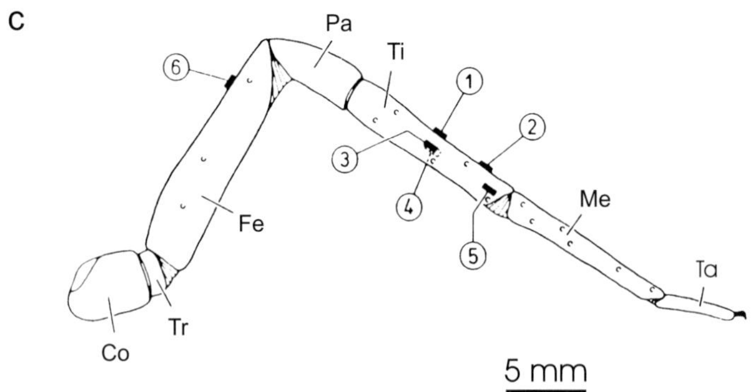

Fig. 5a–c. The measurement of forces and strains. **a** Overview of experimental setup. *1* Walking corridor; *2* lever; *3* electric cables from measuring sites, *4* dummy strain gauges; *5* electronic circuit to compensate for strains due to temperature changes; *6* power supply and amplifier; *7* electronic filter; *8* FM tape recorder; *9* pen recorder; *10* light-emitting diode used as time marker; *11* video camera; *12* video recorder; *13* monitor. **b** Force platform (*left* mechanical design, *right* top view and circuitry). *1* Base; *2* Plexiglas rings; *3* bronze springs with strain gauges attached; *4* platform the spider leg contacts during locomotion; *F* ground reaction force; *XYZ* strain gauges in Wheatstone bridge for measurement along three axes (+ and – indicate connectivity in bridge circuit). **c** Measuring points on the walking-leg exoskeleton chosen for strain measurement; anterior view of leg 2 of *Cupiennius salei*; *1* tibia – center of dorsal side; *2* tibia at attachment site of flexor metatarsi bilobatus; *3* tibia – center of anterior side; *4* tibia – center of posterior side; *5* tibia – close to joint on latero-posterior side; *6* femur – at attachment site of flexores patellae; *Co* coxa; *Tr* trochanter; *Fe* femur; *Pa* patella; *Ti* tibia; *Me* metatarsus; *Ta* tarsus. (Brüssel 1987)

The methods employed by Andreas Brüssel (1987) were very similar to those previously developed by Reinhard Blickhan (Blickhan and Barth 1985). A force platform was set into the floor of a channel in which the spider walked, as shown in figure 5b. It was extremely sensitive: horizontally, x axis 1175 mV/mN, y axis 505 mV/mN; vertically (z axis) $\geq$ 221 mV/mN. The resonant frequency of the platform was above 300 Hz. The three plastic rings clearly visible in the figure were also connected to one another by semiconductor strain gauges, an important element in achieving the very high sensitivity. Movement episodes were recorded on video tape so that they could be correlated with the mechanical measurement signals.

To obtain quasipunctate strain measurements, foil strain gauges were positioned at the sites of the spider leg shown in figure 5c. With one exception (femur) they were on the tibia, because in terms of functional morphology and the physiological properties of the lyriform organs, this is by far the most thoroughly studied leg segment. In order to get an idea of the distribution of strains within the tibia, the measurement sites should be as far apart as possible and also, if at all possible, be exposed to maximal loading with various kinds of loads. The sites 1 and 2 on the dorsal tibia (in the middle and at the insertion point of the metatarsal flexor muscle) are particularly exposed to dorsally directed bending moments and, of course, to the forces exerted directly by the flexor muscle (on site 2). In contrast, the sites 3 and 4 in the lateral midregion of the tibia are affected especially by axial forces and laterally directed bending moments. Site 5 is in the region of lyriform organs and serves for comparison with sites 1 to 4, in regions lacking lyriform organs. The site 6 on the femur is above the insertion of the patellar flexor muscles, and hence can be compared with site 2, which is likewise over a muscle insertion point. Because the strains in the skeleton resulting from bending load are greatest in the direction of the long axis of the leg (Szabo 1977, 1984), the measurements were made in this direction.

Ground Reaction Forces

The forces exerted on the substrate by the individual legs, normalized to the mean body weight of adult female *Cupiennius salei* (3 g), were measured by Andreas Brüssel for a spider standing on a horizontal and a sloping substrate as well as while walking straight (velocity 10 cm/s) on a horizontal surface. The most important findings were as follows.

Standing on Horizontal or Sloping Substrate. The force vectors (Fig. 6) measured for the various legs of *Cupiennius salei* and also for a bird spider tell us that stationary spiders should be regarded as very nearly rigid structures. On the other hand, normally not all joints are kept stiff; we know this because not only vertically directed forces occur (Cruse 1976). Because the vertical force components are mirror-symmetric, we can infer that the movable joint is in the plane of symmetry, at the places where the legs are connected to the prosoma. This is also the location of the body's center of gravity. That *Cupiennius* can adjust to variations in the substrate and also stand still as a substantially rigid structure is evident when the spider is set onto mercury: its posture remains stable even though the low viscosity of the mercury excludes horizontal ground reaction forces. In a less exotic, indeed thoroughly "biological" situation the same thing is achieved by *Dolomedes triton*, which normally hunts for prey on the surface of a body of water (see Section 5; Barnes and Barth 1991).

Horizontally directed forces that act on all eight legs stabilize body posture. This applies both to spiders that stand on a horizontal surface and those standing on a slope. Both positions are equally relevant to *Cupiennius* – when it is waiting for prey, for instance, at night on a bromeliad.

Examined a little more closely, the distribution of force vectors shows that the body weight is borne mainly by the third pair of legs. Both the vertical (z) and the lateral (y) force component is greater for the third pair than for any of the other legs. As we know (Schüch and Barth 1985), this also plays a role in courtship: the female introduces her vibratory answer to the male into the plant mainly by way of the third legs. While these stay in firm contact with the substrate, the other legs are repeatedly raised during signalling.

When the spider is standing at an angle, on a planar surface tilted by 45°, as expected the ground reaction forces change considerably (Fig. 6b, Table 1). In a downward-facing position (the typical one for *Cupiennius*) all the forces parallel to the body long axis (x) point forward; they decrease distinctly from the 4th to the 1st pair of

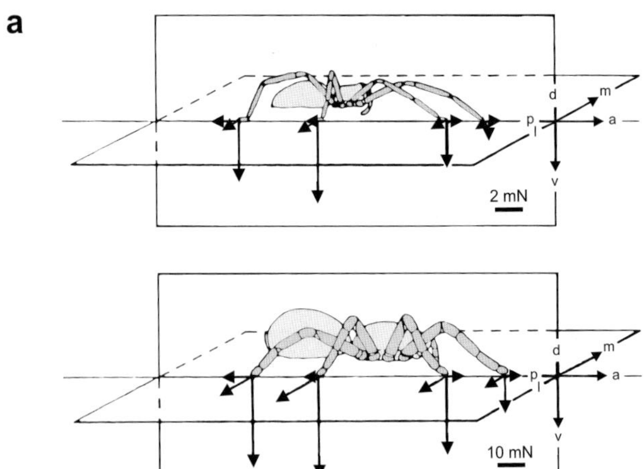

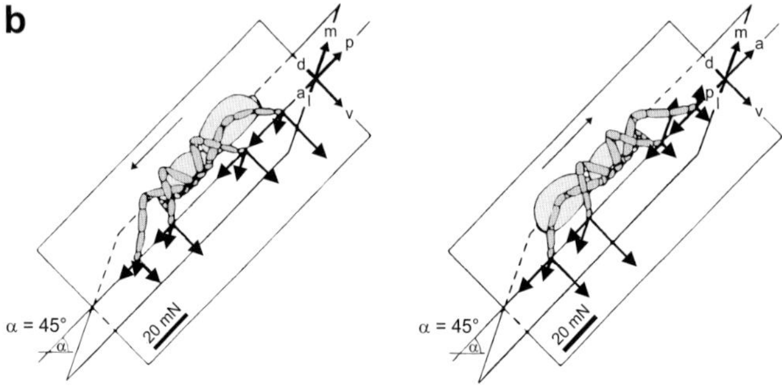

Fig. 6 a, b. Ground reaction forces at the different legs of *Cupiennius salei* and of bird spiders when standing on horizontal and inclined platform. **a** Horizontal platform (*top Cupiennius*, *bottom* bird spider); force components along body long axis (*a* anterior, *p* posterior), along transverse axis (*m* medial, *l* lateral) and along vertical axis (*d* dorsal, *v* ventral); mean values normalized for a body mass of 3 g in the case of *Cupiennius* (N=5, n≥23) and a mass of 20 g in the case of bird spiders (N=4, n≥18). **b** Inclined platform (inclination angle 45°, bird spiders) with the animals looking downward and upward; mean values normalized for body mass of 20 g (N=4, n≥21). (Brüssel 1987)

Table 1. Ground reaction forces (mN) of individual legs of bird spiders standing on inclined platform (see Fig. 6) and looking downward (top) and upward (bottom). Inclination angle 45°, mean values ±SD (N=4, n≥21) normalized for body mass of 20 g. (Brüssel 1987)

	x-axis +: anterior	y-axis +: laterad	z-axis +: ventrad
Downward			
Leg 1	+10.1 (±2.7)	+5.0 (±1.3)	+12.4 (±4.3)
Leg 2	+10.9 (±3.1)	+9.1 (±2.4)	+17.0 (±3.3)
Leg 3	+14.3 (±4.2)	+10.6 (±4.1)	+17.8 (±5.7)
Leg 4	+18.6 (±2.9)	+7.2 (±2.1)	+25.3 (±7.2)
Upward			
Leg 1	−14.2 (±4.3)	−6.2 (±1.2)	+6.9 (±3.4)
Leg 2	−8.3 (±2.8)	−17.7 (±4.7)	+6.2 (±1.5)
Leg 3	−12.9 (±3.1)	+13.3 (±2.0)	+26.8 (±8.2)
Leg 4	−18.1 (±5.3)	+7.5 (±1.9)	+21.4 (±9.7)

legs. The spider is hanging on by its hindlegs. The change from the horizontal situation is least for the laterally directed forces (y), while the vertical ground reaction forces (z) are now greatest at the hindlegs. When the spider is facing upward, the 4th leg pair again has an especially important supporting function: the forces (x) acting in the direction of the body long axis are greatest here. Now, however, the forces at the forelegs are almost as large. That is, the spider is not only bracing itself against the hindlegs, it is also hanging by the first ones. The forces (z) acting perpendicular to the substrate are considerably larger for legs 3 and 4 than for legs 1 and 2.

During a Straight-Line Walk. When a spider walks, it stabilizes its body posture just as it does

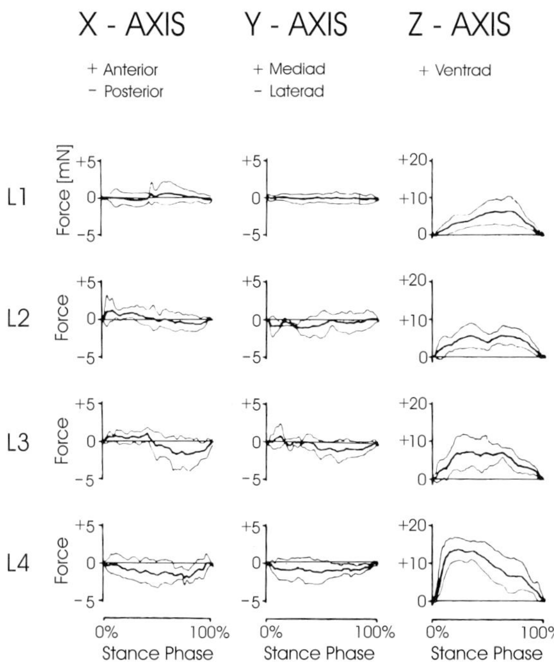

Fig. 7. Ground reaction forces of individual legs of *Cupiennius salei* during quiet locomotion (10 cm/s). Mean values (normalized to body mass of 3 g; ±SD, n≥18) of forces exerted by left legs *L1–4* during the stance phase in the direction of the body long axis (*X*), transverse axis (*Y*) and vertical axis (*Z*). (Brüssel 1987)

when standing, by means of forces directed laterally at all legs, forward at the forelegs and backward at the hindlegs (Fig. 7). Not unexpectedly, the forces parallel to the body exerted by the 3rd leg pair during a stance phase are directed first forward and then backward. During this phase the 3rd legs rotate about an axis transverse to the animal, thus helping to support the body weight of *Cupiennius salei* even while it is walking.

Strains

Strains

What strains appear in the skeleton of the second leg at the six measurement sites (Fig. 5c) – during walking on a horizontal or vertical substrate or on a slope, during starting, stopping and jumping?

(i) Walk on Horizontal Substrate. As had previously been measured on the distal tibia near the joint (Blickhan and Barth 1985; see Chapter VII), during slow walking (v=5 cm/s) the strain changes in the rhythm of the movement at all the sites considered here, on the tibial exoskeleton as well as the femoral site. Whereas the swing phase is associated with only slight, near-zero strains, during the stance (load-bearing) phase they are considerable both dorsally, in the middle of the tibia and at the insertion of the flexor muscle, and laterally at the joint to the metatarsus: of the order of −70 με to −80 με. Laterally in the middle of the tibia and on the femur, the strain amplitudes are only about 20 με and 40 με, respectively, and the time courses are bimodal, with positive and negative values. As the walking speed increases (v=10–30 cm/s) so, in general, does the load, especially in the joint region (Figs. 8, 9; Table 2), where at v=30 cm/s strain values up to −360 με have been measured during the stance phase. During a walk on a horizontal substrate it makes no appreciable difference, as far as strain is concerned, whether *Cupiennius* is upright or hanging from the undersurface.

(ii) Walk on a Vertical Plane. Now consider walking on a vertical plane, which is entirely normal for *Cupiennius*. As it walks upward, the strain values in the dorsal midregion of the tibia are larger than during a horizontal walk, and when it walks downward they are smaller. In the joint region upward walks increase the load by 2- to 3-fold and produce conspicuously irregular strain time courses with strain amplitudes that are similar for all walking speeds tested, though there is considerable scatter (Fig. 10). During a downward walk the same region of cuticle exhibits a different strain time course for each velocity, with strain amplitudes that differ in detail. Rapid downward walks (v=20 cm/s) load the joint region considerably more severely than walks on a horizontal substrate. On the lateral tibial surfaces (measurement sites 3 and 4) the strains are largely independent of whether the substrate is horizontal or vertical.

(iii) Rapid Starts and Stops. As would be expected, during rapid starting, stopping and

v = 5 cm/s

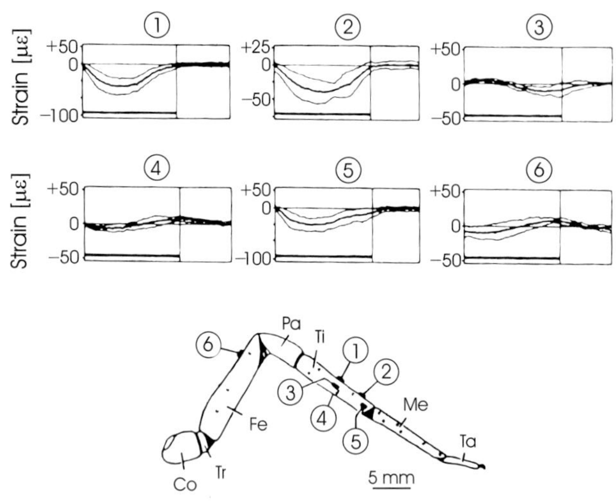

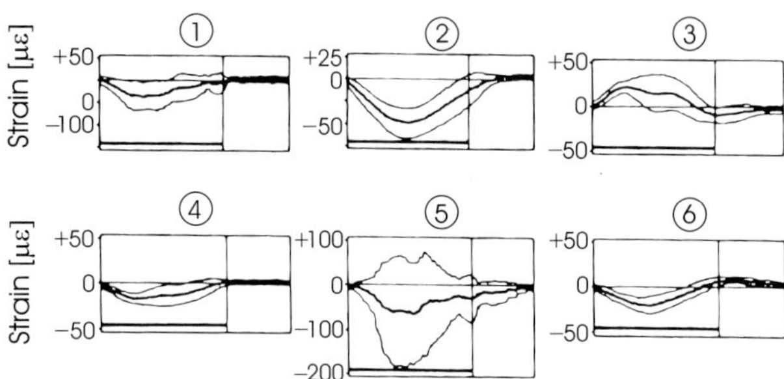

Fig. 8. Strains in the exoskeleton of femur and tibia of walking leg 2 of *Cupiennius salei* during locomotion in the horizontal plane at different speeds (5 cm/s and 30 cm/s). Mean values (±SD) of strains occurring at measuring points *1* to *6* within one stepping cycle. (Brüssel 1987)

Table 2. Mean positive and negative strain amplitudes (in $\mu\varepsilon \pm$ SD) measured at the sites shown in Figs. 5c and 8 on femur and tibia of *Cupiennius salei* during locomotion in a horizontal plane at different speeds. Direction of measurement always parallel to leg long axis. (Brüssel 1987)

[cm/s]	Strain [$\mu\varepsilon$]					
	Tibia ①	Tibia ②	Tibia ③	Tibia ④	Tibia ⑤	Femur ⑥
5	+5.7 (±4.3) -73.8 (±19.3)	5.4 (±4.2) -84.9 (±17.1)	+9.6 (±3.6) -15.5 (±3.6)	+8.7 (±3.6) -11.6 (±3.9)	+9.7 (±7.5) -73.4 (±31.7)	+16.5 (±7.7) -22.7 (±10.2)
10	-7.3 (±3.8) -69.4 (±19.5)	+7.3 (±6.7) -84.3 (±20.9)	+14.0 (±6.0) -19.1 (±6.3)	+7.9 (±4.8) -24.7 (±2.6)	+14.8 (±21.3) -69.3 (±43.8)	+21.1 (±12.3) -26.2 (±10.9)
20	+6.7 (±3.2) -68.0 (±33.2)	+15.7 (±6.0) -108.7 (±35.8)	+9.6 (±4.4) -9.7 (±3.7)	+3.1 (±2.3) -12.6 (±2.5)	+5.7 (±23.1) -21.5 (±76.8)	+21.9 (±7.7) -53.3 (±18.3)
30	+10.1 (±5.4) -81.3 (±27.9)	+21.4 (±11.2) -84.4 (±43.0)	+28.6 (±9.9) -19.7 (±7.0)	+4.6 (±2.5) -21.5 (±6.5)	+42.5 (±19.5) -93.0 (±107.0)	+19.3 (±10.4) -48.4 (±17.9)

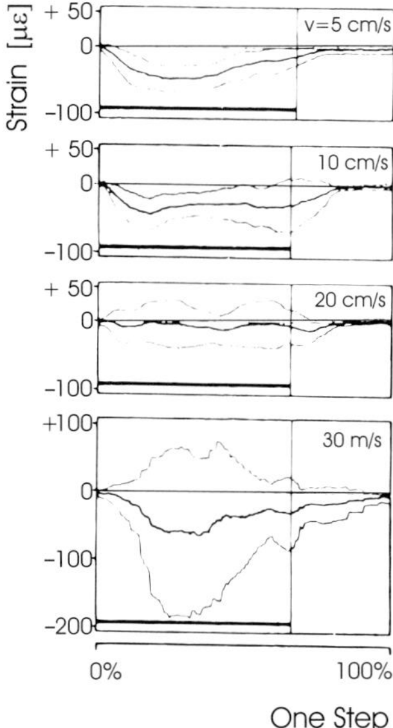

Fig. 9. Strains in the tibia of *Cupiennius salei* close to joint with metatarsus (see measuring point ⑤, Fig. 5c) and during locomotion on horizontal plane at different speeds. *Black bars* indicate duration of stance phase (N=4, n≥23). (Brüssel 1987)

jumping movements the strain values are high (Fig. 11). This situation can be described briefly as follows: the high accelerations impose loads practically equivalent to those measured during the most rapid walks in our sample (with the same orientation of the spider to the direction of gravity).

The loading of the femur at the insertion of the patella flexor muscles in *Cupiennius* is only about half as great as at the corresponding site on the tibia. On the basis of our current understanding, it is practically impossible to interpret this finding reliably. Is the femur as a whole under less load? Are other areas more like those on the tibia? We need more measurements if we are to answer such questions. Interestingly, it is also the case in the water strider (*Gerris lacustris* L.) that the femora are only half as heavily loaded as the tibiae (Darnhofer-Demar 1977).

Position of Slit Sensilla

What has this complicated investigation of the strains in the skeleton taught us about the significance of the position of the slit sensilla, the

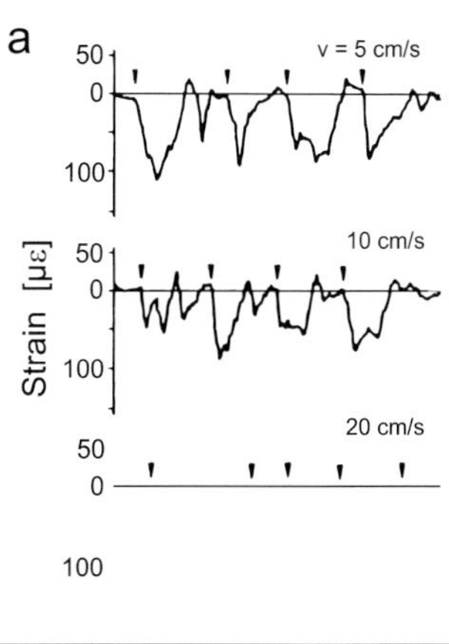

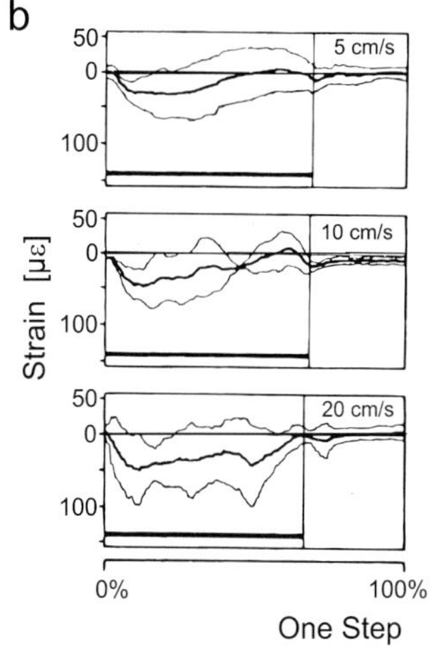

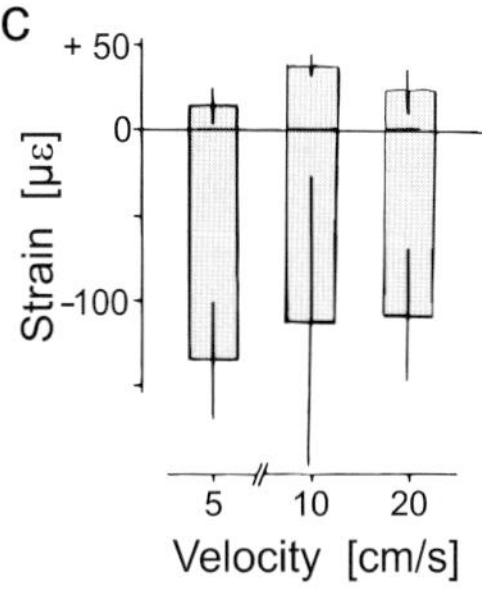

Fig. 10a–c. Strains in the tibia at same site as in Fig. 9 but in spiders (*Cupiennius salei*) walking upward on a vertical plane at different speeds. **a** Original recording; *arrowheads* indicate instant when leg was just touching the substrate. **b** Mean strains (±SD) during one step; *bar* indicates duration of stance phase (N=4, n≥20). **c** Mean amplitudes (±SD) of strains recorded within one step (N=4, n≥20). (Brüssel 1987)

Horizontal

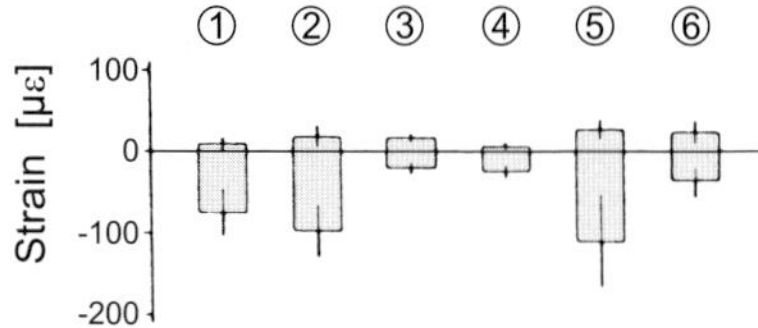

Upward

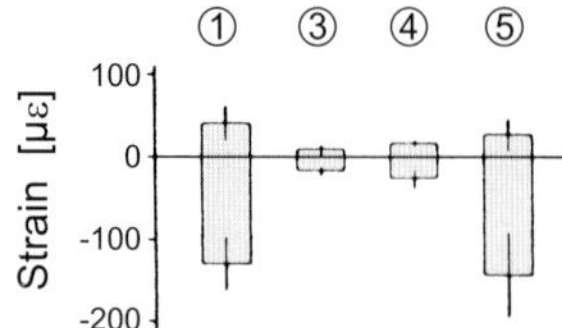

Downward

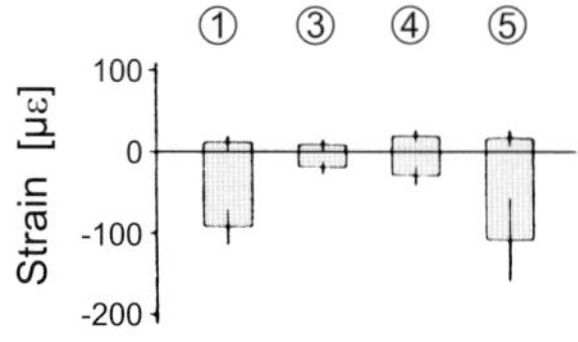

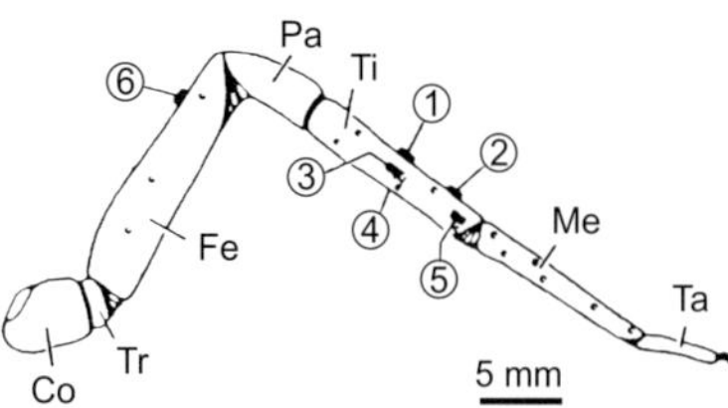

Fig. 11. Strain amplitudes in $\mu\varepsilon$ (means ± SD) measured during starts, stops and jumps of *Cupiennius salei* at the six measuring points on femur and tibia shown in Fig. 5c. Movements in the *horizontal* plane and upward and downward movement in the *vertical* plane. (Brüssel 1987)

biological strain gauges? First: at least the tibia of *Cupiennius* is an object of equally distributed strength under the conditions of slow, uniform movement. During rapid movements, however, it is not; the loading becomes nonuniform and concentrated in the distal joint region. Accordingly, the lyriform organs are located in a region with widely varying strains and high peak loadings. As was discussed in Chapter VII, the various lyriform organs of the tibia are well able to monitor just such variable load and movement states, because of the differing details of their position and orientation and also because of their physiological properties. They would thus be an excellent means of keeping track of the load on the tibia-metatarsus joint and of counteracting dangerous load peaks by triggering synergic reflexes.

The single, isolated slits, which are – typically – a long way away from the joint (see also Fig. 1 in Chapter VII), are evidently situated where relatively regular mechanical events occur that reliably identify a particular kind of load. Because the single slits situated dorsolaterally on the tibia are oriented parallel to the long axis of the leg, they are very selectively excited only by those strains that are associated with activity of the flexor muscles. Hence they are optimally suited to signal the stance phase of a step and the moments of flexion actively generated in the joint. Remember, though, that the sequence of movements in free locomotion is maintained after the lyriform organs on the legs have been inactivated. It follows that additional sensors are involved, and/or that there is a strong influence of central nervous control (see also Chapter XXI on kinesthetic orientation).

Safety Factor

Finally, another biomechanical detail. If we know the modulus of elasticity E of the material and the maximal strains, compressive stress can be calculated. Assuming the values 18 GPa for E (Blickhan and Barth 1985) and −160 $\mu\varepsilon$ for strain (dorsal surface of tibia), then we find at most ca. 2.9 MPa compressive stress, which would correspond to 6.5 MPa for the distal tibia in the joint region. The insect cuticle breaks at a stress of 30–100 MPa when under a tensile load and at ca. 160 MPa when bent (Hepburn and Joffe 1976, Nachtigall 1982). Taking this as a guideline, we might conclude that the tibia of *Cupiennius* has a high safety factor (dorsal tibia 10–55, joint region 5–25). However, in view of Curry's (1967) finding that in the proximal leg segments of another spider (*Pholcus* sp.) the buckling stress amounted to 5.8 MPa, it would appear that the safety factor is only 2 for the middle of the dorsal tibia of *Cupiennius* (this is the only region at risk of buckling). In the water strider (*Gerris*) the safety factors with respect to permanent damage have been found to be 16.6 for the femur and 8.3 for the tibia (Nachtigall 1982).

4 Energetics of Walking

It isn't quite correct to call *Cupiennius*' straight-ahead walking on a horizontal substrate "uniform". The velocity of locomotion varies from moment to moment, by as much as ±25% about a mean (for example, v = 10 cm/s). The variation is sinusoidal with a period corresponding to half the step length (Fig. 12). The walking speed is always highest when all the third and fourth legs are on the ground, and minimal when only one of these legs is down. When the time course of variation of the total ground reaction force during a step is calculated from the ground reaction forces generated by the individual legs, again periodic oscillations appear in the rhythm of the leg movement, for forces in the x, y and z directions:

- Along the body long axis (x): fluctuations with a period half that of the stepping cycle; backward-directed acceleratory forces during almost the whole step, and more propulsion than braking force. Assuming a constant speed of locomotion (no acceleration), it must be concluded that these propulsive forces serve to overcome friction.
- Along the transverse axis (y): force varies with a period corresponding to the step duration; forces alternately directed toward left and right, with zero-crossing when the fourth legs are set down.
- Along the vertical axis (z): the largest of all force fluctuations, corresponding to about 40% of the spider's weight; period equal to half the step duration; in the stance phase, force exceeds body weight, whereas during

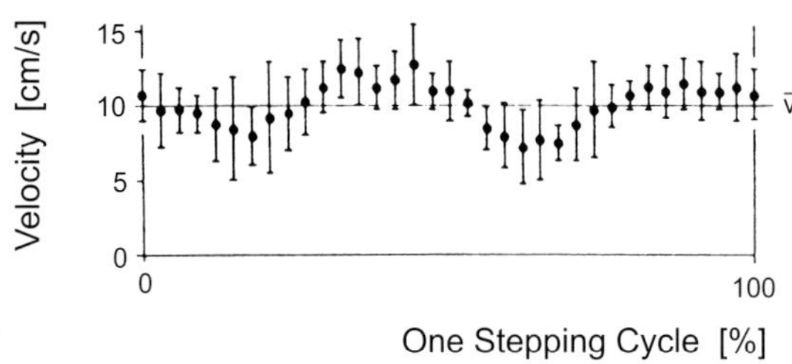

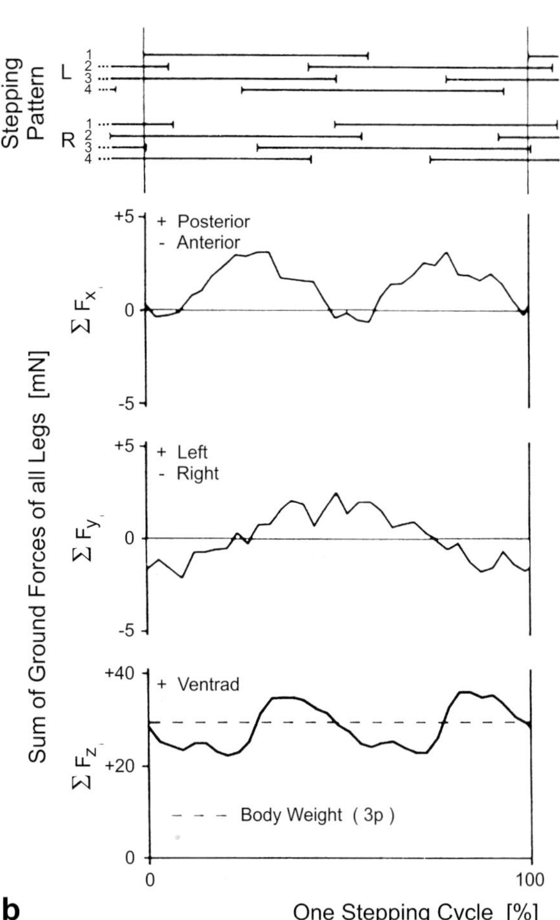

Fig. 12a,b. Instantaneous velocity and ground reaction forces. **a** Instantaneous velocity of spider (*Cupiennius salei*) locomotion during one step of quiet walking at a constant mean speed of 10 cm/s (N = 4, n = 13). **b** The summed ground forces of all legs occurring during the step shown in **a** and measured in all three directions of space (*from top to bottom* parallel to body long axis, parallel to transverse axis, parallel to the vertical axis). Note changes coupled to stepping pattern. (Brüssel 1987)

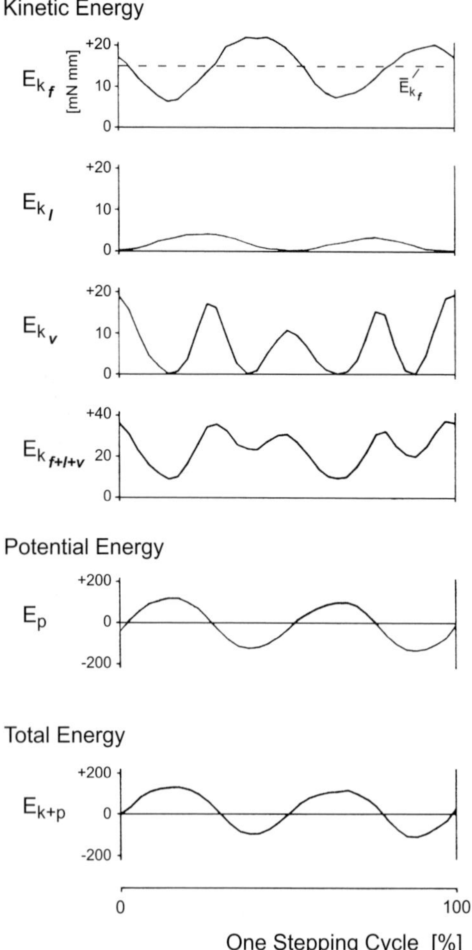

Fig. 13. Time course of kinetic, potential and total mechanical energy of the center of gravity of *Cupiennius salei* during one step of quiet locomotion (V = 10 cm/s). E_{kf}, E_{kl}, E_{kv} kinetic energy of center of gravity in the direction of locomotion and of the transverse and vertical axis. E_{kf+l+v} sum of component kinetic energies; E_p potential energy of center of gravity; E_{k+p} total mechanical energy of center of gravity. (Brüssel 1987)

the transition of the third legs from the swing to the stance phase force equals body weight.

The fluctuations of momentary velocity and of body height, even during "uniform" walking, indicate that *Cupiennius* must always perform mechanical work in order to accelerate itself anew and raise its center of gravity. As in vertebrates, including humans, the kinetic as well as the potential energy of the center of gravity thus oscillates between a maximum and a minimum (Cavagna et al. 1963, 1964, 1977 a, b; Alexander and Vernon 1975; Cavagna and Kaneko 1977; Alexander and Jayes 1980; Alexander et al. 1980). Cavagna et al. (1963) call the work that has to be done for this purpose the "external mechanical work W_{ext} of locomotion". Andreas Brüssel (1987) calculated W_{ext} for *Cupiennius* (v = 10 cm/s) from the ground reaction forces of all legs, according to the method given by Cavagna (1975) (Fig. 13). It turned out that the total mechanical energy is determined substantially by the time course of the potential energy, the absolute values of which exceed those of the kinetic energy by about tenfold. Given a mean walking speed of 10 cm/s, he found a value of 0.49 mJ for W_{ext} in one step cycle, while the mechanical power P_{ext} was 0.78 mW. If the work required to lower and slow down the body is also taken into account, as well as the proportion corresponding to constant propulsion, then W_{ext} increases to 1.07 mJ per step. The phase angle of the kinetic and potential energies of the center of gravity with respect to one another differs characteristically for various gaits (walking, running, galloping). During walking the two kinds of energy are 180° out of phase, which enables energy transfer in the sense of a pendulum mechanism and thus reduces the work needed for acceleration and raising the body by 35–70% (Cavagna et al. 1977 a, b). The locomotion of *Cupiennius* studied by Andreas Brüssel should clearly be classified as "walking". The energy transfer nevertheless amounts only to 8.4%, indicating that the pendulum mechanism in this case produces little reduction of the work to be performed, in comparison to the vertebrates.

Assuming that the value of W_{ext} found for *Cupiennius salei* represents the minimal energy needed for locomotion, because it does not take into account a number of additional energy costs (Brüssel 1987), we can conclude that there is after all some similarity to the energetics of vertebrates and other arthropods. In these the energy consumption, the cost of transport over a specific distance, also depends substantially on the mass of the body (Taylor et al. 1970, 1982) (Fig. 14).

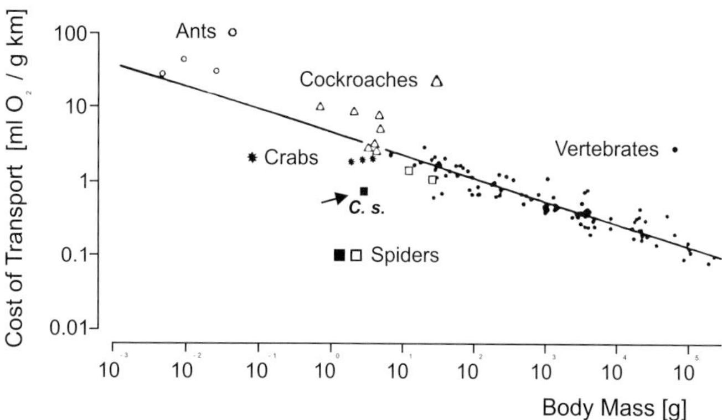

Fig. 14. Transportation costs of locomotion in arthropods and vertebrates as a function of body mass and in comparison to the value calculated for *Cupiennius salei* (*C.s.* ■); ● values referring to vertebrates and regression line (Taylor et al. 1982); ○ ants (Jensen and Holm-Jensen 1980); △ cockroaches (Herreid and Full 1984; Herreid et al. 1986b; Bartholomew and Lighton 1985); * crabs (Full and Herreid 1983, 1984; Herreid and Full 1986); □ bird spiders (Herreid 1981, Andersen and Prestwich 1985). (Brüssel 1987)

5 *Cupiennius* on the Water

I can still remember every detail. We had come from the "rain forest of the Austrians" in the south of Costa Rica, near the Pacific Coast city of Golfito, and were tired, sweaty and glad that we had reached the Rio Bonito. After the long march we settled down with great anticipation to pure astounded observing, which offers so much if only it is given a chance. Among kingfishers, herons, scorpions and all kinds of insects, one animal by a quiet pool at the side of the river attracted our attention more than any of the others: the natives call it "chisbala". The English name is Jesus Christ Lizard and there is actually something biblical about it, or something of the circus, where the impossible becomes possible. The Jesus Christ Lizard is a basilisk (*Basiliscus* sp.), which like a little dinosaur rushes over the water on two legs with the front part of the body raised, leaving the observer astonished and enchanted. Small animals walk on water for distances of 20 m or more, but the large adult males can cover only a few meters. According to the most recent studies (Glasheen and McMahon 1996a,b; 1997), despite its rapid locomotion and large hind feet the animal sinks further into the water than someone watching the quick movements can detect without film analysis. The trick of not quite going under is that the legs generate most of the necessary upward force as soon as they strike the water surface. A fraction of a second later they are already in the swing phase of the step. The foot pushes the water away, creating an air space around itself. For the basilisk this provides additional resistive forces, brought about by the acceleration of the displaced water. The animal is further supported by the pressure difference between the air space above the foot and the hydrostatic pressure below it. But all this would be of little use if the basilisk did not pull its foot out of the air space before the water filled it up again.

It may seem amazing, but if *Cupiennius* is placed on a pool of water, it too neither sinks without trace into the depths nor is incapable of movement. Instead, it adjusts to the new medium and rows over the surface – a Jesus Spider. It is easy to imagine that this ability is useful during the rainy season in its tropical habitat, when downpours are a daily occurrence. But what properties of the substrate cause the spider to convert from one kind of locomotion to the other? Although multiple gaits are known to exist in aquatic beetles and bugs, in water striders and even in migratory locusts thrown onto a water surface (Hughes 1958, Bowdan 1978, Pflüger and Burrows 1978, Wendler et al. 1985), the question of how the switch is made remained open for other arthropods as well as spiders until Jon Barnes came from the University of Glasgow to Frankfurt, in order to pursue it in our laboratory. The results were published in 1991 as Chapter 10 in a book about the neuronal mechanisms of locomotion (Barnes and Barth 1991).

We analyzed the spider's walking and rowing on water of different depths, on sugar solutions

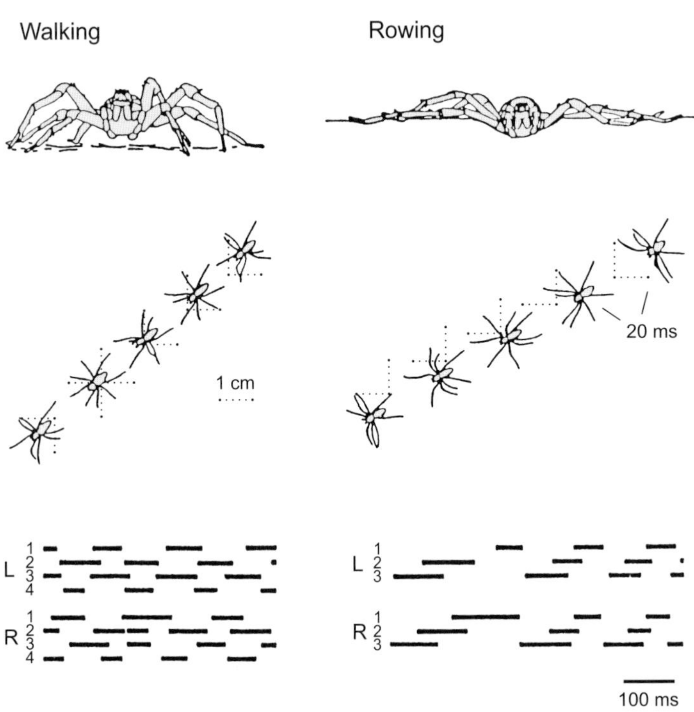

Fig. 15. Walking and rowing in *Dolomedes*. *Top* Characteristic body position taken on land and on the water surface. *Middle* Sequence of video frames (interval between frames 20 ms) with static reference point. *Bottom* Examples of stepping patterns; *bars* indicate duration of stance phase, *gaps* between them swing phase of step; *L*, *R* left and right legs. (Barnes and Barth 1991)

of specified viscosities and on olive oil and mercury, the surface tension of which is lower and higher, respectively, than that of water. To do this Jon used a high-speed film camera (200 frames/s). *Dolomedes fimbriatus*, a semiaquatic spider for which catching prey on the water surface is normal behavior as well as walking on land (see Chapter XVIII), served for comparison with *Cupiennius salei*, for which walking on solid ground is the normal thing to do. *Dolomedes* exhibits the phenomenon of gait switching, which cannot by any means be elicited in all spiders (for instance, not in *Tegenaria*, the ubiquitously familiar house spider with a sheet-like web), and they are particularly inclined to do so (Fig. 15) (see also Shultz 1987, *Dolomedes triton*). While the spider is rowing, its body rests on the water surface, the legs are extended and the hindlegs are not used; that is, they do not push as they do during a walk on land.

Water Depth

Cupiennius seems to put off rowing as long as possible: in water from 3 to 10 mm deep it wades, raising its body and the moving legs high above the water surface and frequently making searching movements with the first pair of legs. Often it also jumps forward. Differences in the behavior of individuals are correlated with the leg length, not with weight (Fig. 16). *Dolomedes* on the other hand, as might be expected from its mode of life, rows as soon as it can: when the water depth is only 0.7 mm they all row, regardless of the length of their legs. Hence when similarly sized individuals of the two species are compared, *Dolomedes* is found to be considerably more likely to engage in rowing than is *Cupiennius* (for example, given a leg length 11–15 mm and water depth 0.5 mm: *Dolomedes* 51%, *Cupiennius* 11%). In both cases the rowing is not elicited until the leg tips have lost contact with the ground.

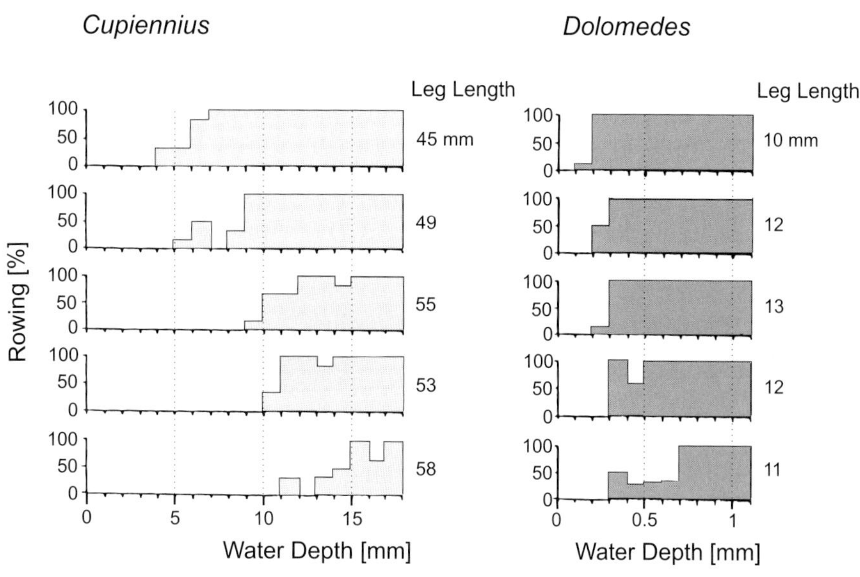

Fig. 16. Histograms showing percentage of trials in which spiders rowed on water of varying depths, for adult female *Cupiennius salei* (mass 2.4–2.9 g) and immature *Dolomedes. 0 mm* represents glass base of experimental chamber. *Cupiennius* 3–6 trials per spider at each depth; *Dolomedes:* 6–15 trials per spider at each depth. (Barnes and Barth 1991)

Viscosity

By raising the kinematic viscosity of the liquid it is possible to test whether the spider can measure the increase in mechanical resistance during the stance phase of the leg movement and use it to initiate the transition from rowing to walking. In low-viscosity conditions rowing predominates, and when the viscosity is high, walking prevails. *Dolomedes* begins to walk at lower viscosities than do *Cupiennius* individuals of comparable size (Fig. 17). These and other observations suggest that resistance is indeed measured and used as proposed above.

Surface Tension

The surface tension acts like a skin on the surface, which supports objects with a density higher than that of the liquid. The greater the surface tension, the heavier the object can be. The hypothesis: the spiders notice whether the substrate is supporting them adequately or not and adjust their gait accordingly.

The result: The surface tension of olive oil is too low (32 mN/m); the spiders sink into it. On water (73 mN/m) *Cupiennius* and *Dolomedes* both row, as we already know, and *Tegenaria* walks as though it were moving on a solid sub-

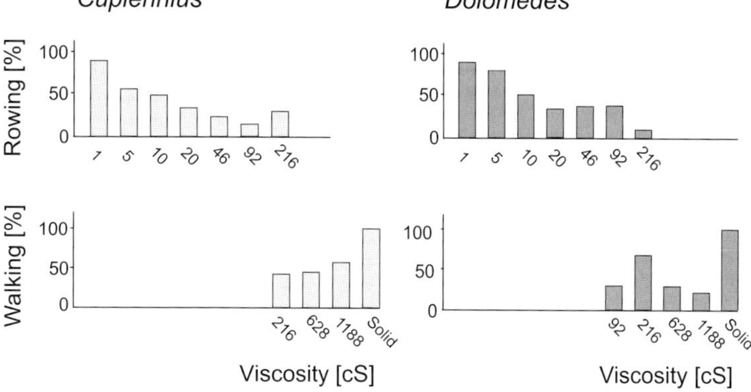

Fig. 17. Percentage of trials in which 12 immature individuals of *Cupiennius salei* (leg length 11–15 mm; 15–29 trials for each type of fluid) and 7 individuals of *Dolomedes* (size like that of *Cupiennius*) rowed on the surface or walked as if on solid ground on sugar solutions varying in kinematic viscosity. *1cS* Water; *5cS* 40% sucrose solution; *1188cS* 74% sucrose solution. (Barnes and Barth 1991)

strate. The rowers distribute their weight between legs and body (see Fig. 15, *Dolomedes* lies on the water surface); on mercury, with its considerably higher surface tension (435 mN/m), this is unnecessary, so all the weight is borne by the legs. However, mercury also has a considerably lower viscosity (0.11 cS) than water. *Cupiennius* usually walks on mercury, whereas *Dolomedes* on the same liquid rows, jumps and walks (in order of decreasing frequency).

The outcome of the study as a whole is thus as follows:

- Because spiders of both species can row both on mercury and in shallow water, the change in locomotor rhythm clearly has nothing to do with water *per se*.
- The absence of a firm substrate is certainly one of the factors that elicits rowing. This was shown particularly clearly by the experiments with *Cupiennius* in water of different depths.
- That increasing resistance of the substrate for the legs during the stance phase causes the transition from rowing to walking is suggested by the experiments with liquids of different viscosities.
- An increase in surface tension has the same effect on *Cupiennius* as does contact with the substrate.
- The occasional occurrence of forms of locomotion intermediate between walking and rowing suggests the existence of a flexible motor program that can generate several output patterns.
- The contact effect is stronger in *Cupiennius* than in *Dolomedes*. This is consistent with the differences in their modes of life and in the ecology of the two spiders, and its consequence is that *Dolomedes* is more readily induced to row.

It will not have escaped the attentive reader that the mechanical principles that enable spiders to locomote on the water surface must be quite different from the "slap and stroke" mechanism used by the basilisk. Like other arthropods in the same habitat, such as the water strider, spiders here locomote mainly by means of *horizontal* leg movements. The propulsive forces in this case are based substantially on the resistive force produced when, during the stance phase, the leg moves backward *along with* the indentation it has produced in the water surface. This resistive force is proportional to the density of the medium and to the square of the movement velocity (Suter et al 1997).

Both the surface tension and the bow waves are associated with resistances considerably less important for locomotion. The frictional force generated by the feet must *a priori* be smaller on the water surface than during walking on land: the cuticular surfaces of the spider are hydrophobic and the interaction with the water molecules is correspondingly slight. The mechanical resistance necessary for horizontal locomotion must primarily come from somewhere else – namely from the resistive force produced when the leg is pulled through the water together with its water-surface indentation (Suter et al. 1997). That the surface tension of the water bears the weight of the spider, by providing the necessary vertical resistive force, is independent of this effect.

The basilisk, the spider ... and Jesus? "As for humans, they have nothing to learn from the lizards except to stay ashore: an 80-kilogram person would have to run 30 meters per second (65 miles an hour) and expend 15 times more sustained muscular energy than a human being has the capacity to expend" (Glasheen and McMahon 1997). So that's it, then: either circus or divine.

XXV Swinging to a New Plant: the Dispersal of the Spiderlings

1 Drop and Swing

Spiders do not fly and in this respect differ – with quite a few consequences – from the insects, a group with countless masters of this art. But in a certain sense spiders can fly, and what some of them achieve here is remarkable. Young spiders can be carried away on the wind, which serves to disperse the species. Willis Gertsch (1979), one of the great American arachnologists and curator for insects and spiders at the American Museum of Natural History in New York, describes this so important moment in the young life of a spider with almost poetic affection: "Up and up they climb, to the tips of the tall grass stems and the summits of the leafless shrubs which mark the meadow site of the egg sac. Straight toward the sun they climb until they can climb no higher, impelled by a strong urge to throw silken threads out upon the soft breezes." This remarkable behavior is called "ballooning", with complete disregard of the physical differences between this and what keeps a balloon in the air.

Once the spider has reached a peak in its surroundings, it turns so that it is facing downwind, extends its legs as far as it can, bends its opisthosoma upward, spins some threads and takes off, as soon as the frictional forces acting on the threads and its body exceed its weight. JAC Humphrey (1987) has extensively documented the flow-mechanical boundary conditions of this process. Sometimes the winds carry their spider freight up several thousand meters, and Charles Darwin mentioned in his diary (1832; quoted by Gertsch, 1979 p. 30) "vast numbers of small spiders, about one tenth inch in length, and of a dusky red color", which became caught in the rigginig of the research ship Beagle 60 miles off the South American coast.

Cupiennius is certainly not one of the world champions in this kind of sport. There are good biological reasons, to which I shall return below. Still, *Cupiennius* does use similar methods to scatter its abundant progeny (from a single egg sac as many as 2000 spiderlings can hatch) and to avoid overpopulation in the vicinity of the sac.

Our experiments in this connection mostly employed *Cupiennius getazi.* Pilot experiments with *Cupiennius salei* and *Cupiennius coccineus*, the two other large species, showed no differences in behavior, so that *Cupiennius getazi* can perfectly well serve as an exemplary case. After the female has carried her egg sac around with her for about 25 days, she attaches it to the plant on which she sits and loosens the densely woven wall of the sac so that the spiderlings can leave their nursery, in which they have already undergone three molts. Within a short time the sac is surrounded by a cloud of hundreds of spiders (Plate 10). They hang in the tangle of delicate threads that their mother has previously spun around the sac. Quite in contrast to the adults, the children are friendly and tolerant toward one another for a few days. But after about 9 days their supply of yolk runs low, and they lose what is presumably the most important reason for their tolerance. Now they begin competing for food and living space. If spiders in this stage are kept confined, cannibalism is a normal occurrence.

Under field conditions, however, the small spiders show a behavior that we have called "Drop and Swing Dispersal Behavior" (DASDB), the most important function of which is to allow a spiderling to leave the area densely occu-

Fig. 1. A young *Cupiennius getazi*, 9 days after leaving the egg sac. *Bar* 1 mm. (Barth et al. 1991)

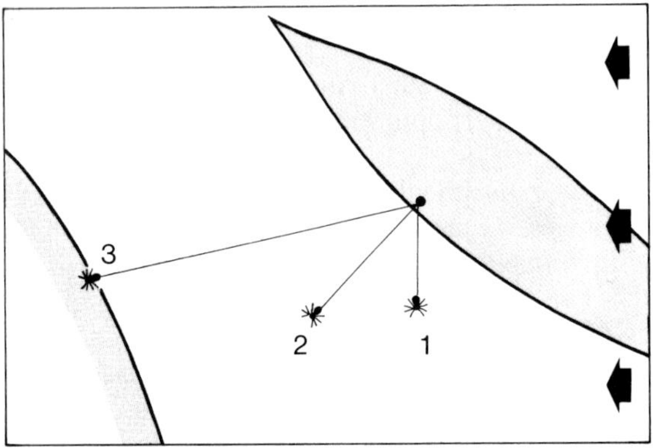

Fig. 2. "Drop and swing" behavior of *Cupiennius*. *Arrows* indicate direction of airflow. Upon exposure to airflow spiderlings drop from the plant and swing in the wind, dangling from their gradually lengthening safety thread. On touching the new substrate they attach to it. (Barth et al. 1991)

pied by competitors (Barth et al. 1991). The drop-and-swing behavior is less spectacularly effective for dispersal than ballooning; as a rule only relatively small distances are covered. But these too are vital.

Cupiennius does not take off from a "tiptoe" position at an elevated site as ballooning spiders do. Instead, it allows itself to fall so that it is hanging from the plant by its dragline. The spiderling, with a body length of only about 2 mm (Fig. 1), swings in the wind on its progressively lengthening thread until it feels something solid under its feet – usually another leaf of the same or a nearby plant (Fig. 2). Often the process is then repeated. A similar behavior is also known to occur in other spider species. Coyle (1983, 1985) describes it for two orthognathous spiders, in which it is usually a prelude to the actual ballooning that occurs once the dragline has broken. The same "pre-ballooning" has also been observed in 13 other species of 9 different spider families (Decae 1987).

What particularly interested us was the set of physical conditions that enable this behavior and whether these might be detected by the trichobothria.

2 What Kind of Wind Induces the Behavior?

To answer this question, we put a small *Cupiennius*, along with a plant (*Aechmea fasciata*) on which it could sit, into a wind tunnel 1 m in diameter so that the flow mechanics of the ambient air could be controlled; the wind tunnel was kindly made available by the Federal Institute for Experimentation and Research (Dr. J. Kränke) in Vienna. As soon as the animals are exposed to the wind, the ones that are going to show drop-and-swing behavior begin to run around, touching one another with their first and second leg pairs. However, they do not proceed to a higher place on the plant. Instead, soon they let themselves fall off on a dragline that could be as long as 70 cm under our experimental conditions; with a wind speed of at least 1.3 m/s the dragline is swept out nearly horizontally, if it is at least 30 cm long.

In the ideal case the small spiders are about 9 days old: then most of them (up to 70%) show drop-and-swing behavior. At this age their average mass is 1.26 mg (±0.35 SD, N=30) and their effective area (that is, for the action of the wind; the leg tips are at the circumference of this area, which is treated as though it were solid) is 6.1 mm^2 (±0.7 SD, N=10) in dorsal view and 4.0 mm^2 (±0.5 SD, N=10) as viewed from the side.

The number of responses given by the animals is strongly dependent on both the wind speed and the degree of turbulence of the air-

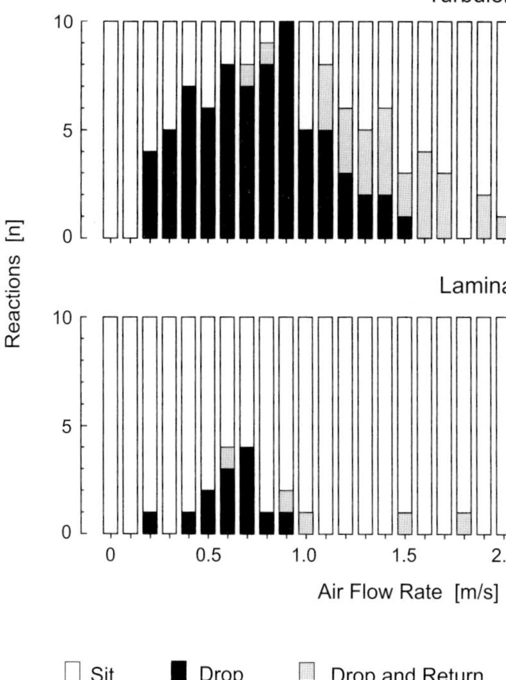

Fig. 3. Types of behavior shown by spiderlings (*Cupiennius getazi,* 9 days of age) in the wind tunnel when stimulated by airflow of different speeds and different degrees of turbulence; ten animals for each experimental situation. At the beginning of the experiment all spiderlings sat motionless on their bromeliad (*Aechmea fasciata*). *Top* Highly turbulent airflow (degree of turbulence 35%). *Bottom* Nearly laminar airflow (degree of turbulence 5%). (Barth et al. 1991)

stream. Responses can be triggered by wind at speeds as low as 0.2 m/s, but above 1.5 m/s the wind is ineffective; if a spider has already dropped when it reaches that speed, it climbs back to the starting point. As the turbulence increases, the number of responses rises considerably and the most effective mean velocity changes from ca. 0.7 m/s (quasi-laminar, 5% turbulence) to ca. 0.9 m/s (35% turbulence) (Fig. 3). Figure 4 shows the difference between the two kinds of airflow.

The degree of turbulence is calculated by $Tu = 100 \times c'/c$, where c is the mean velocity and c′ the average "root mean square" of the velocity. It can be measured with a hot-wire anemometer or a laser Doppler anemometer.

The effective wind speeds found for *Cupiennius getazi* are similar to those that, according to Vugts and von Wingerden (1976) and Richter (1970, 1971), elicit attempts at ballooning in other spider species. *Erigone arcticer* (Linyphiidae) initiates its pre-ballooning behavior at wind speeds above 3 m/s (horizontal component). *Pardosa purbeckensis*, a wolf spider (Lycosidae), shows the typical body-raising behavior ("standing on tiptoe") most frequently when the wind speed is between 0.35 m/s and 1.7 m/s; at speeds above 3 m/s it stops altogether.

Our wind measurements in the biotope of *Cupiennius* in Central America indicate that during this spider's activity period, after sunset, the wind speed is typically low: up to 0.4 m/s but

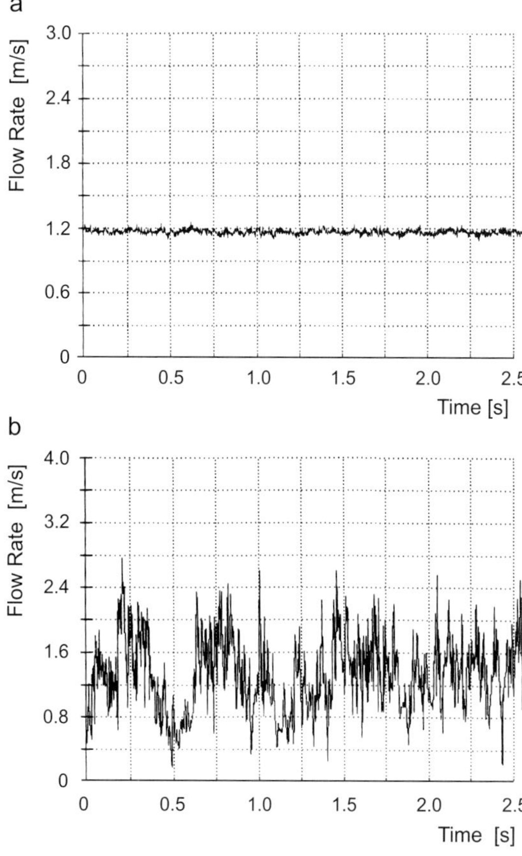

Fig. 4. Typical velocity patterns of airflows used in the experiments, monitored with a hot-wire anemometer; degrees of turbulence 5 and 25%, respectively. (Barth et al. 1991)

usually around 0.1 m/s (cf. Chapters III and XIX). Under such conditions – particularly with a level of fluctuation below 15% r.m.s. – DASDB is likely to be elicited only rarely; however, we have observed some responses in wind at only 0.2 m/s.

As we know from Chapter XIX, the trichobothria are all excited by such winds and, because of their markedly phasic discharge, give particularly vigorous responses to turbulent airflow. The small spider should thus not lack information about the air movement, even when the relevant sensory equipment amounts to only 6 or 7 trichobothria per walking leg, as opposed to almost 100 in adult animals (Fig. 5). It would be interesting to see what happens when trichobothria are removed, but such experiments have not yet been done. It is not a simple matter to operate on such a tiny spider without causing other, unintended damage. The prediction is that after such an operation DASDB, and in other spiders pre-ballooning, would be much less readily elicited.

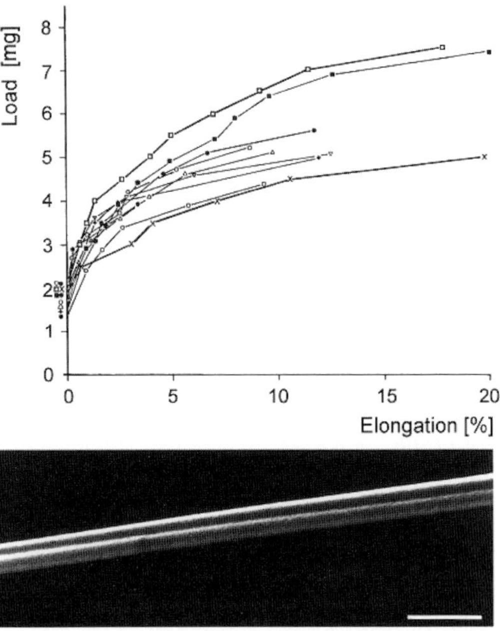

Fig. 6. Elongation of individual dropping lines of spiderlings of *Cupiennius getazi* with increasing mass added to their free end. The maximum elongation given is close to the breaking point of the thread; breakage immediately followed the further addition of 0.5 mg or 1 mg, respectively. Scanning electron micrograph shows example of draglines actually used for drop-and-swing behavior. Bar/μm (Barth et al. 1991)

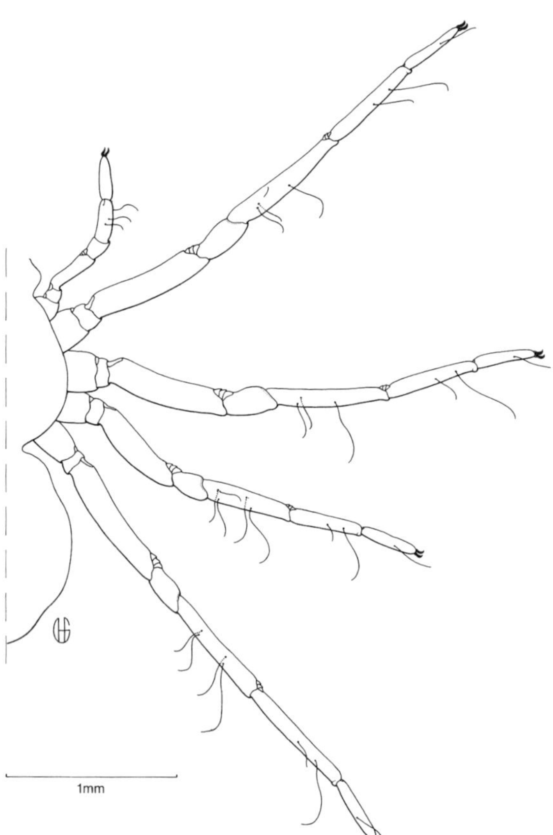

Fig. 5. Trichobothria of *Cupiennius salei* at the age of 9 days

3 When Does the Dragline Break?

The dragline is 0.5 to 1.0 μm thick and is composed of several threads. In the scanning electron microscope there is no discernible difference between the dragline used for dropping and those present when the spider is walking around normally. On average it can be statically loaded with a mass of 6.05 mg (±1.77 SD, N=27) before breaking – that is, with about five times the mass of a 9-day-old spider. Under this load, it lengthens by 11.6% (±5.8 SD, N=19). The modulus of elasticity in the first part of the stress-strain curve is of the order of 10^3 MPa (Fig. 6).

4 The Physical Model

Figure 7 summarizes the physical model. It is basically quite simple, showing the forces and moments that should be considered. T stands for the force with which the dragline is attached to the plant. D_s is the force acting on the spider, D_f the force acting per unit length of the dragline, and L the length of the dragline. The product M·g stands for the weight of the spider, M for its mass and g for gravity. p and n denote parallel and normal (that is, perpendicular) to the dragline. V is the mean horizontal velocity component moving the thread together with the spider. Z denotes the coordinate direction along the safety thread. The equations that specify this system quantitatively are somewhat more complicated than the drawing.

The moment applied at the site of the spider:

$$T_n L = \int_0^L D_{fn} Z dZ \,. \tag{1}$$

The moment applied at the point of attachment to the plant:

$$\int_0^L D_{fn} Z dZ = L\,M \cdot g \sin(\theta) - L D_s \,. \tag{2}$$

The force equilibrium parallel to the dragline:

$$T_p = D_s \sin(\theta) M \cdot g \cos(\theta) + \int_0^L D_{fp} dZ \,. \tag{3}$$

The forces applied to the spider, which is approximated by a sphere (Humphrey 1987):

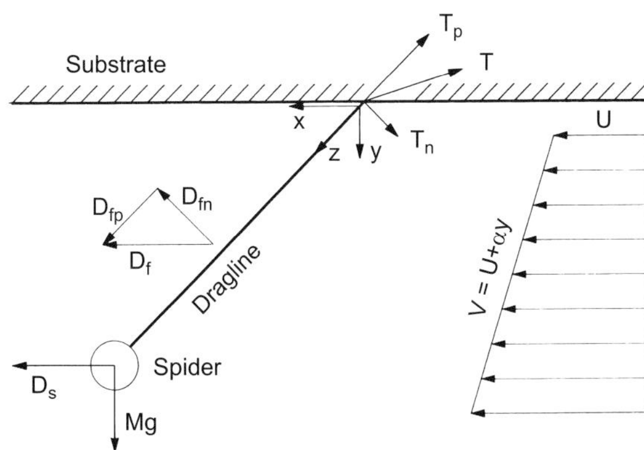

Fig. 7. The physical model: forces acting upon the idealized spider-dragline system including definitions of the coordinate system and general shape of the wind-velocity profile. *T* Force of attachment of dragline to the substrate; D_s drag force on the spider; D_f drag force per unit length on the dragline; $M \times g$ the spider's weight (*M* mass, *g* gravitational acceleration), *V* average horizontal component of air velocity, subscripts *p* and *n* denote parallel to and normal to the dragline, respectively. (Barth et al. 1991)

$$D_s = \frac{24}{Re_D}\left[1 + 0.125\, Re_D^{0.72}\right] \frac{1}{2}\, \varsigma\, V^2 \pi \left(\frac{D}{2}\right)^2 . \tag{4}$$

where $Re_D = DV/\nu$ is the Reynolds number of the spider, based on its effective diameter, and ς is the density and ν the kinematic viscosity of air.

The forces for the dragline, which is approximated as a long, rigid cylinder (Happel and Brenner 1965):

(i) component normal to the thread

$$D_{fn} = \frac{8\pi}{Re_{dn} \ln\left(\frac{7.4}{Re_{dn}}\right)} \cdot \frac{1}{2}\, \varsigma V_n^2 d \,. \tag{5 a}$$

(ii) component parallel to the thread

$$D_{fp} = \frac{4\pi}{Re_{dp}\left[\ln\left(\frac{2L}{d}\right) - 0.72\right]} \cdot \frac{1}{2}\, \varsigma V_p^2 d \,. \tag{5 b}$$

where $Re_{dn} = dV_n/\nu$ und $Re_{dp} = dV_p/\nu$ are the Reynolds numbers expressed with reference to

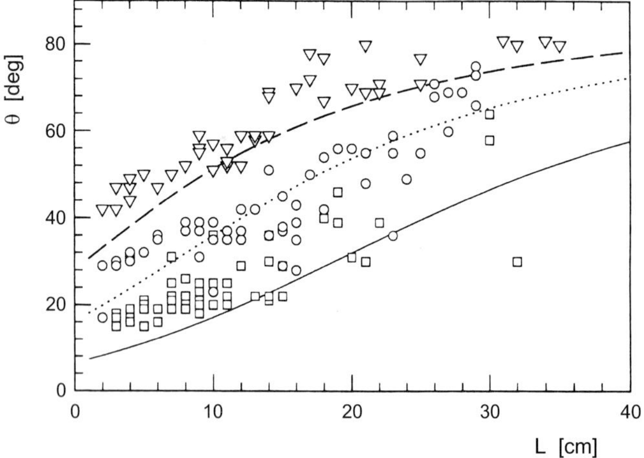

Fig. 8. Filament angle θ versus filament length L at different airflow velocities. *Continuous line and squares* V = 0.5 m/s; *dotted line and circles* 0.9 m/s; *broken line and triangles* 1.3 m/s. Whereas the curves are calculated, the symbols represent experimentally determined values. (Barth et al. 1991)

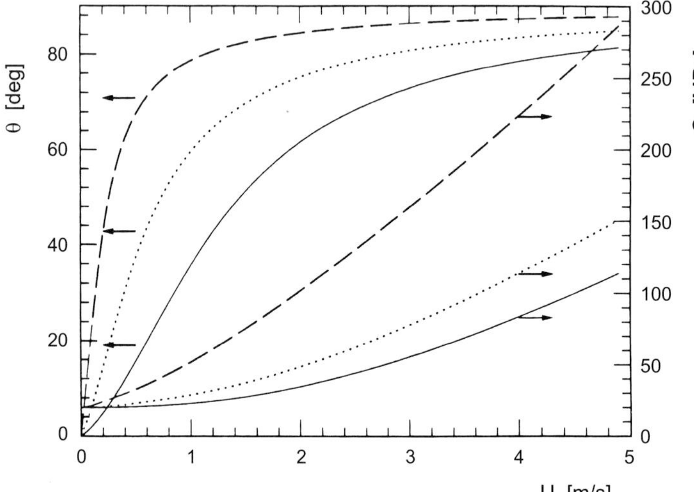

Fig. 9. Plot of filament angle θ and tension Q as a function of wind speed V for three filament lengths L. *Continuous line* $L = 0.25$ m; *dotted line* $L = 1.0$ m; *broken line* $L = 4.0$ m. Calculation conditions correspond to the absence of a velocity gradient and a constant dragline cross-section under the applied tension force. (Barth et al. 1991)

the diameter of the dragline and the normal and parallel velocity components of the airflow.

Regarding the procedure needed to solve these equations, the reader wanting to learn more about the biomechanics is referred to the original literature (Barth et al. 1991). What is of more interest here is the result of their application.

In brief: the agreement between model and experimental findings can be evaluated by plotting the deflection angle θ of the dragline for various wind speeds against the length of the dragline (Fig. 8). It is very close when the diameter D of the spider is measured not only as the length of the body itself but also includes the legs; while swinging, the spiderlings hold their legs outspread and thus considerably increase the force acting on them. It is also important to take into account a velocity gradient (instead of assuming uniform flow with no shear), which is to be expected under natural conditions. The model also makes it possible to estimate the wind speed at which the applied forces exceed the stability in terms of tensile strength (S_{ts}) of the dragline – that is, the breaking point. As the velocity of airflow increases, both the angle θ and the tensile stress Q increase more rapidly in long lines than in short ones (Fig. 9). The average value for the breaking strength of the dragline was found experimentally to be S_{ts}= 133 MPa. From figure 9 it can be seen that a spider-bearing dragline will not break as long as the wind speed is less than ca. 2.5 m/s and the line is 4 m long, or with a more realistic line length of 1 m, when the wind speed is below ca. 4.5 m/s. The experiments in the wind tunnel had shown that the spiders already stop showing DASDB when the wind speed exceeds ca. 1.5 m/s. This explains very nicely why breakage of a dragline was so rarely observed.

5 Ecological Consequences

The lack of true ballooning behavior must have considerable ecological consequences for *Cupiennius*. The dispersal of *Cupiennius* would be expected, and indeed is observed in the field, to be a short-range phenomenon. The spiders ordinarily land no further than 1 m away from the starting point. This may be one of the reasons why the dispersal of *Cupiennius getazi*, like that of *Cupiennius salei* and *Cupiennius coccineus*, is found in the field to be very local and stable, at least over a few years, and why the spiders in general tend to be rare rather than common.

We conclude that *Cupiennius getazi* and its close relatives *Cupiennius salei* and *Cupiennius coccineus* are among those tropical animal species with stable and specialized habitats (Barth, Seyfarth et al. 1988), which they cannot leave without serious risk (Robinson 1980, 1982). In such cases it is less important for the species to become widely distributed than to be in the appropriate habitat. The risky exploration of more distant fields is left to the few *Cupiennius getazi* that are carried away by the wind when their dragline breaks.

In general the unpredictability of a habitat is regarded as a basic factor in the evolution of migratory behavior in animals. Experiments by Greenstone (1982) on ballooning in two species of wolf spiders support this view, even in the present context. Although no long-term studies of the conditions in the habitat of *Cupiennius* have been carried out, it is most likely that the plants preferred by these spiders, which are available all year in the tropical climate, are a fairly predictable habitat. If this interpretation is correct, it provides another argument for the "site fidelity" of *Cupiennius*.

That *Cupiennius* has travelled from its Central American home into research laboratories as far away as Munich, Frankfurt, Jena, Bern, Cologne, Edmonton, Halifax and Vienna is another story.

Epilogue

This, then, is the story about our spider. It is a story with an open end, with the many unsolved problems it has brought to our attention. But this is normal in research. Our possibilities are always limited in some way, the severest limitation being time. Our enthusiasm, however, was quite unlimited along large stretches of the path we followed, providing the right background noise in our search for the relevant signals. If the pages between the covers of this book help to generate an interest in continuing our work and thus turning present knowledge into future discoveries, then to introduce the reader to *A Spider's World* was worth the effort.

Admittedly, I still do not consider myself an arachnologist in the classical sense but primarily a sensory and neurobiologist. This is not to say, however, that there is no special relationship between spiders and myself. They are wonderful animals indeed and fascinating to study. Their world may be small, with no long-distance senses like our eyes and ears, but *Cupiennius* taught us how rich it is.

It is unfortunate that a spider like *Cupiennius*, which has turned out to be so well suited for neurobiological studies, has no popular name. I would call it "bromeliad spider", if I had a say, alluding to its close relationship to particular plants. I cannot see much significance in the scientific name except in a very personal sense. *Cupiennius* derives from the Latin *cupiens*, which means desirous. *Cupiennius* is the one who desires – indeed, demands – our attention. When first choosing *Cupiennius* as an experimental animal I was far from anticipating that its demands would keep us busy for such a long time. The old dictionary of my own school days says that *cupiens* may also mean "attached to" or "kindly disposed to". In view of the many data and insights we owe *Cupiennius* this more positive translation may indeed be more appropriate.

Now, more than 30 years later, I see a wealth of potentially very rewarding routes for future exploration. A particular treasure chest still to be fully opened is a tiny piece of tissue measuring only 3 mm×1 mm×1 mm. This is the central nervous system, which has command over the rich repertoire of behavior the book talks about. Apart from first glimpses we are far from understanding its functional operations. The extensive neuroanatomical work now available and the several studies already done on interneurons may nevertheless provide a helpful basis for future work. Spiders certainly have a lot to offer for comparative brain biology.

While I have been writing my way through this book, a dedicated crew of Ph. D. students in my lab is trying to learn more about the tactile sense, which has not received the attention it deserves in arthropods in general. *Cupiennius* is covered by several hundreds of thousands of tactile hairs. These come in different shapes, and with different mechanical and physiological properties that are likely to be related to differences in their natural stimuli and associated behaviors. The abundance of the tactile hairs strengthens my impression that one of the aspects future work on the senses should concentrate on is to find out what arrays of hundreds of sensilla can do better and more than the individual sensillum or small groups of sensilla. How are complex spatio-temporal stimulus patterns pictured by the nervous system? The question applies not only to tactile hairs but to trichobothria and slit sensilla as well. Here problems of the ontogeny of sensory systems come in – and, again, the need to bring sensory physiologists together with engineers and physicists who have the adequate computational and informatics background. Another attraction for future research is the visual system. Now that we know something about the neuroanatomical details, to study the physiological properties of the different types of neurons in the visual pathway seems promising, not least from a comparative point of view. This approach seems even more justified given that no such analysis has yet been done even in spiders whose behavior is largely dominated by the visual sense (unlike that of *Cupiennius*). The last example of a future research area that seems particularly attractive from my present viewpoint is efferent control and modulation in the sensory periphery. It is to be hoped that our now much improved knowledge of the fine structural details and of the potential neurotransmitters will be an incentive for the electrophysiological analysis of the complex phenomena probably involved.

Many nice problems lie ahead of those interested in general questions in sensory and behavioral biology as well as in spider-related specializations. Balthasar Gracián, who was given the first word in this book, shall also have the last one: "Not to be amazed is the result of knowledge in the minority; the majority fail to be amazed because they don't notice anything."* I hope that the stories in this book will provide an incentive to take notice of the many interesting questions a zoologist and neurobiologist can ask a spider.

* From: El Criticón 1651–1657; quoted from: Criticón oder über die allgemeinen Laster des Menschen. Rowohlt, Hamburg 1957

References

Adams ME, Carney RL, Enderlin FE, Fu ET, Jarema MA, Li JP, Miller CA, Schooley DA, Shapiro MJ, Venema VJ (1987) Structures and biological activities of three synaptic antagonists from orb weaver spider venom. Biochem Biophys Res Comm 148:678–683

Adams ME, Herold EE, Venema VJ (1989) Two classes of channel-specific toxins from funnel web spider venom. J Comp Physiol A 164:333–342

Alcock J (1989) Animal Behavior. Sunderland, Massachusetts: Sinauer Associates

Alexander RMcN, Jayes AS (1980) Fourier analysis of forces exerted in walking and running. J Biomech 13:383–390

Alexander RMcN, Vernon A (1975) Mechanics of hopping by kangaroos (Macropodidae). J Zool Lond 177:265–303

Alexander RMcN, Jayes AS, Ker RF (1980) Estimates of energy cost for quadrupedal running gaits. J Zool Lond 190:155–192

Altevogt R (1970) Form und Funktion der vibratorischen Signale von *Uca tangeri* und *Uca inaequalis* (Crustacea, Ocypodidae). forma et functio 2:178–187

Altner H, Loftus R (1985) Ultrastructure and function of insect thermo- and hygroreceptors. Annu Rev Entomol 30:273–295

Altner R, Prillinger L (1980) Ultrastructure of invertebrate chemo-, thermo- and hygroreceptors and its functional significance. Int Rev Cytol 67:69–139

Ameismeier F, Loftus R (1988) Response characteristics of cold cell on the antenna of *Locusta migratoria* L. J Comp Physiol A 163:507–516

Anderson JF (1970) Metabolic rates of spiders. Comp Biochem Physiol 17:973–982

Anderson JF, Prestwich KN (1985) The physiology of exercise at and above maximal aerobic capacity in a theraphosid (tarantula) spider, *Brachypelma smithi* (F.O. Pickard-Cambridge). J Comp Physiol A 155:529–539

Anonymous (1996) Necrotic arachnidism. Med Sci Bull 18(12):4

Anton S (1991) Zentrale Projektionen von Mechano- und Chemorezeptoren bei der Jagdspinne *Cupiennius salei* Keys. Dissertation, Universität Wien

Anton S, Barth FG (1993) Central nervous projection patterns of trichobothria and other cuticular sensilla in the wandering spider *Cupiennius salei* (Arachnida, Araneae). Zoomorphology 113:21–32

Anton S, Tichy H (1994) Hygro- and thermoreceptors in tip-pore sensilla of the tarsal organ of the spider *Cupiennius salei*: innervation and central projection. Cell Tissue Res 278:399–407

Anton S, Ehn R, Loftus R, Tichy H (1992) Spider tarsal organ: Hygroreceptors in papillae with pore openings. In: Elsner N, Richter DW (eds) Rhythmogenesis in Neurons and Networks. Thieme, Stuttgart New York, p 459

Arnesen SJ, Olivo RF (1988) The effects of serotonin and octopamine on behavioral arousal in the crayfish. Comp Biochem Physiol 91:259–263

Aschoff J (1979) Circadian rhythms: influences of internal and external factors on the period measured in constant conditions. Z Tierpsychol 49:225–249

Autrum H (1941) Über Gehör- und Erschütterungssinn bei Locustiden. Z vergl Physiol 28:580–637

Autrum H (1984) Leistungsgrenzen von Sinnesorganen. Verh Ges Dtsch Naturforsch Ärzte 113:87–112

Baba Y, Hirota K, Shimozawa T, Yamaguchi T (1995) Differing afferent connections of spiking and nonspiking wind-sensitive local interneurons in the terminal abdominal ganglion of the cricket *Gryllus bimaculatus*. J Comp Physiol A 176:17–30

Babu KS (1975) Post-embryonic development of the central nervous system of the spider, *Argiope aurantia* (Lucas). J Morphol 146:325–342

Babu KS (1985) Patterns of arrangement and connectivity in the central nervous system of arachnids. In: Barth FG (ed) Neurobiology of Arachnids. Springer, Berlin Heidelberg New York Tokyo, pp 3–19

Babu KS, Barth FG (1984) Neuroanatomy of the central nervous system of the wandering spider, *Cupiennius salei* (Arachnida, Araneida). Zoomorphology 104:344–359

Babu KS, Barth FG (1989) Central nervous projections of mechanoreceptors in the spider *Cupiennius salei* Keys. Cell Tissue Res 258:69–82

Babu KS, Barth FG, Strausfeld NJ (1985) Intersegmental sensory tracts and contralateral motor neurons in the leg ganglia of the spider *Cupiennius salei* Keys. Cell Tissue Res 241:53–57

Bacetti B, Bedini C (1964) Research on the structure and physiology of eyes of a lycosid spider. I. Microscopic and ultramicroscopic structure. Archs Ital Biol 102:97–122

Bacon JP, Murphey RK (1984) Receptive fields of cricket giant interneurons are related to their dendritic structure. J Physiol 352:601–623

Barnes WJP, Barth FG (1991) Sensory control of locomotor mode in semi-aquatic spiders. In: Armstrong DM, Bush

BMH (eds) Locomotor Neural Mechanisms in Arthropods and Vertebrates. Manchester Univ Press, Manchester, chap 10, pp 105–116

Barnes WJP, Spirito CP, Evoy WH (1972) Nervous control of walking in the crab *Cardisoma guanhumi*. II. Role of resistance reflexes in walking. Z vergl Physiol 76:16–31

Barth FG (1967) Ein einzelnes Spaltsinnesorgan auf dem Spinnentarsus: seine Erregung in Abhängigkeit von den Parametern des Luftschallreizes. Z vergl Physiol 55:407–449

Barth FG (1969) Die Feinstruktur des Spinnenintegumentś. I. Die Cuticula des Laufbeins adulter häutungsferner Tiere (*Cupiennius salei* Keys.). Z Zellforsch 97:137–159

Barth FG (1970) Die Feinstruktur des Spinnenintegumentś. II. Die räumliche Anordnung der Mikrofasern in der lamellierten Cuticula und ihre Beziehung zur Gestalt der Porenkanäle (*Cupiennius salei* Keys., adult, häutungsfern, Tarsus). Z Zellforsch 104:87–106

Barth FG (1971 a) Der sensorische Apparat der Spaltsinnesorgane (*Cupiennius salei* Keys. Araneae). Z Zellforsch 112:212–246

Barth FG (1971 b) Ein einzelnes Spaltsinnesorgan auf dem Spinnentarsus: seine Erregung in Abhängigkeit von den Parametern des Luftschallreizes. Z vergl Physiol 55:407–449

Barth FG (1972a) Die Physiologie der Spaltsinnesorgane. I. Modellversuche zur Rolle des cuticularen Spaltes beim Reiztransport. J Comp Physiol 78:315–336

Barth FG (1972b) Die Physiologie der Spaltsinnesorgane. II. Funktionelle Morphologie eines Mechanoreceptors. J Comp Physiol 81:159–186

Barth FG (1973) Laminated composite material in biology. Microfiber reinforcement of an arthropod cuticle. Z Zellforsch 144:409–433

Barth FG (1976) Sensory information from strains in the exoskeleton. In: Hepburn HR (ed) The Insect Integument, pp 445–473. Elsevier Sci Publ Co, Amsterdam Oxford New York

Barth FG (1978) Slit sense organs: "Strain gauges" in the arachnid exoskeleton. Symp zool Soc Lond 1977 42:439–448

Barth FG (1981) Strain detection in the arthropod exoskeleton. In: Chapter 8; Laverack MS, Cosens D (eds) The Sense Organs, pp 112–141. Blacky, Glasgow

Barth FG (1982) Vibratory communication in a spider. Joint Symp Neurobiology and Strategies of Adaptation. FG Barth (ed) 1–9; Universitätsdruck, Frankfurt am Main

Barth FG (1985 a) Neuroethology of the spider vibration sense. In: Barth FG (ed) Neurobiology of Arachnids. Springer, Berlin Heidelberg New York Tokyo, pp 203–229

Barth FG (1985 b) Slit sensilla and the measurement of cuticular strains. In: Barth FG (ed) Neurobiology of Arachnids. Springer, Berlin Heidelberg New York Tokyo, pp 162–188

Barth FG (1986 a) Vibrationssinn und vibratorische Umwelt von Spinnen. Naturwissenschaften 73 (9):519–530

Barth FG (1986 b) Zur Organisation sensorischer Systeme: die cuticularen Mechanorezeptoren der Arthropoden. Verh Dtsch Zool Ges 79:69–90

Barth FG (1993 a) Sensory guidance in spider pre-copulatory behavior. Comp Biochem Physiol 104A:717–733

Barth FG (1993 b) Balz und vibratorische Kommunikation der Spinne *Cupiennius salei* (Ctenidae). Film C2318 of ÖWF, Vienna; Österr. Bundesinstitut für den wiss. Film (German or English Commentary, Color, 25 min)

Barth FG (1997) Vibratory communication in spiders: Adaptation and compromise at many levels. In: Lehrer M (ed) Orientation and Communication in Arthropods. Birkhäuser, pp 247–272

Barth FG (1998) The vibrational sense of spiders. In: Hoy RR, Popper AN, Fay RR (eds) Springer Handbook of Auditory Research. Comparative Hearing: Insects. Springer, New York, pp 228–278

Barth FG (2000) How to catch the wind: Spider hairs specialized for sensing the movement of air. Naturwissenschaften 87:51–58

Barth FG, Cordes D (1998) *Cupiennius remedius* (Araneae, Ctenidae), a new species in Central America, and a key for the genus *Cupiennius*. J Arachnol 26(2):133–141

Barth FG, Blickhan R (1984) Mechanoreception. In: Bereiter-Hahn J, Matoltsy AG, Richards R (eds) Biology of the Integument. Springer, Berlin, pp 554–582

Barth FG, Bohnenberger J (1978) Lyriform slit sense organ: threshold and stimulus amplitude ranges in a multi-unit mechanoreceptor. J Comp Physiol 125:37–43

Barth FG, Geethabali (1982) Spider vibration receptors. Threshold curves of individual slits in the metatarsal lyriform organ. J Comp Physiol A 148:175–185

Barth FG, Höller A (1999) Dynamics of arthropod filiform hairs. V. The response of spider trichobothria to natural stimuli. Phil Trans R Soc Lond B 354:183–192

Barth FG, Libera W (1970) Ein Atlas der Spaltsinnesorgane von *Cupiennius salei* Keys. Chelicerata (Araneae). Z Morph Tiere 68:343–369

Barth FG, Pickelmann H-P (1975) Lyriform slit sense organs. Modelling an arthropod mechanoreceptor. J Comp Physiol 103:39–54

Barth FG, Schmid A (eds) (2001) Ecology of Sensing. Springer, Berlin Heidelberg New York, pp 341

Barth FG, Schmitt A (1991) Species recognition and species isolation in wandering spiders (*Cupiennius* spp., Ctenidae). Behav Ecol Sociobiol 29:333–339

Barth FG, Seyfarth E-A (1971) Slit sense organs and kinesthetic orientation. Z vergl Physiol 74:326–328

Barth FG, Seyfarth E-A (1979) *Cupiennius salei* Keys. (Araneae) in the highlands of central Guatemala. J Arachnol 7:255–263

Barth FG, Stagl J (1976) The slit sense organs of arachnids. A comparative study of their topography on the walking legs. Zoomorphology 86:1–23

Barth FG, Wadepuhl M (1975) Slit sense organs on the scorpion leg (*Androctonus australis, L. Buthidae*). J Morph 145(2):209–227

Barth FG, Ficker E, Federle H-U (1984) Model studies on the mechanical significance of grouping in compound spider slit sensilla. Zoomorphology 104:204–215

Barth FG, Bleckmann H, Bohnenberger J, Seyfarth E-A (1988) Spiders of the genus *Cupiennius* SIMON 1891

(Araneae, Ctenidae). II. On the vibratory environment of a wandering spider. Oecologia 77:194–201

Barth FG, Humphrey JAC, Wastl U, Halbritter J, Brittinger W (1995) Dynamics of arthropod filiform hairs. III. Flow patterns related to air movement detection in a spider (*Cupiennius salei* Keys.). Phil Trans R Soc Lond B 347, 397–412

Barth FG, Komarek St, Humphrey JAC, Treidler B (1991) Drop and swing dispersal behavior of a tropical wandering spider: experiments and numerical model. J Comp Physiol A 169:313–322

Barth FG, Nakagawa T, Eguchi E (1993) Vision in the ctenid spider *Cupiennius salei*: spectral range and absolute sensitivity (ERG). J Exp Biol 181:63–79

Barth FG, Seyfarth E-A, Bleckmann H, Schüch W (1988) Spiders of the genus *Cupiennius* Simon 1891 (Araneae, Ctenidae). I. Range distribution, dwelling plants, and climatic characteristics of the habitats. Oecologia 77:187–193

Barth FG, Wastl U, Humphrey JAC, Devarakonda R (1993) Dynamics of arthropod filiform hairs. II. Mechanical properties of spider trichobothria (*Cupiennius salei* Keys.). Phil Trans R Soc Lond B 340:445–461

Bartholomew GA, Lighton JRB (1985) Ventilation and oxygen consumption during rest and locomotion in a tropical cockroach, *Blaberus giganteus*. J Exp Biol 118:449–454

Bässler U (1972) Der Regelkreis des Kniesehnenreflexes bei der Stabheuschrecke *Carausius morosus*: Reaktionen auf passive Bewegungen der Tibia. Kybernetik 12:8–20

Bässler U (1983) Neural Basis of Elementary Behavior in Stick Insects. Studies of Brain Function, vol 10. Springer, Berlin Heidelberg New York

Battelle BA, Calman BG, Andrews AW, Grieco FD, Mleziva MB, Callaway JC, Stuart AE (1991) Histamine: a putative afferent neurotransmitter in *Limulus* eyes. J Comp Neurol 305:527–542

Baurecht D, Barth FG (1991) Vibratory communication in spiders: receptor response to synthetic vibratory signals. In: Elsner N, Penzlin H (eds) Synapse – Transmission, Modulation. Proc 19th Göttingen Neurobiol Conf. Thieme, Stuttgart New York, p 133

Baurecht D, Barth FG (1992) Vibratory communication in spiders. I. Representation of male courtship signals by female vibration receptor. J Comp Physiol A 171:231–243

Baurecht D, Barth FG (1993) Vibratory communication in spiders. II. Representation of parameters contained in synthetic male courtship signals by female vibration receptor. J Comp Physiol A 173:309–319

Becherer C (1993) Immunhistochemische Nachweise möglicher Neurotransmitter im Zentralnervensystem von Arachniden. Diplomarbeit, Universität Wien

Beck L (1972) Zur Tagesperiodik der Laufaktivitäten von *Admetus pumilio* C. Koch (Arach., Amblypygi) aus dem neotropischen Regenwald. II. Oecologia 9:65–102

Bertkau Ph (1878) Versuch einer natürlichen Anordnung der Spinnen nebst Bemerkungen an einzelnen Gattungen. Arch Naturgesch 44:351–410

Bertkau Ph (1886) Beiträge zur Kenntniss der Sinnesorgane der Spinnen. I. Augen der Spinnen. Arch mikrosk Anat 27:589–631

Bettini S (ed) (1978) Arthropod Venoms. Springer, Berlin Heidelberg New York

Bickeböller C, Kadel M, Seyfarth EA (1991) Coxal muscle c2 in spiders: identification of motoneurons, joint receptors, and their role in body raising behavior. In: Elsner N, Penzlin H (eds) Synapse – transmission, modulation. Proc 19th Göttingen Neurobiology Conference. Thieme, Stuttgart New York, p 59

Bicker G, Menzel R (1989) Chemical codes for the control of behavior in arthropods. Nature 337:33–39

Biederman-Thorson M, Thorson J (1971) Dynamics of excitation and inhibition in the light adapted *Limulus* eye in situ. J Gen Physiol 58:1–19

Bieman CFM (1986) Acoustic differentiation and variation in planthoppers of the genus *Riautodelphax* (Homoptera, Delphacidae). Neth J Zool 36:461–480

Bishop CA, Krouse ME, Wine JJ (1991) Peptide cotransmitter potentiates calcium channel activity in crayfish skeletal muscle. J Neurosci 11(1):269–276

Bleckmann H (1982) Reaction time, threshold values and localization of prey in stationary and swimming surface feeding fish *Aplocheilus lineatus* (Cyprinodontidae). Zool Jb Physiol 86:71–81

Bleckmann H (1985a) Discrimination between prey and non-prey wave signals in the fishing spider *Dolomedes triton* (Pisauridae). In: Kalmring K, Elsner N (eds) Acoustic and Vibrational Communication in Insects. Paul Parey, Berlin, pp 215–222

Bleckmann H (1985b) Perception of water surface waves: how surface waves are used for prey identification, prey localization, and intraspecific communication. In: Ottoson O (ed) Progress in Sensory Physiology 5. Springer, New York, pp 147–166

Bleckmann H (1988) Prey identification and prey localization in surface-feeding fish and fishing spiders. In: Atema J, Fay RR, Popper AN, Tavolga WN (eds) Sensory Biology of Aquatic Animals. Springer, New York, pp 619–641

Bleckmann H (1994) Reception of hydrodynamic stimuli in aquatic and semiaquatic animals. Progress in Zoology 41. Fischer, Stuttgart Jena New York

Bleckmann H, Barth FG (1984) Sensory ecology of a semiaquatic spider (*Dolomedes triton*) II. The release of predatory behavior by water surface waves. Behav Ecol Sociobiol 14:303–312

Bleckmann H, Bender M (1987) Water surface waves generated by the male pisaurid spider *Dolomedes triton* during courtship behavior. J Arachnol 15:363–369

Bleckmann H, Rovner JS (1984) Sensory ecology of a semiaquatic spider. I. Roles of vegetation and wind-generated waves in site selection. Behav Ecol Sociobiol 14:297–301

Bleckmann H, Borchardt M, Horn P, Görner P (1994) Stimulus discrimination and wave source localization in fishing spiders (*Dolomedes triton* and *Dolomedes okefinokensis*). J Comp Physiol A 174:305–316

Blest AD (1978) The rapid synthesis and destruction of photoreceptor membrane by a dinopid spider: a daily cycle. Proc R Soc Lond B 200:463–483

Blest AD (1985a) The fine structure of photoreceptors in relation to function. In: Barth FG (ed) Neurobiology of Arachnids. Springer, Berlin Heidelberg New York Tokyo, pp 79–102

Blest AD (1985b) Retinal mosaics of the principal eyes of jumping spiders (Salticidae) in some neotropical habitats: optical trade-offs between sizes and habitat illuminances. J Comp Physiol A 157:391–405

Blest AD, Day WA (1978) The rhabdomere organisation of some nocturnal pisaurid spiders in light and darkness. Phil Trans R Soc Lond B 283:1–23

Blest AD, Land MF (1977) The physiological optics of *Dinopis subrufus* L. Koch: a fish-lens in a spider. Proc R Soc Lond B, 196:198–222

Blickhan R, Barth FG (1979) Dehnungen und Spannungen im Außenskelett von Arthropoden. GESA-Symp, Exp Spannungsanal, Braunschweig, 21 p

Blickhan R, Barth FG (1985) Strains in the exoskeleton of spiders. J Comp Physiol A 157:115–147

Blickhan R, Barth FG, Ficker E (1982) Biomechanics in a sensory system. Strain detection in the exoskeleton of arthropods. VIIth Int Conf Exp Stress Anal, Haifa, pp 223–233

Blickhan R, Weber W, Barth FG (1984) Strain at the site of biological strain detectors in the exoskeleton of spiders. Proc V Intern Congr Exp Mech, Montreal, pp 37–47

Blumenthal H (1935) Untersuchungen über das "Tarsalorgan" der Spinnen. Z Morphol Ökol Tiere 29:667–719

Bodian D (1937) A new method for staining nerve fibers and nerve endings in mounted paraffin sections. Anat Rec 69:153–162

Bodnar DA, Miller JP, Jacobs GA (1991) Anatomy and physiology of identified wind-sensitive local interneurons in the cricket cercal sensory system. J Comp Physiol A 168:553–564

Boevé J-L (1991) The injection of venom by a spider (*Cupiennius salei*, Ctenidae) and the weight of an insect prey (*Acheta domesticus*, Gryllidae). Bull Soc Neuchâtel Sci nat 116(1):41–47

Boevé J-L (1994) Injection of venom into an insect prey by the free hunting spider *Cupiennius salei* (Araneae, Ctenidae). J Zool (Lond) 234:165–175

Bohnenberger J (1979) Das Übertragungsverhalten eines zusammengesetzten Spaltsinnesorgans auf dem Spinnenbein. Dissertation, Universität Frankfurt am Main

Bohnenberger I (1981) Matched transfer characteristics of single units in a compound slit sense organ. J Comp Physiol A 142:391–401

Bohnenberger J, Seyfarth E-A, Barth FG (1983) A versatile feedback controller for electro-mechanical stimulation devices. J Neurosci Methods 9:335–341

Bonnet P (1945) Bibliographia Araneorum. Analyse méthodique de toute la litterature aranéologique jusqu'en 1939. Tome Les Frères Douladoure, Toulouse

Bonsel W (1912) Die Biene Maja und ihre Abenteuer. In: Bonsel R-M (Hrsg) Bonsel W – Wanderschaft zwischen Staub und Sternen. Gesamtwerk. Albert Langen, Georg Müller-Verlag, München Wien 1980

Bowdan E (1978) Walking and rowing in the water strider, *Gerris remigis*. I. A cinematographic analysis of walking. J Comp Physiol A 123:43–49

Bowerman RF (1972a) A muscle receptor organ in the scorpion postabdomen. I. The sensory system. J Comp Physiol 81:133–146

Bowerman RF (1972b) A muscle receptor organ in the scorpion postabdomen. II. Reflexes evoked by MRO stretch and release. J Comp Physiol 81:147–157

Bowers CW, Phillips HS, Lee P, Jan YN, Jan LY (1987) Identification and purification of an irreversible presynaptic neurotoxin from the venom of the spider, *Hololena curta*. Proc Natl Acad Sci 84:3506–3510

Braitenberg V (1970) Ordnung und Orientierung der Elemente im Sehsystem der Fliege. Kybernetik 7:235–242

Branton WD, Kolton L, Jan YN, Jan LY (1987) Neurotoxins from *Plecytreurys* spider venom are potent presynaptic blockers in *Drosophila*. J Neurosci 7:4195–4200

Brants A (1838) Bijdragen tot de kennis van de eenvoudige oogen der gelede dieren, Articulata, Cuv. Tijdschr Nat Geschied 4:135–153

Brenner HR (1972) Evidence for peripheral inhibition in an arachnid muscle. J Comp Physiol 80:227–231

Brescovit AD, von Eickstedt VRD (1995) Ocorrência de *Cupiennius* Simon na América do Sul e redescrição de *Cupiennius celerrimus* Simon (Araneae, Ctenidae). Revta bras Zool 12(3):641–646

Brittinger W (1998) Trichobothrien, Medienströmungen und das Orientierungsverhalten von Jagdspinnen (*Cupiennius salei* Keys). Dissertation, Universität Wien

Brodführer PD, Hoy RR (1990) Ultrasound sensitive neurons in the cricket brain. J Comp Physiol A 166:651–662

Brownell P, Farley RD (1979) Detection of vibrations in sand by tarsal sense organs of the nocturnal scorpion, *Paruroctonus mesaensis*. J Comp Physiol A 131:23–30

Brüssel A (1987) Belastungen und Dehnungen im Spinnenskelett unter natürlichen Verhaltensbedingungen. Dissertation, Universität Frankfurt, Frankfurt am Main

Brüssel A, Gnatzy W (1985) A somatotopic organization of leg afferents in the spider *Cupiennius salei* Keys. (Araneae, Ctenidae). Experientia 41:468–470

Burrows M (1975) Monosynaptic connections between wing stretch receptors and flight motoneurones of the locust. J Exp Biol 62:189–219

Burrows M (1996) The Neurobiology of an Insect Brain. Oxford Univ Press, Oxford New York Tokyo

Bush BMH (1965) Leg reflexes from chordotonal organs in the crab *Carcinus maenas*. Comp Biochem Physiol 15:567–587

Callaway JC, Stuart AE (1989) Biochemical and physiological evidence that histamine is the transmitter of barnacle photoreceptors. Visual Neurosci 3:311–325

Calman BG, Battelle BA (1991) Central origin of the efferent neurones projecting to the eyes of *Limulus polyphemus*. Visual Neurosci 6:481–495

Camhi JM, Levy A (1989) The code for stimulus direction in a cell assembly in the cockroach. J Comp Physiol A 165:83–97

Camhi JM, Tom W, Volman S (1978) The escape behavior of the cockroach *Periplaneta americana* L. II. Detection of natural predators by air displacement. J Comp Physiol A 128:203–212

Carlson AD, Jalenak M (1986) Release of octopamine from the photomotor neurones of the firefly lanterns. J Exp Biol 122:453–457

Carricaburu P, Muñoz-Cuevas A, Ortega-Escobar J (1990) Electroretinography and circadian rhythm in *Lycosa tarentula* (Araneae, Lycosidae). Acta Zool Fennica 190:63–67

Cavagna GA (1975) Force plates as ergometers. J Appl Physiol 39:174–179

Cavagna GA, Kaneko M (1977) Mechanical work and efficiency in level walking and running. J Physiol Lond 268:467–481

Cavagna GA, Heglund NC, Taylor CR (1977a) Mechanical work in terrestrial locomotion: two basic mechanisms for minimizing energy expenditure. Am J Physiol 233:R243–R261

Cavagna GA, Heglung NC, Taylor CR (1977b) Walking, running and galloping: Mechanical similarities between different animals. In: Pedley TJ (ed) Scale Effects in Animal Locomotion. Academic Press, London New York, pp 111–125

Cavagna GA, Saibene FP, Margaria R (1963) External work in walking. J Appl Physiol 18:1–9

Cavagna GA, Saibene FP, Margaria R (1964) Mechanical work in running. J Appl Physiol 19:249–256

Chabaud F (1990) Developpement du système nerveux de l'araignée *Cupiennius salei* Keyserling (Ctenidae). Travail de Diplom, Université Genf

Chapman DM, Mosinger JL, Duckrow RB (1979) The role of distributed viscoelastic coupling in sensory adaptation in an insect mechanoreceptor. J Comp Physiol 131:1–12

Christian UH (1971) Zur Feinstruktur der Trichobothrien der Winkelspinne *Tegenaria derhami* (Scopoli), (Agelenidae, Araneae). Cytobiology 4:172–185

Christian UH (1972) Trichobothrien, ein Mechanorezeptor bei Spinnen. Elektronenmikroskopische Befunde bei der Winkelspinne *Tegenaria derhami* (Scopoli), (Agelenidae, Araneae). Verh Dtsch Zool Ges 66:31–36

Clark DL, Uetz GW (1993) Signal efficacy and the evolution of male dimorphism in the jumping spider, *Maevia inclemens*. Proc Natl Acad Sci USA 90:11954–11957

Cloudsley-Thompson JL (1968) The water-relations of scorpions and tarantulas from the Sonoran desert. Entomol Mon Mag 103:217–220

Cloudsley-Thompson JL (1970) Terrestrial invertebrates. In: Whittow GC (ed) Comparative Physiology of Thermoregulation, Vol I. Academic Press, New York London, pp 15–77

Cloudsley-Thompson JL (1973) Entrainment of the 'circadian clock' in *Buthotus minax* (Scorpiones: Buthidae). J Interdiscipl Cycle Res 4:119–123

Cloudsley-Thompson JL (1975) Entrainment of the 'circadian clock' in *Babycurus centrurimorphus* (Scorpiones: Buthidae). J Interdiscipl Cycle Res 6:185–188

Cloudsley-Thompson JL (1978) Biological clocks in arachnida. Bull Brit Arachnol Soc 4:184–191

Cloudsley-Thompson JL (1987) The biorhythms of spiders. In: Nentwig W (ed) Ecophysiology of Spiders. Springer, Berlin Heidelberg New York London Paris Tokyo, pp 371–379

Coddington JA, Levi IT (1991) Systematics and evolution of spiders. Ann Rev Ecol Syst 22:565–592

Coelho P (1996) Der Alchimist. Diogenes, Zürich

Comstock JH (1913) The Spider Book. Doubleday, New York

Corbière-Tichané G (1971) Structure nerveuse énigmatique dans l'antenne de la larve du *Speophyes lucidulus* Delar. (Coléoptère cavernicole de la sous-famille des Bathysciinae). Étude au microscope électronique. J Microscopie 10:191–202

Corbière-Tichané G, Loftus R (1983) Antennal thermal receptors of the cave beetle, *Speophyes lucidulus* Delar. II. Cold receptor response to slowly changing temperature. J Comp Physiol A 153:343–351

Coyle FA (1983) Aerial dispersal by mygalomorph spiderlings (Araneae, Mygalomorphae). J Arachnol 11:283–286

Coyle FA (1985) Ballooning behavior of *Ummidia* spiderlings (Araneae, Ctenizidae). J Arachnol 13:137–138

Cremer L, Heckl M, Ungar EE (1973) Structure-borne Sound. Structural Vibrations and Sound Radiation at Audio Frequencies. Springer, Berlin Heidelberg New York

Cruse H (1976) The function of the legs in the free walking stick insect, *Carausius morosus*. J Comp Physiol 112:235–262

Cull-Candy SG, Neal H, Usherwood PNR (1973) Action of black widow spider venom on an aminergic synapse. Nature 241:353–354

Currey JD (1967) The Failure of Exoskeletons and Endoskeletons. J Morphol 123:1–16

Dahl F (1883) Über die Hörhaare bei den Arachniden. Zool Anz 6:267–270

Dahl F (1908) Die Lycosiden oder Wolfsspinnen Deutschlands und ihre Stellung im Haushalt der Natur. Nach statistischen Untersuchungen dargestellt. N Act Acad Caes Leop-Carol 88:175–678

Dahl F (ed) (1925) Die Tierwelt Deutschlands. VEB G Fischer, Jena

Dambach M (1989) Vibrational responses. In: Huber F, Moore TE, Loher W (eds) Cricket Behavior and Neurobiology. Cornell Univ Press, Ithaca NY, pp 178–197

Damen WGM, Tautz D (1998) A Hox class 3 orthologue from the spider *Cupiennius salei* is expressed in a Hox-gene-like fashion. Dev Genes Evol 208:586–590

Damen WGM, Hausdorf M, Seyfarth E-A, Tautz D (1998) A conserved mode of head segmentation in arthropods revealed by the expression pattern of Hox genes in a spider. Proc Natl Acad Sci USA 95:10665–10670

Darnhofer-Demar B (1977) Funktionsmorphologie der Mittelbeine von Wasserläufern der Gattung *Gerris*. Fortschr Zool 24:115–122

Davies ME, Edney EB (1952) The evaporation of water from spiders. J Exp Biol 29:571–582

Davis H (1965) A model for transducer action in the cochlea. Cold spring Harb Symp quant Biol 30:181–190

Decae AE (1987) Dispersal: ballooning and other mechanisms. In: Nentwig W (ed) Ecophysiology of Spiders. Springer, Berlin Heidelberg New York Tokyo, pp 348–356

Den Otter CJ (1974) Setiform sensilla and prey detection in the bird-spider *Sericopelma rubronitens* Ausserer (Araneae, Theraphosidae). Neth J Zool 24:219–235

Denny M (1976) The physical properties of spider's silk and their role in the design of orb webs. J Exp Biol 65:483–506

Devarakonda R, Barth FG, Humphrey JAC (1996) Dynamics of arthropod filiform hairs. IV. Hair motion in air and water. Phil Trans R Soc Lond B 351:933–946

Devetak D, Gogala M, Čokl A (1978) A contribution to the physiology of vibration receptors in the bugs of the family Cydnidae (Heteroptera). Biol Vestn 26:131–139

DeVoe RD (1972) Dual sensitivities of cells in wolf spider eyes at ultraviolet and visible wavelengths of light. J Gen Physiol 59:247–269

DeVoe RD, Small RJW, Zvargulis JE (1969) Spectral sensitivities of wolf spider eyes. J Gen Physiol 54:1–32

Dickinson MH, Götz K (1997) The wake dynamics and flight forces of the fruit fly *Drosophila melanogaster.* J Exp Biol 199:2085–2104

Dierkes S (1992) Der Mechanismus der Erzeugung vibratorischer Balzsignale bei der Spinne *Cupiennius getazi* (Arachnida: Araneae). Dissertation, Universität Wien

Dierkes S, Barth FG (1995) Mechanism of signal production in the vibratory communication of the wandering spider *Cupiennius getazi* (Arachnida, Araneae). J Comp Physiol A 176:31–44

Dijkgraaf (1947) Über die Reizung des Ferntastsinnes bei Fischen und Amphibien. Experientia 3:206–208

Doherty JA (1985) Trade-off phenomena in calling song recognition and phonotaxis in the cricket, *Gryllus bimaculatus* (Orthoptera, Gryllidae). J Comp Physiol A 156:787–801

Drewes CD, Bernard RA (1976) Electrophysiological responses of chemosensitive sensilla in the wolf spider. J Exp Zool 198:423–428

Dugès A (1836) Observations sur les aranéides. Ann Sci nat (2), Zool 6:159–218

Dumpert K (1978) Spider odor receptors: electrophysiological proof. Experientia 34:754–756

Duncker PM (1992) Vorkommen und Verteilung biogener Amine im Zentralnervensystem der Jagdspinne *Cupiennius salei* Keyserling (Ctenidae, Araneae, Arachnida). Diplomarbeit, Universität Wien

Dusenbery DB (1992) Sensory Ecology. How Organisms Acquire and Respond to Information. WH Freeman and Company, New York

Eberhard WG (1985) Sexual Selection and Animal Genitalia. Harvard Univ Press, Cambridge

Eckweiler W (1983) Topographie von Propriorezeptoren, Muskeln und Nerven im Patella-Tibia- und Metatarsus-Tarsus-Gelenk des Spinnenbeins. Diplomarbeit, Fachbereich Biologie, Goethe-Universität Frankfurt am Main

Eckweiler W, Seyfarth E-A (1988) Tactile hairs and the adjustment of body height in wandering spiders: behavior, leg reflexes, and afferent projections in the leg ganglia. J Comp Physiol A 162:611–621

Eckweiler W, Hammer K, Seyfarth E-A (1989) Long smooth hair sensilla on the spider leg coxa: Sensory physiology, central projection pattern, and proprioceptive function. Zoomorphology 109:97–102

Eguchi E (1982) Retinular fine structure in compound eyes of diurnal and nocturnal sphingoid moths. Cell Tissue Res 223:29–42

Ehn R (1995) Functional properties of a spider thermoreceptor. Dissertation, Universität Wien

Ehn R, Tichy H (1994) Hygro- and thermoreceptive tarsal organ in the spider *Cupiennius salei.* J Comp Physiol A 174:345–350

Ehn R, Tichy H (1996a) Response characteristics of a spider warm cell: temperature sensitivities and structural properties. J Comp Physiol A 178:537–542

Ehn R, Tichy H (1996b) Threshold for detecting temperature changes in a spider thermoreceptor. J Neurophysiol 76:2608–2613

Elste A, Koester J, Shapiro E, Panula P, Schwartz JH (1990) Identification of histaminergic neurons in *Aplysia.* J Neurophysiol 64:736–744

Engelhardt W (1964) Die mitteleuropäischen Arten der Gattung *Trochosa* Cl Koch, 1848 (Araneae, Lycosidae). Morphologie, Chemotaxonomie, Biologie, Autökologie. Z Morphol Ökol Tiere 54:219–392

Erber J (1989) Serotonin, octopamine and FMRFamide modulate visual and olfactory antennal reflexes in the honeybee. In: Erber J, Menzel R, Pflüger HJ, Todt D (eds) Neural Mechanisms of Behavior. Georg Thieme, Stuttgart New York, p 234

Erber J, Homberg U, Gronenberg W (1987) Functional roles of the mushroom bodies in insects. In: Gupta AP (ed) Arthropod Brain. Wiley, New York Chichester Singapore, pp 485–511

Evans PD, O'Shea M (1978) The identification of an octopaminergic neurone and the modulation of a myogenic rhythm in the locust. J Exp Biol 73:235–260

Fabian R, Seyfarth E-A (1997) Acetylcholine and histamine are transmitter candidates in identifiable mechanosensitive neurons of the spider *Cupiennius salei*: an immunocytochemical study. Cell Tissue Res 287:413–423

Fabian R, Volknandt W, Seyfarth E-A (1997) Peripheral synapses at identifiable mechanosensory neurons in spiders: synapsin-I immunocytochemistry. In: Elsner N, Wässle H (eds) Proc 25th Göttingen Neurobiology Conference vol II, Thieme, Stuttgart New York, p. 738

Fabian R, Höger U, Seyfarth E-A, Meinertzhagen IA (1998) Three-dimensional reconstruction of the peripheral synaptic connections of identified spider mechanosensory neurons. In: Elsner N, Wehner R (eds) Proc 26th Göttingen Neurobiology Conference vol II,. Thieme, Stuttgart New York, p 585

Fabian-Fine R, Meinertzhagen IA, Seyfarth E-A (2000) Organization of efferent synapses at mechanosensory neurons in spiders. J Comp Neurol 420:195–210

Fabian-Fine R, Volknandt W, Seyfarth E-A (1999) Peripheral synapses at identifiable mechanosensory neurons in the spider *Cupiennius salei*: synapsin-like immunoreactivity. Cell Tissue Res 295:13–19

Fabian-Fine R, Höger U, Seyfarth E-A, Meinertzhagen IA (1999) Peripheral synapses at identified mechanosensory neurons in spiders: three-dimensional reconstruction and GABA immunocytochemistry. J Neuroscience 19(1):298–310

Felber R (1994) The phylogenetic relationship of spiders in the genus *Cupiennius* deduced from mitochondrial DNA sequences. Diploma thesis, University of Vienna

Fincke T, Paul R (1989) Book lung function in arachnids III. The function and control of spiracles. J Comp Physiol B 159:433–441

Fleissner G, Fleissner G (1985) Neurobiology of a circadian clock in the visual system of scorpions. In: Barth FG (ed) Neurobiology of Arachnids. Springer, Heidelberg Berlin New York Tokyo, pp 351–375

Fletcher NH (1978) Acoustical response of hair receptors in insects. J Comp Physiol 127:185–189

Foelix RF (1970) Chemosensitive hairs in spiders. J Morph 132:313–334

Foelix RF (1975) Occurrence of synapses in peripheral sensory nerves of arachnids. Nature 254:146–148

Foelix RF (1985) Mechano- and chemoreceptive sensilla. In: Barth FG (ed) Neurobiology of Arachnids. Springer, Berlin Heidelberg New York Tokyo, pp 118–137

Foelix RF (1985) Sensory nerves and peripheral synapses. In: Barth FG (ed) Neurobiology of Arachnids. Springer, Berlin Heidelberg New York Tokyo, pp 189–200

Foelix RF (1992) Biologie der Spinnen. Thieme, Stuttgart New York

Foelix RF, Choms A (1979) Fine structure of a spider joint receptor and associated synapses. Eur J Cell Biol 19:149–159

Foelix RF, Chu-Wang I-W (1973a) The morphology of spider sensilla I. Mechanoreceptors. Tissue & Cell 5:451–460

Foelix RF, Chu-Wang I-W (1973b) The morphology of spider sensilla. II. Chemoreceptors. Tissue & Cell 5(3):461–478

Foelix RF, Müller-Vorholt G, Jung H (1980) Organization of sensory leg nerves in the spider *Zygiella x-notata* (Clerck) (Araneae, Araneidae). Bull Br Arachnol Soc 5:20–28

Forster L (1985) Target discrimination in jumping spiders (Araneae: Salticidae). In: Barth FG (ed) Neurobiology of Arachnids. Springer, Berlin Heidelberg New York Tokyo, pp 249–274

Fossier P, Baux G, Tauc L (1990) Activation of protein kinase C by presynaptic FLRFamide receptors facilitates transmitter release at an *Aplysia* cholinergic synapse. Neuron 5:479–486

Fourtner CR, Evoy WH (1973) Nervous control of walking in the crab. *Cardisoma guanhumi*. IV. Effects of myochordotonal organ ablation. J Comp Physiol 83:319–329

Frank U (1957) Untersuchungen zur funktionellen Anatomie der lokomotorischen Extremitäten von *Zygiella x-notata*, einer Radnetzspinne. Zool Jahrb Abt Anat Ontog Tiere 76:423–460

French AS (1988) Transduction mechanisms of mechanosensilla. Ann Rev Entomol 33:39–58

Friedel T, Barth FG (1995) The response of interneurones in the spider CNS (*Cupiennius salei* Keys.) to vibratory courtship signals. J Comp Physiol A 177:159–171

Friedel T, Barth FG (1997) Wind-sensitive interneurones in the spider CNS (*Cupiennius salei*). Directional information processing of sensory inputs from trichobothria on the walking legs. J Comp Physiol A 180:223–233

Friedel T, Nentwig W (1989) Immobilizing and lethal effects of spider venoms on the cockroach and common meatbeetle. Toxicon 27:305–316

Friedrich OC (1998) Tasthaare bei Spinnen. Zur äußeren Morphologie, Biomechanik und Innervierung mechanorezeptiver Haarsensillen bei der Jagdspinne *Cupiennius salei* Keys. (Ctenidae). Diplomarbeit, Universität Wien

Frohlich C, Buskirk RE (1982) Transmission and attenuation of vibration in orb spider webs. J Theor Biol 95:13–36

Füller H, Eckert M, Blechschmidt K (1989) Distribution of GABA-like immunoreactive neurons in the optic lobes of *Periplaneta americana*. Cell Tissue Res 255:225–233

Full RJ, Herreid II CF (1983) Aerobic response to exercise of the fastest land crab. Am J Physiol 244:R530–R536

Full RJ, Herreid II CF (1984) Fiddler crab exercise: the energetic cost of running sideways. J Exp Biol 109:141–161

Gaffin DD, Brownell Ph (1997) Electrophysiological evidence of synaptic interactions within chemosensory sensilla of scorpion pectines. J Comp Physiol A 181:301–307

Gerhardt HC (1978) Temperature coupling in the vocal communication system of the gray tree frog, *Hyla versicolor*. Science 199:992–994

Gertsch WJ (1979) American Spiders (2nd ed) Van Nostrand Reinhold, New York

Gettrup E (1965) Sensory mechanisms in locomotion: The campaniform sensilla of the insect wing and their function during flight. Cold Spring Harb Symp quant Biol 30:587–599

Gettrup E (1966) Sensory regulation of wing twisting in locusts. J Exp Biol 44:1–16

Gingl E (1998) Nachweis der Rezeptoren für das Kontaktpheromon an der Seide der weiblichen Jagdspinne. Diplomarbeit, Universität Wien

Gitter AH, Klinke R (1989) Die Energieschwellen von Auge und Ohr in heutiger Sicht. Naturwissenschaften 76:160–164

Glasheen JW, McMahon TA (1996a) A hydrodynamic model of locomotion in the Basilisk lizard. Nature 380:340–342

Glasheen JW, McMahon TA (1996b) Size dependence of water – running ability in basilisk lizards (*Basiliscus basiliscus*). J Exp Biol 199:2611–2618

Glasheen JW, McMahon TA (1997) Running on water. Sci Am Sept 1997, 68–69

Gnatzy W, Heußlein R (1986) Digger wasp against crickets. I Receptors involved in the antipredator strategies of the prey. Naturwissenschaften 73:212–215

Gnatzy W, Kämper G (1990) Digger wasp against crickets: II. An airborne signal produced by a running predator. J Comp Physiol A 167:551–556

Gnatzy W, Schmidt K (1971) Die Feinstruktur der Sinneshaare auf den Cerci von *Gryllus bimaculatus* (Saltatoria, Gryllidae). Z Zellforsch 122:190–209

Gnatzy W, Tautz J (1980) Ultrastructure and mechanical properties of an insect mechanoreceptor: Stimulus-transmitting structures and sensory apparatus of the cercal filiform hairs of *Gryllus*. Cell Tissue Res 213:441–463

Goodman CS, Spitzer NC (1979) Embryonic development of identified neurons: differentiation from neuroblast to neuron. Nature, Lond 313:385–403

Gorb SN, Barth FG (1994) Locomotor behavior during prey-capture of a fishing spider *Dolomedes plantarius* (Araneae: Araneidae): galloping and stopping. J Arachnol 22, 89–93

Gorb SN, Barth FG (1996) A new mechanosensory organ on the anterior spinnerets of the spider *Cupiennius salei* (Araneae, Ctenidae). Zoomorphology 116:7–14

Gorb SN, Landolfa MA, Barth FG (1998) Dragline-associated behaviour of the orb web spider *Nephila clavipes* (Araneoidea, Tetragnathidae). J Zool Lond 244:323–330

Görner P (1958) Die optische und kinästhetische Orientierung der Trichterspinne *Agelena labyrinthica* (Cl.). Z vergl Physiol 41:111–153

Görner P (1962) Die Orientierung der Trichterspinne nach polarisiertem Licht. Z vergl Physiol 45:307–314

Görner P (1965) A proposed transducing mechanism for a multiply-innervated mechanoreceptor (trichobothrium) in spiders. Cold Spring Harbor Symp Quant Biol 30:69–73

Görner P (1973) Beispiele einer Orientierung ohne richtende Außenreize. Fortschr Zool 21:20–45

Görner P, Andrews P (1969) Trichobothrien, ein Ferntastsinnesorgan bei Webspinnen (Araneen). Z vergl Physiol 64:301–317

Görner P, Claas B (1985) Homing behavior and orientation in the funnel-web spider, *Agelena labyrinthica* Clerck. In: Barth FG (ed) Neurobiology of Arachnids. Springer, Berlin Heidelberg New York Tokyo, pp 275–297

Görner P, Zeppenfeld Ch (1980) The runs of *Pardosa amentata* (Araneae, Lycosidae) after removing its cocoon. Proc Int Congr Arachnol 8:243–248

Goyffon M, Drouet J, Francaz J-M (1980) Neurotransmitter aminoacids and spontaneous electrical activity of the prosomian nervous system of the scorpion. Comp Biochem Physiol C 66:59–64

Gracián B (1637) The Art of Worldly Wisdom. Shambhala Publications Inc, Boston 1993

Graeser K (1973) Die Übertragungseigenschaften des Netzes von *Zygiella x-notata* (Clerck) für transversale Sinusschwingungen im niederen Frequenzbereich und Frequenzanalyse beutetiererregter Netzvibrationen. Diplomarbeit, Universität Frankfurt am Main

Gramoll S (1990) Air stream stimulation of a spider's trichobothria and central nervous responses. In: Elsner N, Roth G (eds) Proc 18th Göttingen Neurobiol Conf. Thieme, Stuttgart New York, p 105

Greenstone MH (1982) Ballooning frequency and habitat predictability in two wolf spider species (Lycosidae: *Pardosa*). Florida Entomol 65:83–89

Grenacher H (1879) Untersuchungen über das Sehorgan der Arthropoden, insbesondere der Spinnen, Insecten und Crustaceen. Vandenhoeck & Ruprecht, Göttingen

Gronenberg W (1986) Physiological and anatomical properties of optical input-fibers to the mushroom body in the bee brain. J Insect Physiol 32:695–704

Gronenberg W (1989) Anatomical and physiological observations on the organization of mechanoreceptors and local interneurones in the central nervous system of the wandering spider *Cupiennius salei*. Cell Tissue Res 258:163–175

Gronenberg W (1990) The organization of plurisegmental mechanosensitive interneurons in the central nervous system of the wandering spider *Cupiennius salei*. Cell Tissue Res 260:49–61

Groome JR, Tillinghast EK, Townley MA, Vetrovs A, Watson WH III (1990) Identification of proctolin in the central nervous system of the horseshoe crab, *Limulus polyphemus*. Peptides 11:205–211

Groome JR, Townley MA, Tschaschell MD, Tillinghast EK (1991) Detection and isolation of proctolin-like immunoreactivity in arachnids: possible cardioregulatory role for proctolin in the orb-weaving spiders *Argiope* and *Araneus*. J Insect Physiol 37 (1):9–19

Grusch M (1994) Feinstruktur und Retinalgehalt der Augen von *Cupiennius salei* (Araneae, Ctenidae). Diplomarbeit, Universität Wien

Grusch M, Barth FG, Eguchi E (1997) Fine structural correlates of sensitivity in the eyes of the ctenid spider *Cupiennius salei* Keys. Tissue & Cell 29(4):421–430

Habermehl G (1976) Gift-Tiere und ihre Waffen. Springer, Berlin Heidelberg New York

Hagen H-O von (1967) Nachweis einer kinästhetischen Orientierung bei *Uca rapax*. Z Morphol Ökol Tiere 58:301–320

Hammer M, Menzel R (1995) Learning and memory in the honeybee. J Neuroscience 15:1617–1630

Hanström B (1921) Über die Histologie und vergleichende Anatomie der Sehganglien und Globuli der Araneen. Kgl Svenska Vet Akad Handl 61:1–39

Hanström B (1923) Further notes on the central nervous system of arachnids: scorpions, phalangids and trapdoor spiders. J Comp Neurol 35:249–272

Hanström B (1925) The olfactory centers in crustaceans. J Comp Neurol 38:221–250

Hanström B (1926) Untersuchungen über die relative Größe der Gehirnzentren verschiedener Arthropoden unter Berücksichtigung der Lebensweise. Z Mikr Anat Forsch 7:139–190

Hanström B (1928) Vergleichende Anatomie des Nervensystems der Wirbellosen Tiere. Springer, Berlin Heidelberg New York

Hanström B (1935) Fortgesetzte Untersuchungen über das Araneengehirn. Zool Jahrb Abt Ontog Tiere Anat 59:455–478

Happel J, Brenner H (1965) Low Reynolds number hydrodynamics with special applications to particulate media. Prentice Hall, New Jersey

Harris DJ (1977) Hair regeneration during moulting in the spider *Ciniflo similis* (Araneae, Dictynidae). Zoomorphology 88:37–63

Harris DJ, Mill PJ (1973) The ultrastructure of chemoreceptor sensilla in *Ciniflo* (Arachnida, Araneida). Tissue & Cell 5:679–689

Harris DJ, Mill PJ (1977a) Observations on the leg receptors of *Ciniflo* (Araneidae: Dictynidae). I. External mechanoreceptors. J Comp Physiol 119:37–54

Harris DJ, Mill PJ (1977b) Observations on the leg receptors of *Ciniflo* (Araneida: Dictynidae). II. Chemoreceptors. J Comp Physiol 119:55–62

Harris-Warrick RM, Flamm RE, Johnson BR, Katz PS (1989) Modulation of neuronal circuits in Crustacea. Amer Zool 29:1305–1320

Hartmann G, Wehner R (1995) The ant's path integration system: a neural architecture. Biol Cybern 73:483–497

Haug T (1986) Struktur, Funktion und Projektion der antennalen Thermo- und Hygrorezeptoren von *Antheraea pernyi* (Lepidoptera: Saturniidae). Dissertation, Universität Regensburg

Haupt J (1982) Hair regeneration in a solpugid chemotactile sensillum during moulting (Arachnida: Solifugae). Wilhelm Roux' Arch Entwicklungsmech Org 191:137–142

Haupt J (1996) Fine structure of the trichobothria and their regeneration during moulting in the whip scorpion *Typopeltis crucifer* Pocock, 1894. Acta Zoologica 77, 2:123–136

Hausen K (1984) The lobula-complex of the fly: structure, function and significance in visual behavior. In: Ali MA (ed) Photoreception and Vision in Invertebrates. Plenum, New York, pp 523–559

Hebets EA, Uetz GW (1999) Female responses to isolated signals from multimodal male courtship displays in the wolf spider genus *Schizocosa* (Araneae: Lycosidae). Anim Behav 57:865–872

Helsdingen PJ van (1965) Sexual behaviour of *Leptyphantes leprosus* (Ohl) (Araneida, Linyphiidae), with notes on the function of the genital organs. Zool Meded 41:15–42

Helversen D von, Helversen O von (1983) Species recognition and acoustic localization in acridid grasshoppers: a behavioral approach. In: Huber F, Markl H (eds) Neuroethology and Behavioral Physiology. Springer, Berlin Heidelberg New York Tokyo, pp 95–107

Helversen D von, Helversen O von (1987) Innate receiver mechanism in the acoustic communication of orthopteran insects. In: Guthrie DM (ed) Aims and Methods in Neuroethology. Manchester Univ Press, Manchester, pp 104–150

Helversen D von, Helversen O von (1990) Pattern recognition and directional analysis: routes and stations of information flow in the CNS of a grasshopper. In: Gribakin FG, Wiese K, Popov AV (eds) Sensory Systems and Communication in Arthropods. Birkhäuser, Basel Boston Berlin, pp 209–216

Hepburn HR, Joffe I (1976) On the material properties of insect exoskeletons. In: Hepburn HR (ed) The Insect Integument. Elsevier, Amsterdam New York, pp 207–235

Hergenröder R, Barth FG (1983a) The release of attack and escape behavior by vibratory stimuli in a wandering spider (*Cupiennius salei* Keys.). J Comp Physiol A 152:347–358

Hergenröder R, Barth FG (1983b) Vibratory signals and spider behavior: How do the sensory inputs from the eight legs interact in orientation? J Comp Physiol A 152:361–371

Herreid II CF (1981) Energetics of pedestrian arthropods. In: Herreid CF, Full CR (eds) Locomotion and Energetics in Arthropods. Plenum, New York London, pp 491–526

Herreid II CF, Full RJ (1984) Cockroaches on a treadmill: aerobic running. J Insect Physiol 30:395–403

Herreid II CF, Full RJ (1986) Energetics of hermit crabs during locomotion: the cost of carrying a shell. J Exp Biol 120:297–308

Herreid II CF, Prawel DA, Full RJ (1981) Energetics of running cockroaches. Science 212:331–333

Hill DE (1979) Orientation by jumping spiders of the genus *Phidippus* (Araneae: Salticidae) during the pursuit of prey. Behav Ecol Sociobiol 5:301–322

Hoffmann C (1967) Bau und Funktion der Trichobothrien von *Euscorpius carpathicus* L. Z vergl Physiol 54:290–352

Höger U (1994) Die postembryonale Entwicklung von Tasthaaren bei Spinnen. Diplomarbeit, Universität Frankfurt am Main

Höger U, Seyfarth E-A (1995) Just in the nick of time: postembryonic development of tactile hairs and of tactile behavior in spiders. Zoology (ZACS) 99:49–57

Höger U, Torkkeli PH, Seyfarth E-A, French AS (1997) Ionic selectivity of mechanically activated channels in spider mechanoreceptor neurons. J Neurophysiol 78:2079–2085

Holman GM, Nachman RJ, Wright MS (1990) Insect neuropeptides. Annu Rev Entomol 35:201–217

Homann H (1928) Beiträge zur Physiologie der Spinnenaugen. I. Untersuchungsmethoden. II. Das Sehvermögen der Salticiden. Z vergl Physiol 7:201–269

Homann H (1931) Beiträge zur Physiologie der Spinnenaugen. III. Das Sehvermögen der Lycosiden. Z vergl Physiol 14:40–67

Homann H (1961) Die Stellung der Ctenidae, Tectricinae und Thoicininae im System der Araneae. Senck biol 42:397–408

Homann H (1971) Die Augen der *Araneae*. Anatomie, Ontogenie und Bedeutung für die Systematik. Z Morphol Tiere 69:201–272

Homberg U (1984) Processing of antennal information in extrinsic mushroom body neurons in the bee brain. J Comp Physiol A 154:825–836

Homberg U, Christensen TA, Hildebrand JG (1989) Structure and function of the deutocerebrum in insects. Annu Rev Entomol 34:477–501

Honegger H-W, Brunniger B, Bräunig P, Elekes K (1990) GABA-like immunoreactivity in a common inhibitory neuron of the antennal motor system of crickets. Cell Tissue Res 260:349–354

Horch KW, Salmon M (1969) Production, perception and reception of acoustic stimuli by semiterrestrial crabs (genus *Ocypode* and *Uca*, family Ocypodidae). forma et functio 1:1–25

Horch KW, Salmon M (1971) Responses of the ghost crab *Ocypode* to acoustic stimuli. Z Tierpsychol 30:1–13

Hoyle G (1983) Forms to modulate tension in skeletal muscles. Comp Biochem Physiol 76A:203–210

Hoyle G (1985) Neurotransmitters, neuromodulators and neurohormones. In: Gilles R, Balthazart J (eds) Neurobiology. Springer, Berlin Heidelberg, pp 264–279

Huber KC, Haider THS, Müller MW, Huber BA, Schweyen RJ, Barth FG (1993) DNA-sequence data indicates the polyphyly of the family Ctenidae (Araneae). J Arachnol 21:194–201

Hudspeth AJ (1985) The cellular basis of hearing: the biophysics of hair cells. Science 230:745–752

Hudspeth AJ (1989) How the ear's works work. Nature 341:397–404

Hughes GM (1958) The co-ordination of insect movements. III. Swimming in *Dytiscus*, *Hydrophylus* and a dragonfly nymph. J Exp Biol 35:567–583

Humphrey JAC (1987) Fluid mechanic constraints on spider ballooning. Oecologia 73:469–477

Humphrey JAC, Devarakonda R, Iglesias I, Barth FG (1993) Dynamics of arthropod filiform hairs. I. Mathematical modelling of the hair and air motions. Phil Trans R Soc Lond B 340:423–444

Humphrey JAC, Devarakonda R, Iglesias I, Barth FG (1998) Errata re. Humphrey et al (1993). Phil Trans R Soc Lond B 352:1995

Humphreys WF (1975) The influence of burrowing and thermoregulatory behaviour on the water relations of *Geolycosa godeffroyi* (Araneae: Lycosidae), an Australian wolf spider. Oecologia 21:291–311

Jackson RR, Blest AD (1982) The distance at which a primitive jumping spider, *Portia fimbriata*, makes visual discriminations. J Exp Biol 97:441–445

Jander P (1970) Ein Ansatz zur modernen Elementarbeschreibung der Orientierungshandlung. Z Tierpsychol 27:771–778

Jensen TF, Holm-Jensen I (1980) Energetic cost of running in workers of three ant species, *Formica fusca* L., *Formica rufa* L., and *Camponotus herculeanus* L. (Hymenoptera, Formicidae). J Comp Physiol A 137:151–156

Juusola M, Seyfarth E-A, French AS (1994) Sodium-dependent receptor current in a new mechanoreceptor preparation. J Neurophysiol 72:3026–3028

Kadel M (1992) Zentralnervöse Korrelate lokaler und plurisegmentaler Muskelreflexe bei Spinnen, physiologische und morphologische Identifizierung von Einzelneuronen. Dissertation, Universität Frankfurt am Main

Kadel M, Seyfarth E-A (1989) Body raising in spiders: neuroethology of a „simple" behavior. In: Erber J, Menzel R, Pflüger H-J, Todt D (eds) Neural Mechanisms of Behavior. Proc 2nd Intern Congr Neuroethology. Thieme, Stuttgart New York, p 32

Kadel M, Seyfarth E-A (1991) Neuronal basis of local and plurisegmental leg reflexes in spiders. In: Elsner N, Penzlin H (eds) Synapse – Transmission, Modulation. Proc 19th Göttingen Neurobiology Conference. Thieme, Stuttgart New York, p 60

Kadel M, Bickelböller C, Seyfarth E-A (1989) Local and plurisegmental correlates of reflex activity in spiders. In: Elsner N, Singer W (eds) Dynamics and Plasticity in Neuronal Systems. Proc 17th Göttingen Neurobiology Conference. Thieme, Stuttgart New York, p 118

Kaestner A (1924) Beiträge zur Kenntnis der Lokomotion der Arachniden. I. Araneae. Arch Naturgesch 90A:1–19

Kaissling K-E, Thorson J (1980) Insect olfactory sensilla: structural, chemical and electrical aspects of the functional organization. In: Hall LM, Hildebrand JG, Satelle DB (eds) Receptors for Neurotransmitters, Hormones and Pheromones in Insects. Elsevier, Amsterdam Oxford New York, pp 261–282

Kalmring K (1983) Convergence of auditory and vibratory senses at the neuronal level of the ventral cord in grasshoppers, its possible importance for behaviour in the habitat. In: Horn E (ed) Multimodal Convergences in Sensory Systems. Fischer, Stuttgart New York, pp 129–141

Kalmring K, Lewis B, Eichendorf A (1978) The physiological characteristics of the primary sensory neurons of the complex tibial organ of *Decticus verrucivorus* L (Orthoptera, Ensifera). J Comp Physiol 127:109–121

Kämper G, Kühne R (1983) The acoustic behavior of the bushcricket *Tettigonia cantans*. II. Transmission of airborne sound and vibration signals in the biotope. Behav Proc 8:125–145

Kämper G, Kleindienst H-U (1990) Oscillation of cricket sensory hairs in a low frequency sound field. J Comp Physiol A 167:193–200

Kanou M, Osawa T, Shimozawa T (1989) Mechanical polarization in the air-current sensory hair of a cricket. Experientia 45:1082–1083

Kanzaki R, Arbas EA, Strausfeld NJ, Hildebrand JG (1989) Physiology and morphology of projection neurons in the antennal lobe of the male moth *Manduca sexta*. J Comp Physiol A 165:427–453

Kaps F (1994) Retinabewegungen bei *Cupiennius salei* (Araneae, Ctenidae). Diplomarbeit, Universität Wien

Kaps F (1998) Anatomische und physiologische Untersuchungen zur Funktion der Retinabewegungen bei *Cupiennius salei* (Araneae, Ctenidae). Dissertation, Universität Wien

Kaps F, Schmid A (1996) Mechanism and possible behavioural relevance of retinal movements in the ctenid spider *Cupiennius salei*. J Exp Biol 199:2451–2458

Karner Ch, Barth FG (1994) Vibrations in the spider web. I. The architecture of a spiders's orb web. 9th Int Meeting on Insect Sound and Vibration, Seggau: 48

Kaston BJ (1935) The slit sense organs of spiders. J Morphol 58:189–209

Kaulen P, Erber J, Mobbs P (1984) Current source-density analysis in the mushroom bodies of the honey bee (*Apis mellifera carnica*). J Comp Physiol A 154:569–582

Keil St (1988) DMS-Produktion in Darmstadt. Eine historische Rückschau aus Anlaß des 50jährigen Jubiläums des Dehnungsmeßstreifens. Mechtech Briefe MTB 24, 2:43–52; Hottinger-Baldwin Meßtechnik GmbH, Darmstadt

Keller LR (1961) Untersuchungen über den Geruchssinn der Spinnenart *Cupiennius salei* Keyserling. Z vergl Physiol 44:576–612

Kenyon FC (1896) The meaning and structure of the so-called "mushroom bodies" of the hexapod brain. Amer nat 30:643–650

Keyserling E (1877) Über amerikanische Spinnenarten der Unterordnung Citigradae. Verh zool-bot Ges Wien 26:609–708

Kirschfeld K (1976) The resolution of lens and compound eyes. In: Zettler F, Weiler R (eds) Neural Principles in Vision. Springer, Berlin Heidelberg New York Tokyo, pp 354–370

Klärner D, Barth FG (1982) Vibratory signals and prey capture in orbweaving spiders (*Zygiella x-notata, Nephila clavipes*; Araneidae). J Comp Physiol A 148:445–455

Klemm N (1985) The distribution of biogenic monoamines in invertebrates. In: Gilles R, Balthazart J (eds) Neurobiology. Springer, Berlin Heidelberg, pp 280–296

Kloppenburg P, Erber J (1989) Octopamine and serotonin modulate the activity of motionsensitive neurons in the honeybee. In: Erber J, Menzel R, Pflüger HJ, Todt D (eds) Neural Mechanisms of Behavior. Georg Thieme, Stuttgart New York, p 235

Kobayashi M, Muneoka Y (1989) Functions, receptors and mechanisms of the FMRFamide-related peptides. Biol Bull 177:206–209

Kondoh Y, Hisada M (1987) The topological organization of primary afferents in the terminal ganglion of crayfish, *Procambarus clarkii*. Cell Tissue Res 247:17–24

Kosok G (1993) Gefäße und Hämolymphversorgung des zentralen Nervensystems von *Cupiennius salei* Keys. (Chelicerata, Araneae). Diplomarbeit, Universität Frankfurt am Main

Krafft B, (1978) The recording of vibratory signals performed by spiders during courtship. Symp Zool Soc Lond 42:59–67

Kravitz EA (1991) Hormonal orchestration of behavior: amines and the biasing of behavioral output in lobsters. In: Elsner N, Penzlin H (eds) Synapse - Transmission, Modulation. Georg Thieme, Stuttgart New York, pp 141–153

Kuhn-Nentwig L, Schaller J, Nentwig W (1994) Purification of toxic peptides and the amino acid sequence of CSTX-I from the multicomponent venom of *Cupiennius salei* (Araneae, Ctenidae). Toxicon 32:287–302

Kuhn-Nentwig L, Schaller J, Streb B, Kämpfer U, Nentwig W (1998) CSTX-4, a novel bactericidal and insecticidal peptide in the venom of *Cupiennius salei*. Toxicon 36:1276–1277

Küppers J (1974) Measurements of the ionic milieu of the receptor terminal in mechanoreceptive sensilla of insects. Rheinisch-Westfäl Akad Wiss 53:387–394

Labhart T, Nilsson D-E (1995) The dorsal eye of the dragonfly *Sympetrum*: Specializations for prey detection against the sky. J Comp Physiol A 176:437–453

Lachmuth U, Grasshoff M, Barth FG (1984) Taxonomische Revision der Gattung *Cupiennius* SIMON 1891 (Arachnida - Araneae - Ctenidae). Senck biol 65 (3/6):329–372

Land MF (1969a) Structure of the retinae of the principal eyes of jumping spiders (*Salticidae: Dendryphantinae*) in relation to visual optics. J Exp Biol 51:443–470

Land MF (1969b) Movements of the retinae of jumping spiders (Salticidae: Dendryphantidae) in response to visual stimuli. J Exp Biol 51:471–493

Land MF (1971) Orientation by jumping spiders in the absence of a visual feedback. J Exp Biol 54:119–139

Land MF (1972) Mechanisms of orientation and pattern recognition in jumping spiders (Salticidae). In: Wehner R (ed) Information Processing in the Visual Systems of Arthropods. Springer, Berlin Heidelberg New York Tokyo, pp 231–247

Land MF (1981) Optics and vision in invertebrates. In: Autrum H (ed) Handbook of Sensory Physiology vol. VII/6B. Springer, Berlin Heidelberg New York Tokyo, pp 471–592

Land MF (1985) The morphology and optics of spider eyes. In: Barth FG (ed) Neurobiology of Arachnids. Springer, Berlin Heidelberg New York Tokyo, pp 53–78

Land MF (1997) Visual acuity in insects. Ann Rev Entomol 42:147–177

Land MF, Barth FG (1992) The quality of vision in the ctenid spider *Cupiennius salei*. J Exp Biol 164:227–242

Landolfa MA, Barth FG (1996) Vibrations in the orb web of the spider *Nephila clavipes*. Cues for discrimination and orientation. J Comp Physiol A 179:493–508

Landolfa MA, Jacobs GA (1995) Direction sensitivity of the filiform hair population of the cricket cercal system. J Comp Physiol A 177:759–766

Landolfa MA, Miller JP (1995) Stimulus-response properties of cricket cercal filiform receptors. J Comp Physiol A 177:749–757

Lang HH (1980a) Surface wave discrimination between prey and nonprey by the back swimmer *Notonecta glauca* L. (Hemiptera, Heteroptera). Behav Ecol Sociobiol 6:233–246

Lang HH (1980b) Surface wave sensitivity of the back swimmer *Notonecta glauca*. Naturwissenschaften 67:204–205

Laughlin S, Blest AD, Stowe S (1980) The sensitivity of receptors in the posterior median eye of the nocturnal spider, *Dinopis*. J Comp Physiol A 141:53–65

Lax E, Synowietz C (1967) D'Ans-Lax Taschenbücher für Chemiker und Physiker, Vol 1. Springer, Berlin Heidelberg New York

Lee HM, Wyse GA (1991) Immunocytochemical localization of octopamine in the central nervous system of *Limulus polyphemus*: A light and electron microscopic study. J Comp Neurol 307:683–694

Lehtinen PT (1980) Trichobothrial patterns in high level taxonomy of spiders. Proc 8th International Conf Arachnol. Egermann, Wien, pp 493–498

Lewis ER, Narins PM (1985) Do frogs communicate with seismic signals? Science 215:187–189

Leydig F (1855) Zum feineren Bau der Arthropoden. Arch Anat Physiol 376–480

Liesenfeld FJ (1956) Untersuchungen am Netz und über den Erschütterungssinn von *Zygiella x-notata* (Cl.) (Araneidae). Z vergl Physiol 38:563–592

Liesenfeld FJ (1961) Über Leistung und Sitz des Erschütterungssinnes von Netzspinnen. Biol Zbl 80:465–475

Linzen B, Gallowitz P (1975) Enzyme activity patterns in muscles of the lycosid spider, *Cupiennius salei.* J Comp Physiol 96:101–109

Linzen B, Angersbach D, Loewe R, Markl J, Schmid R (1977) Spider hemocyanins: Recent advances in their structure and function. In: Bannister JV (ed) Structure and Function of Haemocyanin. Springer, Berlin Heidelberg New York, pp 31–36

Linzen B et al (1985) The structure of arthropod hemocyanins. Science 229:519–524

Loewe R, Linzen B (1975) Hemocyanins in spiders II. Automatic recording of oxygen binding curves, and the effect of Mg^{++} on oxygen affinity, cooperativity and subunit association of *Cupiennius salei* hemocyanins. J Comp Physiol 98:147–156

Loewe R, Linzen B, Stackelberg W (1970) Die gelösten Stoffe in der Hämolymphe einer Spinne, *Cupiennius salei* (Keyserling). Z vergl Physiol 98:147–156

Loftus R (1968) Response of the antennal cold receptor of *Periplaneta americana* to rapid temperature changes and to steady temperature. Z vergl Physiol 59:413–455

Loftus R (1969) Differential thermal components in the response of the antennal cold receptor of *Periplaneta americana* to slowly changing temperature. Z vergl Physiol 63:415–433

Loftus R (1976) Temperature-dependent dry receptor on antenna of *Periplaneta.* Tonic response. J Comp Physiol 111:153–170

Loftus R, Corbière-Tichané G (1981) Antennal warm and cold receptors of the cave beetle, *Speophyes lucidulus* Delar., in sensilla with a lamellated dendrite. I. Response to sudden temperature change. J Comp Physiol 143:443–452

Loftus R, Corbière-Tichané G (1987) Response of antennal cold receptors of the catopid beetles, *Speophyes lucidulus* Delar. and *Choleva angustata* Fab. to very slowly changing temperature. J Comp Physiol A 161:399–405

Lorenz K (1943) Die angeborenen Formen möglicher Erfahrung. Z Tierpsychol 5:235–409

Lorenz K (1973) Die Rückseite des Spiegels. Piper, München

Macmillan DL (1975) A physiological analysis of walking in the American lobster *Homarus americanus.* Phil Trans R Soc (B) 270:1–59

Magni F, Papi F, Savely HE, Tongiorgi P (1962) Electroretinographic responses to polarized light in the wolf spider *Arctosa variana* C.L. Koch. Experientia 18:1–3

Magni F, Papi F, Savely HE, Tongiorgi P (1965) Research on the structure and physiology of the eyes of a lycosid spider. III. Electroretinographic responses to polarized light. Arch Ital Biol 103:146–158

Malli H, Vapenik Z, Nentwig W (1993) Ontogenetic changes in the toxicity of the venom of the spider *Cupiennius salei* (Araneae, Ctenidae). Zool Jb Physiol 97:113–122

Manira AE, Rossi-Durand C, Clarac F (1991) Serotonin and proctolin modulate the response of a stretch receptor in crayfish. Brain Research 541:157–162

Mann DW, Chapman KM (1975) Component mechanism of sensitivity and adaptation in an insect mechanoreceptor. Brain Res 97:331–336

Maretić Z (1987) Spider venoms and their effect. In: Nentwig W (ed) Ecophysiology of Spiders. Springer, Berlin Heidelberg New York London Paris Tokyo, pp 142–159

Markl H (1962) Borstenfelder an den Gelenken als Schweresinnesorgane bei Ameisen und anderen Hymenopteren. Z vergl Physiol 45:475–569

Markl H (1969) Verständigung durch Vibrationssignale bei Arthropoden. Naturwissenschaften 56:499–505

Markl H (1983) Vibrational communication. In: Huber F, Markl H (eds) Neuroethology and Behavioral Physiology. Springer, Berlin Heidelberg New York, pp 332–353

Markl J (1980) Hemocyanins in spiders XI. The quaternary structure of *Cupiennius* hemocyanins. J Comp Physiol 140:199–207

Masters WM (1984a) Vibrations in the orb webs of *Nuctenea sclopetaria* (Araneidae). I. Transmission through the web. Behav Ecol Sociobiol 15:207–215

Masters WM (1984b) Vibrations in the orb webs of *Nuctenea sclopetaria* (Araneidae). II. Prey and wind signals and the spider's response threshold. Behav Ecol Sociobiol 15:217–223

Masters WM, Markl H (1981) Vibration signal transmission in spider orb-webs. Science 213:363–365

Masters WM, Markl H, Moffat AJM (1985) Transmission of vibrations in a spider's web. In: Shear WA (ed) Spiders: Webs, Behavior, and Evolution. Stanford Univ Press, Stanford CA, pp 49–69

Maxworthy T (1981) The fluid dynamics of insect flight. A Rev Fluid Mech 13:329–350

McClintock WJ, Uetz GW (1996) Female choice and pre-existing bias: visual cues during courtship in two *Schizocosa* wolf spiders (Araneae: Lycosidae). Anim Behav 52:167–181

McIndoo NE (1911) The lyriform organs and tactile hairs of araneids. Proc Acad Nat Sci Philadelphia 63:375–418

Meinertzhagen IA (1975) The organization of perpendicular fibre pathways in the insect optic lobe. Phil Trans R Soc Lond B 274:555–596

Melamed J, Trujillo-Cenóz O (1966) On the fine structure of the visual system of *Lycosa* (*Araneae, Lycosidae*). I. Retina and optic nerve. Z Zellforsch Anat 74:12–31

Melamed J, Trujillo-Cenóz O (1971) Innervation of the retinal muscles in wolfspiders. J Ultrastruct Res 35:359–369

Melchers M (1963a) *Cupiennius salei* (Ctenidae) – Kokonbau und Eiablage. Encyclop Cinematogr, Film E 363

Melchers M (1963b) Zur Biologie und zum Verhalten von *Cupiennius salei* (Keyserling), einer amerikanischen Ctenide. Zool Jb Syst 91:1–90

Melchers M (1967) Der Beutefang von *Cupiennius salei* Keyserling (Ctenidae). Z Morph Ökol Tiere 58:321–346

Menzel R (1979) Spectral sensitivity and color vision in invertebrates. In: Autrum H (ed) Handbook of Sensory Physiology vol VII/6. Springer, Berlin Heidelberg New York Tokyo, pp 503–580

Meßlinger (1987) Fine structure of scorpion trichobothria (Arachnida, Scorpiones). Zoomorphology 107:49–57

Meyer EP, Matute C, Streit P, Nässel DR (1986) Insect optic lobe neurons identifiable with monoclonal antibodies to GABA. Histochemistry 84:207–216

Meyer W, Poehling HM (1987) Regional distribution of putative amino acid neurotransmitters in the CNS of spiders (Arachnida: Araneida). Neurochem Int 11:241–246

Meyer W, Poehling HM, Neuhoff V (1980) Comparative aspects of free amino acids in the central nervous system of spiders. Comp Biochem Physiol 67:83–86

Meyer W, Schlesinger C, Poehling HM, Ruge W (1984) Comparative quantitative aspects of putative neurotransmitters in the central nervous system of spiders (Arachnida: Araneida). Comp Biochem Physiol 78:357–362

Michelsen A, Fink F, Gogala M, Traue D (1982) Plants as transmission channels for insect vibrational songs. Behav Ecol Sociobiol II:269–281

Milde JJ, Seyfarth E-A (1988) Tactile hairs and leg reflexes in wandering spiders: physiological and anatomical correlates of reflex activity in the leg ganglia. J Comp Physiol A 162:623–631

Mill PJ, Harris DJ (1977) Observations on the leg receptors of *Ciniflo* (Araneida, Dictynidae). III. Proprioreceptors. J Comp Physiol 119:63–72

Miller JP, Jacobs GA, Theunissen FE (1991) Representation of sensory information in the cricket cercal system. I. Response properties of the primary interneurones. J Neurophysiol 66:1680–1689

Millot J (1931) Les glandes venimeuses des aranéides. Anm Sci Nat Zool 10:113–145

Millot J, Vachon M (1949) Ordre des scorpions. In: Grassé T (ed) Traîté de Zoologie, vol VI. Masson, Paris, pp 386–436

Mittelstaedt H (1985) Analytical cybernetics of spider navigation. In: Barth FG (ed) Neurobiology of Arachnids. Springer, Berlin Heidelberg New York Tokyo, pp 298–316

Mittelstaedt H, Eggert T (1989) How to transform topographically ordered spatial information into motor commands. In: Arbb M, Ewert J-P (eds) Visuomotor Coordination: Amphibians, Comparisons, Models, and Robots. Plenum, New York, pp 569–585

Mittelstaedt H, Mittelstaedt M-L (1973) Mechanismen der Orientierung ohne richtende Außenreize. Fortschr Zool 21:46–58

Mittelstaedt M-L, Mittelstaedt H (1980) Homing by path integration in a mammal. Naturwissenschaften 67:566

Mitter E (1994) Sitzplatzpflanzenwahl bei der mittelamerikanischen Jagdspinne *Cupiennius* (Araneae). Dissertation, Universität Wien

Mizunami M, Weibrecht J, Strausfeld NJ (1992) A new role for the insect mushroom bodies: place memory and motor control. In: Beer RD, Ritzmann R, McKenna T (eds) Biological Networks in Invertebrate Neuroethology and Robotics. Academic Press, Cambridge, pp 199–225

Müller M, Wehner R (1994) The hidden spiral: systematic search and path integration in desert ants, *Cataglyphis fortis*. J Comp Physiol A 175:525–530

Murphey RK (1973) Mutual inhibition and the organisation of a nonvisual orientation in *Notonecta*. J Comp Physiol A 84:31–69

Murphey RK, Possidente D, Pollack G, Merritt DJ (1989) Modality-specific axonal projections in the CNS of the flies *Phormia* and *Drosophila*. J Comp Neurol 290:185–200

Nachtigall W (1982) Biostatik. In: Hoppe W, Lohmann W, Markl H, Ziegler H (eds) Biophysik. Springer, Berlin Heidelberg New York Tokyo, pp 632–639

Narins PM (1995) Frog communication. Sci Am 273(2):62–67

Narins PM, Lewis ER (1984) The vertebrate ear as an exquisite seismic sensor. J Acoust Soc Am 76(5):1384–1387

Nässel DR (1988) Serotonin and serotonin-immunoreactive neurons in the nervous system of insects. Prog Neurobiol 30:1–85

Nässel DR (1991) Neurotransmitters and neuromodulators in the insect visual system. Prog Neurobiol 37:179–254

Nässel DR, Pirvola U, Panula P (1990) Histamine-like immunoreactive neurons innervating putative neurohaemal areas and central neuropil in the thoraco-abdominal ganglia of the flies *Drosophila* and *Calliphora*. J Comp Neurol 297:525–536

Nässel DR, Waterman TH (1979) Massive, diurnally modulated photoreceptor membrane turnover in crab light and dark adaptation. J Comp Physiol A 131:205–216

Nentwig W (1986) Non-webbuilding spiders: prey specialists or generalists? Oecologia 69:571–576

Nentwig W, Friedel T, Manhart C (1992) Comparative investigations on the effect of the venoms of 18 spider species onto the cockroach *Blatta orientalis* (Blattodea). Zool Jb Physiol 96:279–290

Nicklaus R (1965) Die Erregung einzelner Fadenhaare von *Periplaneta americana* in Abhängigkeit von der Größe und Richtung der Auslenkung. Z vergl Physiol 50:331–362

Nishikawa M, Yokohari F, Ishibashi T (1991) Deutocerebral interneurons responding to thermal stimulation on the antennae of the cockroach, *Periplaneta americana* L. Naturwissenschaften 78:563:565

Nishikawa M, Yokohari F, Ishibashi T (1992) Response characteristics of two types of cold receptors on the antennae of the cockroach, *Periplaneta americana* L. J Comp Physiol A 171:299–307

O'Shea M (1985) Are skeletal motoneurones in arthropods peptidergic? In: Selverston AI (ed) Model Neural Networks and Behavior. Plenum Press, New York London, pp 401–413

Oldfield BP (1988) Tonotopic organization of the insect auditory pathway. Trends Neurosci 11:267–270

Orchard I (1982) Octopamine in insects: neurotransmitter, neurohormone and neuromodulator. Can J Zool 60:659

Orchard I, Loughton BG (1981) Is octopamine a transmitter mediating hormone released in insects? J Neurobiol 12:143–153

Orchard I, Belanger JH, Lange AB (1989) Proctolin: A review with emphasis on insects. J Neurobiol 20 (5):470–496

Pabst H (1965) Elektrophysiologische Untersuchung des Streckreceptors am Flügelgelenk der Wanderheuschrecke *Locusta migratoria*. Z vergl Physiol 50:498–541

Palmgren A (1948) A rapid method for selective silver staining of nerve fibres and nerve endings in mounted paraffin sections. Acta Zool 29:377–392

Palmgren P (1936) Experimentelle Untersuchungen über die Funktion der Trichobothrien bei *Tegenaria derhami* Scop. Acta Zool Fenn 19:3–27

Palmgren P (1939) Ökologische und physiologische Untersuchungen über die Spinne *Dolomedes fimbriatus* (Cl.). Acta Zool Fennica 24:1–42

Palmgren P (1978) On the muscular anatomy of spiders. Acta Zool Fennica 155:1–41

Parry DA (1954) On the drinking of soil capillary water by spiders. J Exp Biol 31:218–227

Parry DA (1957) Spider leg-muscles and the autotomy mechanism. Q J Microsc Sci 98(3):331–340

Parry DA (1960) The small leg-nerve of spiders and a probable mechanoreceptor. Q J Microsc Sci 101:1–8

Pass G, Sperk G, Agricola H, Baumann E, Penzlin H (1988) Octopamine in a neurohaemal area within the antennal heart of the American cockroach. J Exp Biol 135:495–498

Pasztor VM, Macmillan DL (1990) The actions of proctolin, octopamine and serotonin on crustacean proprioreceptors show species and neurone specificity. J Exp Biol 152:485–504

Paul RJ (1991) Oxygen transport from book lungs to tissues – environmental physiology and metabolism of arachnids. Verh Dtsch Zool Ges 84:9–14

Paul RJ, Fincke T (1989) Book lung function in arachnids II. Carbon dioxide release and its relations to respiratory surface, water loss and heart frequency. J Comp Physiol B 159:419–432

Paul RJ, Fincke T, Linzen B (1989) Book lung function in arachnids I. Oxygen uptake and respiratory quotient during rest, activity and recovery – relations to gas transport in the hemolymph. J Comp Physiol B 159:409–418

Paul RJ, Bergner B, Pfeffer-Seidl A, Decker A, Efinger R, Storz H (1994a) Gas transport in the hemolymph of arachnids I. Oxygen transport and the physiological role of hemocyanin. J Exp Biol 188:25–46

Paul RJ, Colmorgen M, Hüller S, Tyroller F, Zinkler D (1994b) Optophysiologische Untersuchung der Atmungsregulation bei kleinen Metazoen (*Daphnia magna, Folsomia candida, Tubifex tubifex*). Verh Dtsch Zool Ges, p 188

Paul RJ, Pfeffer-Seidl A, Efinger R, Pörtner HO, Storz H (1994c) Gas transport in the hemolymph of arachnids II. Carbon dioxide transport and acid-base balance. J Exp Biol 188:47–63

Pauls M, Schürmann FW (1993) Structural evidence for synaptic interactions in tracts of insect ganglia. In: Elsner N, Heisenberg M (eds) Gene-Brain-Behaviour. Thieme, Stuttgart New York, p 522

Paulus HF (1979) Eye structure and the monophyly of the arthropoda. In: Gupta AP (ed) Arthropod Phylogeny. Van Nostrand Reinhold, New York, pp 299–383

Pearson KG (1972) Central programming and reflex control of walking in the cockroach. J Exp Biol 56:173–193

Penzlin H (1989) Neuropeptides – occurrence and functions in insects. Naturwissenschaften 76:243–252

Peters W, Pfreundt C (1986) Die Verteilung von Trichobothrien und lyraförmigen Organen an den Laufbeinen von Spinnen mit unterschiedlicher Lebensweise. Zool Beitr N F 29:209–225

Peterson RE (1966) Stress Concentration Design Factors. John Wiley & Sons Inc, New York London Sydney

Pfeiffer-Linn C, Glantz RM (1991) GABA-mediated inhibition of visual interneurons in the crayfish medulla. J Comp Physiol A 168:373–381

Pflüger H-J, Burrows M (1978) Locusts use the same motor pattern in swimming as in jumping and kicking. J Exp Biol 75:81–93

Pflüger HJ, Bräunig P, Hustert R (1981) Distribution and central projections of mechanoreceptors in the thorax and proximal leg joints of locusts. II The external mechanoreceptors: hair plates and tactile hairs. Cell Tissue Res 216:79–96

Pickard-Cambridge FO (1897–1905) Biologia Centrali-Americana. Arachnida, Araneida and Opiliones 2:1–610

Pickles JO, Corey DP (1992) Mechanotransduction by hair cells. Trends in Neurosci 15:254–259

Pittendrigh CS (1960) Circadian rhythms and the circadian organization of living systems. Cold Spring Harbor Symp Quant Biol 25:159–184

Pollack I, Hofbauer A (1991) Histamine-like immunoreactivity in the visual system and brain of *Drosophila melanogaster*. Cell Tissue Res 266:391–396

Pollard SD, Macnab AM, Jackson RR (1987) Communication with chemicals: pheromones and spiders. In: Nentwig W (ed) Ecophysiology of Spiders. Springer, Berlin Heidelberg New York London Paris Tokyo, pp 133–141

Porterfield W (1759) A treatise on the eye. The manner and phenomena of vision. Edinburgh

Prell GD, Green JP (1986) Histamine as a neuroregulator. Ann Rev Neurosci 9:209–254

Pringle JWS (1938) Proprioception in insects. III. The function of the hair sensilla at the joints. J Exp Biol 15:467–473

Pringle JWS (1955) The function of the lyriform organs of arachnids. J Exp Biol 32:270–278

Pringle JWS (1963) The proprioceptive background to mechanisms of orientation. Ergebnisse der Biologie 26:1

Pulz R (1986) Patterns of evaporative water loss in tarantulas: transpiration and secretion (Araneae, Theraphosidae). Proc 9th Int Congr Arachnol Panama 1983, pp 197–201

Pulz R (1987) Thermal and water relations. In: Nentwig W (ed) Ecophysiology of Spiders. Springer, Berlin Heidelberg London Paris Tokyo, pp 26–55

Rádl E (1912) Neue Lehre vom Zentralen Nervensystem. Engelmann, Leipzig

Rane SG, Gerlach PH, Wyse GA (1984) Neuromuscular modulation in *Limulus* by both octopamine and proctolin. J Neurobiol 15:207–220

Rathmayer W (1966) Die Innervation der Beinmuskeln einer Spinne, *Eurypelma hentzi* Chamb. (Orthognatha, Aviculariidae). Verh Dtsch Zool Ges, Fischer, Stuttgart 1965:505–511

Rathmayer W (1967) Elektrophysiologische Untersuchungen an Propriorezeptoren im Bein einer Vogelspinne (*Eurypelma hentzi* Chamb.). Z vergl Physiol 54:438–454

Rathmayer W, Koopmann J (1970) Die Verteilung der Propriorezeptoren im Spinnenbein. Untersuchungen an der Vogelspinne *Dugesiella hentzi* Chamb. Z Morphol Tiere 66:212–223

Reißland A, Görner P (1978) Mechanics of trichobothria in orb-weaving spiders (Agelenidae; Araneae). J Comp Physiol 123:59–69

Reißland A, Görner P (1985) Trichobothria. In: Barth FG (ed) Neurobiology of Arachnids. Springer, Berlin, pp 138–161

Richter CJJ (1970) Aerial dispersal in relation to habitat in eight wolf spider species (*Pardosa*: Araneae: Lycosidae). Oecologia 5:200–214

Richter CJJ (1971) Some aspects of aerial dispersal in different populations of wolf spiders, with particular reference to *Pardosa amentata* (Araneae: Lycosidae). Misc Papers, Landbouwhogeschool. Wageningen 8:77–88

Richter K, Stürzelbecher J (1971) Zur Wirkung von Neurohormon D aus *Periplaneta americana* (L.) (Insecta) auf das Herz und das Herzganglion von *Tegenaria atrica* CL. Koch und *Coelotes atropos* Walckenaer (Arachnida-Araneae). Zool Jb Physiol 76:64–79

Rick R, Barth FG, Pawel A von (1976) X-ray microanalysis of receptor lymph in a cuticular arthropod sensillum. J Comp Physiol 110:89–95

Ridgel AL, Frazier SF, DiCaprio RA, Zill SN (2000) Encoding of forces by cockroach tibial campaniform sensilla: implication in dynamic control of posture and locomotion. J Comp Physiol A 186:359–374

Riechert SE (1985) Decisions in multiple goal contexts: habitat selection of the spider, *Agelenopsis aperta* (Gertsch). Z Tierpsychol 70:53–69

Ritzmann RE (1993) The neuronal organization of cockroach escape and its role in context dependent orientation. In: Beer RD et al (eds) Biological Neural Networks in Invertebrate Neuroethology and Robotics. Academic Press, Boston, pp 113–137

Ritzmann RE, Pollack AJ (1990) Parallel motor pathways from thoracic interneurones of the ventral giant interneurone system of the cockroach, *Periplaneta americana*. J Neurobiol 21:1219–1235

Robinson MH (1980) The ecology and behavior of tropical spiders. In: Gruber J (ed) Proc 8th Int Arachnol Congr Vienna. Egermann, Wien, pp 13–32

Robinson MH (1982) The ecology and biogeography of spiders in Papua New Guinea. In: Gressitt JL (ed) Monogr Biol 42:557–581. Junk, The Hague

Robinson MH, Mirick H (1971) The predatory behavior of the golden-web spider *Nephila clavipes* (Araneae: Araneidae). Psyche 78:123–139

Robinson MH, Robinson B (1976) A tipulid associated with spider webs in Papua New Guinea. Entomol Mon Mag 112:1–4

Roemer van de A (1980) Eine vergleichende morphologische Untersuchung an dem für die Vibrationswahrnehmung wichtigen Distalbereich des Spinnenbeins. Diplomarbeit, Universität Frankfurt am Main

Roland C, Rovner JS (1983) Chemical and vibratory communication in the aquatic pisaurid spider *Dolomedes triton* (Araneae: Pisauridae). J Arachnol 11:77–85

Römer H (1987) Representation of auditory distance within a central neuropil of the bushcricket *Mygalopsis marki*. J Comp Physiol A 161:33–42

Rovner JS (1968) Territoriality in the sheet web spider *Linyphia triangularis* (Clerck) (Araneae, Linyphiidae). Z Tierpsychol 25:232–242

Rovner JS, Barth FG (1981) Vibratory communication through living plants by a tropical wandering spider. Science 214:464–466

Ruhland M, Rathmayer W (1978) Die Beinmuskulatur und ihre Innervation bei der Vogelspinne *Dugesiella hentzi* (Ch.) (Araneae, Aviculariidae). Zoomorphologie 89:33–46

Rupprecht R (1968) Das Trommeln der Plecopteren. Z vergl Physiol 59:38–71

Rupprecht R (1975) Die Kommunikation von *Sialis* (Megaloptera) durch Vibrationssignale. J Insect Physiol 21:305–320

Salmon M (1965) Waving display and sound production in *Uca pugilator*, with comparisons to *U. minax* and *U. pugnax*. Zoologica (NY) 50:123–150

Schäfer S, Bicker G (1986) Distribution of GABA-like immunoreactivity in the brain of the honeybee. J Comp Neurol 246:287–300

Schenberg S, Pereira Lima FA (1978) Venoms of Ctenidae. In: Bettini S (ed) Arthropod Venoms. Springer, Berlin Heidelberg New York, pp 217–245

Scheuring L (1914) Die Augen der Arachnoideen. II. Zool Jahrb Anat 37:369–464

Schildberger K (1981) Some physiological features of mushroom body linked fibers in the house cricket brain. Naturwissenschaften 67:623

Schimitschek E (1968) Insekten als Nahrung, in Brauchtum, Kult und Kultur. In: Helmcke J-G, Starck D, Wermuth H (eds) Handb Zool II Insecta. deGruyter, Berlin, pp 1–62

Schlichting H (1979) Boundary Layer Theory. McGraw-Hill, New York

Schmid A (1997) A visually induced switch in mode of locomotion of a spider. Z Naturforsch 52c:124–128

Schmid A (1998) Different functions of different eye types in the spider *Cupiennius salei*. J Exp Biol 201:221–225

Schmid A, Duncker M (1993) Histamine immunoreactivity in the central nervous system of the spider *Cupiennius salei*. Cell Tissue Res 273:533–545

Schmidt JM, Smith JJB (1987) The external sensory morphology of the legs and hairplate system of female *Tri-*

chogramma minutum Riley (Hymenoptera: Trichogrammatidae). Proc R Soc Lond B 232:323–366

Schmitt A, Schuster M, Barth FG (1990) Daily locomotor activity patterns in three species of *Cupiennius* (Araneae: Ctenidae). The males are the wandering spiders. J Arachnol 18, 3:249–255

Schmitt A, Schuster M, Barth FG (1992) Male competiton in a wandering spider (*Cupiennius getazi*, Ctenidae). Ethology 90:293–306

Schmitt A, Schuster M, Barth FG (1994) Vibratory communication in a wandering spider (*Cupiennius getazi*, Ctenidae). Female and male preferences of various features of the conspecific male's releaser. Anim Beh 48:1155–1171

Schnorbus H (1971) Die subgenualen Sinnesorgane von *Periplaneta americana*. Histologie und Vibrationsschwellen. Z vergl Physiol 71:14–48

Schüch W, Barth FG (1985) Temporal patterns in the vibratory courtship signals of the wandering spider *Cupiennius salei* Keys. Behav Ecol Sociobiol 16:263–271

Schüch W, Barth FG (1990) Vibratory communication in a spider: female responses to synthetic male vibrations. J Comp Physiol A 166:817–826

Schulz S, Toft S (1993) Identification of a sex pheromone from a spider. Science 260:1635–1637

Schulz S, Papke M, Tichy H, Gingl E, Ehn R (2000) Identification of a new sex pheromone from silk dragline of the tropical hunting spider *Cupiennius salei*. Angew Chem (in Druck)

Schwarz HF (1921) Spider myths of the American indians. Natural History (magazine of the American Museum of Natural History) 21:382–385

Seitz K-A (1966) Normale Entwicklung des Arachniden-Embryos *Cupiennius salei* (Keyserling) und seine Regulationsbefähigung nach Röntgenbestrahlung. Zool Jb Anat 83:327–447

Seitz K-A (1967) Untersuchungen über dynamische Vorgänge im Ei der Spinne *Cupiennius salei* (Ctenidae) mittels Röntgen-Barrieren. Zool Jb Anat 84:343–374

Seitz K-A (1970) Embryonale Defekt- und Doppelbildungen in der Spinne *Cupiennius salei* (Keyserling) (Araneae, Ctenidae). Zool Jb Anat 87:588–639

Seitz K-A (1971) Licht- und elektronenmikroskopische Untersuchungen zur Ovarentwicklung und Oogenese bei *Cupiennius salei* (Keyserling) (Araneae, Ctenidae). Z Morph Tiere 69:283–317

Seitz K-A (1972) Zur Histologie und Feinstruktur des Herzens und der Hämocyten von *Cupiennius salei* Keys. (Araneae, Ctenidae). I. Herzwandung, Bildung und Differenzierung der Hämocyten. Zool Jb Anat 89:351–384

Seitz K-A (1975) Licht- und elektromikroskopische Untersuchungen an den Malpighischen Gefäßen der Spinne *Cupiennius salei* (Keyserling) (Ctenidae, Araneae). Zool Jb Anat 94:413–440

Seitz K-A (1976) Zur Feinstruktur der Häutungshämocyten von *Cupiennius salei* (Keyserling) (Araneae, Ctenidae). Zool Jb Anat 96:280–292

Sellick PM, Patuzzi R, Johnstone BM (1982) Measurement of basilar membrane motion in the guinea pig using the Mössbauer technique. J Acoust Soc Am 72:131–141

Seyfarth E-A (1978a) Lyriform slit sense organs and muscle reflexes in the spider leg. J Comp Physiol 125:45–57

Seyfarth E-A (1978b) Mechanoreceptors and proprioceptive reflexes: lyriform organs in the spider leg. Symp Zool Soc Lond 42:457–467

Seyfarth E-A (1980) Daily patterns of locomotor activity in a wandering spider. Physiol Entomol 5:199–206

Seyfarth E-A (1985) Spider proprioception: Receptors, reflexes and control of locomotion. In: Barth FG (ed) Neurobiology of Arachnids. Springer, Berlin, pp 230–248

Seyfarth E-A (1993) Taktiles Körperanheben – neuronale Grundlagen eines einfachen Verhaltens bei Spinnen. In: Abschlußbericht SFB 45. Unpubl manuscript, pp 21

Seyfarth E-A, Barth FG (1972) Compound slit sense organs on the spider leg: mechanoreceptors involved in kinesthetic orientation. J Comp Physiol 78:176–191

Seyfarth E-A, Bohnenberger J (1980) Compensated walking of tarantula spiders and the effect of lyriform slit sense organ ablation. Proc Int Congr Arachnol 8:249–255

Seyfarth E-A, Pflüger HJ (1984) Proprioreceptor distribution and control of a muscle reflex in the tibia of spider legs. J Neurobiol 15:365–374

Seyfarth E-A, Bohnenberger J, Thorson J (1982) Electrical and mechanical stimulation of a spider slit sensillum: outward current excites. J Comp Physiol A 147:423–432

Seyfarth E-A, Eckweiler W, Hammer K (1985) Proprioceptors and sensory nerves in the legs of a spider, *Cupiennius salei* (Arachnida, Araneida). Zoomorphology 105:190–196

Seyfarth E-A, French A (1994) Intracellular characterization of identified sensory cells in a new spider mechanoreceptor preparation. J Neurophysiol 71, 4:1422–1428

Seyfarth E-A, Gnatzy W, Hammer K (1990) Coxal hair plates in spiders: physiology, fine structure, and specific central projections. J Comp Physiol A 166:633–642

Seyfarth E-A, Hammer PM, Grünert U (1990) Serotonin-like immuno-reactive cells in the CNS of spiders. Verh Dtsch Zool Ges 83:292

Seyfarth E-A, Hammer K, Spörhase-Eichmann U, Hörner M, Vullings HGB (1993) Octopamine immunoreactive neurons in the fused central nervous system of spiders. Brain Res 611 (2):197–206

Seyfarth E-A, Hergenröder R, Ebbes H, Barth FG (1982) Idiothetic orientation of a wandering spider: compensation of detours and estimates of goal distance. Behav Ecol Sociobiol 11:139–148

Shaw SR (1994) Re-evaluation of the absolute threshold and response mode of the most sensitive known „vibration" detector, the cockroach's subgenual organ: a cochlea-like displacement threshold and a direct response to sound. J Neurobiol 25:1167–1185

Shepherd D, Kämper G, Murphey RK (1988) The synaptic origins of receptive field properties in the cricket cercal sensory system. J Comp Physiol A 162:1–11

Sherman RG (1985) Neural control of the heartbeat and skeletal muscle in spiders and scorpions. In: Barth FG (ed) Neurobiology of Arachnids. Springer, Berlin Heidelberg New York Tokyo, pp 319–336

Shimizu I, Barth FG (1996) The effect of temperature on the temporal structure of the vibratory courtship signal of a spider (*Cupiennius salei* Keys). J Comp Physiol A 179:363–370

Shimozawa T, Kanou M (1984a) Varieties of filiform hairs: range fractionation by sensory afferents and cercal interneurons of a cricket. J Comp Physiol A 155:485–493

Shimozawa T, Kanou M (1984b) The aerodynamics and sensory physiology of range fractionation in the cercal filiform sensilla of the cricket *Gryllus bimaculatus*. J Comp Physiol A 155:495–505

Shimozawa T, Kumagai T, Baba Y (1998) Structural scaling and functional design of the cercal wind-receptor hairs of cricket. J Comp Physiol A 183:171–186

Shimozawa T, Murakami J, Kumagai T (1998) Cricket wind receptor cell detects mechanical energy of the level of *k*T of thermal fluctuation. Abstract 112, International Society of Neuroethology Conference, San Diego

Shultz JW (1987) Walking and surface film locomotion in terrestrial and semi-aquatic spiders. J Exp Biol 128:427–444

Simon E (1891) Description de quelques arachnides de Costa Rica communiqués par Getaz MA (de Genève). Bull Soc zool France 16:109–112

Simon E (1897) Histoire naturelle des araignées 2(1):1–192

Simon E (1898) Histoire naturelle des araignées 2(2):193–380

Sivian LJ, White SD (1933) On minimum audible sound fields. J Acoust Soc Am 4:288–321

Snodgrass RE (1965) A Textbook of Arthropod Anatomy. Hafner, New York London

Sommerfeld A (1970) Vorlesungen über theoretische Physik II. Mechanik der deformierbaren Medien. Akad Verlagsges, Leipzig

Speck J, Barth FG (1982) Vibration sensitivity of pretarsal slit sensilla in the spider leg. J Comp Physiol A 148:187–194

Speck-Hergenröder J (1984) Vibrationsempfindliche Interneurone im Zentralnervensystem der Spinne *Cupiennius salei* Keys. Dissertation Universität Frankfurt am Main

Speck-Hergenröder J, Barth FG (1987) Tuning of vibration sensitive neurons in the central nervous system of a wandering spider, *Cupiennius salei* Keys. J Comp Physiol A 160:467–475

Speck-Hergenröder J, Barth FG (1988) Vibration sensitive hairs on the spider leg. Experientia 44 (1):13–14

Städler E, Hansen FH (1975) Olfactory capabilities of the "gustatory" chemoreceptors of the tobacco hornworm larvae. J Comp Physiol 104:97–102

Starrat AN (1979) Proctolin, an insect neuropeptide. TINS, Jan 79:15–17

Steinbrecht RA (1994) The tuft organs of the human body louse, *Pediculus humanus corporis* – cryofixation study of a thermo-hydrosensitive sensillum. Tissue & Cell 26:259–275

Steinbrecht RA, Müller B (1991) The thermo-/hygrosensitive sensilla of the silkmoth, *Bombyx mori*: morphological changes after dry- and moist-adaptation. Cell Tissue Res 266:441–456

Stokes GG (1851) On the effect of the internal friction of fluids on the motion of pendulums. Trans Camb Phil Soc 9:8 ff (reprinted in: Mathematical and Physical Papers, vol III, pp 1–141. Cambridge Univ Press 1901)

Stout JF, McGhee RW (1988) Attractiveness of the male *Acheta domestica* calling song to females. II. The relative importance of syllable period, intensity, and chirp rate. J Comp Physiol A 164:277–287

Stout JF, DeHaan CH, McGhee RW (1983) Attractiveness of the male *Acheta domestica* calling song to females. I. Dependence on each of the calling song features. J Comp Physiol 153:509–521

Strand E (1910) Eine neue cteniforme Spinne aus Guatemala. Soc ent 25:14

Stratton GE, Uetz GW (1983) Communication via substratum coupled stridulation and reproductive isolation in wolf spiders (Araneae: Lycosidae). Anim Behav 31:164–172

Strausfeld NJ, Barth FG (1993) Two visual systems in one brain: Neuropils serving the secondary eyes of the spider *Cupiennius salei*. J Comp Neurol 328:43–62

Strausfeld NJ, Nässel DR (1980) Neuroarchitecture of brain regions that subserve the compound eye of crustacea and insects. In: Autrum H (ed) Handbook of Sensory Physiology, Vol VII/6B: Comparative Physiology and Evolution of Vision in Invertebrates. Springer, Berlin Heidelberg New York, pp 1–132

Strausfeld NJ, Weltzien P, Barth FG (1993) Two visual systems in one brain: neuropils serving the principal eyes of the spider *Cupiennius salei*. J Comp Neurol 328:63–75

Strausfeld NJ, Hansen L, Li Y, Gomez RS, Ito K (1998) Evolution, discovery, and interpretations of arthropod mushroom bodies. Learning & Memory 5:11–37

Suter RB, Rosenberg O, Loeb S, Wildman H, Long JH (1997) Locomotion on the water surface: propulsive mechanisms of the fisher spider *Dolomedes triton*. J Exp Biol 200:2523–2538

Szabo I (1972) Höhere technische Mechanik. Springer, Berlin Heidelberg New York

Szabo I (1975) Einführung in die technische Mechanik. Springer, Berlin Heidelberg New York

Szabo I (1977) Höhere technische Mechanik. Springer, Berlin Heidelberg New York Tokyo

Szabo I (1984) Einführung in die technische Mechanik. Springer, Berlin Heidelberg New York Tokyo

Tarsitano MS, Jackson RR (1993) Influence of prey movement on the performance of simple detours by jumping spiders. Behaviour 123:106–120

Tarsitano MS, Jackson RR (1997) Araneophagic jumping spiders discriminate between detour routes that do and do not lead to prey. Anim Behav 53:257–266

Tautz J (1977) Reception of medium vibration by thoracal hairs of caterpillars of *Barathra brassicae* L. (Lepidoptera, Noctuidae). I. Mechanical properties of the receptor hairs. J Comp Physiol 118:13–31

Tautz J (1978) Reception of medium vibration by thoracal hairs of caterpillars of *Barathra brassicae* L. (Lepidoptera, Noctuidae). II. Response characteristics of the sensory cell. J Comp Physiol 125:67–77

Tautz J (1979) Reception of particle oscillation in a medium – an unorthodox sensory capacity. Naturwissenschaften 66:452–461

Tautz J, Markl H (1978) Caterpillars detect flying wasps by hairs sensitive to medium vibration. Behav Ecol Sociobiol 4:101–110

Taylor CR, Heglund NC, Maloiy GMO (1982) Energetics and mechanics of terrestrial locomotion. I. Metabolic energy consumption as a function of speed and body size in birds and mammals. J Exp Biol 97:1–21

Taylor CR, Schmidt-Nielsen K, Raab JL (1970) Scaling of energetic cost of running to body size in mammals. Am J Physiol 219:1104–1107

Telionis DP (1981) Unsteady Viscous Flows. Springer Series in Computational Physics. Springer, New York

Theunissen FE, Miller JP (1991) Representation of sensory information in the cricket cercal system. II. Information theoretic calculation of system accuracy and optimal tuning-curve widths of four primary interneurones. J Neurophysiol 66:1690–1703

Thornhill R (1979) Male and female sexual selection and the evolution of mating strategies in insects. In: Blum MS, Blum NA (eds) Sexual Selection and Reproductive Competition in Insects. Academic Press, London, pp 81–121

Thorson J, Biederman-Thorson M (1974) Distributed relaxation processes in sensory adaptation. Science 183:161–172

Thorson J, Weber T, Huber F (1982) Auditory behavior of the cricket. II. Simplicity of calling-song recognition in *Gryllus*, and anomalous phonotaxis at abnormal carrier frequencies. J Comp Physiol 146:361–378

Thurm U (1974) Basics of the generation of receptor potentials in epidermal mechanoreceptors in insects. In: Schwarzkopff J (ed) Mechanoreception. Rheinisch-Westfäl Acad Wiss 53:355–385

Thurm U (1982) Grundzüge der Transduktionsmechanismen in Sinneszellen. Mechano-elektrische Transduktion. In: Hoppe W, Lohmann W, Markl H, Ziegler H (eds) Biophysik. Springer, Berlin, pp 681–696

Thurm U, Küppers J (1980) Epithelial physiology of insect sensilla. In: Locke M, Smith D (eds) Insect Biology in the Future. Academic Press, London New York, pp 735–763

Tichy H (1979) Hygro- and thermoreceptive triad in antennal sensillum of the stick insect, *Carausius morosus*. J Comp Physiol A 132:149–152

Tichy H (1987) Hygroreceptor identification and response characteristics in the stick insect *Carausius morosus*. J Comp Physiol A 160:43–53

Tichy H, Loftus R (1996) Hygroreceptors in insects and a spider: humidity transduction models. Naturwissenschaften 83:255–263

Tichy H, Gingl E, Ehn R, Papke M, Schulz S (2000) Female sex pheromone of a wandering spider (*Cupiennius salei*): identification and sensory reception. J Comp Physiol A (eingereicht)

Tietjen WJ (1977) Dragline-following by male lycosid spiders. Psyche 84:165–178

Tietjen WJ, Rovner JS (1980) Physico-chemical trail following behaviour in two species of wolf spiders: sensory and etho-ecological concomitants. Anim Behav 28:735–741

Tietjen WJ, Rovner JS (1982) Chemical communication in lycosids and other spiders. In: Witt PN, Rovner JS (eds) Spider Communication. Mechanisms and Ecological Significance. Princeton Univ Press, Princeton NJ, pp 249–279

Trivers RL (1972) Parental investment and sexual selection. In: Campbell B (ed) Sexual Selection and the Descent of Man, 1871:1971. Aldine, Chicago, pp 136–179

Trujillo-Cenóz O, Melamed J (1967) The fine structure of the visual system of *Lycosa* (Araneae: Lycosidae) 2: primary visual centers. Z Zellforsch Mikrosk Anat 76:377–388

Uexküll J von (1909) Umwelt und Innenwelt der Tiere. Berlin

Uexküll J von (1920) Theoretische Biologie. Berlin

Usherwood PNR, Duce IR, Boden P (1984) Slowly-reversible block of glutamate receptor-channels by venoms of the spiders, *Argiope trifasciata* and *Araneus gemma*. J Physiol (Paris) 79:241–245

Vest DK (1987) Necrotic arachnidism in the northwest United States and its probable relationship to *Tegenaria agrestis* (Walckenaer) spiders. Toxicon 25(2):175–184

Vest DK (1993) Protracted reactions following probable hobo spider (*Tegenaria agrestis*) envenomation. American Arachnology (abstract) 48:10

Vogel H (1923) Über die Spaltsinnesorgane der Radnetzspinnen. Jena Z Med Naturwiss 59:171–208

Vollrath F (1979a) Behavior of the kleptoparasitic spider *Argyrodes elevatus* (Araneae, Theridiiae). Anim Behav 27:515–521

Vollrath F (1979b) Vibrations: their signal function for a spider kleptoparasite. Science 205:1149–1151

Vrijer PWF de (1986) Species distinctiveness and variability of acoustic calling signals in the planthopper genus *Javesella* (Homoptera: Delphacidae). Neth J Zool 36:162–175

Vugts HF, Wingerden WKRE van (1976) Meteorological aspects of aeronautic behavior of spiders. Oikos 27:433–444

Wainwright SA, Biggs WD, Currey JD, Gosline JM (1976) Mechanical Design in Organisms. Unwin, London

Walcott Ch, Kloot WG van der (1959) The physiology of the spider vibration receptor. J Exp Zool 141:191–244

Walker TJ (1957) Specificity in the responses of female tree crickets (Orthoptera, Gryllidae, Oecanthinae) to calling songs of the males. Ann Entomol Soc Am 50:626–636

Walla P, Barth FG, Eguchi E (1996) Spectral sensitivity of single photoreceptor cells in the eyes of the ctenid spider *Cupiennius salei* Keys. Zoological Science 13:199–202

Walrond JP, Wiens TJ, Govind CK (1990) Inhibitory innervation of a lobster muscle. Cell Tissue Res 260:421–429

Walter, Lieth (1967) Klimadiagramm – Weltatlas. Fischer, Jena

Wang CY (1968) On high-frequency oscillatory viscous flows. J Fluid Mech 32:55–68

Watson PJ (1986) Transmission of female sex pheromone thwarted by males in the spider *Linyphia litigosa* (Linyphiidae). Science 233:219

Wehner R (1992) Arthropods. In: Papi F (ed) Animal Homing. Chapman and Hall, London, pp 45–144

Wehner R (1995) The ant's path integration system: a neural architecture. Biol Cybern 73:483–497

Wehner R, Wehner S (1986) Path integration in desert ants. Approaching a long-standing puzzle in insect navigation. Monitore zool ital (N.S.) 20:309–331

Wehner R, Wehner S (1990) Insect navigation: use of maps or Ariadne's thread? Ethology Ecology & Evolution 2:27–48

Weltzien P (1988) Vergleichende Neuroanatomie des Spinnengehirns unter besonderer Berücksichtigung des „Zentralkörpers". Dissertation, Fachbereich Biologie, Universität Frankfurt am Main

Weltzien P, Barth FG (1991) Volumetric measurements do not demonstrate that the spider brain "central body" has a special role in web building. J Morphol 207:1–8

Wendler G (1964) Laufen und Stehen der Stabheuschrecke *Carausius morosus*: Sinnesborstenfelder in den Beingelenken als Glieder von Regelkreisen. Z vergl Physiol 48:198–250

Wendler G, Teuber H, Jander JP (1985) Walking and swimming and intermediate locomotion in *Nepa rubra*. In: Gewecke M, Wendler G (eds) Insect Locomotion. Paul Parey, New York, 103–110

Weygoldt P (1985) Ontogeny of the arachnid central nervous system. In: Barth FG (ed) Neurobiology of Arachnids. Springer, Berlin Heidelberg New York Tokyo, pp 20–34

Weygoldt P, Paulus HF (1979) Untersuchungen zur Morphologie, Taxonomie und Phylogenie der Chelicerata. I. Morphologische Untersuchungen. Z zool Systematik Evolutionsforsch 17:85–116

Wiese K (1974) The mechanosensitive system of prey localization in *Notonecta*. J Comp Physiol A 92:317–325

Wikgren M, Reuter M, Gustafsson MKS, Lindros P (1990) Immunocytochemical localization of histamine in flatworms. Cell Tissue Res 260:479–484

Williams DS, McIntyre PD (1980) The principal eyes of a jumping spider have a telephoto component. Nature 288:578–580

Wilson DM (1967) Stepping patterns in tarantula spiders. J Exp Biol 47:133–151

Wilson EO (1984) Biophilia. The Human Bond with Other Species. Harvard Univ Press, Cambridge MA

Wirth E (1984) Die Bedeutung von Zeit- und Amplitudenunterschieden für die Orientierung nach vibratorischen Signalen bei Spinnen, Diplomarbeit, Universität Frankfurt am Main

Wirth E, Barth FG (1982) Forces in the spider orb web. J Comp Physiol A 171:359–371

Witthöft W (1967) Absolute Anzahl und Verteilung der Zellen im Hirn der Honigbiene. Z Morph Tiere 61:160–184

Wittman Th, Schwegler H (1995) Path integration – a network model. Biol Cybern 73:569–575

Wong RKS, Pearson KG (1976) Properties of the trochanteral hair plate and its function in the control of walking in the cockroach. J Exp Biol 64:233–249

Wuttke W (1966) Untersuchungen zur Aktivitätsperiodik bei *Euscorpius carpathicus* L. (Chactidae). Z vergl Physiol 53:405–448

Yamada H (1970) Strength of biological materials. In: Evans FG (ed) Strength of Biological Materials. Williams & Wilkins, Baltimore

Yamashita S (1985) Photoreceptor cells in the spider eye: Spectral sensitivity and efferent control. In: Barth FG (ed) Neurobiology of Arachnids. Springer, Berlin Heidelberg New York Tokyo, pp 103–117

Yamashita S, Tateda H (1981) Efferent neural control in the eyes of orb weaving spiders. J Comp Physiol A 143:477–483

Zeigler DD, Stewart KW (1986) Female response thresholds of two stonefly (Plecoptera) species to computer-simulated and modified male drumming calls. Anim Behav 34:929–931

Zill SN, Moran DT (1981a) The exoskeleton and insect proprioreception. I. Responses of tibial campaniform sensilla to external and muscle-generated forces in the American cockroach, *Periplaneta americana*. J Exp Biol 91:1–24

Zill SN, Moran DT (1981b) The exoskeleton and insect proprioreception. III. Activity of tibial campaniform sensilla during walking in the American cockroach, *Periplaneta americana*. J Exp Biol 94:57–75

Zill SN, Moran DT, Varela FG (1981) The exoskeleton and insect proprioreception. II. Reflex effects of tibial campaniform sensilla in the American cockroach, *Periplaneta americana*. J Exp Biol 94:43–55

Zill SN, Underwood MA, Rowley IC, Moran DT (1980) A somatotopic organization of groups of afferents in insect peripheral nerves. Brain Res 198:253–269

Zimmermann B (1991) Differentiation of the thermo-/hygrosensitive (no-pore) sensilla on the antenna of *Antheraea pernyi* (Lepidoptera, Saturniidae): a study of cryofixed material. Cell Tissue Res 266:427–440

Zottl S (1994) Synapses in sensory tracts and neuropils in the wandering spider *Cupiennius salei* (Arachnida, Araneae). Diplomarbeit, Universität Wien

Appendix

Identification Key

This key also contains data regarding the typical coloration of living representatives of the various species of *Cupiennius*. In the case of the large species, even subadults can be identified by this criterion. Material preserved in alcohol, however, usually loses most or all of its color. Then the shapes of the epigynum, the vulva and the sclerites of the bulb become the most important distinguishing characters. In particular, classification of the smaller species, which have inconspicuous and variable coloration, is possible only by dissecting the vulva (females) or closely examining the bulb (males). The key makes use of all the characters of the genitalia used previously by Lachmuth et al. (1984). However, it includes *C. remedius* and *C. celerrimus* and adds additional characters such as coloration, body size and further features of the genitalia. For definition of the color terminology, we employed a color table for colored pencils designed by Faber-Castell, Germany.

Females:

1. Large spider (carapace length >9 mm); legs and/or body with conspicuous markings or color patterns 2
- Medium-sized spider (carapace length <9 mm); legs and/or body uniformly brown or with comparatively indistinct or variable markings 4

2. (1) Legs brown with conspicuous dark markings 3
- Femora I to IV bright carmine-red ventrally; prosoma and opisthosoma medium to dark brown dorsally with a darker median band; ventral opisthosoma without any dark markings (Fig. A1); epigynum with narrow median septum, widening distally; distal part of septum with strongly sclerotized hook (Fig. A2) (Plate 3) ... **coccineus**

3. (2) Femora I to IV with distinct black annular patterns; prosoma dorsolaterally with light grayish-brown pattern contrasting with the

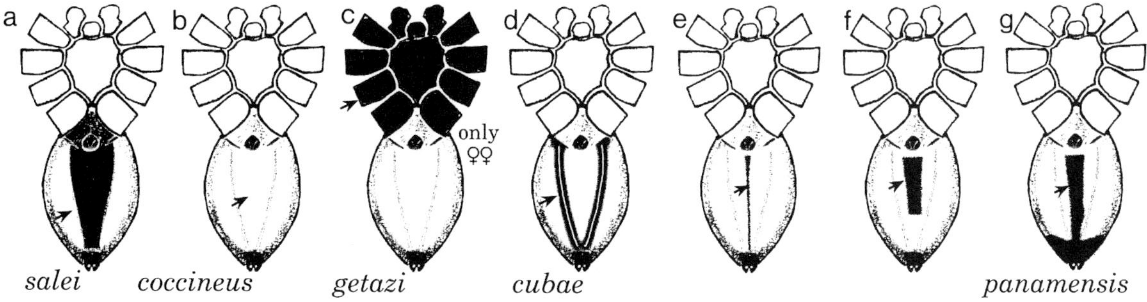

Fig. A1. Ventral views of different species of *Cupiennius* to show difference in patterning. For the three small species the extent of variation is indicated. (Barth and Cordes 1998, after Lachmuth et al. 1984; Brescovit and von Eickstedt 1985)

darker median band; coxae densely covered with terra-cotta red hairs ventrally; ventral opisthosoma always with broad black median stripe (Fig. A1 a); in some specimens pairs of yellowish to whitish spots disto-laterally on both sides of the cardiac mark; epigynum with narrow median septum of uniform width (Fig. A2); body length up to 45 mm (largest species, Fig. 3, Chapter II) (Plate 1) ..... **salei**

- Femora I to IV on the ventral side with many small black spots; either sternum or sternum and coxae (variable) dark brown to black (Fig. A1 c); dorsally, body coloration distinct and species-specific: median dark band on prosoma, colored areas laterally on the body; dark cardiac mark (opisthosoma); dark inverse V-shaped stripes, distal to cardiac mark; ventral opisthosoma light brown (populations from Barro Colorado Islands and from Panama were observed to have only a dark median ventral opisthosomal band and no speckled femora). A grayish morph and an orange morph exist. Epigynum with broad median septum of roughly uniform width, but widening distally; distal part of septum with sclerotized nose-like process (Plate 2) .... **getazi**

4. (1) Epigynal plate oval or trapezoid 5
- Epigynal plate distinctly triangular (Fig. A2); median septum of epigynum strongly widened distally forming a sphere; seminal receptacle I cone-like; body color in general uniformly grayish to brownish, ventral opisthosoma with outlines of a dark median band, consisting of a series of short dark reddish hairs (Fig. A1 d) (Plate 6) ... **cubae**

5. (4) Lateral plate of epigynum directly connected to the median septum forming a loop (Fig. A2 b) 6
- Lateral plate of epigynum not directly connected to the median septum and extending to the anterior-lateral border of the epigynal plate (Fig. A2 a) 8

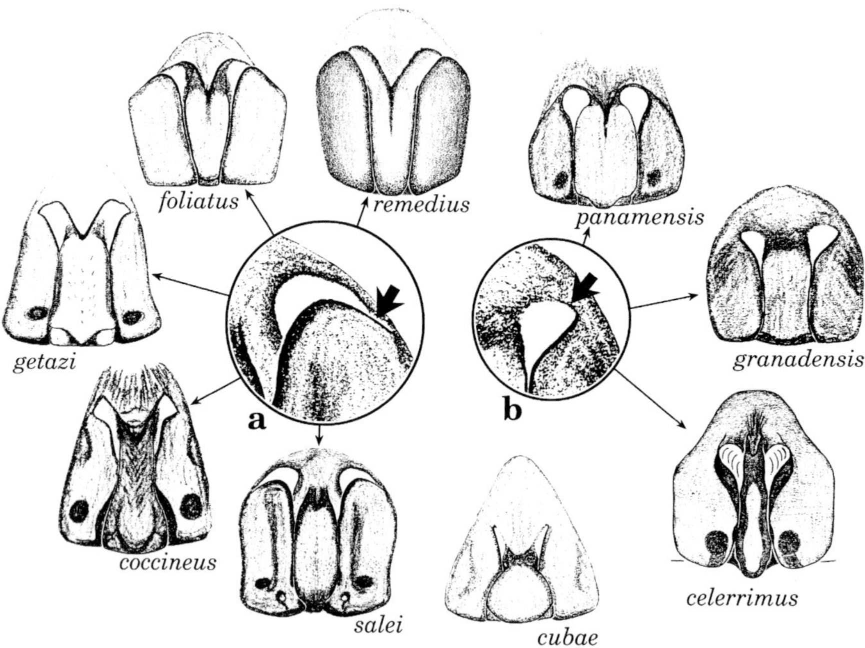

Fig. A2. Ventral view of the epigynums of all nine species of *Cupiennius*. Groups *a* and *b* differ with respect to the connection between the lateral plates and the epigynal plate. (Barth and Cordes 1998, after Lachmuth et al. 1984; Brescovit and von Eickstedt 1995)

6. (5) Epigynum with narrow median septum, seminal receptacles I with seminal ducts of different shapes: S-shaped, twisted, winding or rolled 7

- Epigynum wider than long (Fig. A2b); median septum broad and leaf-like; vulva: seminal receptacles I ball-shaped with seminal ducts sturdy and slightly curved laterally (Fig. A3g); prosoma light brown; opisthosoma darker brown, with narrow dark-shaded median band ventrally (Fig. A1); smallest species (Fig. 3, Chapter II) (Plate 8) .... **panamensis**

7. (6) Median septum with parallel borders, distally ending broad, and with a small hook (Fig. A2); vulva: seminal receptacles I with distinctly S-shaped seminal ducts (Fig. A3e) .... **granadensis**

- Median septum long, narrow and slightly widening distally (Fig. A2); vulva: seminal receptacles large and ball-shaped, seminal ducts rolled dorsoventally (Fig. A3f); body orange to brown with darker brown median band, legs I to IV yellow ventrally on coxae and femora .... **celerrimus**

8. (5) Lateral plates of epigynum ending rounded before connecting to the epigynal plate (Fig. A2a), median septum of epigynum narrow and continuously narrowing distally; vulva with ball-shaped seminal receptacles, seminal duct strongly winding (Fig. A3h); medium-large spider (carapace length 7–8 mm); annular patterns on femora, and body remarkably spotted; tarsi of the legs I to IV with long dark hairs both dorsally and ventrally (Plate 8) .... **remedius**

- Lateral plates of epigynum ending as indicated in Fig. A2a before connecting to the anterior-lateral end of the epigynal plate, median septum of epigynum as in figure A2a; seminal receptacles I ball-shaped, seminal ducts as in figure A3h; spider smaller (carapace length up to 7 mm); body without distinct color pattern or with a series of dark spots along the cardiac mark on the opisthosoma .... **foliatus**

Males:

1. Large spider (carapace length >9 mm). Legs with conspicuous markings (except one case,

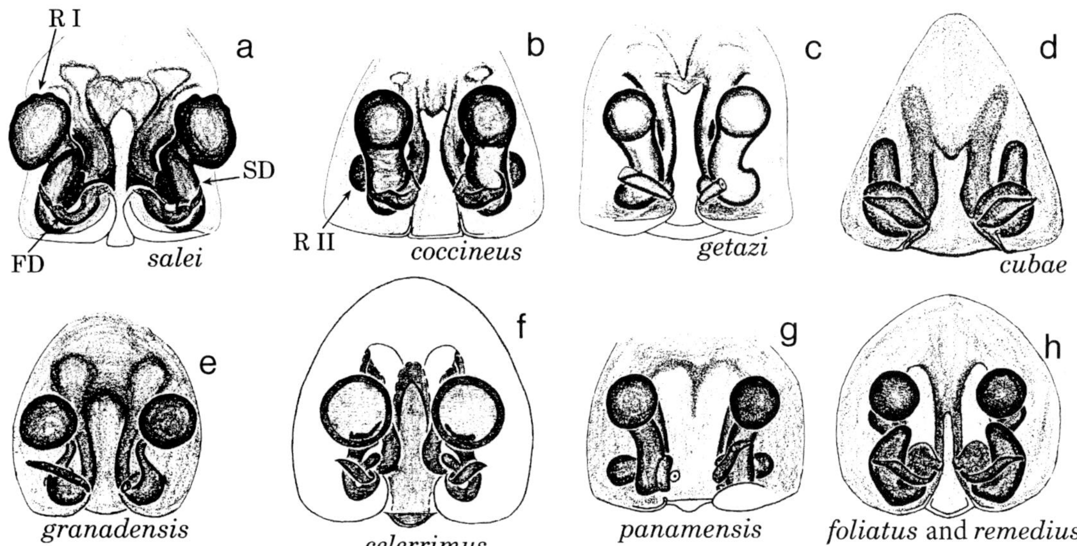

Fig. A3. Dorsal view of the epigynums of all nine species of *Cupiennius*. *RI, RII* Receptacula seminis; *SD* seminal duct; *FD* fertilization duct. (Barth and Cordes 1998, after Lachmuth et al. 1984; Brescovit and von Eickstedt 1995)

see 2); body light gray, light brown to medium brown or bright orange dorsally; ventral opisthosoma with or without broad dark median stripe (Fig. A1) **2**

- Medium-sized spider (carapace length <9 mm); legs and/or body uniformly brown or with indistinct markings, or pro- and opisthosoma with variable arrangement of more or less isolated dark dots and lines; opisthosoma light ventrally or with a narrow dark median stripe (Fig. A1) **4**

2. (1) Legs and/or body with conspicuous markings **3**

- Legs without conspicuous coloration; legs and body gray-brown with median band on dorsal prosoma consisting of thin dark lines; light opisthosoma with dark cardiac mark, lacking dark markings ventrally; bulb with terminal apophysis bent downward, embolic apophysis strongly curved and twisted (Fig. A4b) (Plate 3) ..... **coccineus**

3. (2) Femora I to IV with distinct black annular patterns ventrally; body grayish dorsally with dark lines along the length of the prosoma (=median band); sternum and coxae grayish, opisthosoma with broad dark median band ventrally; bulb with terminal apophysis large and bent downward, embolic apophysis robust and curved (Fig. A4a); body length up to 30 mm (largest species, Plate 1)....... **salei**

- Femora I to IV with many small black spots ventrally; sternum and coxae dark

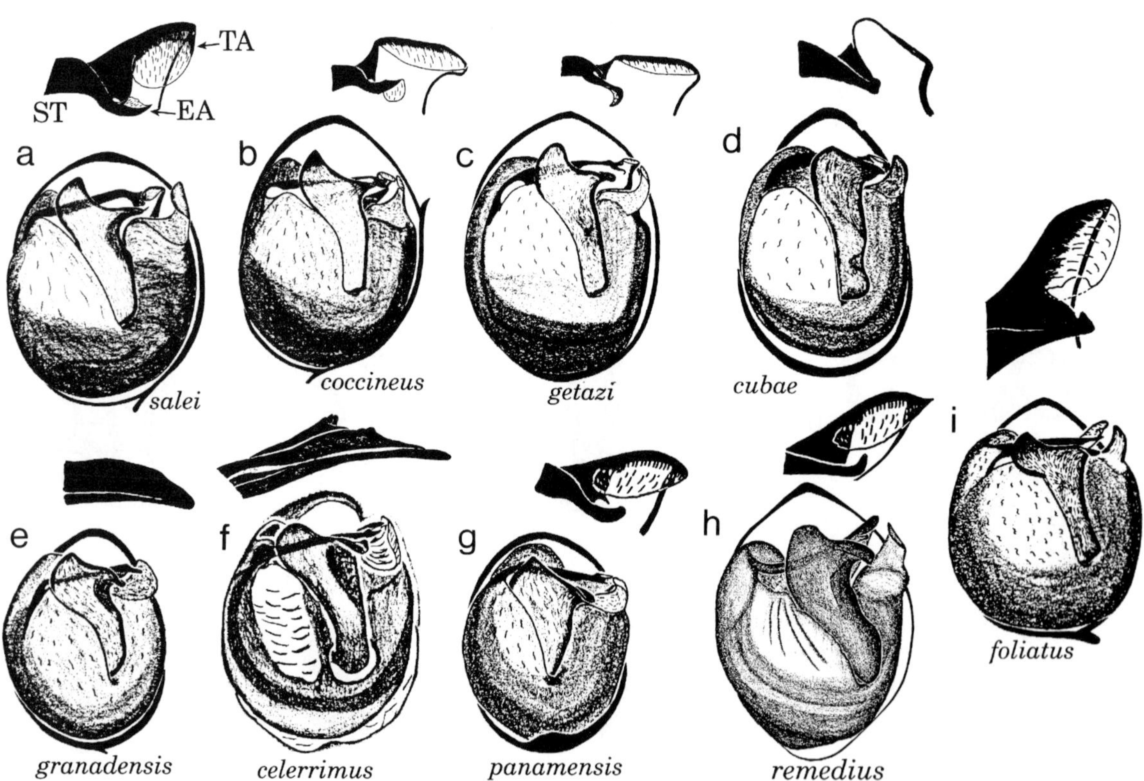

Fig. A4. Genital bulbs and terminal structures of the embolus of the males of all nine species of *Cupiennius*. *TA* Terminal apophysis; *ST* stipes embolus; *EA* embolic apophysis. (Barth and Cordes 1998, after Lachmuth et al. 1984; Brescovit and von Eickstedt 1995)

brownish (variable); conspicuous species-specific body coloration: a dark median band dorsally on prosoma and opisthosoma bordered by light areas laterally; dark cardiac mark dorsally on opisthosoma, and dark inverse V-shaped stripes posterior to it; two morphs with either grayish or orange basic coloration. Bulb with terminal apophysis bent downward, embolic apophysis strongly curved and twisted (Fig. A4c) (Plate 2) .... **getazi**

4. (1) Opisthosoma with narrow dark median stripe ventrally (Fig. A1) or without ventral markings **5**
- Opisthosoma only with dark reddish outlines of the ventral median stripe (Fig. A1d); bulb with median apophysis comparatively straight and notched in the proximal third of its length, distal process and lateral shovel-like process very small, terminal apophysis strongly domed and extending over the short embolic apophysis. Body grayish or brownish .... **cubae**

5. (4) Bulb with embolic base (stipes-embolus) massive (Fig. A4e,f), terminal and embolic apophysis not distinct **6**
- Embolic base (stipes-embolus) with distinct terminal and embolic apophysis (Fig. A4a–d,g–i) **7**

6. (5) Embolic base (stipes-embolus) bill-shaped and folded forming one furrow (Fig. A4e); body light yellow-brown with a sparse coverage of hairs; prosoma with median line markings dorsally .... **granadensis**
- Embolic base (stipes-embolus) strongly folded forming two furrows; embolic tip appears severed with a pair of short processes (Fig. A4f); body and legs orange-brown with a brown median band on pro- and opisthosoma; ventral surface of coxae and femora yellow .... **celerrimus**

7. (5) Terminal apophysis level with the embolic base (stipes-embolus) (Fig. A4h,i) **8**
- Terminal apophysis elevates at an angle of approximately 45° at the embolic base (Fig. A4i) and covers the embolic apophysis; opisthosoma with a variable line of spots along the border of the cardiac mark .... **foliatus**

8. (6) Carapace length about 8 mm; body with spotted coloration pattern dorsally; legs long (sexual-dimorphic), covered with a "brush" of long and thin hairs along the tibia and metatarsus and with the longest hairs at the proximal part of the tibia-metatarsus joint; median apophysis with an elevation near the lateral process, tegulum with deep furrows ventrally (Fig. A4h) .... **remedius**
- Carapace length about 5 mm; body without distinct coloration pattern dorsally; dorsal opisthosoma darker than prosoma and with a small dark median band ventrally, widening towards the posterior part of the opisthosoma (Fig. A1g) .... **panamensis**

I gratefully acknowledge the permission granted by the following publishers and societies to use already published material (mostly in modified form). The respective references are given at the end of the figure legends.

American Physiological Society, Bethesda: Chapter VII, Fig. 10.
Birkhäuser Verlag, Basel: Chapter VIII, Fig. 7; Chapter XV, Fig. 1.
Blackwell Science, Oxford: Chapter IV, Figs. 1–3.
Elsevier Science, Oxford: Chapter V, Fig. 2; Chapter XVII, Figs. 1, 3; Chapter XX, Figs. 14, 25.
Forschungsinstitut und Naturmuseum Senckenberg, Frankfurt am Main: Chapter II, Fig. 2; Appendix Figs. 1–4.
Harcourt Publishers, London: Chapter V, Fig. 3; Chapter VII, Fig. 3, Chapter XX, Fig. 29; Chapter XXIV, Figs. 3, 4.
John Wiley & Sons Inc., New York: Chapter X, Figs. 1, 14; Chapter XIV, Fig. 6; Chapter XVI, Figs. 1–16.
Journal of Arachnology: Chapter II, Figs. 3, 4, 6a; Chapter IV, Fig. 4; Appendix Figs. 1–4.
Springer Verlag, Heidelberg: Chapter III, Figs. 1, 2, 4, 5; Chapter VII, Figs. 1, 3–5, 7, 8, 11, 13–24; Chapter VIII, Figs. 1, 5, 6, 8, 9; Chapter X, Figs. 2, 4–7, 9, 10–13, 15; Chapter XI, Fig. 9; Chapter XIII, Figs. 1–7, 8, 13, 14; Chapter XIV, Figs. 1–5, 7–9; Chapter XV, Figs. 2–6, 8; Chapter XVIII, Figs. 1–13, 15, 16; Chapter XIX, Figs. 23–26; Chapter XX, Figs. 3, 6–13, 15–24, 26–28; Chapter XXI, Figs. 1–9; Chapter XXII, Figs. 1–4, 6–10; Chapter XXIV, Fig. 2; Chapter XXV, Figs. 1–4, 6–9.
The Company of Biologists, Cambridge: Chapter XI, Figs. 3–6; Chapter XXII, Figs. 1, 3–5.
The Royal Society, London: Chapter IX, Figs. 1–4, 10–24; Chapter XIX, Figs. 1–13.
Urban & Fischer Verlag, Jena: Chapter V, Fig. 1; Chapter IX, Figs. 6, 7; Chapter X, Fig. 16.
Verlag Zeitschrift für Naturforschung, Tübingen: Chapter XXII, Fig. 2.
Zoological Science, Sapporo: Chapter XI, Fig. 8.

Subject Index

Bold numbers refer to pages with figures.